SEMICONDUCTOR MODELING

SEMICONDUCTOR MODELING
For Simulating Signal, Power, and Electromagnetic Integrity

Roy G. Leventhal
Leventhal Design and Communications

Lynne Green, Ph.D.
Green Streak Programs

Contributing author:

Darren J. Carpenter
BT Exact

Springer

Roy Leventhal
Leventhal Design & Communications
Arlington Heights, IL, USA

Lynne Green
Green Streak Programs
Woodinville, WA, USA

Modeling Semiconductors: For Simulating Signal, Power, and Electromagnetic Integrity

Library of Congress Control Number: 2006924288

Additional material to this book can be downloaded from http://extras.springer.com.

ISBN 0-387-24159-0 e-ISBN 0-387-24160-4
ISBN 978-0387-25870-6

Printed on acid-free paper.

Printed in the United States of America.

9 8 7 6 5 4 3 2 1

springer.com

Dedication

Dedicated to my wife Charlotte. Her skill and patience as editor helped us organize the content and express our ideas.
–Roy Leventhal

Dedicated to my husband Kelly, for endless support and patience throughout this project.
–Lynne Green

CONTENTS

PREFACE

Why Read This Book?

Semiconductor Modeling is written to give designers and engineers the broadest possible view of using semiconductor models to design high-speed circuit boards. Most issues in the field of designing high-frequency circuits, from the beginning of modeling and simulation using EDA tools to future trends, are discussed and integrated. The authors focus on key ideas in each area.

The main technical focus of *Semiconductor Modeling* is on the use of simulation to help solve practical engineering problems. It draws on over 90 years of diverse technical experience and brings together theory, tools, components, software, and experience to help engineers create successful electronic products.

This book is written to help engineers understand how to go about problem solving while designing high-speed circuits. In particular, it helps engineers in the following areas:

- What to do about non-existent, incomplete, and erroneous semiconductor circuit models.
- How to apply EDA tools to achieve practical designs despite model limitations.
- How a grasp of basic high-speed digital, RF, and EMC design concepts guides design intent.
- How to apply EDA tools to cutting-edge digital technology.
- How to work with suppliers, CAD team members, and others.
- How processes and procedures enable and enhance the design process.

Who Should Read This Book?

This book is written for electronics engineering students and professionals working with models and simulation. This means choosing, obtaining, simulating, analyzing, and verifying models. The authors' goal is to organize and present the material in a way that is helpful to circuit and product design engineers who have little or no specialized knowledge of high-speed digital design or semiconductors.

To get the most from this book, the reader should have:

- A basic familiarity of analog electronic circuit design, Ohm's Law, and signal analysis in time and frequency domains.
- Exposure to different types of semiconductor devices.
- A Bachelor of Science in Electrical or Electronics Engineering (or its equivalent) or several years of practical experience.

What Models Are Covered?

The following information about models is covered:

- Their inner workings
- Their suitability
- Their limitations
- Their design benefits

Specifically, the book covers IBIS models in detail. They are the most useful models for simulation of high-speed digital designs. For the sake of completeness, other models are also included. They are:

- 2-Port ("black-box") matrix models. They form an underpinning to 2-Port Scattering-Parameter models and find application inside many of the EDA programs used for simulation.
- SPICE models.
- Scattering-Parameter models.
- TCAD device physics models.

Writing Approach Used in This Book

This book draws on a vast body of technical literature. The authors cite references wherever possible for readers who want to pursue derivations, proofs, and in-depth coverage. The analog high-frequency effects of high-speed digital switching are also addressed.

What the authors attempt to present is a technical area so broad, and so deep, that it would take several shelves of college textbooks to cover the subjects in depth. The authors intend to:

- Explain the technical areas in enough technical detail to provide a clear description of the subject.

- Provide derivations and proofs to the extent that they help an understanding of the subject.
- Place all of the material within a framework of:
  - How modeling for high-speed circuitry is done.
  - How different technical specialists work together in this field.
  - How theoretical tools, EDA software, physical observations, and practical experience combine for a successful product design.

The authors' points of view are essentially that of an end user of EDA tools for purposes of designing high-speed electronic circuits. The challenge is to be successful despite limitations in theory, software, work environment, component suppliers, or previous condition of practical experience.

Discrete Semiconductors Are Used as Simple Examples

To promote understanding, this book starts with simple, fundamental concepts and then advances to more complex systems and structures. The discussion starts with the physical-electrical operation of discrete semiconductors. After considering simple gates, the authors switch to IC chips that may contain from tens to hundreds of millions of individual gates.

Individual transistor I/O behavior is similar to the I/O behavior of an IC chip. The IC chip's physical construction at its I/O causes its electrical behavior as a driver or receiver. For most high-speed digital design simulations, the IBIS model is a balanced compromise between speed and complexity. IBIS model characteristics are similar to that of a discrete transistor's input or output. IBIS ignores what happens between the IC input and IC output to the PCB world. Instead, the authors focus on the electrical performance of the interconnections between a driver (generator or output) and a receiver (load or input).

Throughout this book, the authors re-use the same example transistor properties to explain how the properties operate through several levels of use and association.

How Is This Book Organized?

The basic parts of the book are:

1. Where models and simulation fit into product development.
2. Generating 2-Port, Scattering Parameter, SPICE, and IBIS models.
3. Selecting appropriate components for the product and suitable models to simulate the design.
4. Information about the IBIS model file: format, data, and quality control.

5. Managing IBIS models for simulation.
6. Checking and verifying IBIS models.
7. The future of IBIS and related modeling techniques.
8. Glossary, Bibliography, Index, and CD-ROM

About the CD-ROM

The companion CD-ROM includes software selected by the authors to help users learn how to extract models and simulate with them. Sonnet® Lite from Sonnet Software and Visual IBIS Editor® from Mentor Graphics are provided.

The CD-ROM also includes example data sheets, sample model files, IBIS syntax guides, presentations with practical model-building laboratory exercises (including input, intermediate, and output files), and other documentation. The appendixes on the CD include:

- Appendix A, Sample Data Sheets
- Appendix B, Sample IBIS Model Files
- Appendix C, Sample SPICE Models
- Appendix D, Sample S-Parameter Model
- Appendix E, Key Concepts of the IBIS Model
- Appendix F, List of Websites
- Appendix G, List of Software Providers
- Appendix H, Production Realities
- Appendix I, IBIS Quality Checklist
- Appendix J, Device Physics
- Appendix K, Logic Selection Guides
- Appendix L, Training Presentations

Feedback

The authors hope this book helps the reader simulate and design high-speed digital circuits and would greatly appreciate the reader's feedback. Please send suggestions and comments by email to:

Roy Leventhal
Training and Support in Signal Integrity, EMI, and Semiconductor Modeling
Leventhal Design & Communications
Roy.Leventhal@ieee.org

ACKNOWLEDGMENTS

Extraordinary thanks go to Charlotte Leventhal, Roy's wife and partner at Leventhal Design and Communications. She was coach, mentor, guide, and editor during the three long years spent on writing this book, contributing countless hours to improve its readability and usability.

Exceptional thanks go to:

- Darren Carpenter, of BT Exact (UK) who contributed "Chapter 23, Using Probability: The Ultimate Future of Simulation."
- Dr. S. Mahfuz Aziz of the School of Electrical and Information Engineering, University of South Australia
- Dr. Sonia Ben Dhia and Dr. Etienne Sicard of the Institut National des Sciences Appliquées of Toulouse (INSA-Toulouse, France)
- Bob Ross of Teraspeed
- Todd Westerhoff of Cisco Systems
- Ken Willis of Cadence Design Systems

Special thanks go to the following individuals who made significant contributions to this book:

- John Barnes of dBi Corporation
- Ron Brewer of Instrument Specialties
- Dr. Alistair Duffy, School of Engineering and Technology, De Montfort University (UK)
- IBIS Committee
- Dr. Randy Jost of Electrical and Computer Engineering, Utah State University
- Arpad Muranyi of Intel Corporation
- George Opshal of Clearbrook Systems

The following companies and organizations generously provided technical information and software licenses:

- 3Com, Inc.
- Cadence Design Systems, Inc.
- EIA/IBIS Committee
- Fairchild Semiconductor, Inc.
- Mentor Graphics Corporation
- Philips Semiconductor, Inc.
- Sonnet Software, Inc.

- Texas Instruments Semiconductor, Inc.
- TDA Systems, Inc.
- TechSmith Corporation
- XiLinx, Inc.

Many thanks go to the following individuals who contributed to portions of the book:

- Dr. Gino Antonini of the Dept. of Electrical Engineering, University of L'Aquila (Italy)
- Dr. Harvey Brodsky of Beta-Tronics
- Bill Hargin of HyperLynx/Mentor Graphics
- John Horner, Dennis Nagle, and Lance Wang of Cadence Design Systems
- Wolf Josenhans of US Robotics
- Jay Katzman, Dr. Phillip Leventhal, and Bob Milstein
- Scott McMorrow of Teraspeed
- Jim Nadolny of Samtec
- Dr. James Rautio of Sonnet Software
- DC Sessions of Philips Semiconductor
- Dr. Matthias Troescher of SimLab (Germany)
- Joseph Ziolkowski of GE Transportation Systems

Thanks also go to the members of the SI reflector (si-list@freelists.org) and members of the IBIS reflectors (ibis-users@eda.org; ibis@eda.org), and the many helpful individuals who took the time to discuss technical topics and provide additional insight.

Working with this diverse group of contributors and reviewers was a privilege and would have been impossible without the enabling power of the Internet.

—Roy Leventhal, Leventhal Design and Communications
—Lynne Green, Green Streak Programs

PART 1: INTRODUCTION

Chapter 1

HOW THE WORKPLACE SUPPORTS SUCCESSFUL DESIGN

For modeling and simulation to be successful, workplace support is required

Abstract: High-speed digital design is a deep, broad, and complex technical area. To meet the challenges of designing world-class products, especially on a tight schedule, engineers must be technically specialized. Design projects require a team of specialists to work together; good interpersonal skills; a strong sense of what is physically possible in a design; and good judgments about competing design objectives. By leveraging corporate-wide experience and knowledge, the company can facilitate good design processes and procedures. Skill with EDA modeling and simulation tools is essential.

1.1 HIGH-SPEED DIGITAL DESIGN IS CHALLENGING

High-speed digital design (HSDD) is intellectually rewarding and challenging. As technology advances, HSDD is continually becoming more complex. Shrinking component size and increasing switching speed are fundamentally changing the nature of the design and development process. In the 1990s, ideas were tried out in a design prototype by soldering in parts and poking about with test instrument probes. Today, that usually cannot be done on HSDD circuit boards. The aid of EDA software tools has become essential when responding to the increased complexity and difficulty of breadboarding and prototype debug. An even smarter approach is to prototype and debug the *virtual design* while it still exists only in the computer.

Figure 1-1 shows how the traditional design process compares to the newer process incorporating EDA tools, modeling, and simulation. With modeling and simulation tools, 90% of the time a new product works as designed when the first prototype is built.[1]

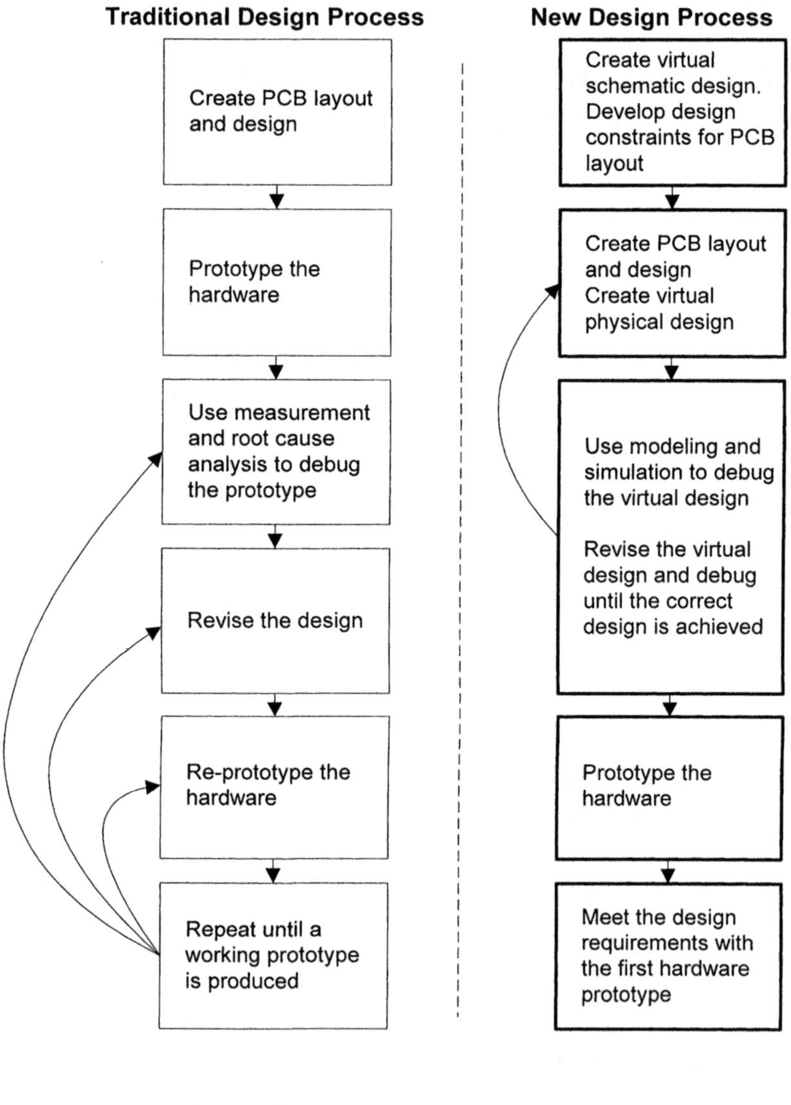

Figure 1-1. Traditional process compared to modeling and simulation process

[1] The authors assume that the models and the simulation tool are calibrated in the design process. Calibrated means that measured and simulated data match well.

Getting a physical prototype to work the first time built requires a change in the design process. In the past, the debug phase was consciously left until after a prototype was built. With the new process, more time and effort are spent early in the design process, while the design still exists only as a *virtual* design. Building the hardware prototype is delayed until after the virtual design works. The traditional method takes much longer to develop a working hardware prototype because the build→debug→re-prototype process takes longer than manipulating a virtual design in the computer. It is becoming impossible to probe all the relevant signals in the hardware prototype because PCB feature sizes are shrinking. In many PCBs, passive components are part of, and buried in, the layers of the board and are not even accessible for measurement.

In the 1960s, RF PCBs operating at a few hundred MHz and beyond were constructed with discrete transistors and leaded passive components. Probing with instruments to measure signals was tricky. However, it could be done on an improvised basis for debugging purposes. Inserting dedicated test points was done mainly to verify that ICs assembled on a board were functional. Real estate on today's boards is too scarce to do much insertion of test points for routine debugging.

Today, success in passing functional and Signal Integrity requirements with the first physical prototype is achieved routinely. After passing the functional and Signal Integrity testing, the prototype must pass EMI/EMC Regulatory testing. For this, the physical prototype gets submitted to a Regulatory test cell. Then, 70% to 90% of the first physical prototype [43] of new designs fail in their first try to pass Regulatory testing.

The difference in success is because Signal Integrity is modeled and simulated routinely, whereas EMI/EMC is not. EMI/EMC can achieve better results, but this will require a change in design process similar to the revolution that Signal Integrity experienced. Fostering a can-do, positive attitude towards EMI/EMC modeling and simulation is one reason this book was written.

EMI/EMC engineers know how to pass Regulatory testing. But the fixes to radiation and emission problems are usually applied late in a program, are expensive, and delay the product release. To avoid radiation and emission problems, EMI engineers would like a more cost-effective and timely process. Modeling and simulation offers a cost-effective and timely process.

The need to design high-speed circuits has fueled the growth of engineering software programs. Without Electronic Design Automation (EDA) tools, analysis and simulation would be too difficult. Instead of designing with lab breadboarding and debugging, engineers now often prefer to perform virtual design on a computer using modeling and simulation.[2] Computer Aided Engineering (CAE) software facilitates the new methods. A computer database (for virtual design of the product) facilitates coordinating, combining, and tracking the contributions of members of the design team. A virtual design facilitates the rapid testing of many what-ifs and is an essential part of the design process.

Computer modeling and simulation are not the whole story behind successful designs. The design process starts with intuition and inspiration about how to solve a problem [17]. Then the analytical powers of our technical training take over to massage the idea into a practical solution. The design process takes advantage of four important tools:

- Rules of thumb[3]
- Analytic approximations[4]
- Numerical simulation tools[5]
- Measurements[6]

Continued improvements in engineering software programs and component modeling have benefited from the advances in computer hardware and software engineering. Let us hope that need and results continue to lift each other in continuing cycles of advancement.

1.2 NEEDS FOR TECHNICAL SPECIALIZATION

As challenges, solutions, and technology advance, the need for technical specialization increases. Design engineers need to master complex EDA software tools as well as to apply specialized knowledge of modeling and simulation.

Engineering knowledge is usually very specialized. For instance, the design of the most advanced connectors, IC packages, and PC boards may require dedicated, highly specialized engineers. They all involve the design of passive interconnections and the same basic physical laws. To be first-class at designing a particular type of component, the engineer may need to focus exclusively on it. This is because the best materials, structure, scale issues, and specialized design techniques can vary by type of component.

---

[2] Something is lost when prototype breadboard and debug are no longer a common experience. Sensory experiences stimulate learning and abstract thinking [99].

[3] Leverage experience and knowledge and avoid "death by analysis."

[4] Early answers aid good strategic choices.

[5] Massage the idea into a practical solution.

[6] Verify our analysis and correct for mistaken modeling and simulation.

It's a rare individual who has the expert skills in all the necessary technical specialties to produce an entire product. These specialties include logic design, Signal Integrity design, Computer Aided Design (CAD) of Printed Circuit Boards (PCBs), design of Electromagnetic Control (EMC) and Regulatory Engineering. Specialization and a division of labor allow individuals to get really good at a subset of these specialties. Therefore, to design electronic products, most engineers work as part of a team of specialists.

How well a team works together is just as important as their individual technical skills. In addition, all members of the design team need to have a minimum knowledge of each other's specialties so that they can communicate effectively and understand the tradeoffs that cross their specialty areas.

Today's complex, high-performance electronic products need teams of specialists who work well together while using the best EDA tools available. At the core of the HSDD team are usually a logic-design and product-design engineer, a signal-integrity and EMI engineer, and a mechanical engineer and printed circuit board (PCB) designer.

When something is hard to do, people often get tense about it. Technical specialization involving teams of people puts more of a premium on working well with others. When a team of accomplished individuals works well together, great things can be accomplished.

Most importantly, technical specialists who apply engineering software tools need to have a keen understanding of physical reality. Computers are great enabling tools in the hands of someone who understands the limitations of both the tool and the models. Computer-generated data should be judged against physical reality. Computers execute only the instructions they are given. When they are given the wrong instructions, they produce the wrong results.

1.3 THE ROLE OF PROCESSES AND PROCEDURES

An additional requirement for success is using the lessons of experience to do the right things at the right time in the design process. Learning from past experiences should help avoid repeating the same mistakes. When projects are busy, complex, and on a short schedule, it is easy to forget some things. For instance, remembering to simulate at min and max conditions, and not just at typical conditions.

Following good design procedures and processes help engineers create quality products by reminding them to do the right things at the right time. Incorporating corporate-wide experience and knowledge into the design processes and procedures makes good sense. Incorporating EDA simulation tools and project management tools also makes good sense.

1.4 USING JUDGMENT WHEN MAKING DESIGN TRADEOFFS

Just as important as technical skills, people skills, and procedures are design imagination and judgment. Because of inherent uncertainties, limitations, and simplifications of modeling and simulation, engineers need to use common sense and judgment. Common sense and judgment are hard to teach. There are no formulas to apply. Fortunately, people learn common sense and judgment from their own experience and the experience of others.

In forming the judgment of how far to trust modeling and simulation calculations, design engineers must bring the following skills to the task:

- *Imagination*
  Computers are not very good at problem solving; neither are processes and procedures. People are good at problem solving and at imagining new and better ways to do something. When imagination is enhanced by experience and training, the result is beneficial.
- *Judgment and understanding*
  The engineer asks: "Does the data make sense? Can the data be explained by matching theory with observation?" The best engineers possess both theoretical and observational (bench) skills. This balance of theoretical and observational skills is most critical when pushing theory into new areas, or when modeling large systems.
- *Ability to make wise design choices early enough to avoid problems*
  One example of applying judgment is choosing a digital logic part with appropriate switching speeds and edge rates, rather than designing with complex termination and timing schemes. A good design choice is better than demanding absolute modeling accuracy and absolute tolerance control on parts used in production.
- *Ability to integrate design rules of thumb with comprehensive modeling and simulation*
  The wise engineer recognizes which portions of the design are robust enough to work properly even when designed using the common rules of thumb. At the same time, he or she pays close attention to the sections of the design that are critical to proper operation by employing more sophisticated modeling and simulation techniques.

Note: Technology is advancing too rapidly for engineers to develop rules of thumb that have been thoroughly tested against experience. The application of the old rules is breaking down. Since many more high frequency phenomena are occurring simultaneously, old rules can easily be misapplied. Simultaneous phenomena are causing conflicts between the rules of thumb. *Computation,* and the use of computers, helps engineers keep track of all these phenomena and give them their proper weight.

- *Ability to verify modeling and simulation with measurements*
  Instead of calling it ability, we could call it a passion to verify results with measurements. Modeling and simulation requires many approximations, compromises, and assumptions along the way. Plus, verification is particularly important when pushing the envelope on technology. Many unexpected issues arise in the design and manufacture of products. Whenever a design, product, supplier, or application is new, engineers need to use some caution.

1.5 HSDD NEEDS THE HELP OF EDA TOOLS

Part of the modeling and simulation challenge is the complexity of the product itself and its manufacture. People are *not* good at keeping track of complexity and detail. They are prone to careless mathematical mistakes. Computers are *good* at keeping track of complexity and detail, and usually do not make mathematical mistakes unless there is an error in the code.

Engineers should gain a background in manipulating EDA software tools by attending classes offered by the software providers.

1.6 HSDD NEEDS A TEAM THAT EXTENDS BEYOND THE COMPANY

To be effective, the design team may need individuals and companies to provide products and services, including:

- Engineering software tools used to design products
- Technical training in electronics and software tools
- Manufactured parts used in products
- Simulation models

The virtual company, a new work paradigm, is increasingly evident in the workplace. The usual understanding of this term is that a company that brands and owns the intellectual property for a product may subcontract all or part of the actual work. Communications are usually maintained through secure Internet connections. The mini-team composed of a product design engineer, a signal integrity engineer, and a PCB designer may never meet face-to-face. They may only know each other, and communicate through their computer terminals. Today, email, phone conferencing, and video conferencing are valuable team tools.

There is an added challenge when communicating across technical specialties and cultures, especially when English is the second language for some participants. To be effective, design teams and supporting companies need to deal with several issues, such as:

- Reducing the lines of communication, both physically and organizationally in every way possible.
- Coaching individuals in the skills of personally interacting in a positive way with people from many cultures.
- Negotiating common objectives and getting team acceptance of them.
- Ensuring that team members possess adequate technical and managerial skills to deliver the necessary performance.
- Ensuring that the team shares a common authority-responsibility reporting structure.

1.7 HSDD TEAM MEMBERS OFTEN HAVE THEIR OWN AGENDAS

Some companies that participate in the success of the design team may have their own products and services to sell. Sometimes there is a divergence of objectives between the design team and the supporting companies.

- *Engineering software companies* have one main objective: namely, to sell software. As required, they teach users how to run and manipulate their tool. Users should make sure the companies provide good training, service, and support. Engineers should recognize that software companies might not want to be a source of information on how well semiconductor companies do in delivering models.
- *University professors* have a different objective: namely, to provide knowledge to students about theory at a deep level. Engineers should recognize that professors might not like to cover the practical workaday world of designing and manufacturing hardware in great detail.

- *Part suppliers*, primarily semiconductor companies, have another objective: namely, to make and sell the best IC components available. They usually do not get paid[7] for making models of the parts sold. Most semiconductor suppliers see modeling—particularly circuit-oriented modeling—as an added expense with no Return-On- (their) Investment (ROI), as well as a possible legal liability in today's litigious environment.
- *Part users*, primarily the designers working for an Original Equipment Manufacturer (OEM) designing electronic products, need to bring together components and theory plus figure out how to make them all work together under extreme competitive pressure. These engineers would like to avoid spending a lot of time on:

  - Making models work, or obtaining them.
  - Making EDA software work, or learning how to use it.
  - Becoming experts in many very deep technical areas to be able to use modeling and simulation.

1.8 HSDD SIMULATIONS PERFORMED IN THE WORKPLACE

Throughout the design process, HSDD modeling and simulation are used to resolve the following signal noise issues:

- *Signal Integrity (SI)* involves noise generated on digital signals because of impedance mismatches, signal reflections, and crosstalk.
- *Power Integrity (PI)* involves noise generated by digital signals due to unbalanced power-supply impedances and ground-voltage bounce.
- *Electro-Magnetic (Interference) Integrity (EMI)* involves noise generated on digital signals because of conducted, magnetically coupled, and radiated electrical sources.

To solve these signal noise problems, digital engineers need some related insights from analog and RF circuit analysis. Noise on a digital signal is mainly an analog high-frequency effect of high-speed or high-current switching.

---

[7] A recent exception is licensed IP cores used increasingly today in modern digital and embedded system design.

1.9 MODELING AND SIMULATION VERSUS PROTOTYPE AND DEBUG

Practitioners of HALT (Highly Accelerated Life Test) and HASS (Highly Accelerated Stress Screening) caution that design and analysis are not good enough to produce durable and reliable products [52]. They maintain that reality is simply too complex to model and simulate. Sometimes their warnings should be heeded, especially when a product has to be durable. Therefore, their advice is to limit the use of modeling and simulation, and use HASS and HALT instead.

If that approach is adopted, it will simply be a fact-hurdling leap to an erroneous conclusion. The rebuttal is: modeling and simulation is a defense against errors in the design process. Modeling and simulation alone do not guarantee success, but their absence almost surely leads to serious problems.

Leaving the application of effort, knowledge, and engineering to the end of the development cycle is a bad idea. The earlier we can apply effort, knowledge, and engineering, the more options we have. We can more easily optimize among performance, cost, time, and other factors.

More to the point, finding errors at the physical prototype (hardware) stage has some basic disadvantages:

- Fixing problems at the hardware stage is hundreds of times to thousands of times more expensive and time consuming than the cost of avoiding problems predicted at an early, virtual design stage [15, 51, 62, 95].
- Testing (modeling and simulating) a virtual design on a computer takes minutes to hours. Diagnosing root cause, designing a fix, and verifying it in hardware take days to many weeks.
- Too many problems at any stage tend to overwhelm the procedures set up to detect and eliminate them. It is always prudent to eliminate potential future problems as early as possible. Too many problems left undetected until the HALT and HASS stage will overwhelm these tests and their excellent methodologies. There will certainly be test escapes of known problems to the field and to the customer. It is simply prudent and cost-effective to eliminate potential problems at the earliest design stage possible.

Quality, reliability, and excellence must be *designed in* from the very start and re-iterated continuously during a product's development.

1.10 TEN TIPS FOR MODELING AND SIMULATION

Figure 1-2 lists ten important principles to keep in mind as you begin the study and practice of modeling and simulation.

1. Pick the right component and the best model.
2. Simulate early and often.
3. Understand where the numeric results come from, and depend on computers to keep track of the numbers.
4. Balance model simplification and detail to achieve speed and accuracy.
5. Correlate the simulation with laboratory measurements.
6. Develop advanced skills for analyzing and measuring.
7. Understand models at a physical level.
8. Avoid designing with tight tolerances.
9. Work in a team of specialists.
10. Leverage the company and supplier component libraries, plus other resources.

Figure 1-2. Ten tips for effective modeling and simulation

1.11 SUMMARY

High-speed digital design involves many complex technical areas. To compete with cutting-edge products on a global scale, management must form effective teams of experts. Management must also supply engineers with good EDA tools and a supportive company culture.

The old practice of delaying the debug phase until the prototype was available is no longer competitive. It has been replaced by a new, improved design process involving early simulation, debug of the virtual prototype, and the use of many effective EDA tools.

Chapter 2

INTRODUCTION TO MODELING CONCEPTS

Basic terms and concepts: models, systems, and distributed systems

Abstract: Clear and consistent definitions of model and system terms can help engineers understand what modeling and simulation are all about. This chapter introduces modeling for simulation and discusses top-down and bottom-up concepts, sources of noise on digital signals, and limitations of models.

2.1 MODELING AND SIMULATION FOR ALL SCALES OF SYSTEM SIZE

Modeling components, especially semiconductors, is a fascinating and important activity. It can begin at atomic energy levels and extend to the effects of output-switching energy on radiated emissions from an enclosure.

We begin our look at modeling by defining basic simulation terms and concepts. These will be our working definitions and we will explain how they are used.

2.2 COMMUNICATING ACROSS SPECIALTIES

A primary objective of this book is making semiconductor modeling and simulation as simple and understandable as possible. The authors believe that communicating across engineering specialties is important and is a bigger challenge than most people realize. For instance, think about the lack of communication between silicon manufacturers struggling to get more die from a wafer and logic designers dealing with parts that electrically switch on and off much faster than the logic speed requires. The faster the signals switch, the worse the noise generation and ringing problems become. The extra switching speed brings no benefit—only potential problems. The missed communication between supplier and user is not only due to opposing goals but also to misunderstanding each other's technical terminology, culture, and needs.

This chapter begins by defining model and system. For help with other technical terms and acronyms relevant to modeling and simulation, see the Glossary near the end of the book.

2.3 WHAT IS A MODEL?

Models are essential for circuit simulation software to make predictions about system behavior. Specifically, these simulation programs need models of components, modules, and subsystems.

To facilitate cross-disciplinary communication, we must first precisely define the term *model.*

> MODEL DEFINITION:
> A *model* is a mathematical representation of a physical thing, and is used to predict its behavior. The mathematical representation can be either a formula, or data to be used in a formula.

Consider a schematic symbol for a resistor:

A particular physical resistor, or resistors in general, can be represented by this symbol. Then the connections between resistors and other types of parts can be drawn as a schematic. Thus a system (circuit) can be represented in a drawing called a *schematic*. Whenever a system is to be electrically simulated and analyzed, there is an implied model attached to each of its symbols. The models represent the electrical behavior. For a resistor, this model comes readily to mind:

$$I = V/R \tag{2-1}$$

In Figure 2-1 the generator supplies the voltage V across the resistance R and the result is that a current I flows through the R. Because charge and energy are conserved, the current completes the journey around the circuit loop.

2.3.1 Model as Measured Behavior

One model can predict *multiple* behaviors. On the other hand, different models can predict the *same* behavior. Models can represent individual parts of the design, or represent the whole design. Engineers can use the model relationship in these two ways:

- To predict an output from given inputs, and
- To solve for an input from a given output.

2.3.2 Model as an Equation

Given some input x, where a model or a system *operates* on the input according to some function f in the variable x, or $f(x)$ (that is, f of x), we can determine the output of the system y.

> For a simple system, the equation might be: $y = f(x)$
> In this case, we say that $y=f(x)$ *models* the system.

This equation $y = f(x)$ is just *one* general equation; there are nearly an infinite number of specific instances. The equation $y = f(x)$ represents a simple system, where the dependent variable y depends on only one independent variable x.

For a slightly more complex system, the dependent variable y can depend on several independent variables $x_1, x_2, x_3, \ldots$

> For a more complex system, the formula might be:
> $y= f(x_1, x_2, x_3, \ldots)$

Note that $f(x)$ can be any of the following:

- A mathematical function: linear, non-linear, continuous, discontinuous, or state-variable.
- An electrical, mechanical, chemical, thermal, finite, or distributed model.
- A circuit diagram or a mechanical mechanism.
- A vector, a matrix, or matrixes of vectors for multiple interaction effects.
- A system model, representing a combination of models.

Here are two instances of models as formulas:

- The transfer function (analog behavioral model) of a particular electronic circuit.
- The equation that computes the value of Beta for a transistor.

During an electronic system's development, models are exercised with signals to simulate the system's behavior.

2.3.3 Model as Data

The mathematical representation is often *not* a formula. Instead, it is often data used in a formula to make predictions. For example, the value of R in equation R =V/I can be understood in this way. This data representation can also contain the basic structure and characteristics of the design object. If you look at SPICE, Matrix, Scattering-Parameter, or IBIS models, you will not see formulas. You will see data.

In another example, the SPICE model of a 10-ohm resistor has a variable *r* with a value of 10. The formula uses a value when Ohm's Law is used to compute a result, for example *voltage = resistance times current.*

2.3.4 Model as a Structural File

A model can be described as a structural file in the same way that a SPICE netlist is a structural file. The netlist, which follows a structural format, can pull in components, sub-models, and model files as elements. If we want to analyze a net on a PCB, the simulation software usually extracts RLGC transmission line models from the board physical structure and layout of the net. The simulation software also obtains IC models from a SPICE, IBIS, or other component level model. Then with these model elements, the structured netlist forms the *structured* model. *Structured* as used here means that the netlist conforms to a defined format and syntax.

Using a structured model does not mean that we are using a 2D or 3D CAD board layout, plus stackup *directly.* Nor would a 2D or 3D semiconductor structural or device physics file of an IC be used directly.

2.4 WHAT IS A SYSTEM?

Circuit simulation software makes predictions about systems, specifically systems of models. Even when the system analyzed is a single component, the component may be composed of multiple sub-parts. The question always seems to arise about how big is a system? At what size do we become engaged in system analysis?

Again, to facilitate cross-disciplinary communication, we must precisely define the term *system*. Here is how we define system:

SYSTEM DEFINITION:
A *system* is a group of objects operating together.

The Sun and its planets comprise a solar system. Their motions, masses, and distances from each other influence, and cause them to operate on, or orbit (operate) together. A collection of rocks sitting in the backyard is not considered a system, as the rocks have no such detectable influence on each other, even though the same forces (gravitational, electromagnetic, nuclear, etc) operate between the rocks.

For example, in Figure 2-1, consider a schematic (symbolic representation) of a simple system, which is composed of a voltage generator driving a resistive load.

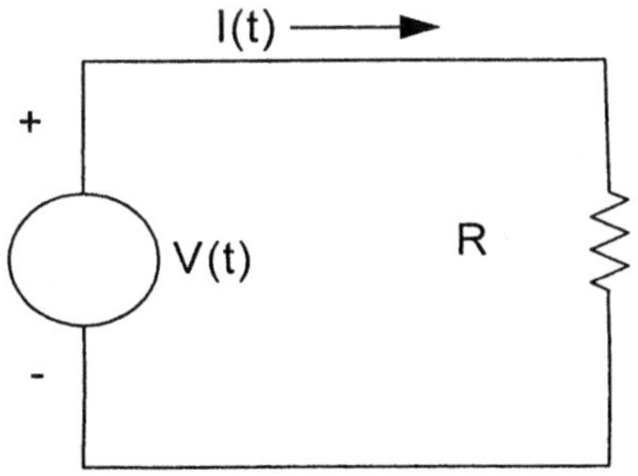

Figure 2-1. A schematic of a simple system

Now let's visualize how this simple system operates. The equation that expresses the system behavior is (Ohm's Law):

$$I(t) = V(t)/R \qquad (2\text{-}2)$$

This system can be the simple representation of an IBIS[1] output (generator) driving a signal over some interconnections to the input (load) of a receiver.

An electronic product or component can be composed of one or more systems as distinguished from a collection of parts. The size of a system is

1 IBIS: Acronym for Input-output Buffer Information Specification.

unimportant. We can start with a system-on-chip, combine chips on a board, combine boards in a chassis, combine several chassis into groups of racks, and scale up to a national telephone system. We can reverse the process, and break the units apart into sub-systems.

In large systems composed of many sub-parts, sub-systems, and providing many functions, we often want to control which parts will influence other parts. We often want to keep various sub-functions from affecting one another accidentally. The design activity includes both making some parts operate together *and* keeping other parts from operating together. The same laws of physics apply in both cases. The designer has to ensure which signals should be passed between parts and which should not.

2.5 NEEDS FOR MODEL ACCURACY CHANGE AS A DESIGN PROGRESSES

Models are essential for circuit simulation software programs to operate and make predictions. Without models of components, modules, and subsystems, these programs cannot simulate system behavior.

Since models are mathematical representations of physical things, we need to know: how accurate are they and how accurate do they need to be? They will probably be less accurate and less detailed than the real item and have a more limited range of conditions over which they are valid. That is all right as long as they are accurate, detailed, and valid enough to predict performance within an engineering approximation. We cannot afford the expense of building something before we have any clue about whether it will work. We also cannot afford the luxury of waiting for the perfect model before we make any predictions about whether a design will work.

There is a tradeoff between design robustness and the need for detail and accuracy in the model. The more robust a design is, the better it tolerates process variation and stress conditions, and the less need for accurate and complex models. Choices that designers make can result in a robust system, or one that is too sensitive to component variation. The need for model accuracy and complexity varies at different stages of product development.

At the early stages of product development, there is usually less need for model accuracy and complexity. At this stage, major choices of technology, product design, and cost are yet to be made. If ballpark answers are available quickly, we can move forward. Early indications of inability to meet specifications should generate immediate action. For instance, if we are faced with amplifying multiple data channels, we can end with a parts count that is too high to meet predicted reliability goals. This would occur if each data channel was amplified and processed in an individual hardware circuit.

A smarter choice might be to multiplex the data and amplify it in a single channel. The right approach is simply a design choice that avoids problems.

In addition, high-speed circuits (digital or RF) are often dominated by the interconnections. Until we have at least a trial topology, layout, and routing, the issue of component models' accuracy may be premature.

As the design nears completion, it is important to give more attention to the modeling accuracy of the most sensitive, state-of-the-art circuits where predicted safety margins are small. If early simulation predictions are based on models with a high level of uncertainty, efforts to get more accurate models may be required. This can happen when working with both:

- Very *new* technologies of which little is known.
- Very *old* technologies of which little is remembered.

For example, some logic device types were developed before simulation came into common use, and some no longer have (or never had) the original SPICE model available. For both cases, we need to pay more attention to lab verification and qualification of the physical product.

The chief criterion in deciding whether more accurate models are needed is the amount of predicted safety margin in meeting design objectives.

Complex systems, which are common, involve many variables. These variables take random values. As a result, performance and predictions of performance involve some uncertainty. To guard against performance failure, designers try to provide some *safety margin* to their predictions. Thus, if a receiver tolerates a noise voltage of 250 mV before falsely switching, the circuit noise budget could be targeted to stay below 150 mV of noise with a 100 mV *safety margin* on noise.

For model accuracy questions, it is important to understand the *physical basis of model parameter variability.* By doing so, designers gain crucial insight into the accuracy and usability of the models. Questions to ask about a physical product include:

- Will all units from a population of devices work?
- Or, will some fail? Under all conditions or some conditions?

For example, a resistor whose parameter resistance value is 10 ohms may vary from +10% to -10% in tolerance due to manufacturing variations. In addition, resistance varies with temperature, causing an additional, physical source of variation that widens the spread from minimum to maximum.

2.6 THERE ARE MANY KINDS OF MODELS AND SIMULATIONS

Model and modeling terms are used in a variety of overlapping ways. When we try to compare and contrast the models and modeling terms, the uninitiated may get confused. It's a bit like looking into a bowl of alphabet soup. Models can be classified in the following ways:

- *Speed* of simulation versus detail of results.
- *Abstraction* versus physical detail level.
- *Analog versus logic operation equation-based models for digital circuits.*
- *Physical detail.*
- *Interconnection analysis* in the frequency domain versus the time domain.

Figure 2-2 is a visual aid presenting the idea that there are many models and many modeling terms.

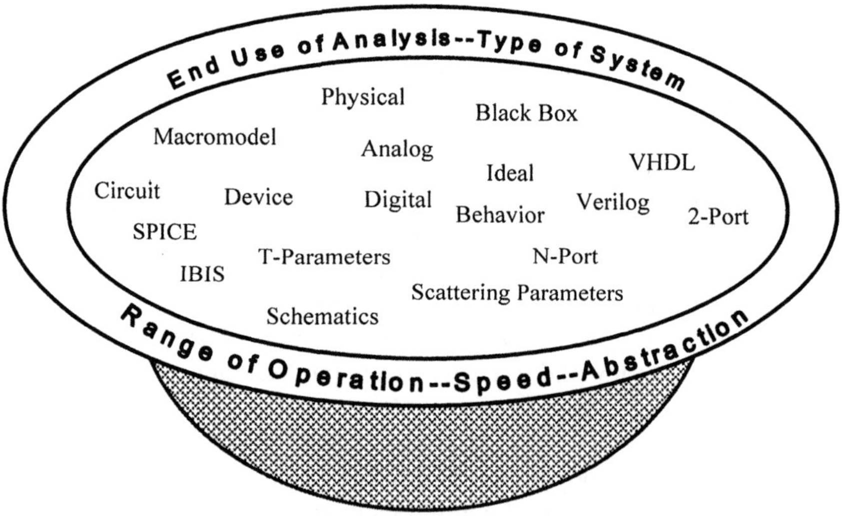

Figure 2-2. The alphabet soup of model types

Some of the confusion over which model is best arises because each model has advocates with enthusiasm that approaches the religious. These advocates believe that the model they support provides all things to all users and does everything best. The authors' position is that each model has a few things that it provides best. And yes, it can also provide some other things, but at reduced efficiency compared to an alternate model.

This leaves model users at a disadvantage because life would be simpler if they could choose one model and stick to it. But as in other things, diversity has value in modeling too. The authors will offer some ideas on living with and profiting from this diversity.

In summary:

- A *single* model can be effective in a variety of user design activities.
- A *variety* of models can be used in a single design activity, and can be more or less effective in their simulation.

2.7 MODELING AND SIMULATION FOR SYSTEMS

Today, much attention is being paid to modeling and simulation. Traditional methods of building→debugging→re-prototyping new products are no longer as successful or competitive compared to incorporating modeling and simulation into the design process. To facilitate the design processes, we must use computers. Computers give us the flexibility and speed of developing a virtual design—one that exists as a computer database—by building →debugging→and re-prototyping a *virtual* prototype.

2.7.1 Components Assembled into Complex Systems

As components get assembled into systems, the systems get physically larger and more complex. As physical size gets larger, so does electrical size. As the electrical size increases, so does the dominance of different electrical effects.

During the assembly of systems, modeling and simulation can be applied at many different levels of the design process. The techniques, electrical properties, and models change depending on the nature of the activity. But the same laws of nature apply at every level.

2.7.2 Electromagnetic Signal Noise Propagates Through Systems

At each level of system integration, the designer needs to consider how electrical noise might interfere with the intentional electrical signals in the circuits. The possible noise sources involve the same mechanisms at each level of integration. These possible undesirable noise sources include:

- Signal reflections on a circuit.
- Signal pickup (coupling) from a nearby circuit.

- Noise generated in power-supply and return circuits, which is then conducted into signal lines.
- Signals radiated to and received from other electrical equipment.

As signals become faster, electrical wavelengths become shorter. As products become faster and more complex, components and structures look "electrically larger." Electrically larger means that for the same physical size, but shorter wavelengths, more wavelengths can be fit within a given dimension of the part. This means that the structure is "seen" as being larger.

2.7.3 Power Consumption Problems

Power consumption grows as more functions get built into a product. Power dissipation, and the temperature rise that it causes, affects electrical performance. In response, part suppliers are busy shrinking the size and power consumption of their components. All this changes the nature of the design challenge in many ways—usually making the design more difficult and complex.

2.8 BOTTOM-UP AND TOP-DOWN DESIGN

The design process influences the modeling and simulation activity. How soon do we need models? How accurate and when? But more than that, when should engineers:

- Build up and use macromodels of the circuitry?
- Disassemble large functional-behavioral blocks into smaller, simpler functional-behavioral blocks?

We next introduce the top-down design process and the bottom-up design process. We also discuss the advantages and disadvantages of both, which are summarized in Table 2-1.

2.8.1 Bottom-Up Design

When individual parts are combined into a larger system, it is called bottom-up design. It is also referred to as system integration. The strength of bottom-up design is that each part (model) is built from known facts and behavior. The weakness of bottom-up design is that the combination of all the parts may not generate the overall performance (functionality) desired.

The design and simulation of the system usually occur late in the development cycle—well after strategic choices are pretty much fixed.

2.8.2 Top-Down Design

When a large, more complex system is disassembled into smaller parts, it is called top-down design. This disassembly may proceed down to the microscopic (micro) level of individual transistor gates. The strength of top-down design is that we start with the overall performance (functionality) we want.

Modeling and simulation begin early in the development cycle and are a great help in making strategic design choices. The weakness of top-down design is that—when we decompose each function into its sub-functions (and finally into its individual components)—we may not get parts and sub-functions that are practical or even physically possible to build.

2.8.3 Top-Down Design with Bottom-Up Verification

The best approach is to start with top-down design, but follow it with bottom-up verification at each major decomposition of the system-subsystem transfer function. Table 2-1 summarizes the advantages, disadvantages, and timing of bottom-up and top-down design.

Table 2-1. Summary of bottom-up design and top-down design

Design Type	Description	Advantage	Disadvantage	Timeline
Bottom-up	Individual parts are combined into a larger system	Each part (model) is built from known facts and behavior	The combination of all the parts may not give the performance (functionality) desired.	The system design and simulation occur late in the development cycle after strategic choices are fixed.
Top-down	Large, more complex system is disassembled into small parts	We start with the overall performance (functionality) we want	Decomposing each function into its sub-functions (and components) may not result in sub-functions that are practical or even physically possible to build.	Modeling and simulation begin early in the development cycle and are a great help in making strategic design choices.

At each major step of the development cycle, it is important to do a decompose→design→verify→adjust feedback loop, proceeding from the general system functions to the specific hardware designed to provide the functionality.

Figure 2-3 illustrates top-down and bottom up design. The concentric boxes show going from a single transistor inside an IC (inside box) to a large system of many parts (outside box), or vice-versa. The top-down design arrow flowing in from the left indicates going from the large and overall to the small and specific. The bottom-up design arrow flowing out to the right indicates going from the small and specific to the large and overall.

A top-down design with bottom-up verification of a digital circuit would follow this sequence:

1. Start with a timing analysis and simulation of the overall logic function.
2. Break down the logic function into smaller blocks that can be made into schematic diagrams of specific logic blocks. In this process, the design targets are derived for each block. But the actual performance possible from available parts and board real estate may differ from the specification. So the overall performance may need to be verified or modified.

   Once a block is designed as a high-level schematic, it contains data on the parts, their electrical connectivity, and the timing and noise margin constraints that apply. To this information, engineers can add some additional targets for network path lengths, physical connectivity, and characteristic impedances. This is called a *network topology*.
3. Simulate the network topology and design it for the desired timing and noise margins. Add terminating resistors and change path lengths as necessary. The actual performance possible from available parts and board real estate may differ from the design targets. So overall system performance may need to be verified or modified.
4. Pass the accumulated design information and constraints on to the PCB designer who will actually design the placement of the parts, and the routing of their connections. Work with this designer to confirm what is physically realizable and/or has to be modified in the constraints. Verify additional desired performance dependent on the physical structure, such as coupled and radiated signal level constraints.

Continue this process of designing to target specs and verifying — correcting the spec as the board goes from:

functional
→ virtual
→ place-and-route
→ prototype and beyond
→ . . .

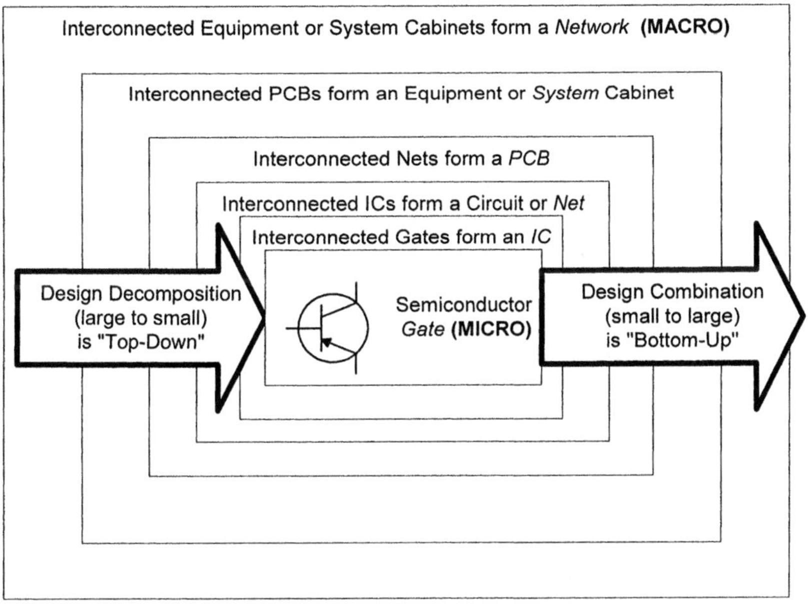

Figure 2-3. Top-down design and bottom-up design

2.9 ANALOG ISSUES IN DIGITAL DESIGN

2.9.1 Frequency, Wavelength, Size, and Distributed Systems

Frequency, wavelength, and size are analog issues. So why should digital designers be concerned about them? Simply put, digital design is no longer just about ones, zeros, and time delay. The issues digital designers face are not just digital logic functionality and timing. Frequency and wavelength are also important because physical size and wavelength interact together to affect signal noise. As switching gets faster, frequencies present in the switching waveforms increase and wavelengths get shorter. The physical size of components and interconnections relative to wavelength size affects how electrical signals behave. More discussion follows:

- When a signal generator sends an electromagnetic wave out of its output terminals, the energy sent out has to match the energy absorbed by the load it is driving. When the time delay between generator and load is short, we are not aware of any effects due to mismatches between energy sent out and energy absorbed. However, as wavelengths shrink, energy reflections, measured as signal reflections, may occur if the energy sent out by the generator is different than what the load can absorb.

- The reflections begin to happen when a physical distance becomes larger than about 1/10th of a wavelength.[2] The signal conduction path then begins to act like a *transmission line.* The load that the generator sees at its terminals is the characteristic impedance of the transmission line. As an example, let us assume that a 5-volt signal is launched onto a 50-ohm transmission line. The current launched will then be 100mA and the energy in the E field and H field will be coupled together.
- If the load impedance at the receiving end is different than 50 ohms, then the current cannot be 100mA when the 5-volt signal arrives there. Therefore, there is an energy mismatch at the receiving end, and part of the energy must be reflected back toward the source (driver) to maintain energy balance.
- Propagation delay becomes significant because the signal starts to change faster than the change can travel to the other end of the circuit. If there is an energy mismatch at the receiving end and the source end, an energy wave begins to travel back and forth between the source (driver) and receiver.
- Signal carrying structures, such as interconnections, begin to radiate when wavelength shrinks to where the structure begins to look like a half wavelength long (or multiples thereof).
- As structures begin to look a half wavelength long, they can begin to support unwanted resonances.
- As frequencies increase, coupling and crosstalk increase.

To put more functions into an IC and into PCBs, semiconductor suppliers are making smaller component sizes. To dissipate less heat, their operating voltages have been decreasing. This is good in relation to the bulleted list above. But there are some cautions:

- Switching currents, particularly di/dt, have increased especially as the numbers of gates in ICs have grown. This is to deliver more functions to the users.
- Along with shrinking size and power levels, device fragility and sensitivity to signal noise have increased.

This last bullet means that noise levels that could have been ignored in the past no longer can be. Switching threshold voltages have decreased along

---

[2] Rule of thumb

with decreased operating voltages. Most electrical engineers and technicians know that increased switching speeds and power supply currents are challenges. However, they may not be as aware that signal-to-noise ratios are shrinking. *We can no longer say* that low noise levels have no discernable influence on our logic operation.

Therefore, component and circuit size matters. Table 2-2 shows what happens to wavelength (represented by lambda, λ) as frequency increases. The authors' purpose is to get engineers to visualize and think in terms of wavelengths. We cannot overemphasize how important this viewpoint is to those that are trying to understand and model circuit components and boards.

Component sizes are shown in Figures 2-4 to 2-6. The photos and Table 2-2 are courtesy of Dr. Randy Yost, Electrical and Computer Engineering, Utah State University.

Table 2-2. Frequency related to wavelength (size)

	Frequency				
	100 MHz	**1 GHz**	**3 GHz**	**10 GHz**	**20 GHz**
λ (Metric)	3 m	30 cm	10 cm	3 cm	1.5 cm
λ (English)	≈ yard	≈ foot	≈ 4 in	≈ 1.2 in	≈ 0.6 in
Application	FM radio	Cell phone	CPU speeds	X-band radar	K-band radar
Physical	NA	NA	NA	← Circuit elements →	
size range	NA	NA	← Circuit boards →		
of variation	← Circuit enclosures →				

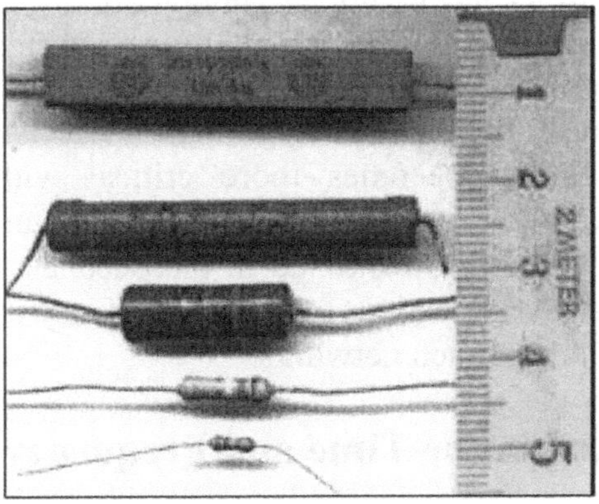

Figure 2-4. Discrete power supply component sizes

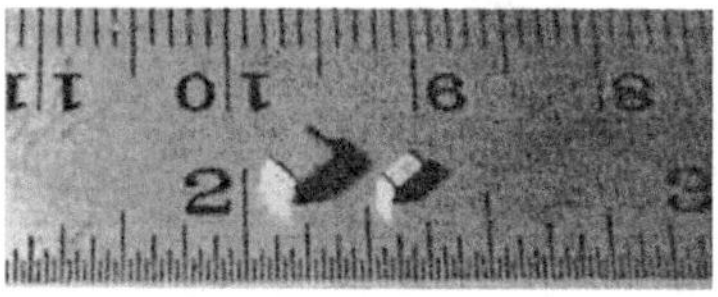

Figure 2-5. SMD digital component sizes

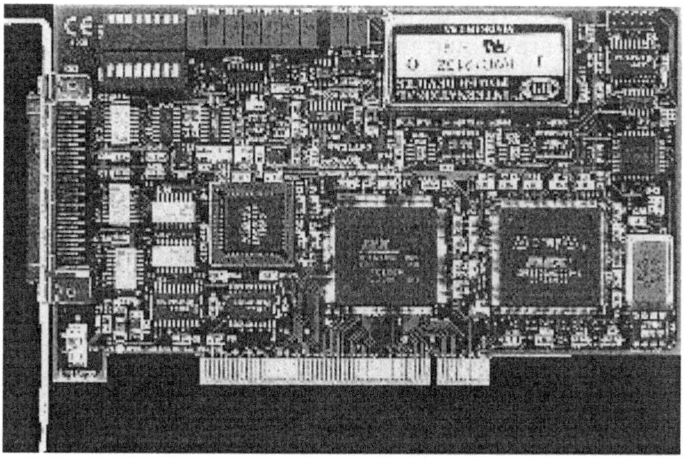

Figure 2-6. Digital PCB (near life size) with placed components

Let us consider clock rates before continuing our discussion of the effects of frequency, wavelength and size, especially as affected by rise time and edge rate. Many books on Signal Integrity edge rate effects emphasize them very strongly because of their analog signal effects. But it is worth remembering that timing becomes more critical with increased *clock frequency*. Timing margins for parallel bus architectures disappear, and propagation delays have to be reckoned in picoseconds. The impracticality of such close tolerances has led to the ascendancy of serial data bus architectures for the fastest such networks.

2.9.2 Thinking in the Time and Frequency Domains

Amplitude versus time electromagnetic (EM) waveforms can be measured when an observer with an oscilloscope probes a point on a circuit. This EM wave propagates along the interconnections from its point of injection to other points in the system.

We do not sensually experience the frequency domain (or dimension) directly as we do with space (distance) and time. But to gain understanding about traveling EM waves, transmission lines, and how the IBIS model is applied, we often need to convert our thinking to the frequency domain. The mathematics that helps us make the transition back and forth between a time waveform and its frequency spectrum is called Fourier analysis and transform. A time waveform can be thought of as a set of discrete frequencies (sine and cosine time waveforms) of various amplitudes and phases.

Once we understand this conceptual and mathematical transformation, we can start to better answer how reactive impedance, characteristic impedance, skin effect losses, dielectric dispersion, wavelength, reflections, electromagnetic radiation and interference, and many other properties behave with frequency. Having found these answers in the frequency domain, we can construct what is going on in the time domain by doing an inverse Fourier transform and reconstructing the time domain behavior.

2.9.3 Time, Distance, and Frequency: Lumped and Distributed Circuits

The velocity of an electromagnetic (EM) wave in air, or in free space is (denoted by):

$$C = 3 * 10^{10}\ cm/sec = v_p \tag{2-3}$$

The EM wave may be partially reflected and partially absorbed (transmitted, received) when it reaches another medium or discontinuity, as in radar and microwaves.

If the reflected wave arrives back at the signal source in a very short time, the time element across a circuit drops out except in terms of charging capacitors, establishing currents in inductors, and turning devices on and off. Everything happens "instantaneously" around a given circuit and the mathematics for analyzing what is happening simplifies. Ohm's Law and Kirchhoff's Laws become the basis for thinking about what is happening. Such circuits are called "lumped" because their properties and behaviors can be modeled in terms of *lumped* elements with connections between them as ideal, property-less wires. The lumped elements are resistance (R), inductance (L), conductance (G), and capacitance (C).

If propagation delay, t_p, across a circuit becomes significant, propagation velocity, v_p, and other issues must be considered:

$$t_p = D/v_p \quad \text{where D = distance} \tag{2-4}$$

For any reflected wave to arrive back at its source, it takes twice t_p to make the round trip.

At low frequencies, EM waves prefer to propagate through conductors where material conductivity is high. The EM waves do not normally propagate into a dielectric (non-conductive) medium surrounding the conductor. To do so, the source energy on the conductor begins acting like an antenna. This begins to occur when the dimensions of the source and conduction path become 1/10 or more of a wavelength, λ.

The length of an electromagnetic wave, λ, in a propagating medium is:

$$\lambda = v_p / f \quad \text{where f = frequency} \tag{2-5}$$

For low to moderate frequencies, electromagnetic propagation is usually confined to guiding conducting wires and the surrounding dielectric medium. R, L, G, and C elements characterize this medium. At higher frequencies (radio, and microwave), wave propagation begins to be characterized by radiation of the wave from antenna structures. At these higher frequencies, losses in conducting elements become high, and it is better to not have conducting or high dielectric constant elements. Energy can propagate through dielectrics and/or be guided by dielectric structures very effectively until the x-ray frequencies.

The region of chief interest to Signal Integrity engineers lies in between the lumped element region and the pure radio and microwave region. In this region, electromagnetic energy is normally guided along conductive paths, but is strongly affected by the properties of the surrounding dielectric medium. The connections that guide the energy become known as *transmission lines*. The EM wave energy penetrates only a short distance into the conductor (skin effect), and the electric and magnetic fields appear chiefly in the surrounding dielectric. Transmission lines can be modeled as lossless at the lower frequency end of this range and lossy at its higher end.

Propagation velocity in a high dielectric constant medium is slower than in air. This fits intuitively because the electric property (per unit volume) of the dielectric capacitance is increased substantially over that of air. And there is a time delay associated with energizing and de-energizing this element. Expressing this in a dielectric, velocity becomes lower:

$$v_{p'} = C/(\varepsilon)^{1/2} \tag{2-6}$$

where ε = dielectric constant, and we may have situations where $\varepsilon = \varepsilon_{r'}$ where $\varepsilon_{r'}$ is effective relative dielectric constant of a microstrip that is partially in air and partially in PCB dielectric.

Effective, $\varepsilon_{r'}$ refers to whether the field lines are in dielectric (stripline) or partially in dielectric and partially in air (microstrip), and relative, ε_r, refers to dielectric constant relative to air. Time delay increases because:

$$t_{p'} = D/v_{p'} = (D/v_p) * (\varepsilon)^{1/2} = t_p * (\varepsilon)^{1/2} \quad (2\text{-}7)$$

Working out the numbers:

$$t_{p'} = 85 * (\varepsilon_{r'})^{1/2} \text{ picoseconds per inch} \quad (2\text{-}8)$$

And wavelength becomes smaller, as seen in equation (2-5). Or as we should really consider for our dielectric medium:

$$\lambda' = v_{p'}/f = v_p/(f * (\varepsilon_{r'})^{1/2}) \quad (2\text{-}9)$$

Thus a high dielectric constant material around our conductor will make it behave like a transmission line at a lower frequency than a conductor in air. Our discussion of time, distance, frequency and wavelength leads to this question: what is the *electrical length* of a waveform feature that will start to interact according to transmission line behavior with the physical length of an interconnection? The electrical length of a feature, for example t_r, is given by:

$$L = t_r / t_{p'} \quad (2\text{-}10)$$

Systems small enough for all points to act in near unison to an EM wave traveling on them are lumped systems. Systems larger than that are transmission line systems. Reflections and re-reflections that travel back and forth on these systems can become a problem because they introduce noise on the waveform being transmitted. The rule of thumb [60] is that systems with:

$$D = (\text{waveform feature electrical length})/6 \quad (2\text{-}11)$$

are at the transition point for transmission line behavior. Assuming a stripline with $\varepsilon_{r'} = 4.6$, we get:[3]

$$t_{p'} = 182.3 \text{ picoseconds/inch} \quad (2\text{-}12)$$

With a $t_r = 1$ nanosecond, we get $L \approx 5.5$ inches and $D \approx 0.92$ inches as the transition to transmission line behavior. At rise times of 100 picoseconds,

[3] ε_r for FR4 material

D ≈ 0.1 inch and transmission line behavior begins to occur within IC packages themselves. Vias connecting the IC to signal planes in a board begin to behave like monopole antennas with no canceling image currents on supply planes.

We briefly mentioned that as frequency increases, the EM wave energy is less able to penetrate a conductor. This is known as *skin effect* and it results in significant increases in conductor resistance and conduction losses. Secondly, most PCB dielectrics begin to behave differently. Their dielectric constant begins to roll off. Wave propagation speeds up at the higher frequencies, and the frequency components of the wave begin to arrive at different times at a receiver. This effect, called dispersion, adds to time domain waveform distortion. A third effect is the increased dielectric conduction losses (characterized by the G parameter of the transmission line model) as frequencies enter the GHz range.

2.10 NOISE MODELING ON ELECTRICAL SIGNALS

2.10.1 Signals Passing Through a Circuit Can Get Noisy

In previous topics, we discussed component parts getting assembled into circuits. We also discussed frequency, wavelength, size, and their effects on electrical performance. It should now be apparent that modeling and simulation in both the time and frequency domain are important skills for digital designers of logic circuits.

We now turn our attention to the electrical signals that propagate through parts and circuits. We further discuss how these clean ones and zeros can get noisy, distorted, and unusable.

When analyzing electrical signals, we must consider what kind of analysis to perform at any level of system integration from the microscopic to the macroscopic. Consider an electrical circuit schematic. It changes in no essential way as we go from the insides of an IC to the outside of, say, a large computer. The same theoretical circuit principles and laws hold and can be applied at any level. But the models change to incorporate new effects. One of our concerns is the sending and receiving of good electrical signals—signals uncorrupted by having picked up too much noise.

2.10.2 Noise Sources and Pathways

The four major pathways for electromagnetic noise signals to interfere with a desired signal are caused by:

- Conduction
- Coupling
- Radiation
- The transient response of the power supply system

Note: The power delivery system gets noisier when the equipment demands greater dI/dt and larger quiescent current capabilities. This noise gets transferred onto the signal output circuitry. This is because the output circuit and the power source-sink share a common conduction path.

Conduction, coupling, and radiation are laws of nature. These mechanisms operate at every level, from microscopic to macroscopic. They can be significant or negligible in their effects depending on electrical circuit properties and signal levels. Unwanted noise can be generated and transferred at each level of system integration by one or many of these mechanisms.

This book's focus is noise on digital electrical signals. Analyzing noise on digital-electrical signals involves modeling some analog effects.

For there to be a problem with noise on digital-electrical signals, there must be a noise source, a pathway, and a receiver. Analyzing source-pathway-receiver is circuit analysis. Circuit analysis is not the focus of this book but does require some discussion. We will focus more on the *models* used in analyzing noise on digital electrical signals.

2.10.3 Noise Problems and Solution

Discussion of Signal Integrity, Power Integrity, and EMI/EMC is necessary. So far we have discussed some of the sources of electrical noise and the technical disciplines applied to their mitigation. As signal speeds increase, so do concerns over noise in electronic circuits. SI, PI and EMI analysis now extends down into the interconnection of individual gates within ICs, as well as upward into all encompassing systems like the Internet.

In Table 2-3, we summarize some of the noise generation and injection concerns and issues of high-speed design.

- Pathology is a disease mechanism.
- Indication is a symptom that is observed.
- Prognosis is the problem caused by the disease.
- Therapy is what to do to avoid the potential problems.

Table 2-3. From noise problem to solution for high-speed design

Pathology	**Indication (symptom)**	**Prognosis (untreated)**	**Therapy**
Conducted Transmission Line discontinuities	Reflections	Poor SI = noise on signal, interference with switching logic	Impedance match. Lower edge rate. Conductor length $< 2t_R$.
Coupled signal line-line, mostly inductive	Crosstalk	Poor SI = noise on signal, interference with switching logic	Increase spacing, less parallelism. Shielding.
Conducted transient voltage sag and ringing	Ground and Power Bounce	Poor SI and PI= noise on signal, interference with switching logic plus EMI effects	Lower the edge rate. Add bypass capacitors. Add series termination to lower the current.
Conducted-coupled-radiated interference and susceptibility	EMI	Regulatory failure, plus SI effects	All the above techniques, plus shield, enclose and source suppress.

2.11 ADDITIONAL DESIGN ISSUES TO MODEL AND SIMULATE

The handling of parts as a result of manufacturing, shipping, assembly, and testing may produce a number of parts with latent defects, which may cause them to fail prematurely. Electrostatic discharge damage (ESD) is a common concern. The design itself and environmental stresses may exercise parts with latent defects in ways that will hasten their demise. The latent defects then contribute to *infant mortality.* Additionally, a circuit application may simply overstress a part or cause it to *wear out.* The traditional *bathtub curve* of part failure rates is usually explained as an infant mortality early period; followed by a low, flat random environmental overstress period, followed by a wearout period.

High power dissipation may require a part to be assembled on a board with a heat sink if it is not to immediately begin wearing out and failing at a high rate. Attaching a heat sink will have immediate implications for SI, PI, and EMI performance. Attaching a heat sink to a component may cause a change in how it should be modeled and simulated.

Variation, latent defects, and environmental stress are issues traditionally dealt with by the disciplines of Quality and Reliability. But process variation causes some device to pass or not pass a timing or Signal Integrity performance requirement. This has immediate quality and reliability implications. Before the physical product has been built, its variations can be modeled, simulated, and designed to meet requirements.

2.11.1 Modeling Limitations: Simplifications and Assumptions

When we simplify a model, the result is often a gain in understanding and speed of analysis. This is because we first want to see the big picture. However, we must be careful to not use a simplified model beyond its range of definition and assumptions. Otherwise, we might run into trouble. We can get into trouble even faster when we assume that we understand a model when we really do not.

Consider a series R-L-C circuit with a voltage generator driving it. For this example, we will assume that time delay around the circuit is negligible. Looking into the circuit in Figure 2-7, we ask, "What is the load impedance across the generator?" A simple application of Ohm's Law and Kirchhoff's Laws will yield our solution.

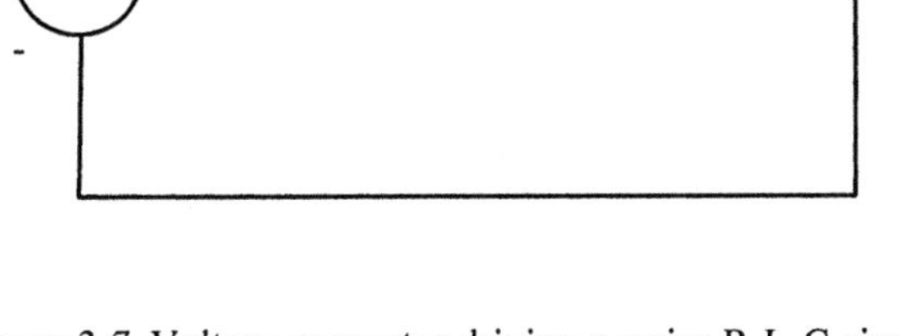

Figure 2-7. Voltage generator driving a series R-L-C circuit

To keep things simple, we'll assume zero initial voltage (charge) on the capacitor and zero initial current (magnetic field) in the inductor.

Also, we have assigned $R_{(total)} = R_{generator} + R_{network}$ (2-13)

By inspection, the time domain response of this network is:

$$V(t) = I(t)[R^2 + (X_L - X_C)^2]^{1/2} \tag{2-14}$$

Solving for $Zin = V(t)/I(t) = [R^2 + (X_L - X_C)^2]^{1/2}$ (2-15)

where $X_L = 2\pi fL$ and $X_C = 1/2\pi fC$ (2-16)

In this case, the impedance, especially near resonance, is sensitive to the magnitude of R. Figure 2-8 shows what happens with I-V relations when we look at typical buffer I/O behavior:

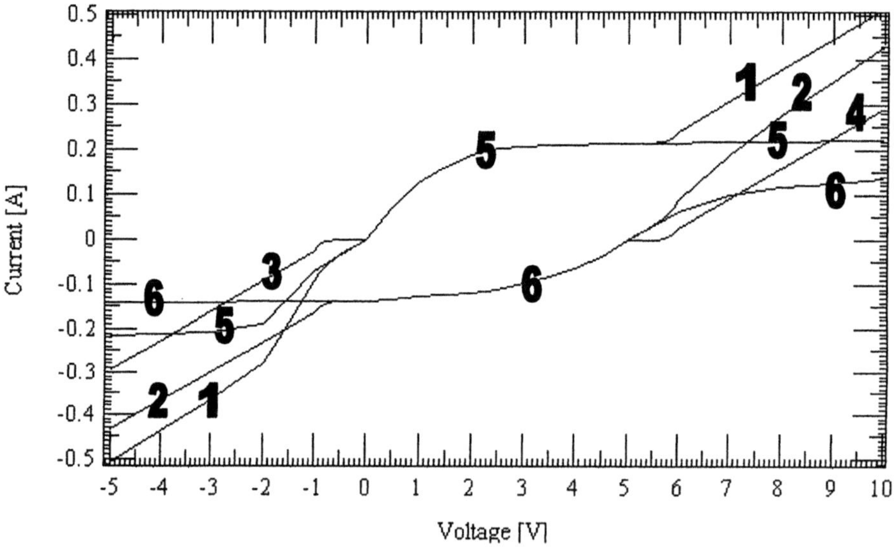

LEGEND

1=Fall Edge Composite Typical
2=Rise Edge Composite Typical
3=Ground Clamp Typical
4=Power Clamp Typical
5=Pull Down Typical
6=Pull Up Typical

Figure 2-8. I-V curves of a CMOS driver

What happens when we replace our generator (having a linear, small magnitude internal resistance of perhaps an ohm or less) by a non-linear output driver, including clamps, of a CMOS switching transistor? A typical set of I-V curves would be like the one shown in Figure 2-8. Here we see conductance (I versus V) directly instead of its inverse, resistance.

This driver might average 25 ohms output impedance as a linear approximation in its active region, and a few ohms in the clamping regions. We see very different results near resonance than for our ideal generator. We also see very different voltage—current waveforms than for our original circuit. This is especially true if we insert a transmission line and drive the clamps on and off with mismatch reflections.

All of these illustrate the need to be aware of our modeling assumptions and model limitations.

2.11.2 Types of Models Presented

In this book on modeling and simulation for HSDD, we primarily cover the IBIS model. This is because it has become industry's model of choice. SPICE gives better results with respect to the effects of temperature, bias, and load variation. Scattering Parameters give better results with respect to very high-frequency effects. But compared to SPICE and Scattering Parameters, IBIS is best for the great majority of HSDD simulation tasks. This is because:

- IBIS is sufficiently accurate for most tasks, and more consistently provides process corner data, not just typicals.
- IBIS is much more available because it is behavioral, and thus cannot readily be reverse engineered.
- IBIS runs much faster in simulators, because it is behavioral.
- IBIS runs in many more simulators, and is portable across them, because of its agreed-to industry standard.

Most of the book's emphasis is on models for semiconductor devices. Models for passive components are discussed as they apply to circuits using semiconductors. Likewise, transmission-line models are discussed as they apply to circuits using semiconductors. Also discussed in the book are:

- Model parameter variability due to component process variability.
- Choosing the best type of model and simulation for a design activity.
- Dependencies of model characteristics on physical device characteristics and design.

AHDL and RF modeling are covered for the sake of completeness and because of some convergence between microwave and serial Gbit and GHz design techniques. SPICE has one of the best reputations and largest followings of any of these modeling methods, but be aware that SPICE comes in many different flavors. To name a few: Berkeley SPICE, BSIM3, BSIM4, HSPICE, PSPICE, AIMSPICE, and EKV. Parameters are not interchangeable between models, or between simulators. For example, HSPICE Level 28 and Level 53 MOS models have different equations and different parameters. The BSIM3 Level 3 parameter values (and even parameter names) are different for different simulators. In addition, parameters for a 0.13-micron salicide process are different from those without salicide.

2.11.3 Judging a Model's Usefulness

By adding our experience to the information in Table 2-3, we can extract the measures of a model's worth, which is summarized in Figure 2-9.

How useful is our model ...

- For analyzing large signal switching?
- For wideband frequency response from a few MHz to many 10s of GHz?
- For modeling input/output impedance of drivers and receivers, including dynamic changes in their magnitude and the effect of clamping diodes?
- For describing model parameter population variability?
- For describing the spread in model parameter values due to variation in voltage, temperature and other environmental conditions?
- In "allowing" a simulator to converge on a solution?
- For modeling the power-ground effects on output-input signals?
- For modeling IC die and package parasitics and couplings between pins?
- For handling ICs with variable drive outputs, series switching and more?
- For the ability to provide model information easily compatible with analysis in the time domain or the frequency/wavelength domain?
- For simulating fast?
- For providing only results appropriate to the analysis at hand?
- In terms of its availability?
- In terms of its acceptance and understanding in the design technical community?
- Also, when and where does the model fail?
- And, does the model fail "gracefully" or suddenly?

Figure 2-9. Questions to use for assessing models

2.12 USING EDA TOOLS FOR SEMICONDUCTORS

Figure 2-10 shows how EDA tools impact the phases of manufacturing and using semiconductors.

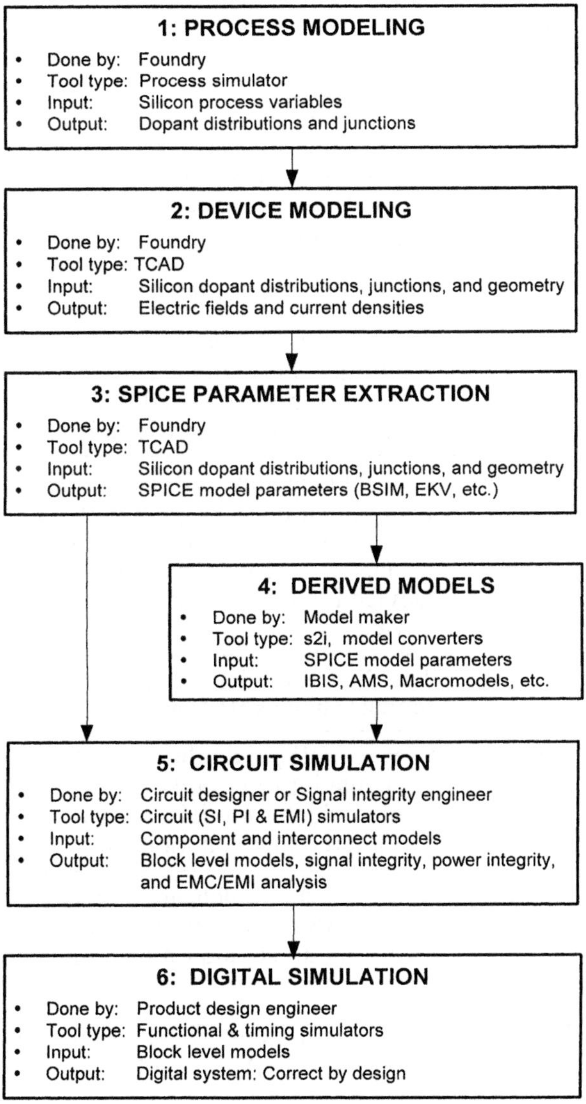

Figure 2-10. The uses of EDA tools in the design, manufacture, and application of semiconductors

Electrical circuit modeling of semiconductors begins with semiconductor material properties and structures. Semiconductors, as used on PC boards, can ultimately be modeled at the circuit and system level.

Figure 2-10 shows the use of EDA tools to model and simulate from input variables to output results. Each box shows who does the process, the type of EDA tool used, the model input to the tool, and the tool's output. The following table explains the flow:

Box in Figure 2-10	Description
1. Process Modeling	Deposition times, drive-in temperatures, dopant species, substrate crystal properties, and much more, are used to find junction depths, dopant gradients and other quantities that specify the structure and materials properties of the semiconductor devices.
2. Device Modeling	The dopant distributions, electron (hole) mobilities, bias voltages, junctions, geometry and much more are used to find energy bands, current densities, electric fields and more.
3. SPICE (transistor-level) Parameter Extraction	Current densities, drift velocities, carrier recombination rates, geometry and much more are used to extract the SPICE model (see "Chapter 3, Model Properties Derived from Device Physics Theory") properties of the device.
4. Derived Model extraction	The SPICE model is used to derive other models such as IBIS and S-Parameters.
5. Circuit Simulation	SPICE, IBIS, and other models are used to find the behavior of circuits composed of the semiconductor models, passive device models, and interconnection models.
6. Digital Simulation	System level models are extracted, using VHDL-AMS and similar methods, and system level functional and timing simulations are run.

In "Chapter 22, Future Trends in Modeling," we discuss the availability of EDA tools to accomplish the various modeling and simulation activities.

2.13 USING EDA TOOLS FOR BOARD INTERCONNECTIONS

Figure 2-11 shows the involvement of EDA tools in the manufacture of PCBs and interconnections and their use in electronic systems.

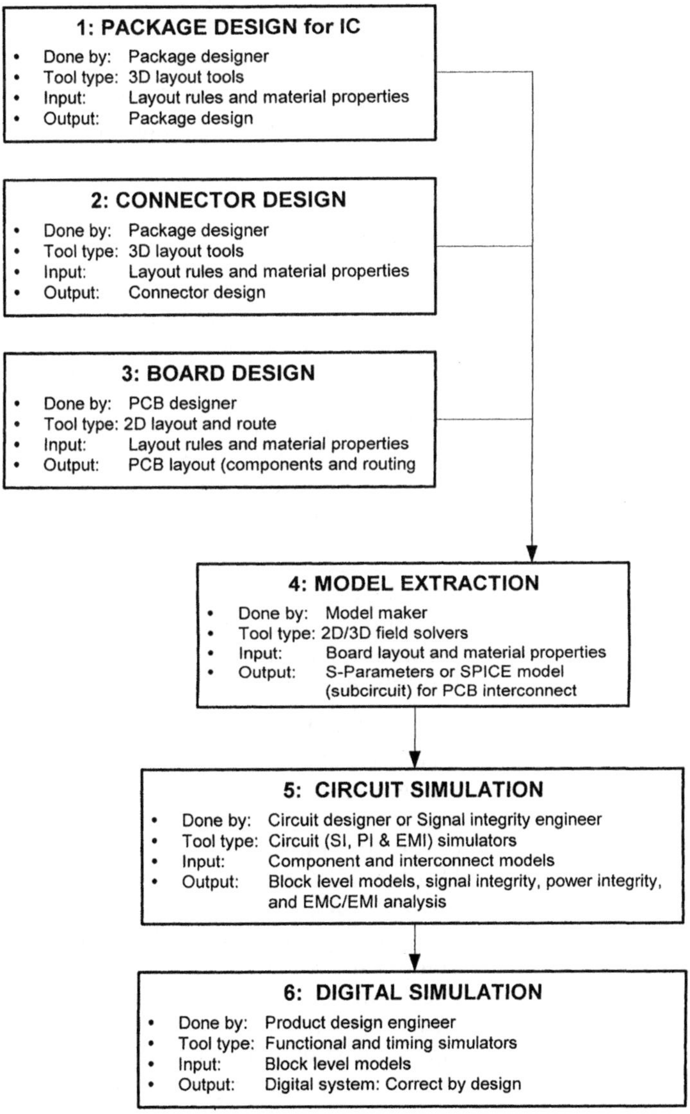

Figure 2-11. The uses of EDA tools in the design, manufacture, and application of interconnection, components, and PCBs

Modeling a modern high-speed digital board must include connectors, IC packages, the PC boards, enclosures, cables, and systems composed of these elements. Passive component models and models of cables and connectors often come from manufactures of those components. Simple devices are easy to add to most circuit simulators. But where do the interconnection models (Box 5 of Figure 2-10) come from? Mostly, the modeling of board interconnections, connectors, and IC packages begins with 2D and 3D EM field solvers.

Figure 2-11 shows the use of EDA tools to model and simulate from input variables to output results. Each Box shows who does the process, the type of EDA tool they use, the model input to the tool, and the tool's output. The following table explains the flow:

Box in Figure 2-11	Description
1. Package Design 2. Connector Design 3. Board Design	CAD tools are used to construct 3D models (and sometimes 2½ D[4] models) of the component to be modeled. Materials properties are also entered into the CAD database.
4. Model Extraction	A 2D or 3D field solver is used to extract model properties. SPICE, RLGC matrices and S-Parameters are the common model types extracted.
5. Circuit Simulation	SPICE, IBIS, and other models are used to find the behavior of circuits composed of the semiconductor models, passive device models, and interconnection models.
6. Digital Simulation	System level models are extracted, using VHDL-AMS, C/C++, or any other language. System level functional, timing simulations, and power consumption are run without describing the structure inside the models. Digital simulation can include anything. It must include functionality (input/output relationships, legal and illegal state detection). It could include cycle-based or nsec timing, load-dependent timing, power consumption, output switching characteristics, and anything else commonly used in modeling digital systems.

[4] 2½ D implies geometry that lies in a 2-dimensional plane with added properties like propagation velocity along the plane layer thicknesses, and more.

Note: Sometimes the division of labor between PCB board layout designer and Signal Integrity engineer is not exact. This is also true for the division of labor between Signal Integrity engineer and Product Design engineer. Most often these are different individuals because of the specialized knowledge required.

For more information on circuit simulation and computational algorithms, see [94, 131].

2.14 LOOKING AHEAD IN THE BOOK

There is more than one way to organize the subject matter of *Semiconductor Modeling.* The chapters are organized in the following general sequence:

1. Where models and simulation fit into product development.
2. Generating 2-port, Scattering Parameter, SPICE, and IBIS models.
3. Selecting appropriate components for the product, and suitable models to simulate the design.
4. Information about the IBIS model file: format, data, and quality control.
5. Managing IBIS models for simulation.
6. Checking and verifying IBIS models.
7. The future of IBIS and other modeling techniques.

This organization follows how a circuit design engineer might proceed with the task of understanding and using models in simulating circuits. The authors started with basic concepts and developed the discussion from there.

In the next few chapters, we look at the specifics: how to measure model parameters, how to characterize devices and models, what a data sheet tells us, and how to simulate model parameters from basic semiconductor physics.

2.15 SUMMARY

The basic concepts about modeling and simulation as presented in this chapter include:

- Models are mathematical representations of the behavior of devices and systems of devices.
- Systems are groups of objects operating together.

- Mathematical expressions for the behavior of systems can be combined from their constituent parts (bottom-up design). Also a mathematical transfer function equation for the system can get decomposed (top-down design) into constituent parts.
- Modeling and simulation is done at all size scales. The same laws of physics apply at all scales. But those physical effects that are significant change with size, switching speed, and other factors.
- Switching and the propagation of signals physically occur in the time domain. High-frequency analog effects of fast switching are sometimes better analyzed and understood in the mathematics of the frequency domain.
- Electrical signal noise is a prime physical consequence of high-speed digital circuit switching. As switching speeds have increased, high-frequency signal attenuation has become more significant. The added noise and loss can interfere with the functioning of logic circuits and has to be modeled and simulated.
- Different types of models exist and help engineers do some analysis jobs well and other jobs not so well.
- Modeling and simulation with the aid of EDA tools is done at every step of the design process, from the construction of ICs to the construction of large electronic systems.

PART 2: GENERATING MODELS

Chapter 3

MODEL PROPERTIES DERIVED FROM DEVICE PHYSICS THEORY

Using basic physics to extract model parameters for modeling and simulation

Abstract: TCAD modeling tools are used to extract circuit model properties for SPICE from the structure, material, and electrical properties of semiconductors. Device design determines electrical performance. TCAD provides a link between semiconductor device design and the electrical behavior represented by models for those devices. Throughout this chapter, discrete semiconductors are used as examples in the discussion of the link between device design and model parameters.

3.1 INTRODUCTION

Simulators are wonderful, almost magical tools. But if simulators get inappropriate model parameters, they produce garbage–*not* useful answers. To ensure that useful output is generated, engineers need knowledge, understanding, and insight to help them avoid making mistakes. The authors recognize that understanding basic concepts and where the numbers come from are important in appreciating the simulation results. This chapter provides insight into semiconductor modeling for those engineers interested in these matters.

This chapter explains how model parameters are calculated from semiconductor device physics and the limitations imposed by both device physics and the simulator implementation of those models. Together, chapters 3, 4, and 5 cover fundamental knowledge that will help users avoid making mistakes in modeling and simulation. If the reader is in a hurry to pick a component model and simulate it, the reader may wish to skip to

"Chapter 6, Using Selection Guides to Compare and Contrast Components." They can return to these chapters later for more detailed understanding of the relationship between datasheet parameters and model parameters.

3.2 WHY DEEP SUB-MICRON TECHNOLOGY IS COMPLEX

CMOS models and equations (3-23) to (3-26) emphasize modeling from a physical perspective. These models are simpler and easier to follow than a full development for deep-sub-micron CMOS. There are currently about a dozen different high-level deep-sub-micron CMOS models, each involving over 70 parameters. Some models incorporate as many as 200 parameters. Most of these models and parameters are not totally portable from one simulation platform to another. Much of the modeling is proprietary and unavailable to the general public. Many of these models are based on UC-Berkeley models [117].

This topic is an overview of modeling methods. Therefore, what is the purpose of presenting CMOS models in full detail to an audience unlikely to specialize in model extraction? The answer is that the authors want circuit designers to understand the physical basis of SPICE models.

For today's deep-sub-micron IC lateral MOSFET devices, modeling assumption require the realization that horizontal spacing is comparable to vertical spacing, The horizontal BJT formed as part of a MOSFET in a substrate well becomes available as a significant circuit element. This BJT can be used as a low-gain element in a design. The opposite side of this convenience occurs when the BJT shows up as an unwanted parasitic element.

For deep sub-micron CMOS, higher-order effects that were negligible have become significant, to the point of sometimes dominating over the first-order effects. In the early days of SPICE modeling, it was assumed that most of the current flow was across a flat area, and that the effects of perimeters and corners could be neglected. For instance, sidewall capacitance is insignificant in a large-area discrete (vertical) BJT. But it predominates other capacitances in a deep sub-micron lateral IC BJT transistor. Later models added some perimeter effects.

A typical fine emitter finger on a 2N3904 (or a 2N918 RF transistor), developed in the early 1950s, might be a mil or two (25 to 50 microns) across. Today's advanced IC technology is working with geometry features below 0.1 micron. The modeling of today's integrated circuit BJTs and MOSFETs is a whole new situation from when SPICE was first developed

By today's standards, the small signal 10x15 mils to 30x30 mils discrete device described in "Chapter 7, Using Data Sheets to Compare and Contrast Components" may seem huge. However, consider some high-power giant BJT transistors that were being sold in the 1970s. Such devices sometimes took an entire 3½-inch diameter wafer per device (the largest wafer then in common use). Today wafer diameters are at 12 inches and larger. We still use power transistors of similar size to handle high-power and high current. Small-signal transistors remain in use as well, typically offered in 1/8 W to 1 W packages. For comparison, today's integrated circuit transistors are made in CMOS, with each transistor handling around 50 micro Watts.

Figure 3-1 shows a single transistor, n-channel, packaged, Insulated Gate Bipolar Transistor (IGBT), the MG300Q1US51. This component is currently sold by Toshiba Semiconductors. The package is about 10 x 6 x 2.5 centimeters. The device handles 300 A, 1200 V, and 2500 W.

Figure 3-1. A high-power silicon IGBT from Toshiba Semiconductor

A 10x15 mil device occupied 150 mils^2 of wafer. A 3½-inch wafer provides 9,621,127 mils^2 of surface. That is an area ratio of 64,140 times larger for the giant transistor than our 40V, 200mA small signal ½-watt device. The giant power BJT is able to handle 100V, 100A, and 243 watts. There are also power devices handling 150V, 200A, and 3000 watts.

3.3 MODELS EXTRACTED FROM SEMICONDUCTOR DESIGN THEORY

One way to model a semiconductor device is to extract its model properties from the device's structure and materials physics. When we take this model extraction approach, it is possible to model a device without ever building it.[1]

There are important advantages to being able to extract model properties from theory. One advantage is being able to develop models on new devices while concurrently designing a new PCB. Another advantage is being able to develop model parameter distributions. These simulated distributions represent the predicted variability of a device's population. They can be generated very early in the device's life cycle.

Often usage rates and the amount of product produced are small. Therefore, gathering significant population statistics from measured units is not possible. But with simulation we can still predict the likely *range* of product variation. This predicted range can be used in simulating a circuit.

The computer aided engineering programs that provide the ability to extract SPICE device models from semiconductor structure and physics are generically called (semiconductor) Technology Computer Aided Design (TCAD) programs.[2] These TCAD programs also aid semiconductor device design engineers and process engineers to design the structure, materials, and processing that yield the desired electrical properties.

In this chapter, we use Bipolar Junction Transistor (BJT) technology to explain how structure determines performance. We also use it to show the interrelationship between many modeling ideas. In the past, BJT technology was a major player, but today it is only a niche player.[3] However, BJT technology is still likely to be familiar to most readers so it makes a good starting point in our discussions. The BJT device and circuit models are used to introduce some device physics and modeling concepts. BJTs are also relevant to those working with BiCMOS circuits.

Deep submicron CMOS FETs have parasitic BJT action, so understanding some basic BJT operation is necessary. Indeed, nearly all FETs can potentially have a parasitic BJT associated with their structure. And nearly all BJTs can potentially have a parasitic surface inversion FET associated with their structure. When the transistor was first invented, its

---

1 This is parallel to modeling and simulating a circuit on a computer before prototyping it.

2 References to read are [11, 32, 35, 38, 40, 44, 90, 97, 100, 117, 132, 140].

3 For readers interested in learning more about current CMOS technology, see [11, 40, 77, 127, 140].

inventors were looking for surface inversion transistor action, instead they found a parasitic BJT and recognized that it was a new type of transistor.[4]

In summary, if we want to understand and use device physics to our advantage, we need to understand some BJT behavior, which is a good starting point for explaining semiconductors.

CMOS technology has taken over most applications, even differential signaling. But with today's sub-sub-micron technologies, deep submicron CMOS is far more complex to explain. The author's purpose is to illustrate some relationships between device structure and model properties. It is important to understand that the relationships both enable and limit device behavior.

3.4 EXAMPLE OF THE BJT PROCESS

Semiconductor device modeling starts with semiconductor device design. Process technology determines device design. Throughout this chapter, the authors use as an example a discrete BJT, which uses planar epitaxial double-diffused technology.

The manufacture of this BJT starts with a crystal ingot of semiconductor material. This material is typically silicon but not exclusively; it is grown and sliced into thin wafers that will serve as a *substrate* for semiconductor devices. A pure silicon crystal is a semiconductor but not a very good one. Dopant atoms are added to the semiconductor crystal to alter its electrical properties in a controlled way. Depending on type, dopant atoms can contribute either an extra electron or an extra "hole," thereby altering conductivity, current carrier injection, and other properties in particular regions of the crystal.

In an actual semiconductor manufacturing process, we create many identical devices (chips) using printing lithography across a wafer. Dopant placement and diffusion are done through a set of features printed on the surface of the semiconductor wafer. At the end of the printing and dopant placement process, individual devices (chips) are separated from each other by scribing and cracking them apart.

The upper half of Figure 3-2 shows a cross section through a vertical BJT pnp device. The lower half shows an idealized one-dimensional model of the device. We refer to this one-dimensional model later in this chapter.

For a complete description of the process of creating semiconductor device chips, see "Appendix J, Device Physics."

---

[4] The high level of surface ionic contaminants, due to the semiconductor processing of the day, masked the surface inversion action they were looking for.

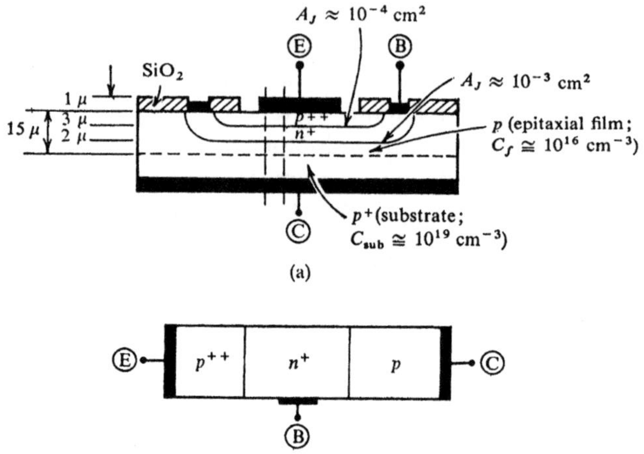

Figure 3-2. BJT cross section [47]

LEGEND:

E=Emitter
B=Base
C=Collector

3.5 HOW BJT AND FET CONSTRUCTION AFFECT THEIR OPERATION

3.5.1 BJT Construction and Operation

A BJT (bipolar junction transistor) is a device with a p-n (or n-p) semiconductor junction similar to that found in a diode. When the junction is forward-biased, it *emits* (injects) current carriers from one side of the junction to the other. This effect is usually enhanced by the doping concentration profile of the junction and base region. The injecting (emitter) side usually has a much higher doping concentration than the receiving side. The region into which the current carriers get injected is called the *base*.

On the other side of the base region is another semiconductor junction, n-p to correspond with p-n (or p-n to correspond with n-p). This junction is reverse biased (Vcb). Current carriers are normally not injected into the base region from it. This effect is usually enhanced by the doping concentration profile of the junction. The base region usually has a much lower doping concentration than both the emitter and collector regions.

This p-n-p (or n-p-n) arrangement is similar to bringing two forward facing diodes together, pointing at each other in a series connection. This arrangement is not very exciting because the injected carriers tend to recombine in the base region before they get very far.

Figure 3-3 shows an energy band diagram scanning across our BJT from and to emitter-base-collector.

- E_c = conduction band energy level
- E_f = Fermi energy level
- E_v = valence band energy level

The top part of Figure 3-3 shows the energy band diagram without externally applied bias voltage. The bottom part of Figure 3-3 shows the energy-band diagram with externally applied bias voltage. Forward bias applied emitter-base lowers that energy band and enhances the injection of current carriers into the base. Reverse bias applied collector-base raises that energy band and discourages the injection of current carriers into the base. An n-emitter injects electrons into a p-base. A p-emitter injects holes into an n-base.

What happens when the base region is made so narrow that some carriers arrive at the reverse-biased junction? The minority carriers have arrived on the wrong side of a reverse-biased junction and the electric field attracts them across the junction into the collector. The junction is still reverse-biased to carriers on the collector side. But it is forward-biased to carriers of the same polarity on the base side. Carriers of that polarity would normally not be there had they not been injected from the emitter and then diffused across the base.

The result is that the carriers get swept up, or *collected*, into the p (or n) region. Conduction from the emitting-to-collecting region can be controlled with a low-voltage, low-energy, and low-impedance emitter-base circuit. Large currents can be made to flow in a higher-voltage, higher-energy, higher-impedance base-collector circuit. This is an interesting and useful result. Many applications take advantage of the current, voltage, and power gains in a bipolar transistor.

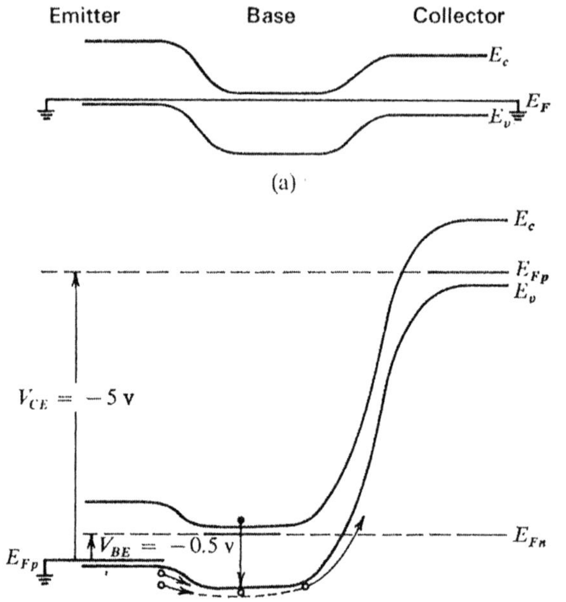

Figure 3-3. Energy band diagram for a pnp BJT transistor [47]

Minority current carriers that do recombine in the base (fall out of the conduction or valence band) and re-attach at a particular atomic site would cause a net, imbalanced charge, un-neutralized at the atomic level to build up in the base. When the current carrier is a conduction-band electron (hole) that looses energy and re-attaches to the semiconductor lattice, it does so at the site of an atom with a hole (electron) that can accept it. Before that event, the atom is electrically neutral. That is, the electric charge of the nucleus is balanced by the charge of the electrons in shells around it. When the extra electron (hole) injected from the emitter attaches to the atom, it unbalances the charge on the atom.

In essence, unless we supplied the means to drain off the extra charges in the base via base current, the accumulated extra charges would quickly repel any further injection from the emitter. The base current that we must supply, Ib, represents the inefficiency of conducting carriers across the base region. The view of a BJT as a *current gain* device can be a bit misleading. It is the built-in junction voltages (due to the energy band and modifying effects of temperature and dopants), plus the applied external voltage biases across those junctions that control injection and collection of carriers across the base. Base current *does not get multiplied* and flow out of the collector.

Figure 3-4 represents the construction of a *lateral* BJT done as part of an integrated circuit (IC). Here, each device is isolated from its neighbors and the backside substrate by sitting in a well formed by *isolation diffusions.* Contact to the buried collector is done through the topside of the IC — collector metal-contact[5] to (collector) contact-diffusion to buried-collector.

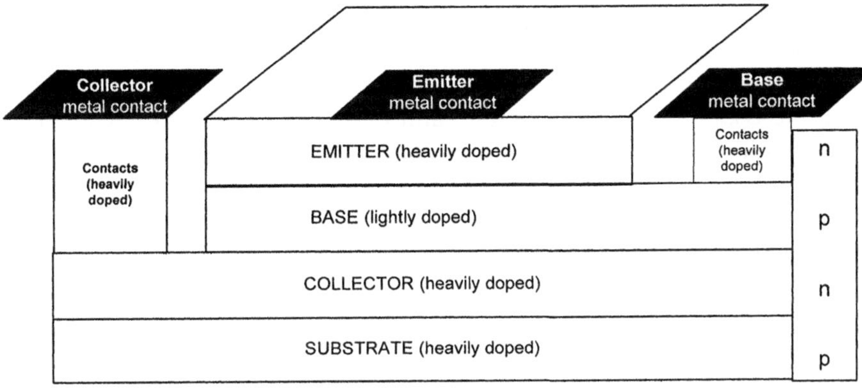

Figure 3-4. Typical lateral BJT construction for an individual transistor

3.5.2 Two Types of FET

The BJT is a semiconductor device dominated by minority current carrier and p-n/n-p junction effects. A Field Effect Transistor (FET) is a semiconductor device dominated by majority current carrier and channel conductance effects. A FET can be constructed in two forms:

- The Junction FET (JFET) transistor. Channel conductance is controlled by space-charge depletion layer changes.
- The Metal-Oxide-Semiconductor (MOSFET) transistor. Channel conductance is controlled by surface-inversion layer changes.

The MOSFET contact pads and field plate can be implemented in metal or in heavily doped polysilicon (semicrystalline silicon). In today's CMOS chips, the field plate is always implemented in polysilicon.

---

[5] Throughout the book, metal to die contacts are usually non-rectifying, non-Schottky Barrier diode contacts.

3.5.3 JFET Construction and Operation

JFETs are interesting and useful devices. They are interesting because they illustrate certain space-charge layer aspects of semiconductor behavior. However, JFETs are even smaller niche players than BJTs.

In the JFET, a conduction *channel* of n or p material inside a semiconductor is controlled by the surrounding material (*gate*) of the opposite material polarity (p for n-channel, n for p-channel). Conduction from a *source* end to a *drain* end of the channel is established by imposing a voltage across the two ends. Source and drain are easily interchanged. The channel region is lightly doped, while the gate, source, and drain regions are more heavily doped.

The p-n (or n-p) junction formed under the gate is reverse biased and is used to control the channel conduction width by varying the width of the space-charge depletion region width. This is why the electrode attached to the surrounding channel width-control-region is called the *gate*. The gate is always reverse biased and thus supplies only leakage current. The leakage current is *much* (orders of magnitude) smaller than the base current.

3.5.4 MOSFET Construction and Operation

Modern MOSFET transistors are capable of performing nearly all switching and amplifier tasks and performing them better than alternative technologies such as BJTs. Deep submicron CMOS technology is certainly the technology of choice in high-speed logic and low-level RF amplifier circuits.

The MOSFET is constructed by placing a metal (or polysilicon) electrode over a lightly doped, semiconductor conduction channel; an insulating oxide layer separates them. This forms a layered Metal-Oxide-Semiconductor (MOS) structure at the surface of a semiconductor crystal. As with the JFET, the source and drain, located on opposite sides of the gate, are more heavily doped. Leakage current across the oxide is *much* (orders of magnitude) smaller than JFET leakage current. The DC bias current that needs to be supplied by the gate is negligible; the transient current is usually larger, due to the gate capacitance ($I(t) = C*dV/dt$).

In Figure 3-5, we see how an insulating oxide layer covers a surface channel. A metal field electrode called the *gate* sits over this oxide layer. The contact at one end of the conduction channel[6] is made through a *source*; the contact at the other end of the channel is made through a *drain*. The

6 The channel does not exist when no gate bias is applied. It is initially a surface p-layer. Under the influence of a gate bias, the surface p-layer inverts to n-type, and a channel is formed.

source and drain are heavily doped n-regions.[7] The conduction channel between them is a lightly doped p-region. In Figure 3-6 we have the same construction in a complementary pnp device. In both cases, conduction from a *source* end to a *drain* end of the channel is established by imposing a voltage across the two ends. Source and drain are easily interchanged.

How does a MOSFET work? The answer depends on underlying construction, dopant concentrations applied during manufacture, and bias applied during operation. There are two possible modes of operation: *enhancement* mode and *depletion* mode.

3.5.4.1 Enhancement Mode MOSFET

Electrons generated by thermally excited dopant atoms are present in the heavily doped source and drain. However, the electrons cannot flow through the p-surface layer between them with zero bias applied to the gate, and nothing else done to *invert* the surface.

What does it mean to invert the surface and what is the result of such inversion? Let us apply a positive gate voltage to our npn p-surface layer NMOS device. Holes will be repelled from the surface and electrons will be attracted to it. As we continue increasing the positive gate bias voltage, more of this migration of electrons occurs. At some point, the concentration of attracted electrons can become higher than the background concentration of holes supplied by the p material. For a thin layer near the silicon surface, the *apparent* character of the silicon goes from p to n. That is, it *inverts.*

Electrons can then carry current between the source and drain through this inversion layer. The value of V_{GS} at which enough mobile electrons are attracted to the surface region to invert it is called the *threshold voltage*, V_t. Removal of the bias voltage causes the character of the inversion layer to revert back to its original p character and conduction ceases.

In the inversion action just explained, a further increase in applied positive gate bias voltage further *enhances* the conduction process. See Figure 3-8 for the characteristic curves of an enhancement mode MOSFET.

For an enhancement mode p-channel pnp PMOS construction, the role of holes and electrons are swapped, and a negative gate voltage is used to create the inversion layer.

---

[7] High conduction and low resistance. These regions are essentially contacts to the conduction channel.

3.5.4.2 Depletion Mode MOSFET

What does it mean if the surface is already inverted and what is the result of such inversion?[8] In this case, the conduction channel already exists and current flows with zero gate bias applied, but source-to-drain bias applied. Such was the case for the discoverers of the BJT transistor. Surface contamination ions had already inverted the surface of their prototype device.

Let us apply a positive gate voltage to a depletion mode npn n-channel NMOS device. More gate voltage increases the conduction just as before. Now, however, if we apply a negative gate voltage, we repel electrons from the conduction channel, the channel becomes shallower, and its conductivity decreases.

In the decreased conduction action just explained, an increase in applied negative gate bias voltage further *depletes* the conduction carriers. The value of V_{GS} at which enough mobile electrons are repelled out of the surface inversion channel to shut off all conduction–even with applied V_{DS} voltage–is called the *threshold voltage*, V_t. Removal of the gate bias voltage causes the character of the inversion layer to revert back to its original n character and conduction resumes.

For depletion mode p-channel pnp PMOS construction the role of holes and electrons are swapped, and a positive gate voltage is used to deplete the inversion layer. See Figure 3-9 for the characteristic curves of a depletion mode MOSFET.

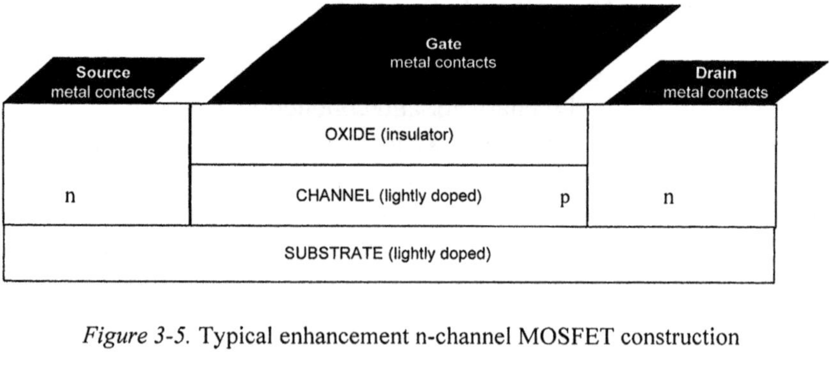

Figure 3-5. Typical enhancement n-channel MOSFET construction

In a FET, conduction does not occur as it does in the BJT with *minority* carriers. In a BJT, the carriers injected from an emitter migrate across a base, and get collected at a collector. In a MOSFET, *majority* carrier conduction

[8] An inversion layer can easily be formed by implanting and diffusing a thin n+ (or p+) layer at the surface under the gate of our npn (or pnp) structure.

occurs when a surface layer gets *inverted* in an enhancement mode device. Inversion by the gate voltage means p material is made to look like n material and vice versa. The inversion layer connects the source and drain. In a depletion mode device the surface is already inverted, can conduct, and can be depleted, that is, shut off by applied gate voltage.

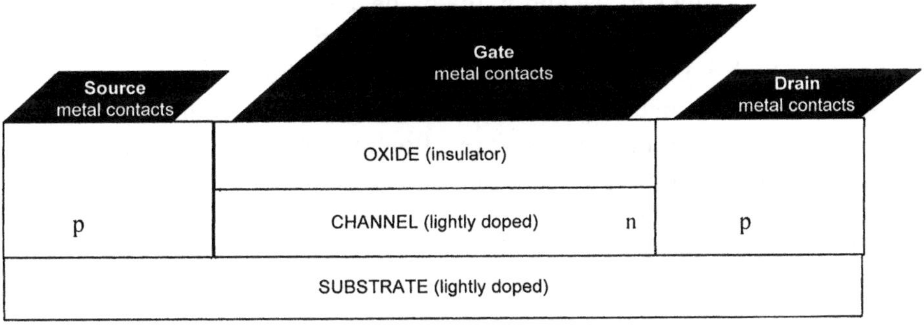

Figure 3-6. Typical p-channel MOSFET construction[9]

Today, most ICs are made using enhancement mode MOSFETs. The gate voltage required to bias the transistor into a conducting condition is called the threshold voltage, Vth, and conduction occurs when Vgs>Vth and for an NMOS enhancement mode device Vds>0V. The conditions for turning on the various types of MOSFETs are summarized in Table 3-1.

Table 3-1. Conditions for turning a MOSFET on

MOSFET Type	Vgs	Conduction
NMOS Enhancement Mode	Vgs > Vth	Yes
NMOS Depletion Mode	Vgs ≤ Vth	Yes
PMOS Enhancement Mode	Vgs < Vth	Yes
PMOS Depletion Mode	Vgs ≥ Vth	Yes

The MOSFET has become the dominant form of transistor for discretes and ICs. In no small part, this trend has been driven by both the accessibility of the semiconductor surface and the advances in lithography, enabling ever-smaller transistor geometries to be fabricated.

3.5.5 BJT Versus MOSFET Comparison

Parasitic BJT action in a MOSFET and parasitic MOSFET action in a BJT are natural and possible occurrences. Device designers and operators

9 To avoid confusion, the contacts are still labeled "metal." However, today's deep sub-micron CMOS contacts will probably be implemented in polysilicon.

must always be aware that they can be present. They must understand both types of transistors.

Consider whether a BJT surface and oxide can be made clean enough to eliminate all possibilities of inversion. Not in a practical sense. First, all semiconductor surfaces can be made to invert. Second, lattice dislocations (especially at the Si-SiO2 interface) due to dopant atoms, and hot-electron state injections at the surface and in the oxide, can alter inversion conditions. Throughout the late 20th century, it was relatively easy to invert only moderately high voltage (40 volts BVceo) pnp BJTs. This meant that few such devices were ever developed without guard or equipotential rings. Guard or equipotential rings are floating rings of highly doped semiconductor material (p+ on a p-surface, n+ on an n-surface) that are difficult to invert. Such rings surround the base-emitter structures on a BJT topside. One example can be seen in Figure 7-22. The guard rings act like an IC-well does in isolating an active device from its neighbors. In this case, we isolate the active BJT from its die edge. The die edge acts like a near-infinite source of low activation energy current carriers. The ring prevents the surface inversion from reaching the die edge and shorting out the device.

Figure 3-7 shows BJT collector characteristic curves that are linearly spaced with Ib, except for low- and high-current non-ideal effects. Figures 3-8 and 3-9 shows MOSFET curves that are spaced as $(Vgs\text{-}Vth)^2$. The BJT curves in Figure 3-7 cover a wider voltage range, resulting in an increase in slope as the BJT approaches breakdown. As collector-base reverse-bias junction voltage is increased the junction depletion region widens further — especially on the lightly doped base side. This narrows the region the minority carriers have to drift across, lessens base recombination and increases beta. Dr. James Early first explained the cause of this behavior in 1952 — thus the name Early effect.

The MOSFET would show a similar upward slope at high voltages if the voltage range were extended. CMOS shows a flatter slope in the normal operating region, and steeper slope above Vcc.[10] The sharp increase in current is caused by parasitic diode turn-on in CMOS. In a BJT a large increase in current occurs because of voltage breakdown, but usually at significantly larger voltages than Vcc.

In a BJT, the Early effect is the name of a base-width-modulation effect due to the reverse-biased Vbc magnitude. In a MOSFET, the equivalent effect is called channel-length modulation (also gate conductivity modulation) due to the reverse-biased drain junction (increasing Vds

---

[10] CMOS breakdown is caused by two effects: parasitic diode turn on, and gate oxide breakdown. The low-voltage oxide breakdown makes ESD protection very important in submicron CMOS parts.

shortens the channel length, and increases current with approximately linear slope).

Both types of devices show approximately linear I-V characteristics below a certain voltage:

- For a BJT, it is Vce ≤ Vce(sat).
- For an n-channel MOSFET, it is Vds ≤ Vgs-Vth.

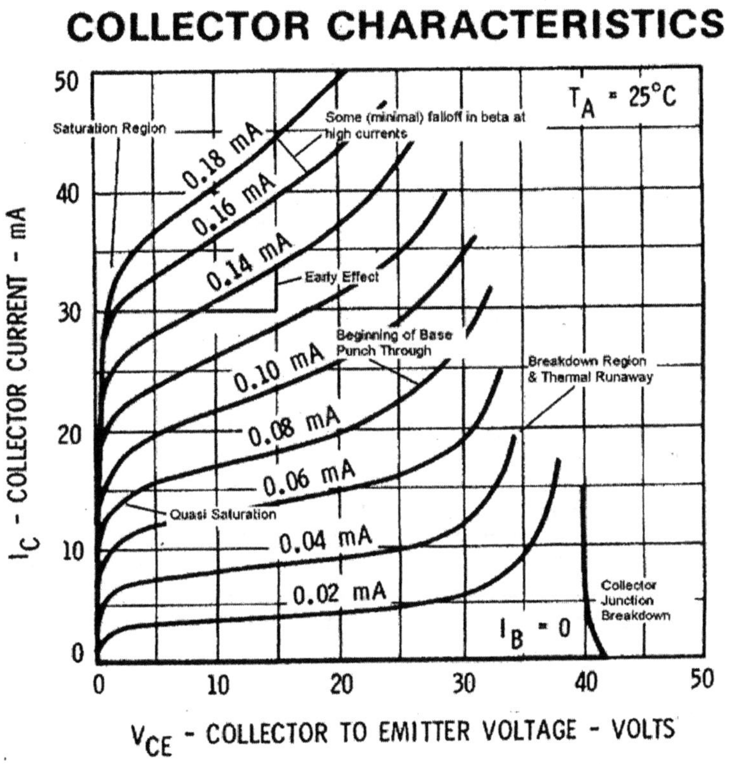

Figure 3-7. Annotated BJT CE collector characteristic curves, 2N3904 transistor

We can make some comparisons of BJT and MOSFET operational regions. For example, the saturation region (see Figure 3-7) for a BJT corresponds to the ohmic region, also known as the triode region (see Figure 3-8 and 3-9) in MOSFET. The MOSFET ohmic region is more linear than the BJT saturation region and it covers a much wider voltage range in MOSFET. The saturation region in MOSFET actually corresponds to the linear operation region in a BJT. These regions of operation, linear in BJT and saturation in MOSFET, provide most of the signal gain used in amplifiers and I/O buffers.

The key difference between the two types is that a BJT draws DC-Ib, while a MOSFET draws only a tiny gate-leakage[11] current (ideally zero). Thus, a BJT consumes significantly more standby power than a MOSFET.

Both BJT and MOSFET can draw relatively large AC currents due to capacitive effects (Cbe, Cgs). Both have an effective feedback capacitance from collector and drain to base and gate (Ccb, Cdg) respectively. A portion of the output signal gets fed back to the input through the feedback capacitance, where the amplifier gain then amplifies the fed back signal. The result makes it appear as though the amplifier gain multiplies the actual capacitance.[12]

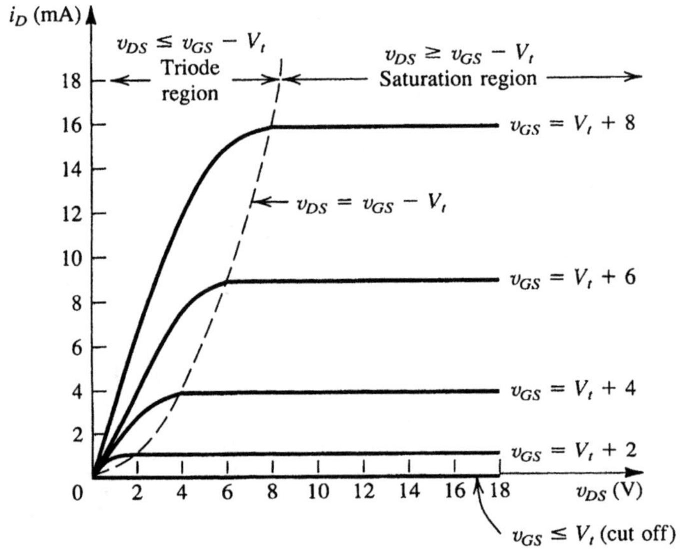

Figure 3-8. MOSFET drain characteristic curves, n-channel enhancement mode transistor [108, page 309]

---

[11] Plus there is some leakage in the parasitic diodes inherent in the IC structure. This leakage is larger then the gate leakage and is becoming a significant contributor to power consumption in handheld (battery operated) products.

[12] This gain multiplication effect was first explained in vacuum tube amplifiers by J. M. Miller [87] and is called Miller Effect Capacitance. One also sees a feedforward effect from gate to drain, which is what causes the initial bump sometimes seen in CMOS V-T curves.

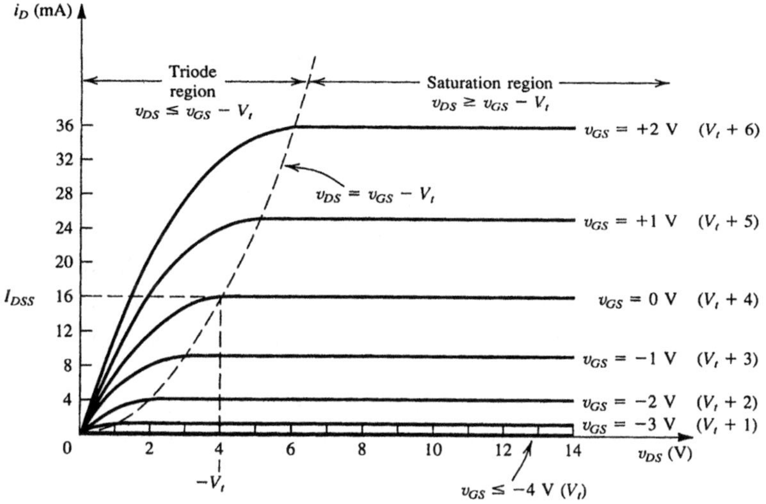

Figure 3-9. MOSFET drain characteristic curves, n-channel depletion mode transistor [108, page 320]

Table 3-2 summarizes the differences between BJT and MOSFET.

Table 3-2. BJT versus MOSFET summary

Property	BJT	MOSFET
Characteristic curves at the output	Evenly spaced with Ib	Evenly spaced as $(Vgs\text{-}Vth)^2$
Gain Modulation at the output due to power supply voltage	Base-width changes and Early effect	Channel length changes
Linear output I-V range	For Vce ≤ Vce(sat)	For Vds ≤ Vgs-Vth
Steady-state power consumption	Relatively large because of Ib	Smaller because gate leakage is small

3.6 CALCULATING DEVICE PHYSICS PROPERTIES

3.6.1 Introduction to TCAD and SPICE

TCAD tools are essential to modern semiconductor design and fabrication. For example, TCAD is often used to optimize chip layout using multiple fingers (emitter or drain) to increase speed (by reducing capacitance and series resistance), and to reduce layout area. This activity is usually done for output transistors on both BJT and MOSFET chips, but is a minor activity for logic and input-stage transistors.

The *T* in TCAD stands for Technology, specifically semiconductor process technology. The *CAD* stands for Computer-Aided-Design. TCAD is comprised of a set of algorithms for computing semiconductor behavioral properties based on device physics. This topic presents examples of how device structure (that is, junction depths, channel lengths, doping profile, and geometry) is used in equations derived for computing some TCAD and SPICE parameters. TCAD is the computer aided engineering implementation of the physics models, process models, and SPICE model extraction.[13] Process models predict such things as implantation depth for dopants and their migration during drive-in diffusion, annealing, and other process steps.

SPICE (Simulation Program with Integrated Circuit Emphasis) is a computer program that was developed to simulate the device physics models as part of a circuit design.[14] Most device models were too complex to readily solve by node and mesh equations. Textbooks on the circuit application of SPICE typically present SPICE parameters as measured properties. Device physics theory is equally as important as measuring the model parameters.

By considering the device-physics equations, we see why SPICE is thought of as a physical model rather than a behavioral model. The device-physics derivations lead to parameters like junction-emission coefficient, and junction-diffusion capacitances, which are related to physical phenomena.

Many SPICE programs are available for device modeling and simulation. Many commercial programs (such as HSPICE®, PSPICE®, Spectre®, Saber®, Aim SPICE®, IntuSoft®, and others) are also available. They are all slightly different from each other. Having no standard SPICE model guarantees that models and simulators will not exchange data seamlessly.

3.6.2 Comparing Different Varieties of SPICE

Different versions of SPICE, SPICE derivatives,[15] and the device physics models on which SPICE is based do not use identical sets of parameters. Default values for parameters can also vary between implementations of models of device types. It is also worth noting that default values are almost always wrong for an actual device – for example, parasitic capacitances defaulted to zero, giving a transistor excessively high bandwidth and speed.

---

[13] Refer to [10, 11, 32, 35, 38, 40, 44, 63, 68, 71, 77, 79, 83, 84, 86, 90, 97, 98, 117, 118, 121, 127, 128, 132, 140].

[14] Refer to [33, 34, 42, 44, 47, 48, 53, 63, 66, 109, 110, 111].

[15] Refer to [10, 35, 40, 63, 68, 69, 79, 83, 84, 86, 100, 103, 118, 132].

3.6.2.1 Capacitor Models

A simple capacitor provides an example of how different versions of SPICE differ. In the original UC Berkeley SPICE, the basic capacitor came in two forms: a fixed capacitor, and a geometric capacitor. The geometric capacitor allowed the user to define the capacitance in terms of physical parameters, such as length, width, sidewall length, and dielectric constant. Neither the fixed nor the geometric capacitor allowed any dependence on voltage or temperature.

In reality, semiconductor capacitance varies with both voltage and temperature. For example, HSPICE allows variation with temperature but not with voltage. IS-SPICE allows variation with temperature or voltage, but not both. Some SPICE simulators, such as HSPICE, IS-SPICE, and PSpice, have added support for user models (equations), making it possible to model capacitor temperature and voltage variations together, as well as modeling non-polynomial variation.

3.6.2.2 BJT SPICE Models

BJT SPICE models have more variation between versions of SPICE than does the simple capacitor. Table 3-3 compares some selected BJT model parameters.[16]

BJT device behavior modeled by these parameters can usually be observed directly on a curve tracer. Phenomena, such as quasi-saturation and high-current beta rolloff, are now incorporated in the models. In addition, the effects of manufacturing and process problems can be seen. Some examples are surface contamination effects on a BJT's low current beta, avalanche breakdown voltages, and surface limited breakdown voltages [74].

Table 3-3 lists only the parameters receiving significant discussion in this chapter. A more complete table is shown in "Appendix J, Device Physics."

High current beta rolloff coefficient is a particularly important example of the issue that not all models have the same parameters. If a SPICE simulator reports "unused parameter," this indicates that there is a mismatch between a parameter set and the model the SPICE simulator is using the parameters in. The user must then either find the correct parameters for the model, or the correct model for the parameter set.

---

[16] Refer to Table 3-8 for MOSFET model parameters.

Table 3-3. Comparison of SPICE with device physics model properties

Parameter	**SPICE 3F3 [97]**	**H-SPICE[17] (1992) [86]**	**P-SPICE (6/2004) [21]**	**Ebers-Moll 2 [38]**	**Gummel -Poon [98]**
Saturation current	IS	IS	IS	IS	I_S
Ideal max forward bias Beta	BF	BF, or BFM	BF	BF	β_F
Forward Early voltage	VAF	VAF+TVAF1, TVAF2 (VA, VBF)	VAF (VA)	VA	V_{AF}
Corner for BF high current rolloff	IKF	IKF+TIKF1, TIKF2 (IK, JBF)	IKF (IK)		I_{KF}
High current beta rolloff coefficient		NKF	NK		
B-E built-in potential	VJE	VJE+TVJE (PE)	VJE (PE)	PE	ϕ_{be}
B-E junction exponential factor	MJE	MJE+TMJE1, TMJE2 (ME)	MJE (ME)	ME	m_{be}
Ideal forward transit time	TF	TF+TTF1, TTF2	TF	TF	τ_F
Energy gap for temperature effect on IS	EG	EG+GAP1, GAP2	EG	EG	eg

3.6.3 Structure Influences BJT Performance

From the original point-contact BJT transistor[18] to modern deep sub-micron ICs, semiconductor process technologies have undergone many developments, evolutions, and revolutions.

The original semiconductor crystals pulled from a crucible of molten material (most likely germanium) averaged barely ½ inch in diameter and a few inches long. Today's crystals are larger than 12 inches in diameter and a few feet long. For many years, silicon has reigned as king of semiconductor materials despite the superb properties of other materials like Gallium

---

[17] Note that parameters following the format TVAF1, TVAF2 mean:
TXXX1= parameter's coefficient for a linear temperature-dependent term.
TXXX2= parameter's coefficient for a quadratic temperature-dependent term.

[18] Point-contact transistors came first. They have the metal wires touching the p or n material, almost creating Schottky junctions. The junction transistor came next, where they put dots of indium or gallium onto the germanium, then melted/diffused it into the germanium at high temperature. Thus an n-emitter (p) and n-collector (p) were diffused into a block of p (n) material. Diffused transistors, where a gas containing the dopant compound would break down at high temperatures, and then diffuse into the wafer, were third-generation transistors.

Arsenide. Once the semiconductor crystal is grown, it is sliced like a loaf of bread into many thin wafers known as substrate wafers.

The substrate wafers, usually heavily doped and with a high conductivity, become a matrix upon which a gas deposits an epitaxial (epitaxial means "grown upon") layer. From this point, we could construct technologies like Mesa Transistors, which are still used in high power discrete transistors. But the dominant technology is the Silicon Planar Process [53]. Semiconductor devices are created in the epitaxial layer. For integrated circuits composed of multiple individual transistors, one or two layers of polysilicon, separated by oxide, are used to form local interconnects. Alternate layers of insulator and metal are grown above the poly layers. Eight or more layers of copper may be used for global interconnects. The resistance of polysilicon is too high to use for global interconnects.

Some device design structural-dopant concentration choices for optimizing different electrical performances are shown in Table J-3 in "Appendix J, Device Physics." These optimizations show up in the electrical performance characteristic curves of devices. These characteristic curves can be used to compare and contrast component performance.

3.6.4 Modern Semiconductor Process Control

Today, TCAD software is used extensively to design devices, generate their SPICE (typically HSPICE, Spectre, or Eldo) model parameters, and control the processes used in their manufacture. A SPICE model library is normally not portable between simulator platforms.[19] When requesting models from a foundry, it is important to specify the simulator, as well as the process name. Foundries typically offer BSIM and the Enz-Krummenacher-Vittoz (EKV) models for these three simulators. As a result, HSPICE is losing its once dominant lead in these applications.

Most semiconductor manufacturers use these TCAD programs to control their processes–not to generate simulation models. Consequently, the SPICE models produced may be less desirable for circuit simulation purposes.

For example, Figure 3-10 shows two sets of I-V data for a diode used in an I/O buffer design. On the left, the SPICE and IBIS models match almost exactly. On the right, the SPICE and test data do not match. The mismatch could have been caused by an incorrect model parameter, or by an underlying model problem internal to SPICE. The causes of this mismatch are related to different understandings of the correct voltage range over which to measure clamp behavior.

---

[19] A model parameter for a different simulator might not cause a particular simulator to crash, it can, and will, generate incorrect currents and voltages.

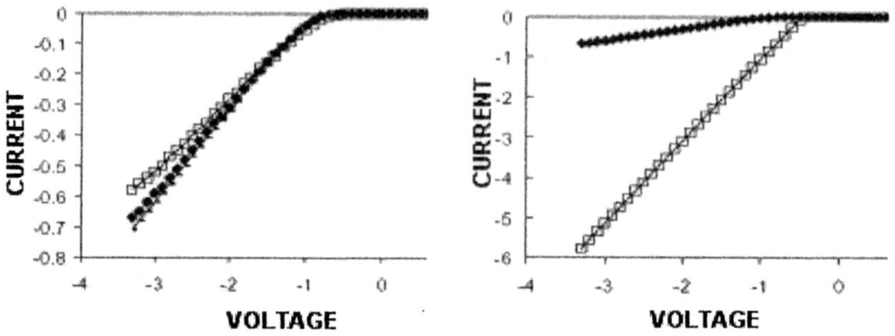

Figure 3-10. SPICE and IBIS models compared to measured data for a clamp diode
Data courtesy of Cypress Semiconductor

3.6.5 More About SPICE and TCAD

In the context of SPICE, the term netlist refers to the list of components and connections for a circuit. Also in the context of SPICE, the term model refers to one or more of the following:

- A schematic of a circuit
- A netlist
- A subcircuit or macromodel
- A parameter set for a particular transistor
- A library of model parameters for a specific technology
- The simulator's equations, coded to use these parameters

Equations are coded differently in different SPICE models–or even the same model in different SPICE programs. For example, HSPICE has over 50 different MOSFET models, each with its own set of parameters. The parameters can be significantly different from one model to the next. For example, the equations for the same transistor model can have:

- Different names for the same parameter
- Different values for the same parameter name
- Different parameters (no name match)

TCAD addresses all aspects of semiconductor design, including such issues as die interconnections. TCAD at the IC chip level sometimes also addresses Signal Integrity, current density, electromigration on the interconnections, EMI, package design, and many other concerns. Some SPICE parameters, like beta and Early effect, are also familiar to SPICE model circuit designers. But there are other device physics TCAD properties that are not used in SPICE. An example is breakdown voltage.[20] This is because SPICE, a program for circuit analysis, uses only a subset of the formulas and properties available to device designers. Also, many SPICE models (parameter sets) do not model reverse-bias operation or second-order effects accurately.

Designing a semiconductor is more than designing its desired behavior as an amplifier or a switch, as addressed in SPICE. For instance, what about its current, voltage, and power handling capabilities? What about electromigration and reliability issues? These issues can be modeled with a combination of equations from device physics and parameters from measurements.

3.7 EXAMPLES OF COMPUTING ELECTRICAL PROPERTIES FROM STRUCTURE

For most IC processes, the fabricator and foundry take care of computing and extracting transistor parameters. This topic of parameter computation and extraction is included to show:

- The connection with modeling
- That there are ways to verify the reasonableness of parameters[21]
- How formulas are used to compute SPICE and TCAD values

3.7.1 Semiconductor Constants and Quantities

Before we discuss the device physics equations, let us examine some of the parameters understood as defined constants and quantities. Some SPICE models calculate parameters from other quantities, and it helps to understand how that is done. Many of the defined constants and quantities can be computed from more fundamental physical properties and semiconductor technology processes. For example, N_{DB} (base doping density) is one such quantity.

Once parameters are defined, they should not have to be re-defined. Table 3-4 lists these defined constants and quantities.

---

[20] Equations for computing collector saturation voltage and breakdown voltage for BJTs are presented in "Appendix J, Device Physics."

[21] The Fabless Semiconductor Association (http://www.fsa.org/) has set up a model quality committee on verifying parameters.

Table 3-4. Parameters in the structure-based equations (3-1) to (3-8) and (3-18) to (3-22)

Parameter	Description
D_{nE}	Electron diffusion coefficient
D_{pB}	Hole diffusion coefficient
N_{DB}	Base doping density
N_{AE}	Emitter doping density
W_E	Width of the emitter region
W_B	Neutral (active) base width
W_{EB}	Width of the emitter-base region
L_{pB}	Diffusion length of minority carriers in the base
τp	Lifetime of holes
τ_o	Excess minority carrier lifetime
V_{EB}	Emitter-Base junction bias voltage
n_i	Equilibrium intrinsic carrier (holes or electrons) concentrations
q	Charge magnitude on the electron = $1.60*10^{-19}$coulomb
$K_b = k_b$	Boltzman's constant: 1.38e-23 Joules/Kelvin
T	Temperature in degrees Kelvin
Eg	Energy band gap of the silicon material
ε_0	Permittivity of free space = $8.86*10^{-19}$F/cm
ε_R	Relative permittivity of silicon = 11.7
μn	Mobility of electrons. Mobility depends on the intrinsic semiconductor material (silicon, germanium, GaAs, etc.) and the doping concentration that has been added to it.
μp	Mobility of holes. Mobility depends on the intrinsic semiconductor material (silicon, germanium, GaAs, etc.) and the doping concentration that has been added to it.

Our discussion about device physics thus far represents a simplified theory without taking into account such things as emitter de-biasing, Early effect, high current injection effects, surface recombination and so on. For the most part, these additional effects are accounted for by other terms in the Gummel-Poon/Ebers-Moll/hybrid-pi (that is, SPICE) model.

For an npn, the base is doped with acceptors (p) and emitter with donors (n). The (n) and (p) subscripts should be swapped for a pnp. Investigating a little deeper into how BJT parameters can be calculated:

$$D_{nE} = \left(\frac{k_b T}{q}\right)\mu_n \tag{3-1}$$

Equation (3-1) is also known as Einstein's relationship.

$$D_{pB} = \left(\frac{k_b T}{q}\right)\mu_p \tag{3-2}$$

Parameters N_{DB} and N_{AE} are found from the physics and chemistry of the doping, ion implant, and diffusion process. That set of processes has its own set of formulas, charts, and constants. Calculating those quantities is usually in the realm of the material scientists and process engineers. Calculating a device's doping profile as a result of its processing is the usual way of finding NDB, NAE, WE, and WB. These parameters are static once set by the starting material, epi growth, oxide growth, doping, diffusion, masking, and other processing of the device.

The emitter-base space-charge region has some dynamic characteristics because its width changes with applied bias. As it changes, the effective widths of the depletion layer and base width also change. This is because depletion regions grow and shrink on either side of a junction due to the applied bias. The change mostly occurs on the lightly doped base side.

Consider the static equilibrium case with no applied bias.

then:

W_0 is the width when Veb = 0V

$$W_0 = \sqrt[3]{12\varepsilon_s\varepsilon_0\,\phi_B / qa} \qquad (3\text{-}3)$$

where:

a = the dopant concentration gradient at the junction[22]

ϕB = the built-in junction potential = ϕFp + | ϕFn |, or the sums of the Fermi potentials.

ϕ_B becomes ϕ_B +/- | V_j | in equation (3-3) when bias is applied. Reverse bias widens the space-charge region, forward bias narrows it. V_j is, of course, V_{EB} in this discussion.

Next:

$$L_{pB} = \sqrt{D_p \tau_p} \qquad (3\text{-}4)$$

where:

Dp is the diffusion constant of holes.[23] The quantity n_i is found from Fermi level statistics and, at thermal equilibrium, we have:

---

22 This is why the SPICE equations fail for many modern devices. The junction can be graded in a non-exponential, non-linear fashion, where the peak in doping is not at the junction, but deeper in another part of the region.

$$n_i^2 = N_c N_v e^{-Eg/k_b T} \tag{3-5}$$

where:
Nc is the number of states at the conduction band
Nv is the number of states at the valence band
and:

$$n_i^2 = np \tag{3-6}$$

where:
n is the number of electrons (carriers) primarily due to doping
p is the number of holes (acceptors) primarily due to doping

Note that np = constant. Doping with Nd increases n and reduces p, and doping with Na increases p and reduces n. In all equations, one could specify n(E) vs. n(B) and p(E) vs., p(B). For example, equation (3-6) is equivalent to:

$$n(E)*p(E) = ni^2 \tag{3-7}$$

$$n(B)*p(B) = ni^2 \tag{3-8}$$

3.7.2 Methods for Refining Models

There are two common modeling approaches to account for extensions to a basic model. The quick (and easy) solution is to change equations and add more parameters until a model fits measured data better. This is the approach used in the series of BSIM MOS models. The other approach is to understand the physics and then modify the equations and add selected parameters as needed. This is the approach is used for the EKV MOS model. The second approach leads to more accurate models that are easier to update as new physical effects become significant.

---

[23] More information on diffusion, recombination lifetimes and hole/electron mobilities can be found in [47, pages 101-112].

3.8 EXAMPLES OF SPICE MODELS AND PARAMETERS

SPICE parameters are used in the diode and MOSFET (CMOS) equations to compute the circuit behavior of those devices. However, the BJT equations show how the BJT SPICE parameters are derived from semiconductor device structure and materials.

3.8.1 The Diode

Semiconductor diodes predate the invention of any form of transistor. The structure of nearly any type of semiconductor inherently includes one or more diodes, intentional or not (parasitic). FET leakage junctions are basically parasitic diodes and their leakage has come to dominate over gate leakage. Clamp diodes are more often designed into digital I/Os, for limiting reflection overshoot and to provide ESD protection, than not.

3.8.1.1 Diode Equivalent Circuit

A diode is a single p-n junction. Under forward bias, electrons move from the n to the p region and holes move from the p to the n region. Under reverse bias, a leakage current flows between the two regions. Most of the reverse current usually results from increased carrier generation and reduced carrier recombination in the widened depletion region, rather than from electrons and holes moving across the depletion region.

A diode has the same breakdown mechanisms as for the collector-base junction described above, including avalanche breakdown, corner and surface breakdown, and reach-through or punch-through-limited breakdown. Diodes, when both sides are highly doped, can also suffer from Zener breakdown, where electrons quantum-mechanically tunnel through the junction, resulting in low breakdown voltages. Interestingly, many so-called Zener diodes are actually low-voltage-breakdown avalanche diodes.

Figure 3-11 shows the circuit symbol for a diode. The p side is labeled NA (the anode) and the n side is labeled NC (the cathode). Figure 3-12 shows a physical cross-section built as a lateral structure, as would be the case where a diode was part of an IC chip, and as opposed to a vertical structure as in the case of discrete diodes.

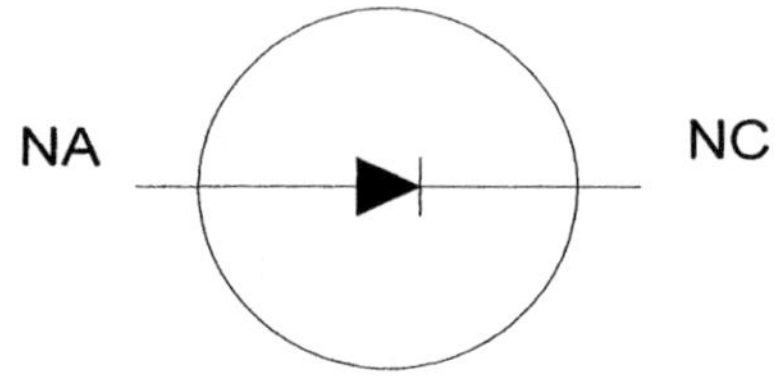

Figure 3-11. Diode schematic symbol [84]

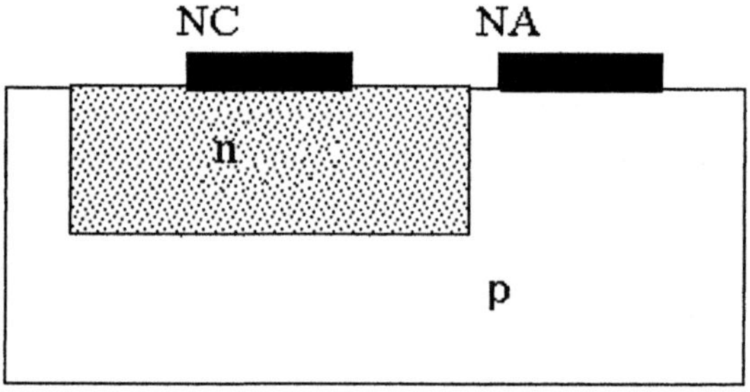

Figure 3-12. Diode cross-section

3.8.1.2 SPICE Diode Parameters

As with the bipolar transistor, the SPICE equations and associated parameters are based on the device physics. The default SPICE parameters are for an ideal small-signal diode having infinite bandwidth.

Unlike the transistor, the diode junction is more likely to have high-injection effects at high forward bias. The junction current depends on three effects: the series resistance due to the bulk silicon, the ideal current that varies as exp(V/Vt), and a second-order effect that varies as exp(V/2Vt). The SPICE diode model has two parameters: RS for the resistance and N for the average of the first and second-order current terms. Diode switching times are determined by two factors:

- Junction capacitance, which is determined by the diode's junction voltage.
- Transit time, which is the time required for a hole or electron to cross the undepleted region.

This second effect is accounted for in the SPICE model by the parameter TT. It can be measured using an oscilloscope. TT corresponds to the time from when the bias voltage is changed to when the current starts to change. This time can be relatively long for high-voltage-breakdown diodes, and is faster for integrated-circuit diodes. The default values for CJ0 and TT are 0.0 in SPICE, which represents an infinite bandwidth. Table 3-5 lists the SPICE diode parameters for equations (3-9) to (3-17).

Table 3-5. SPICE diode model parameters [84]

Name	Parameter	Units	Default Value	Effect of Area
IS	Saturation current	A	1e-14	Multiplies
RS	Ohmic resistance	Ohms	0	Divides
N	Emission coefficient		1	
TT	Transit time	sec	0	
CJO	Zero bias junction capacitance	F	0	Multiplies
VJ	Junction potential	V	1	
M	Grading coefficient		.5	
EG	Activation energy	eV	1.11	
XTI	Saturation current temperature coefficient		3	
KF	Flicker noise coefficient		0	
AF	Flicker noise exponent		1	
FC	Coefficient for forward bias depletion capacitance		.5	
BV	Reverse breakdown voltage		Infinity	
IBV	Current at breakdown voltage		1e-3	

3.8.1.3 Diode SPICE Equations

Forward Voltage:

$$VT = K_b * (T / q) \tag{3-9}$$

= thermal voltage = .025875 electron volts @ room temperature (27°C).

Diode Current:

$$ID = IS * (e^{(VD/(N*VT))} - 1) \tag{3-10}$$

In SPICE, the above equation applies in both forward and reverse operation, although it neglects generation and recombination effects that are actually dominant in reverse bias.

Reverse Breakdown Current:

$$IR = IBV * (e^{(-(VD+BV)/VT)}) \tag{3-11}$$

Reverse-Bias Junction Capacitance:

$$CD = CJO/(1-(VD/VJ))^{M} \tag{3-12}$$

Forward-Bias Junction Capacitance:

$$CD = (CJO/(1-FC)^{(1+M)}) * \{1 - FC(1+M) + M(VD/VJ)\} \tag{3-13}$$

Charge Storage (due to minority-carrier injection):

$$QS = TT * IS * (e^{(VD/(N*VT))} - 1) \tag{3-14}$$

where:

$$TT = TS/\ln(1 + IF/IR) \tag{3-15}$$

where:

IF = the forward current
IR = the reverse current
TS = the storage time

This is not a definition of TT, but a way to calculate TT from measurement of TS, IF, and IR. TT is a SPICE parameter. The value of TS depends on the circuit, so it is not a device parameter. And:

Many versions of SPICE can also calculate signal noise. For example, Shot and Flicker Noise:[24]

$$in^2 / df = 2 * q * ID * (KF / f) * (ID)^{AF} \tag{3-16}$$

where:
in = the spectral noise current
f = the frequency
df = the frequency interval

Effect of Temperature on Parameters:

$$IS@T = IS_{NOM} * (T / T_{NOM})^{(XTI / N)} * e^{(EG /(N*VT))*(T-T_{NOM}) / T_{NOM})} \tag{3-17}$$

It is interesting to note which physical effects are not modeled in the above diode model. Consequently, they have no parameters. In particular, current to the edges and corners (perimeter) of the diode structure are not modeled in this model. The effect of changing depletion width on RS and TT are not modeled. Hyperabrupt junctions (M>1) are not allowed. Increase in leakage due to recombination-generation effects are modeled only through BV and IBV parameters. Self-heating effects are not modeled. The model user must decide which of these effects need to be modeled, and create a macromodel (subcircuit) to add them.

3.8.2 The Bipolar Junction Transistor (BJT)

3.8.2.1 BJT Equivalent Circuit

The bipolar transistor model in SPICE is based on the Gummel-Poon equivalent circuit, shown in Figure 3-13. All of the capacitors and current sources are bias-dependent, as is RB. RC and RE are constant resistance. Also, the default values of the capacitors and base transit time are 0.0, which represents an infinite bandwidth.

The SPICE model accounts for base transit time based on base width but does not fully model base charge storage.[25] The model does not account for

[24] Noise (as used in this equation) is the random and spontaneous fluctuations in the current or voltage signals in the semiconductor device. This type of noise is usually of little consequence in digital switching circuits but is important in analog amplifier circuits. Measured noise is usually classified into thermal, flicker (1/f), or shot noise.

any collector transit time. This mainly affects accuracy for high-voltage switching BJTs. The model assumes that the emitter-base and collector-base junctions do not go into breakdown. These additional physical effects can be addressed by embedding a BJT inside a larger subcircuit that models these effects.

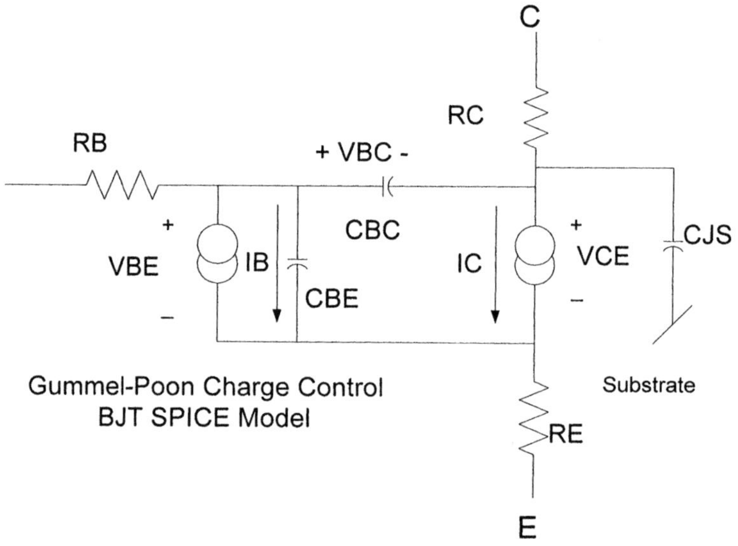

Figure 3-13. BJT Gummel-Poon equivalent circuit [84]

The BJT SPICE parameters and device equations are covered in detail in "Appendix J, Device Physics."

3.8.2.2 SPICE BJT Parameters

As with the diode, the SPICE bipolar transistor equations and associated parameters are based on the device physics. The default SPICE parameters are for an ideal small-signal BJT having infinite bandwidth. Table 3-6 is a partial list of the SPICE BJT parameters. A more complete list of the parameters and their associated equations is found in "Appendix J, Device Physics."

---

[25] This model does not include base storage–transit time effect correctly. Some people actually use the GaAs bipolar model instead, using silicon parameters, of course.

Table 3-6. SPICE - Gummel-Poon BJT model parameters [84]

Name	Parameter	Units	Default Value	Effect of Area
IS	Saturation current	A	1e-16	*
BF	Ideal max forward bias Beta		100	
NF	Forward current emission coefficient		1	
VAF	Forward Early voltage	V	Infinity	
IKF	Corner for BF high current rolloff	A	Infinity	*
ISE	B-E leakage saturation current	A	0	*
NE	B-E leakage emission coefficient		1.5	
BR	Reverse current Beta		1	
RC	Collector resistance	Ohms	0	/
CJE	B-E zero bias depletion capacitance	F	0	*
VJE	B-E built-in potential	V	.75	
MJE	B-E junction exponential factor		.33	
TF	Ideal forward transit time	sec	0	
XTF	Coefficient for base dependence of TF		0	
VTF	Voltage describing VBC dependence of TF	V	Infinity	
ITF	High current parameter for TF	A	0	*
CJC	B-C zero bias depletion capacitance	F	0	*
VJC	B-C built-in potential	V	.75	
MJC	B-C exponential factor		.33	
TR	Ideal reverse transit time	sec	0	

3.8.2.3 BJT Physics Equations

Computing β of a BJT

Beta is a SPICE model parameter. Consider the DC current gain (β) of a Bipolar Junction Transistor (BJT) from [47, page. 220]:

$$\beta = h_{FE} = \frac{\gamma\alpha_T}{1-\gamma\alpha_T} \tag{3-18}$$

In a modern BJT, we want $\gamma\alpha_T$ (γ = emitter efficiency, α_T = base transport factor) to be as close to 1 as possible. This gives a transistor with a large forward β. We can then write:

$$1/\beta \cong 1 - \gamma\alpha_T \tag{3-19}$$

The next step is to formulate a structure and material properties-based equation.

$$\frac{1}{\beta} \approx \underbrace{\frac{N_{DB}W_B}{D_{pB}} * \frac{W_{EB}/\tau_o}{2n_i e^{qV_{EB}/2k_bT}}}_{(a)} + \underbrace{\frac{N_{DB}W_B}{D_{pB}} * \frac{D_{nE}}{N_{AE}W_E}}_{(b)} + \underbrace{\frac{1}{2} * (W_B / L_{pB})^2}_{(c)} \tag{3-20}$$

The first term (a) is a current carrier loss due to the recombination of injecting carriers (from the emitter) within the emitter-base space charge/ depletion/junction region. As W_B decreases due to Early Effect, Beta increases. It also decreases the physical region for recombination in the second term.

The second term (b) is a current carrier loss due to the reverse injection and recombination into the emitter from the base, of carriers of the opposite polarity to the emitter-carrier polarity.

These two terms together make up the emitter injection efficiency, γ.

The third term (c) is a current carrier loss due to the recombination of minority carriers within the neutral base. It is also known as the base transport factor, α_T.

In summary, as the loss factors decrease, the beta of the BJT increases. Early SPICE models did not model the recombination-generation current change with base width, and did not model low- versus high-level injection effects.

The Beta we have just computed from structure in equation (3-20) is the BF in Table 3-6.

Beta Refinements and Non-Ideal Effects

At very low currents, the carrier recombination loss predominates over other losses in the base. The losses of majority carriers injected from the base occur during diffusion into the emitter-base space-charge region and then into the emitter. These losses are fixed. As current increases, they become insignificant. Carrier-recombination losses in the neutral base are a fixed percentage of total current. As current increases, the neutral base current losses become predominant. This gives rise to the classical low-current beta rolloff and rollup of a BJT. This behavior is difficult to see on a scan of collector characteristic curves on a curve tracer—as opposed to a plot of beta versus Ic. Figure 3-14 shows an unusual example of high current carrier losses in the low-current region.

Carrier-recombination losses in the base that are a fixed percentage of total current account for the flat portion of the beta versus Ic curve. The more heavily doped the base region background doping level is, the larger this fixed percentage of collector current recombination becomes. This is because there are more minority-carrier recombination sites.

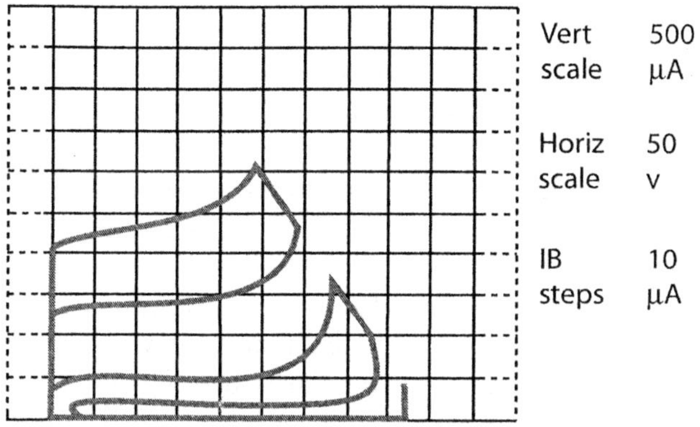

Figure 3-14. Collector characteristics showing pronounced low current beta rolloff [74]

Figure 3-14 shows a curve tracer display of low-current beta rolloff. The vertical axis is Ic @ 10 uA/div and the horizontal axis is Vce @ 50 volts/div. How was it possible to capture this effect? Device selection, curve tracer settings, or some other neat trick?

This picture was taken off of the display of a Tektronix 576 [74] curve tracer. The curves are generated by a stepped series of forced, fixed base current and a swept collector voltage. The resulting collector current is displayed in a series of curves. Notice how the effective β increases as the collector current level increases from about the 5μA level to the 45μA level.

Getting the display to show the low current beta rolloff effect took some manipulation with the curve tracer settings. But the major reason for being able to illustrate the effect is the selection of the device. The device in Figure 3-14 shows a very high amount of low current beta rolloff. The BJT had a very high level of fixed losses in the emitter-base space-charge region. A number of processing quality control issues caused carrier recombination sites to go up, including:

- Lattice stresses, damage, and contaminants.
- Hot carrier site injection under the oxide and other effects of reverse biasing the E-B junction.
- Recombination sites are particularly sensitive to surface contaminants because the E-B space charge region meets the surface of a BJT.

As well, a number of other effects can be seen in a poorly processed device. Especially after life test, as in the above case of a contaminated lot. Among these effects are breakdown voltages that change (push) with applied voltage and surface inversion (channeling) across the E-B junction. Observing these effects and other BJT phenomena is discussed more fully in [74].

Interestingly, the semiconductor process that Bardeen, Brattain and Shockley at Bell Labs had available in 1948 prevented them from first finding a surface FET. The surface FET is what they were originally searching for. A high level of contaminants was already inverting their purported surface-field-effect transistor when they stumbled across the BJT effect and recognized it for what it was. They had set out to observe surface inversion on a semiconductor to attempt to build a useful solid-state switching device. They then realized that they were observing minority-carriers injected from a region of p polarity drifting across a narrow gap of n material to a collecting region of p material, and that a bias voltage on the gap material could control this action.

For more on the discovery and theory of the BJT, see [110, 111].

Computing the Early Effect Voltage

Another example of structure-based parameter computation is computing Early effect. The Early effect, or base-width modulation, is named after Dr. Jim Early, who explained it. It is an extension, or refinement, to the basic model.

In a bipolar transistor, as the collector-base depletion region widens, some of the spreading will be into the base, causing it to narrow. The spreading will be greater with a lightly doped base. The narrowing will be proportionally greater when the base is narrow to begin with. This narrowing of the base increases the effective β. For a pnp transistor with a linearly graded collector-base junction, we have [47, 53, 63]:

$$\partial h_{FE}/\partial V_{CB} = \frac{\left[\frac{W_B}{L_{pb}} + \frac{N_{DB} D_{ne}}{D_{pb} N_{AE} L_{ne}} + \frac{W_{EB} N_{DB} \ell^{-(qV_E/k_b T)}}{2 n_i \tau_o D_{pb}}\right] \Delta V_{CB}}{\left[\frac{W_B}{2L_{pb}^2} + \frac{N_{DB} D_{ne}}{D_{pb} N_{AE} L_{ne}} + \frac{W_{EB} N_{DB} \ell^{-(qV_E/k_b T)}}{2 n_i \tau_o D_{pb}}\right]^2 \left(\frac{1\,\varepsilon_s \varepsilon_o}{q}\right)^{1/3} 3 V_{CB}^{4/3}} \quad (3\text{-}21)$$

The Early Voltage, a mathematical construct, is defined as:

$$V_{AF} \equiv \left[\frac{1}{W_o} * \frac{\partial W}{\partial V_{BC}}\right]^{-1} \tag{3-22}$$

On an I-V curve such as Figure 3-7, the Early voltage can be estimated by extrapolating the linear slopes of the I-V curves to a common point. For more on measuring Early effect, see [63].

The Early Voltage we have just computed from structure in equation (3-22) is the VAF in Table 3-6.

3.8.2.4 Comparison of SPICE Properties for BJTs

Table 3-7 compares some selected non-defaulted SPICE parameters for several discrete BJT transistors, both npn and pnp. As a general rule, npn transistors have higher gain, since electrons have higher mobility in the base region than holes do. The difference in the SPICE parameters for different devices illustrates the importance of having the correct parameters.

All the SPICE parameters for current, such as IS have positive signs. The simulator equations change signs as necessary between npn and pnp models, so that current and junction voltages maintain appropriate physical polarities in all operating regions.

Table 3-7 shows clearly that device construction affects SPICE parameter values. Parameters like mid-current DC current gain, β; low and high current β range, and linearity of β with collector voltage interrelate with each other and with other SPICE related behavior. For the definitions of the parameters in Table 3-7, refer to Table 3-6 and the associated discussion in the text.

Table 3-7. Small-signal general-purpose amplifier and switches

Parameter	2N3904 (npn)	2N4400 (npn)	2N4410 (npn)	MPSA05 (npn)	2N4402 (pnp)
IS	6.734f	26.03f	5.911f	8.324f	.6506f
BF	416.4	7756	413.6	12160	108
VAF	74.03	90.7	62.37	100	115.7
IKF	66.78m	239.7m	12.6m	109.6m	1115m
ISE	6.734f	26.7f	5.911f	73.27f	146.9f
NE	1.259	1.204	1.278	1.368	1.86
BR	0.7371	1.06	1.361	11.1	3.83
RC	1	.5	1.61	.25	.715
CJE	4.493p	24.07p	4.973p	55.61p	19.82p
VJE	0.75	0.75	0.75	0.75	0.75
MJE	0.2593	0.3641	0.4146	0.3834	0.3357
TF	301.2p	573.2p	818.4p	516.1p	761.3p

Table 3-7. (cont.)

Parameter	2N3904 (npn)	2N4400 (npn)	2N4410 (npn)	MPSA05 (npn)	2N4402 (pnp)
XTF	2	0	7	6	1.7
VTF	4	0	4	4	5
ITF	0.4	0	0.35	0.5	0.65
CJC	3.638p	11.01p	4.017p	18.36p	14.76p
VJC	0.75	0.75	0.75	0.75	0.75
MJC	0.3085	0.3763	0.3174	0.3843	0.5383
TR	239.5n	244n	4.749n	72.15n	115.7n

In Tables 6-3, 6-4, and 7-2, we look at this same set of devices and see that as devices grow larger and more rugged (able to handle larger voltages, currents, and power), they become slower in switching speed and bandwidth capabilities. SPICE is used to compute transistor behavior as an amplifier or as a switch. But SPICE is not designed to handle extreme conditions such as self-heating or reversed-biased second breakdown. Voltage, current and power handling capabilities are outside of SPICE's focus. For instance, asking for a device with a particular gain-bandwidth capability has consequences for the magnitude of voltage-standoff capability. For example, a video amplifier chip, used to drive the video gun of a CRT, peaks out at around 100MHz gain-bandwidth and 300V BVceo. Attempts to modify the device design to achieve significantly better gain bandwidth will succeed only by giving up some breakdown voltage capability.

Almost all the shifts in SPICE parameter behavior and power handling behavior can be predicted and studied from TCAD formulas. There is a level of very sophisticated device design techniques that extend device capabilities. For instance, there are device construction techniques that enhance high-current β holdup, voltage standoff, and many other performance factors. These techniques shape electric fields and depletion regions in the silicon, the steepness of doping gradients, and other factors. The discussions of these techniques are a subject for semiconductor device designers and are not relevant to a book on circuit design.

For deep sub-micron CMOS, tradeoff issues would typically involve geometry size, switching speed, current handling, current electromigration resistance, threshold and noise margins, and ESD ruggedness among others.

3.8.3 The MOSFET Transistor

3.8.3.1 Introduction

There are literally dozens of SPICE models for MOSFET transistors. These models cannot be covered completely here. Many additional factors have entered into device modeling, up to and including quantum effects. Quantum effects result from the nature of the extremely small device dimensions in the lateral surface direction. In the past, horizontal dimensions were very much larger than vertical dimensions. Now horizontal dimensions are comparable or even much smaller.

Two excellent resources for studying MOSFET modeling are:

- The Berkeley models:
  http://www-device.eecs.berkeley.edu/~bsim3/get.html [117]
- The Enz, Krummenacher, and Vittoz EKV (Enz, et al., 1995) model:
  http://legwww.epfl.ch/ekv/ [35]
- The Applied Computational Research Society site:
  http://www.cr.org/index.html

BSIM3 and BSIM4 models tend to be based more on curve fitting and binning, while the EKV models are based on physical conservation laws (charge, energy, and fields at boundaries). The EKV models were designed to address MOSFETs operating in low-to-moderate channel inversion, which is characteristic of low-voltage and small-signal integrated circuits. BSIM models, on the other hand, were designed to model digital MOSFETs switching from cutoff to strong inversion.

3.8.3.2 MOSFET Equivalent Circuit

The BSIM MOSFET model in SPICE is based on the equivalent circuit shown in Figure 3-15 and corresponds to SPICE2 from UC Berkeley. All of the capacitors and current sources are bias-dependent; the default values of the capacitors are 0.0, which represents an infinite bandwidth.

The SPICE model does not account for any source and drain transit time; this mainly affects accuracy for high-voltage switching MOSFETs. The model assumes that the gate-oxide layer and source/substrate and drain/substrate junctions do not go into breakdown. These additional physical effects can be addressed by embedding a MOSFET inside a larger subcircuit that models these effects.

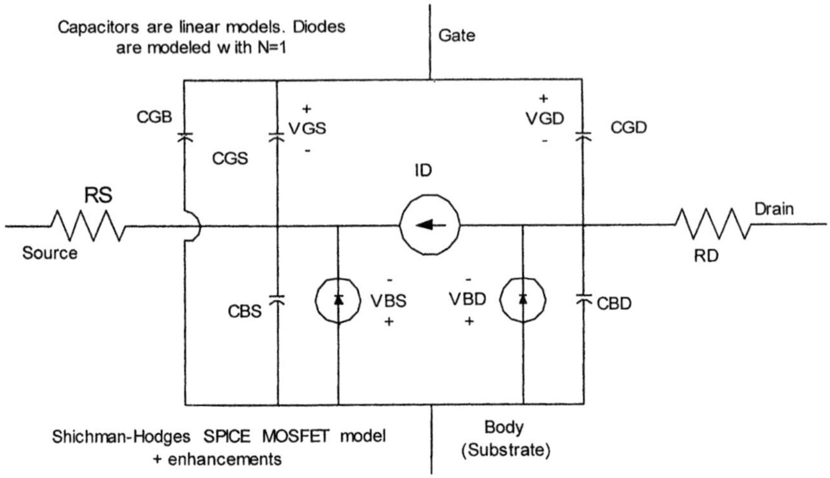

Figure 3-15. MOSFET equivalent circuit [84]

3.8.3.3 SPICE MOSFET Parameters

The MOSFET SPICE parameters and device equations are covered in detail in "Appendix J, Device Physics." Some selected parameters are shown in Table 3-8 for illustration. Equations (3-23) to (3-26) are a sample of the SPICE MOSFET equations that use these parameters. For a complete set of MOSFET SPICE2 parameters refer to "Appendix J, Device Physics." Table 3-8 is based on SPICE2 [152]. Berkeley has released several newer MOSFET models, from MOS1 to BSIM4. Most SPICE versions also support the EKV MOSFET model (which came from a group in Europe).

Table 3-8. SPICE2 MOSFET model parameters from UC Berkeley [84]

Name	Parameter	Units	Default Value
LEVEL	Model index		1
VTO	Zero bias threshold	V	0
KP	Transconductance	A/V^2	2e-5
PHI	Surface potential	V	.6
LAMBDA	Channel length modulation MOS1 and MOS2 only	1/V	0
CBD	Zero bias B-D junction capacitance	F	0
CBS	Zero bias B-S junction capacitance	F	0
IS	Bulk junction saturation current	A	1e-14
CGSO	Gate-Source overlap capacitance	F/m	0
CGDO	Gate-Drain overlap capacitance	F/m	0
CJSW	Zero bias bulk junction sidewall capacitance	F/m	0
TOX	Oxide thickness	m	1e-7

Table 3-8. (cont.)

Name	Parameter	Units	Default Value
NSUB	Substrate doping	cm^{-3}	0
XJ	Metallurgical junction depth	m	0
LD	Lateral diffusion	m	0
UO	Surface mobility	cm/V-s	600
VMAX	Maximum drift velocity of carriers	m/sec	0
DELTA	Width effect on threshold voltage, MOS2 and MOS3		0
THETA	Mobility modulation, MOS3 only	1/V	0
ETA	Static feedback, MOS3 only		0
KAPPA	Saturation field factor, MOS3 only		0.2

3.8.3.4 MOSFET SPICE Equations

A few MOSFET model equations (3-23) through (3-26) are shown for illustration. These equations are level 1 and level 2 with terms for gate modulation specifically included because this effect will be referred to later. Additional equations, governing such things as MOS junction capacitance, can be found in the references [10, 11, 21, 22, 28, 32, 35,38, 40, 42, 47, 63, 68, 77, 84, 86, 90, 97, 98, 100, 109, 117, 118, 121, 127, 132, 133, 140, 152].

For: Forward Region, VDS > 0

$$ID = 0 \tag{3-23}$$

For: VGS - VTO < 0

$$ID=(KP/2)*(W/L)*(VGS-VTE)^2 \tag{3-24}$$

For:0 < VGS - VTO < VDS

$$ID=(KP/2)*(W/L)*VDS*(2*(VGS-VTE)-VDS)*(1+LAMBDA*VDS) \tag{3-25}$$

For: 0 < VDS < VGS – VTO, where:

$$VTE=VTO+GAMMA*(\sqrt{(2PHI-VBS)}-\sqrt{(2PHI)}) \tag{3-26}$$

3.9 MODELING PACKAGING INTERCONNECTIONS

Related to modeling the silicon die is modeling the package and interconnections on and around the chip. The circuit designer needs a model of the whole IC, including interconnections, bond wires, package, and pins. SPICE models in the public domain usually include none of this information. On-die metalization interconnections modeling may be included in some EDA packages and is certainly included by silicon designers as part of their validation and verification process. Today's deep sub-micron switching speeds are tens of picoseconds or less in switching speed edge rates. Signal Integrity, crosstalk, Power Integrity, and EMI/EMC concerns have migrated into the design of the silicon and its packaging. The challenge to the silicon designer and packaging engineer is not lessened by the fact that they do not control the power supplies and power returns—the circuit designer does.

While devices have been shrinking, interconnection technology has also changed. For example:

- Metal and dielectric layer thicknesses have changed. Some of the latest deep-sub-micron technology is working with oxide thicknesses only three atoms wide!
- Horizontal feature size has shrunk dramatically.
- Metal width-to-height ratios have dropped from 3:1 to about 1:3.
- The number of metal layers on a chip has increased from 2 or 3 to over 8. One consequence is that capacitance between adjacent wires is now dominant over trace-to-ground capacitance.

3.9.1 Wavelengths: Interconnections Affect Everything

Package and board design technology have advanced as well. Packages are now being designed with controlled impedance, similar to how high-speed boards have been designed for years. Increasing via density has resulted in more porous ground and power planes, making it more difficult to predict impedance. These changes in package technology have led to the use of 2D and even 3D field solvers to model interconnections at the IC package, and board level. The models are commonly expressed as lumped SPICE subcircuits, RLGC matrices, or S-Parameters.

Field solvers were once rarely used in semiconductor device analysis. They had been mainly used to extract parameters for PCB-level transmission lines, connectors, and cables. For PCB level use, the extraction of parasitics on IC package pins and mutual-coupling effects between pins are sufficient for SI needs.

Now however, with device speeds and edge rates pushing frequency content into the tens of GHz range, field solvers are beginning to see more use. Field solvers are now in widespread use for high-speed ICs. 3D field solvers [113] are mostly used, since there is no ground plane in the IC package itself.

But what about the question of possible EMI radiation and conducted emissions from an IC package? The problem or radiation from IC packages is beginning to receive some attention. The trouble is, little is known about what the issues are. Little has been published on this subject, and many of those doing work in the field treat it as proprietary information.

We do know that IC packages can radiate. Plots of near-field simulations, showing the same board with and without the IC package radiation accounted for, are available. That work was done in close collaboration with one of the EDA supplier's customers. The plots are significantly different. The software provider stated that good correlation between EM-scan results and EMI simulation is not achievable without accounting for the ICs. But the work and the customer's name are proprietary.

How should we model the issues involved? Is the radiation coming from the loop formed by voltage supply rings? Or from via currents that do not have canceling image currents on the power planes because they are perpendicular to the image planes? We suspect that the major effect is from the vias. Or is it due to conducted emissions onto the PCB interconnections? Or some other issue? Packaging engineers seem shocked that anybody thinks their packages radiate. Or else, they see it as entirely the responsibility of the user to figure such things out, because it was the users who determined where to locate the image planes.

Another issue to address in EMI simulation is the ability to analyze the whole problem. EMI software providers have made great strides in improving their products. However, the authors are not aware of any software vendor offering a product that simulates EMI and Signal Integrity issues from the chip to far-field regulatory requirements.

All the same, one author [Leventhal, 85] provided modeling support for a client who needed to simulate near field EMI, to apply source-suppression techniques to pass regulatory requirements. The results correlated well to their lab measurements, enabling the application of useful simulation-based source suppression. In this case, adding an enclosure was not acceptable. Thus, an enclosure with its resonance effects and other coupling and loading effects did not enter into the simulations. So, near-field prediction and hot-spot suppression translated directly into passing regulatory requirements. Another effect of making simulation work is realizing that a designer should:

1. Apply good engineering rules-of-thumb to limit the problems that must be solved by simulation.
2. Apply simulation to sort out conflicts between the rules-of-thumb.

3.9.2 Standards for EMI/EMC Design

Industry and profession society standards committees and organizations have long been involved in developing standards for components and models. Table 3-9 lists some of these organizations now involved in EMI/EMC issues in modeling.

Table 3-9. Standards organizations involved in modeling package parasitics and EMI effects

Committee	Web address
EIA/IBIS Committee	Home page http://www.eigroup.org/ibis/ Input/Output Buffer Information Specification(s): downloadable http://www.eigroup.org/ibis/specs.htm IBIS Interconnections Model (ICM) Specification: draft proposal Intel is pushing this to be used as the proposed IC package specification http://www.eda.org/pub/ibis/connector/
IEC	International Electrotechnical Commission. International standards and conformity assessment for government, business and society for all electrical, electronic and related technologies. Home page http://www.iec.ch/index.html Integrated Circuit Electrical Model and Cookbook: draft proposal http://www.eigroup.org/ibis/specs.htm
IEEE	Institute of Electrical and Electronics Engineers. Home page http://www.ieee.org/portal/index.jsp EMC Society http://grouper.ieee.org/groups/emc/index.html
JEDEC	The JEDEC Solid State Technology Association (Once known as the Joint Electron Device Engineering Council) is the semiconductor engineering standardization body of the Electronic Industries Alliance (EIA), a trade association that represents all areas of the electronics industry. JEDEC was originally created in 1960 as a joint activity between EIA an NEMA, to cover the standardization of discrete semiconductor devices and later expanded in 1970 to include integrated circuits. Home page http://www.jedec.org
JEITA	Japan Electronics and Information Technologies Industries Association. Formed by the merger of EIAJ and JEIDA of Japan. JEITA's mission is to foster a digital network society for the 21st century. Among their many activities are the formulation of component and modeling standards for Japan.
CMC	Compact Model Council. The CMC is a group of Semiconductor Vendor companies and EDA Vendor companies that standardizes any compact model formulation that meets the business needs of its member companies. Currently, there is an effort in MOS, Bipolar, and SOI technology processes. Home page http://www.eigroup.org/cmc/

Much remains to be done in the engineering community regarding EMI/EMC simulation. These industry committees are addressing some of the modeling issues involved. The IBIS and IEC Committees are working on model standards for Power Integrity and SSN/SSO. The IEEE EMC Society is getting deeply involved in the modeling and simulation of EMI/EMC.

3.10 SUMMARY

SPICE was developed to simulate device circuit behavior. Over time, the SPICE program became synonymous with SPICE models. SPICE models can be quite complex, containing more than 200 parameters and 60 circuit nodes in a modern, deep sub-micron CMOS device. Deep-sub-micron CMOS devices are usually simulated with the aid of CAE software programs, collectively known as TCAD programs.

SPICE is the best model for understanding how circuit behavior relates to the internal physical construction of devices. SPICE parameters are also used to monitor and control semiconductor processes because of their often-close relationship to structure and chemistry. The device physics parameters that are useful for controlling semiconductor processes do not coincide exactly with the SPICE parameters that are useful for circuit simulation and analysis.

Engineers should understand the precise SPICE model received from a silicon foundry and the assumptions used to derive that model. In addition, sometimes the model contains errors from a circuit analysis point of view and does not run immediately in a particular simulator.

Not all SPICE parameters have direct physical meaning. Some parameters are purely curve-fitting properties that must be measured rather than be derived from structure.

Chapter 4

MEASURING MODEL PROPERTIES IN THE LABORATORY

Direct measurement of device-model properties with test instruments

Abstract: Modeling is an attempt to represent observations and measurements with mathematical theory. Some model parameters cannot be derived from structure so they *must* be measured. The measurements of model parameters can be used to verify modeling theory. Engineers gain confidence in the model when parameter predictions are verified by laboratory measurements. In this chapter, we introduce how to measure the parameters for 2-Port Matrix, Scattering-Parameter, SPICE, and IBIS models.

4.1 INTRODUCTION TO MODEL MEASUREMENTS

4.1.1 Motivation for Measurements

Careful observation, measurement, and experimentation are the basis of good modeling and simulation–more than the other way around. If we are good at both modeling and measurement, we will be really good at analysis and simulation.

If modeling and simulation predictions derived from theory were perfect and complete, we would have little need for laboratory measurement. If our imagination used to create possible new devices were similarly perfect and complete, we would have even less need for measurement and observation. Alas, such is not the case. Most creativity and accurate understanding of reality comes from experimentation, careful observation, and measurement.

In addition, not all observed semiconductor behaviors have a theoretical construct that allows their prediction from physical structure. Some model parameters are strictly curve-fitting properties and they *must* be measured.

4.1.2 Using Measurements to Predict a Device's Behavior

Modeling is an attempt to explain observations and to predict future behavior based on those observations. We measure a device's behavior and/or properties and use them to predict the device's future behavior when used in similar circumstances. For our purposes, we measure the device's electrical properties and use them to predict electrical performance.

In the case of a resistor, we measure and use the value of its resistance to predict how it will behave in a variety of circuits. This does not require much theory or modeling effort and does not involve much chance of error. Life would be simple if this was all that was required to build successful electronics products. Unfortunately, such is not the case. Without a model of a resistor, we do not know how to manipulate its internal properties for our benefit.

4.1.3 Motivation for Modeling

We can manipulate a device if we model it successfully and can control the physical properties upon which its behavior depends. So, we next try to mathematically explain what we measured electrically in terms of physical parameters, such as the length, area, and conductivity of a conduction channel.

Explaining a device's internal physical behavior takes some extra effort. There is a chance we may misunderstand the physical-electrical relations. But there are benefits to successful modeling. In a physical-electrical model we can use the model for accurate predictions when we change test or use conditions. For instance, what results would happen when we change the voltage across a conduction channel. And, what would happen when we change the length, area, or conduction of the channel (within limits).

An example of when different physical-electrical relations are applied is when very narrow base BJTs were developed. In that case, a new model had to be developed instead of extending the old model. The error in the old model was not discovered by theoretical means. It was discovered by measurements. More recently, the development of deep sub-micron semiconductor technology has resulted in new models and new model parameters.

Model measurements are expensive and time-consuming. At a minimum, they require a well-characterized, representative population sample from a stable and in-control IC product line. The devices themselves are usually electrically fragile and easily damaged by events like electrostatic discharge (ESD). High-speed digital devices require expensive fixturing and instrumentation in the hands of highly skilled practitioners.

Because of measurement expense, most of the supplied IBIS and S-Parameter models are derived from SPICE model simulations. That is, they are simulated from the existing SPICE model.

The SPICE model used to generate derivative IBIS and S-Parameter models may contain errors or bad assumptions. For example, setting clamp diode resistance to zero, or not simulating and measuring I-V over the full –Vcc to +2Vcc range, are common errors. These errors are bad for circuit simulation purposes, but they are not a problem for process control purposes. That is why they still continue to occur in circuit models provided by semiconductor manufacturers.

We have just discussed why laboratory measurements are important. Next we explain how to measure models.

4.2 MATRIX MODELS

This topic presents an overview of matrix models, their notation, and their measurements.

4.2.1 What are Matrix Models?

Matrix models are often thought of as 2-Port models. They can also be generalized to apply to N-Ports. The N-Ports are represented by a set of N equations in N unknowns forming an N-by-N matrix. In "Chapter 8, Selecting the Best Model for a Simulation," we also cover conversions between the various sets of 2-Port matrix models, as well as to other types of models. Measurement of the 2-Port, y-parameter case is presented next.

There are six sets of matrix models:

- z-impedance
- y-admittance
- h-hybrid
- g-inverse hybrid
- s-scattering (also called Scattering-Parameter, s-parameter, and S-Parameter)
- abcd-transmission

The RLGC matrix that an EM field solver extracts for modeling routed etch as a transmission line can also be visualized as a matrix model. Matrix models are easily embedded in a circuit. Modern computer simulation programs for Signal Integrity, EMI, Power Integrity, and crosstalk make extensive internal use of matrix models and matrix mathematics.

A complete matrix model circuit–including active-device matrix models, bias, coupling, tuning, and feedback elements–can readily be solved using Kirchhoff's node-and mesh-equations. Once solved, they yield equations for calculating voltage gain, current gain, input impedance, and output impedance for the complete circuit. By solving for these equations, we can then calculate the total input-to-output port performance, taking into account both the active devices and the external circuit elements. Moreover, the equations allow generalizations about how, for example, amplifier gain performs with respect to the network the transistor is embedded into.

Of the six sets of matrix models, only the Scattering-Parameter models are still widely used today outside of computer matrix manipulations. This is because of the power of SPICE as a physical model, wherein supply voltage, temperature, and other physical parameters can be varied without re-measuring the model parameters. This chapter presents the y-parameter and S-Parameter sets in some detail.

4.2.2 Two-Port Matrix Models

Table 4-1 summarizes the major 2-Port matrix model sets.

Table 4-1. N-Port matrix model parameters

Model Name	Parameters	Description
z-parameter	Impedance	Input-z, output-z, forward and reverse voltage-transfer generators
y-parameter	Admittance	Input-y (y_{11} or y_{fe} or y_{ie} terminology), output-y, forward and reverse current-transfer generators
h-parameter	Hybrid	Input-z, output-y, forward current-transfer generator and reverse voltage-transfer generator
g-parameter	Inverse Hybrid	Inverse hybrid set
S-Parameter or s-parameter	Scattering	Input (s_{11}) and output (s_{22}) reflection coefficients, forward (s_{21}) and reverse (s_{12}) transfer coefficients
ABCD-parameter	Transmission	Less convenient to measure, and less flexible than S-Parameters because of measuring open circuit reflection and transmission. Once characterized they can be convenient for cascaded 2-Port amplifier chains.

Note: Matrix models use lower case to indicate small-signal parameters and UPPER case to indicate large-signal parameters.

For example, both small-signal s-parameters and large-signal S-Parameters are correct notations. In the book we sometimes use them interchangeably but will point out if it is important to distinguish between small- and large-signal analyses. Otherwise, we use S-Parameters throughout to mean either.

4.2.3 Matrix Model Notation

Two-Port models are basically used for linear small signal modeling.

- Notation for *small* signal is y_{11}, alternatively y_{ie}, etc. Sometimes the parameters are measured under large-signal-swing conditions.
- Notation for *large*-signal is Y_{11}, alternatively Y_{ie}, etc. Y_{11} is a general matrix notation. Y_{11} is the input admittance of port 1, most often considered the input port. Y_{ie} specifically means the input admittance of a common-emitter configured amplifier.

Figure 4-1 is a symbol for a black-box component or circuit.

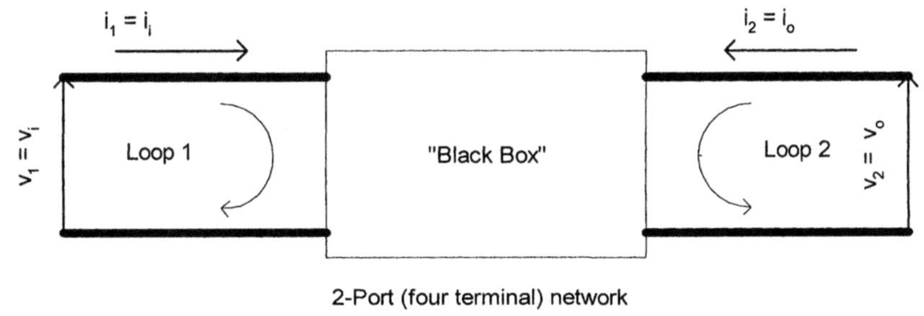

Figure 4-1. The black-box circuit model

In this case it has two 2-terminal ports and called a 2-Port. The 2-Ports are bi-directional. They can be driven from loop 1 (port 1) or loop2 (port 2). In either mode, the parameters can be measured and the response modeled and measured.

For real semiconductors, depending on the port being driven, the actual measured parameter values are usually very different. Transistors are often anything but bi-directional in their physical properties. They could be designed to be bi-directional, but it's more useful to design devices that enhance particular behaviors, such as gain, in one direction.

The current voltage relationships of a y-parameter model for a 2-Port are the equation set (4-1) and (4-2):

$$i_1 = y_{11}v_1 + y_{12}v_2 \quad (4\text{-}1)$$

$$i_2 = y_{21}v_1 + y_{22}v_2 \quad (4\text{-}2)$$

The first subscript on a parameter is the response or dependent node, the second subscript is the driver or independent node. Thus, y_{21} reads as the response at node 2 due to a signal at node 1.

4.2.4 Measurement of 2-Port Parameters

To illustrate the techniques of measurement, it is instructive to consider just one parameter. For a more complete treatment of matrix parameters, see [65, 126].

Example: y_{21}

There are two forms of subscript notation. We have just explained y_{21} in equations (4-1) and (4-2). For a common-emitter amplifier, we can also notate the parameter as y_{fe}, the common-emitter forward transadmittance. Here the f-subscript denotes forward and the e-subscript denotes emitter.

Admittance parameters are always measured into AC short-circuits. As such, they are somewhat easier to measure at higher frequencies than impedance parameters, which are always measured into AC open-circuits. The input to the device-under-test (DUT) is driven with a signal generator. The short- or open-circuit is placed electrically close to the output port. Transmission line effects and parasitic effects outside the DUT are usually ignored.

At high frequencies, the real and imaginary parts of the 2-Port parameters are measured to find the currents, voltages, and their phase shift information. The measurements are usually taken with a vector impedance and admittance meter or a network analyzer (VNA). Measurements are taken over a range of frequencies at a particular collector-emitter voltage and current in the normal class-A amplifier range.

2-Port black-box parameters are simple and easy to measure except under high-frequency conditions beyond about a GHz. They are all measured into open- or short-circuit (AC) conditions except for the Scattering parameter set. Open- and short- circuit conditions can be difficult to establish and maintain at higher frequencies.

Y-parameter models can be measured accurately out to a GHz or so. However, at much higher frequencies, component self-resonances, transmission-line effects, reflections, and other complications make y-parameters too difficult to measure accurately. Instead, measurement and use of Scattering Parameters is preferred.

The definition of y_{21} is:

$y_{21} = i_2/v_1$ with $v_2 = 0$

Open Circuit Impedance Parameter 2-Port Set

The equation set (4-3) defines the impedance parameter set, as shown in Table 4-2:

$$v_1 = z_{11}i_1 + z_{12}i_2$$
$$v_2 = z_{21}i_1 + z_{22}i_2 \qquad (4\text{-}3)$$

Table 4-2. z parameter definitions

Input Impedance	z_{11}	$= v_1/i_1$ when $i_2 = 0$
Forward Transfer Impedance	z_{21}	$= v_2/i_1$ when $i_2 = 0$
Reverse Transfer Impedance	z_{12}	$= v_1/i_2$ when $i_1 = 0$
Output Impedance	z_{22}	$= v_2/i_2$ when $i_1 = 0$

Short Circuit Admittance Parameter 2-Port Set

The equation set (4-4) defines the admittance parameter set, as shown in Table 4-3:

$$i_1 = y_{11}\, v_1 + y_{12}\, v_2$$
$$i_2 = y_{21}\, v_1 + y_{22}\, v_2 \qquad (4\text{-}4)$$

Table 4-3. y parameter set

Input Admittance	y_{11}	$= i_1/v_1$ when $v_2 = 0$
Forward Transfer Admittance	y_{21}	$= i_2/v_1$ when $v_2 = 0$
Reverse Transfer Admittance	y_{12}	$= i_1/v_2$ when $v_1 = 0$
Output Admittance	y_{22}	$= i_2/v_2$ when $v_1 = 0$

Hybrid Parameter 2-Port Set

The equation set (4-5) defines the hybrid parameter set, as shown in Table 4-4:

$$v_1 = h_{11}\, i_1 + h_{12}\, v_2$$
$$i_2 = h_{21}\, i_1 + h_{22}\, v_2 \qquad (4\text{-}5)$$

Table 4-4. h parameter set

Input Impedance	h_{11}	$= v_1/i_1$ when $v_2 = 0$
Forward Transfer Current Ratio	h_{21}	$= i_2/i_1$ when $v_2 = 0$
Reverse Transfer Voltage Ratio	h_{12}	$= v_1/v_2$ when $i_1 = 0$
Output Admittance	h_{22}	$= i_2/v_2$ when $i_1 = 0$

Note: The g-parameters are the inverse of the h-parameters so are not covered here.

4.2.5 Flexibility of Matrix Models

Once in possession of 2-Port models like y-parameters, we can easily analyze the mesh- and node- equations of the circuits in which they are imbedded. We can then derive the voltage and current gain plus input and output impedance of a complete common-emitter amplifier (or whatever other circuit configuration is being analyzed). Feedback elements and oscillator circuits are also easily handled.

Common-emitter parameters can be substituted into a common-base (or common-collector) configuration to derive common-base (or common-collector) closed-form equations with the common-emitter terms.

As another example, we drive what is normally considered the output port (port 2) of a 2-Port with a signal generator, hanging a load on what is normally considered the input port (port 1) of a 2-Port. Once the closed-form equations are developed for current and voltage gain, an isolation amplifier can be designed. In an isolation amplifier, signal transferred in the reverse direction is minimized.

2-Port parameters also lend themselves well to other manipulations:

- For two elements (active devices, and circuit bocks) in *series*, add their Z-parameters directly for the cascaded Z-matrix. Just follow the standard rules for matrix addition and complex numbers.
- For two elements in *parallel*, matrix-add their Y-parameter matrixes directly for the composite Y-matrix.

4.3 SCATTERING-PARAMETER MODELS

This topic presents an overview of scattering-parameter models, their notation, and their measurements.

4.3.1 What is a Scattering-Parameter?

The term scattering parameter comes from the concept of an electronic signal propagating down a transmission line to an input pin, where:

- Some of the signal gets transmitted into the receiving network, and
- Some of the signal gets reflected (scattered) back towards the sending end

The mathematics for 2-Ports can be generalized to N-Ports for all matrix models.

S-Parameters are particularly well suited to measurements at high frequencies because incident and reflected signals are measured relatively easily with directional couplers. Other tools, such as the Vector Network Analyzer (VNA), make the measurement task straightforward. The measurements take place into matched and terminated loads making the transmission line effects easier to control and the measurements easier to calibrate.

At high frequencies, the establishment of open and shorted loads for the measurement of the other matrix parameters can be problematic, due to residual and stray parasitics in non-ideal opens and shorts. Also, active semiconductor devices often oscillate when their output is AC open-circuited or short-circuited.

The forward and backward traveling waves for S-Parameters are measured, looking into port 1 and then into port 2. Transmission lines feed both ports, so it is easy to insert directional couplers, and other equipment to measure the waves without seriously altering the signals involved. To simplify the measurements, it is convenient to match the wave generator and external load to their transmission lines. 50 ohms is a common characteristic impedance. This avoids multiple re-reflections, which complicate the measurements.

4.3.2 Scattering-Parameter Notation

This topic about scattering parameter notation is adapted from [3]. To determine the S-Parameters, a new set of variables (a_i, b_i) is defined. Variables (a_i, b_i) are normalized, complex voltage waves incident on, and reflected from, the ith port of the N-Port. They are defined in terms of the terminal voltage Vi, the terminal current Ii, and an arbitrary reference impedance as follows:

$$a_i = \frac{(V_i + Z_i I_i)}{2\sqrt{|\mathrm{Re}\, Z_i|}} \quad (4\text{-}6)$$

$$b_i = \frac{(V_i - (Z_i^*) I_i)}{2\sqrt{|\mathrm{Re}\, Z_i|}} \quad (4\text{-}7)$$

In equation (4-7), the asterisk, *, denotes the complex conjugate.[1] If Zi is positive and real, this simplifies the terms. Thus, for a 2-Port the independent variables a_1 and a_2 become:

$$a_1 = V_{i1} / \sqrt{Zo} \quad (4\text{-}8)$$

$$a_2 = V_{i2} / \sqrt{Zo} \quad (4\text{-}9)$$

where Zo is the characteristic impedance. Dependent variables b_1 and b_2 are the normalized reflected voltages:

$$b_1 = V_{r1} / \sqrt{Zo} \quad (4\text{-}10)$$

$$b_2 = V_{r2} / \sqrt{Zo} \quad (4\text{-}11)$$

The linear matrix equations for the 2-Port become:

$$b_1 = s_{11} a_1 + s_{12} a_2$$

$$b_2 = s_{21} a_1 + s_{22} a_2 \quad (4\text{-}12)$$

---

[1] For complex variable ($\alpha + j\beta$), its complex conjugate is ($\alpha - j\beta$).

4.3.3 Measurements of Scattering Parameters

The measurement of the S-Parameters is done by using the definitions in Table 4-5.

Table 4-5. S-Parameter measurements

Parameter	**Definition**	**Description**	**Measurement Condition**
s_{11}	b_1/a_1	Input reflection coefficient	Zl = Zo sets $a_2 = 0$
s_{22}	b_2/a_2	Output reflection coefficient	Zg = Zo sets $a_1 = 0$
s_{21}	b_2/a_1	Forward transmission coefficient	Zl = Zo sets $a_2 = 0$
s_{12}	b_1/a_2	Reverse transmission coefficient	Zg = Zo sets $a_1 = 0$

For the actual techniques of measuring S-Parameters, the reader is referred to [1, 2, 124].

4.3.4 A Simple Example of Using Scattering Parameters

To resolve loading effects on complex networks, computational steps can be done with S-Parameters. For instance, what is the effect of a shunt stub on a transmission line? If we let the shunt admittance looking into the stub be Y[2] as in Figure 4-2, we can derive [88, page 279] the resulting S-Parameters. Y usually is a complex number. The results are given in equations (4-13) through (4-15):

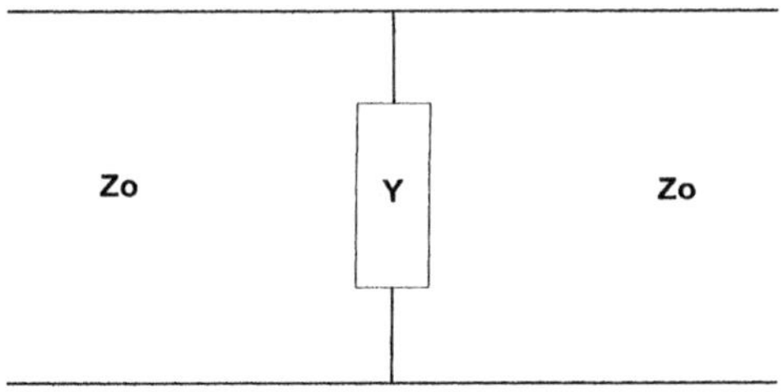

Figure 4-2. Shunt stub admittance S-Parameters

$$S_{11} = S_{22} = -Y/(2Yo+Y) \tag{4-13}$$

$$S_{21} = S_{12} = 2Yo/(2Yo+Y) \tag{4-14}$$

$$Yo = 1/Zo \tag{4-15}$$

[2] Matrix conversions between S, Y and Z parameters are straightforward. The conversions are given in "Chapter 8, Selecting the Best Model for a Simulation." This example merely illustrates some common 2-Port matrix manipulations for finding S-Parameters.

4.4 SPICE MODELS

This topic presents an overview of SPICE models, their notation, and their measurements.

4.4.1 Measuring SPICE-Gummel-Poon BJT Model Parameters

Showing how to measure SPICE parameters for BJTs, MOSFETs and other devices would in itself take a book this size. Plus, it would take the aid of a software program. That is not the purpose of this topic, which is to give readers a sample of how SPICE parameters are measured. The measurements of a few important BJT parameters are presented. The reference [63] presents the measurement of SPICE parameters in full for both the BJT and the MOSFET transistor. The following list presents the BJT parameters and where to start reading for an explanation of making the measurement. For those readers who really need to set up SPICE parameter measurements, see reference [63] and the discussion of specialized parametric analyzers in the Tektronix website: http://www.tek.com/

Once again, we use the BJT for illustration. The parameters in Table 4-6 are found in [63, 84].

Table 4-6. SPICE - Gummel-Poon BJT model properties that can be measured

Parameter Name	**Parameter Description**	**Starting page in [63]**
IS	Saturation current	112
BF	Ideal max forward bias Beta	112
NF	Forward current emission coefficient	112
VAF	Forward Early voltage	118
IKF	Corner for BF high current rolloff	116
ISE	B-E leakage saturation current	115
NE	B-E leakage emission coefficient	115
BR	Reverse current beta	107
NR	Reverse current emission coefficient	107
VAR	Reverse Early voltage	119
IKR	Corner for reverse beta rolloff	117
ISC	B-C leakage saturation current	117
NC	B-C emission coefficient	117
RB	Zero bias base resistance	121
IRB	Current where RB falls to 1/2 min	121
RBM	Min RB at high current	121
RE	Emitter resistance	119
RC	Collector resistance	120
CJE	B-E zero bias depletion capacitance	122
VJE	B-E built-in potential	122

Table 4-6. (cont.)

Parameter Name	**Parameter Description**	**Starting page in [63]**
MJE	B-E junction exponential factor	122
TF	Ideal forward transit time	124
XTF	Coefficient for base dependence of TF	124
VTF	Voltage describing VBC dependence of TF	124
ITF	High current parameter for TF	124
PTF	Excess phase @ frequency 1/(2πTF)Hz	124
CJC	B-C zero bias depletion capacitance	122
VJC	B-C built-in potential	122
MJC	B-C exponential factor	122
XCJC	Fraction of CJC connected to internal base node	123
TR	Ideal reverse transit time	124
CJS	Zero bias collector substrate capacitance	122
VJS	Substrate junction built-in potential	122
MJS	Substrate junction exponential factor	122
XTB	Forward and reverse Beta temperature coefficient	
EG	Energy gap for temperature effect on IS	
XTI	Temperature coefficient for effect on IS	
KF	Flicker noise coefficient	
AF	Flicker noise exponent	
FC	Coefficient for forward bias depletion coefficient	122

The most basic parameters of the BJT are IS (saturation current), BF (forward beta), and NF (forward base-emitter emission coefficient). From device physics:

The collector current equation:

$$Ic = Is * [\exp(q * Vbe / NF * k * t) - 1] \qquad (4\text{-}16)$$

The base current equation:

$$Ib = Ic/BF \qquad (4\text{-}17)$$

We need the DC I-V characteristic of the transistor. The most commonly used are the common-emitter (CE) configuration characteristic curves of Ic versus Vce and Ib.

These curves, for the 2N3904 transistor, are shown in Figure 4-3. This display could have as easily been taken off a curve tracer. A curve tracer amounts to a set of programmable voltage and current supplies, and corresponding voltage and current measurement instruments. In this display, Ic is measured and displayed as a function of Vce and Ib.

From this family of curves (Figure 4-3), we can easily find BF = I_c/I_b for a whole range of $V_{ce.}$. Note that BF increases as V_{ce} increases. This is called Early effect, after the device physicist who first was able to model it. To find forward Early effect, draw a straight line extending the linear portion of a collector characteristic curve to the negative x-axis. The intersection is the value – VAF.[3] If we do this extraction process to the entire set of curves, they will all supposedly intersect the negative x-axis at a single point. Early effect is a non-ideal effect like many other BJT phenomena.

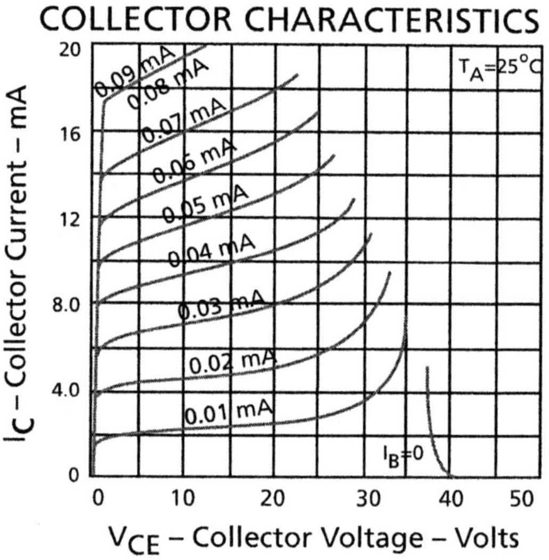

Figure 4-3. Bipolar family of curves, 2N3904 transistor

Figure 4-4 shows the measurement setup for this family of curves.

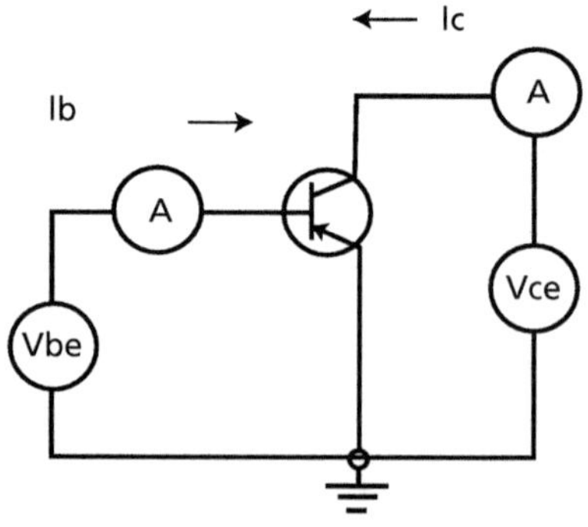

Figure 4-4. Measurement setup of CE collector characteristic curves

[3] VAF, forward Early voltage, is a non-structure based parameter that quantifies an effective increase in beta. This increase is due to narrowing the neutral base region as the collector space charge region widens with increased collector bias.

Note that the intersection of the slopes of the collector characteristic curves at a single point on the negative voltage axis is an idealized model. One author [Leventhal] observed many exceptions where there was a small spread in the value of VAF. Non-ideal effects are labeled on the 2N3904 collector characteristics shown in Figure 4-5.

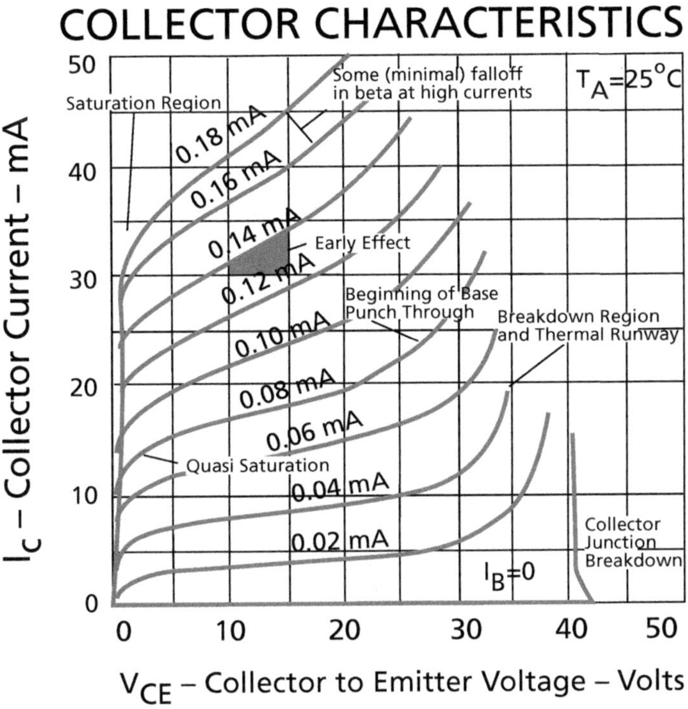

Figure 4-5. BJT CE collector characteristic curves with non-ideal effects labeled

Most standard device physics text books discuss these non-ideal BJT effects [36, 38, 44, 47, 74, 121, 132]. Common BJT non-ideal effects are:

- Collector current saturation [34]
- Collector quasi-saturation [66]
- Thermal runaway [110, 111]
- High current and low current BF falloff [64, 110, 111]
- Collector voltage breakdown
- Base punch-through limited breakdown
- Early effect [33]

These effects and other non-ideal effects cause for the BJT SPICE model having forty plus parameters instead of just a half-dozen or so. For more discussions of non-ideal BJT effects, refer to "Chapter 3, Model Properties

Derived from Device Physics Theory" and "Chapter 7, Using Data Sheets to Compare and Contrast Components."

The family of curves in Figure 4-5 does not completely describe the BJT's I-V relationships. A better plot is obtained from a Gummel plot, Figure 4-6, which is measured using a parametric analyzer, a fancy curve tracer. The Gummel plot shows the base (I_b) and collector (I_c) currents versus the base-emitter voltage (V_{be}) with the base-collector voltage (V_{cb}) held at zero. The natural logs (ln) of I_c and I_b versus V_{be} are plotted.

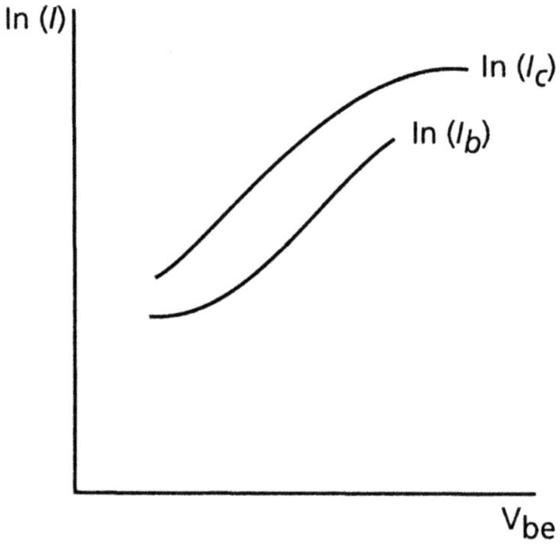

Figure 4-6. BJT Gummel Plot [63]

The measurement setup for this family of curves is shown in Figure 4-7.

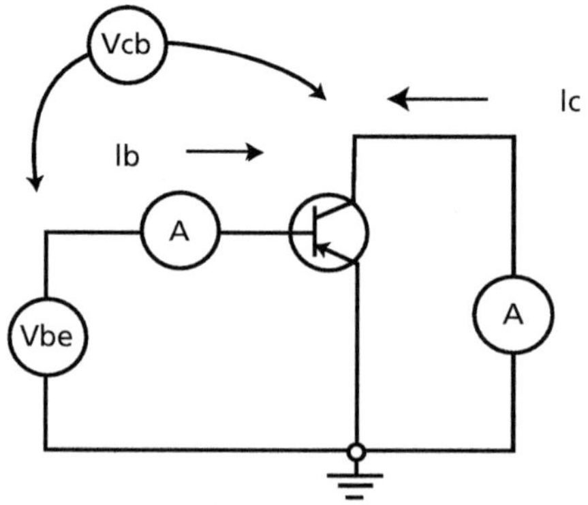

Figure 4-7. BJT CE Gummel Plot measurement setup

Vcb is held at zero volts. Extract IS, BF, and NF from the plots by:

- Drawing a straight-line approximation through the linear portion of the ln(Ic) curve. The intersection of this line with the y-axis is ln(IS). Or:

$$IS = exp(y\text{-}intercept\ of\ ln(Ic)) \quad (4\text{-}18)$$

- Drawing a straight vertical line between ln(Ic) and ln(Ib) at their point of maximum vertical separation. Then:

$$BF = exp(ln(Ic) - ln(Ib)) \quad (4\text{-}19)$$

- Measuring the slope of the straight-line approximation to ln(Ic). Then:

$$NF = 1/(slope*(kT/q)) = 1/(slope*0.0255) \quad (4\text{-}20)$$

At room temperature, these measurements are illustrated on the next Gummel plot, Figure 4-8:

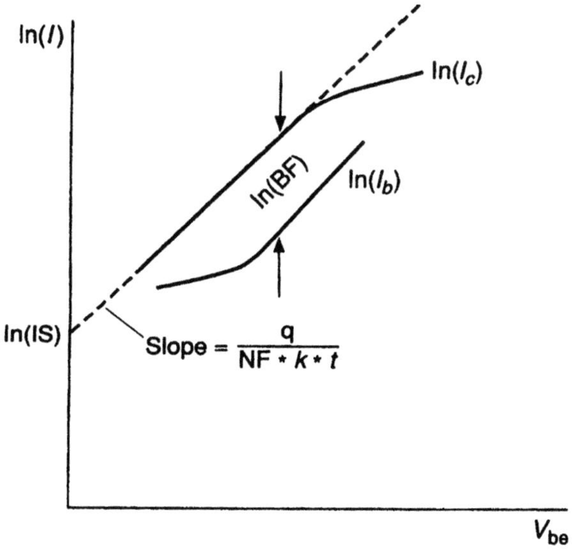

Figure 4-8. IS, BF, NF extraction from the Gummel Plot [63]

In later chapters, we talk about IBIS model extraction. There also I-V curves are central to the model. But they are used as-is and may include clamping diode effects and other effects. But in SPICE, we relate the terminal behavior (for example, I-V curves) to internal physical elements and phenomena inside the transistor under test.

4.4.2 Measuring SPICE MOSFET Model Parameters

Table 4-7 shows reference material for measuring MOSFET model parameters.

Table 4-7. SPICE MOSFET model properties that can be measured

Parameter Name	Parameter Description	Starting page in [63]
LEVEL	Model index	184
VTO	Zero bias threshold	186-191, 200-214
KP	Transconductance	186-191, 200-214
GAMMA	Bulk threshold parameter	
PHI	Surface potential	
LAMBDA	Channel length modulation MOS1 and MOS2 only	184
RD	Drain ohmic resistance	186-191, 200-214
RS	Source ohmic resistance	186, 192, 200-214
CBD	Zero bias B-D junction capacitance	198-199
CBS	Zero bias B-S junction capacitance	198-199
IS	Bulk junction saturation current	192
PB	Bulk junction potential	
CGSO	Gate-Source overlap capacitance	198-199
CGDO	Gate-Drain overlap capacitance	198-199
CGBO	Gate-Bulk overlap capacitance	198-199
RSH	Drain and Source diffusion sheet resistance	
CJ	Zero bias bulk junction bottom capacitance	198-199
MJ	Bulk junction bottom grading coefficient	
CJSW	Zero bias bulk junction sidewall capacitance	
MJSW	Bulk junction sidewall grading coefficient	
JS	Bulk junction saturation current density	
TOX	Oxide thickness	
NSUB	Substrate doping	
NSS	Surface state density	
NFS	Fast surface state density	
TPG	Type of gate material: 1 = opposite of substrate, -1 = same as substrate, 0 = Al gate	
XJ	Metallurgical junction depth	
LD	Lateral diffusion	
UO	Surface mobility	
UCRIT	Critical field for mobility degradation MOS2 only	
UEXP	Critical field exponent for mobility degradation MOS2 only	
UTRA	Transverse field coefficient in mobility degradation MOS2 only	
VMAX	Maximum drift velocity of carriers	
NEFF	Total channel charge coefficient (fixed and mobile) MOS2 only	

Table 4-7. (cont.)

Parameter Name	Parameter Description	Starting page in [63]
XQC	Thin oxide capacitance flag and coefficient of channel charge attributable to drain (0-.5)	
KF	Flicker noise coefficient	
AF	Flicker noise exponent	
FC	Coefficient for forward bias depletion capacitance	
DELTA	Width effect on threshold voltage, MOS2 and MOS3	
THETA	Mobility modulation, MOS3 only	
ETA	Static feedback, MOS3 only	
KAPPA	Saturation field factor, MOS3 only	

Note: For MOSFET Ciss, Coss and Crss, see [63, pages 193-198]. For Ld, Ls, Lg and Rg, see [63, page 199].

4.4.3 SPICE Models from Data Sheet Curves

Software for converting transistor data sheet characteristic curves into a SPICE model is available at these Symmetry Design Systems websites:

- http://www.symmetry.com/
- http://www.i-t.com/engsw/symm/symm.htm

The product for doing the conversion was called MODPEX. Data can come from scanned-in curves or actual measurements. An optimizer can simulate the model and dither the parameters to match the data.

4.4.4 SPICE Models from Software Controlled Automated Measurement Equipment

Computer controlled test equipment is available for taking model measurements. Many types of computers can be programmed to set up the test equipment and for the desired measurement conditions, then measure the responses of the devices-under-test. The computer also checks and adjusts calibration for the test equipment and fixtures. It stores the results in a database, manipulates them to extract parameters and population statistics. Some programs simulate the devices based on the extracted parameters. Then the model parameters are dithered to bring measured and simulated data into closer agreement. One such company supplying this market is

Agilent and their product is called 85190A IC-CAP® Parameter Extraction and Device Modeling Software.

4.5 IBIS MODELS

Acknowledgement
Most of this topic is provided by IBIS version 4.1. The authors' contribution is to arrange parts of the IBIS Specification concerning measurements with a little more focus. Only measurement conditions, I-V tables, Ramp data and V-T table data are addressed because they are of central importance to the IBIS model. There are many other parameters available for data exchange under the IBIS format. But they are embellishments of the basic data of pin connections to models, model behavioral data, and pin parasitics. For other parameters, refer to the IBIS Specification.

This topic presents an overview of IBIS models, their notation, and their measurements.

4.5.1 Format of the IBIS Data

IBIS is basically a format for the exchange of data. Therefore, IBIS is quite inflexible regarding syntax and rules, such as the characters allowed in a data field and the arrangement of the data in a file. For example, the way data is shown in an IBIS file follows a rationale derived from requiring only typical (typ) data, but providing a format for also including minimum (min) and maximum (max) data. When the file is parsed, the data is read left-to-right, top-to-bottom. Typical data is entered in the first column. If the data was arranged min-typ-max, the first field could possibly be empty. This is not good for many computer programs. Therefore, the data is entered typ-min-max left-to-right, top-to-bottom. A short example follows:

```
[Pulldown]
  | voltage    I(typ)         I(min)         I(max)|
  -2.5000      -88.8600mA     -60.2580mA     -0.1064A
  -2.3000      -88.8600mA     -60.2580mA     -0.1064A
  -2.1000      -88.8600mA     -60.2580mA     -0.1064A
```

4.5.2 Measurement Conditions Temperature and Voltage

Measurement conditions apply to all of the measured properties in this topic. Table 4-8 shows temperature and voltage measurement conditions.

Table 4-8. Temperature and voltage measurement conditions

Keyword	Required?	Description
[Temperature Range]	Yes, if other than the preferred 0, 50, 100 degree Celsius range	Defines the die temperature range over which the model is to operate
	Usage Rules List the actual die temperatures (not percentages) in the typ, min, and max format. NA is allowed for min and max only	
	Other Notes The [Temperature Range] keyword also describes the temperature range over which the various I-V tables and ramp rates were derived. Refer to "Notes on Data Derivation Methods" in the IBIS Specification for rules on which temperature values to put in the min and max columns	
[Voltage Range]	Yes, if [Pullup Reference], [Pulldown Reference], [POWER Clamp Reference], and [GND Clamp Reference] are not present	Defines the power supply voltage tolerance over which the model is intended to operate. It also specifies the default voltage rail to which the [Pullup] and [POWER Clamp] I-V data is referenced
	Usage Rules Provide actual voltages (not percentages) in the typ, min, and max format. NA is allowed for the min and max values only.	
	Other Notes If the [Voltage Range] keyword is not present, then all four of the keywords described below must be present: [Pullup Reference], [Pulldown Reference], [POWER Clamp Reference], and [GND Clamp Reference]. If the [Voltage Range] is present the other keywords are optional and may or may not be used as required. It is legal (although redundant) for an optional keyword to specify the same voltage as specified by the [Voltage Range] keyword.	

Examples:

```
| variable              typ     min     max
[Temperature Range]     50.0    -50     180.0
[Voltage Range]          5.0V   4.5V    5.5V
```

The required typ column for all data represents typical operating conditions. For most [Model] keyword data, the min column describes slow, weak performance, and the max column describes the fast, strong performance. It is permissible to use slow, weak components or models to derive the data for the min column, and to use fast, strong components or models to derive the data for the max column under the corresponding voltage and temperature derating conditions for these columns. It is also permissible to use typical components or models derated by voltage and temperature and optionally apply proprietary x% and y% factors (described later) for further derating. This methodology has the nice feature that the data can be derived from either semiconductor vendor proprietary models, or typical component measurement over temperature and voltage.

The voltage and temperature keywords, and optionally the process model, controls the conditions that define the typ, min, and max column entries for all:

- I-V table keywords [Pulldown], [Pullup], [GND Clamp], [POWER Clamp]
- [Ramp] subparameters dV/dt_r and dV/dt_f
- Waveform table keywords
- Subparameters [Rising Waveform], [Falling Waveform], V_fixture, V_fixture_min, and V_fixture_max.

The voltage keywords that control the voltage conditions are [Voltage Range], [Pulldown Reference], [Pullup Reference], [GND Clamp Reference], and [POWER Clamp Reference]. The entries in the min columns contain the smallest magnitude voltages, and the entries in the max columns contain the largest magnitude voltages.

The optional [Temperature Range] keyword contains the temperature, which causes or amplifies the slow, weak conditions in the min column and the temperature, which causes or amplifies the fast, strong conditions in the max column. Therefore, the min column for [Temperature Range] contains the lowest value for bipolar models (TTL and ECL) and the highest value for CMOS models. Default values described later are assumed if temperature is not specified.

The min and max columns for all remaining keywords and subparameters contain the smallest and largest magnitude values. This applies to the [Model] subparameter C_comp as well, even if the correlation to the voltage, temperature, and process variations are known. Information about such correlation is not available in all cases and C_comp is considered an independent variable. This is because C_comp includes bonding pad capacitance, which does not necessarily track fabrication process variations. The conservative approach to using IBIS data associates large C_comp values with slow, weak models, and the small C_comp values with fast, strong models.

4.5.3 I-V Data Table Measurement Methods

Table 4-9 shows the IBIS I-V data table (curve) measurement methods.

Table 4-9. I-V data measurement methods

Keyword	Required?	Description
[Pulldown], [Pullup], [GND Clamp], [POWER Clamp]	Yes, if they exist in the model	The data points under these keywords define the I-V tables of the pulldown and pullup structures of an output buffer and the I-V tables of the clamping diodes connected to the GND and the POWER pins, respectively. Currents are considered positive when their direction is into the component

Usage Rules In each of these sections, the first column contains the voltage value, and the three remaining columns hold the typical, minimum, and maximum current values. The four entries, Voltage, I(typ), I(min), and I(max) must be placed on a single line and must be separated by at least one white space. All four columns are required under these keywords. However, data is only required in the typical column. If minimum and/or maximum current values are not available, the reserved word NA must be used. NA can be used for currents in the typical column, but numeric values MUST be specified for the first and last voltage points on any I-V table. Each I-V table must have at least 2, but not more than 100, rows.

Other Notes The I-V table of the [Pullup] and the [POWER Clamp] structures are 'Vcc relative', meaning that the voltage values are referenced to the Vcc pin. (Note: Under these keywords, all references to 'Vcc' refer to the voltage rail defined by the [Voltage Range], [Pullup Reference], or [POWER Clamp Reference] keywords, as appropriate.) The voltages in the data tables are derived from the equation: Vtable = Vcc -Voutput. Therefore, for a 5 V model, -5 V in the table actually means 5 V above Vcc, which is +10 V with respect to ground; and 10 V means 10 V below Vcc, which is -5 V with respect to ground. Vcc-relative data is necessary to model a pullup structure properly, since the output current of a pullup structure depends on the voltage between the output and Vcc pins and not the voltage between the output and ground pins. Note that the [GND Clamp] I-V table can include quiescent input currents, or the currents of a 3-stated output, if so desired. When tabulating data for ECL models, the data in the [Pulldown] table is measured with the output in the 'logic low' state. In other words, the data in the table represents the I-V characteristics of the output when the output is at the most negative of its two logic levels. Likewise, the data in the [Pullup] table is measured with the output in the 'logic one' state and represents the I-V characteristics when the output is at the most positive logic level. Note that in BOTH of these cases, the data is referenced to the Vcc supply voltage, using the equation: Vtable = Vcc - Voutput.
See the Monotonicity Note below.

Monotonicity Note[4] To be monotonic, the I-V table data must meet any one of the following 8 criteria:
The CURRENT axis either increases or remains constant as the voltage axis is increased.
The VOLTAGE axis either decreases or remains constant as the current axis is decreased.
An IBIS syntax-checking program shall test for non-monotonic data and provide a maximum of one warning per I-V table if non-monotonic data is found.

For example: "Warning: Line 300, Pulldown I-V table for model DC040403 is non-monotonic! Most simulators will filter this data to remove the non-monotonic data." It is also recognized that the data may be monotonic if currents from both the output stage and the clamp diode are added together as most simulators do. To limit the complexity of the IBIS syntax checking programs, such programs will conduct monotonicity testing only on one I-V table at a time. It is intended that the [POWER Clamp] and [GND Clamp] tables are summed

4 The Monotonicity Note is quite lengthy. It serves no real purpose to repeat all of it here.

together and then added to the appropriate [Pullup] or [Pulldown] table when a buffer is driving high or low, respectively.

From this assumption and the nature of 3-statable buffers, it follows that the data in the clamping table sections are handled as constantly present tables and the [Pullup] and [Pulldown] tables are used only when needed in the simulation. The clamp tables of an Input or I/O buffer can be measured directly with a curve tracer, with the I/O buffer 3-stated. However, sweeping enabled buffers results in tables that are the sum of the clamping tables and the output structures. Based on the assumption outlined above, the [Pullup] and [Pulldown] tables of an IBIS model must represent the difference of the 3-stated and the enabled buffer's tables. (Note that the resulting difference table can demonstrate a non-monotonic shape.) This requirement enables the simulator to sum the tables, without the danger of double counting, to arrive at an accurate model in both the 3-stated and enabled conditions.

For a non tri-statable buffer, this difference table cannot be generated through lab measurements (because the clamping tables cannot be measured alone), the [Pullup] and [Pulldown] tables of an IBIS model can contain the sum of the clamping characteristics and the output structure. In this case, the clamping tables must contain all zeroes, or the keywords must be omitted.

Examples for monotonicity note:
[Pulldown]

|Voltage	I(typ)	I(min)	I(max)
-5.0V	-40.0m	-34.0m	-45.0m
-4.0V	-39.0m	-33.0m	-43.0m
- - -	- - -	- - -	- - -
0.0V	0.0m	0.0m	0.0m
- - -	- - -	- - -	- - -
5.0V	40.0m	34.0m	45.0m
10.0V	45.0m	40.0m	49.0m

[Pullup] |Note: Vtable = Vcc - Voutput

|Voltage	I(typ)	I(min)	I(max)
-5.0V	32.0m	30.0m	35.0m
-4.0V	31.0m	29.0m	33.0m
- - -	- - -	- - -	- - -
0.0V	0.0m	0.0m	0.0m
- - -	- - -	- - -	- - -
5.0V	-32.0m	-30.0m	-35.0m
10.0V	-38.0m	-35.0m	-40.0m

The stated voltage ranges for I-V tables cover the most common, single-supply cases. When multiple supplies are specified, the voltages shall extend similarly to values that handle practical extremes in reflected wave simulations.

For the [Ramp] subparameters, the default test load and voltages are provided. However, the test load can be entered directly by the R_load subparameter. The allowable test loads and voltages for the waveform keywords are stated by required and optional subparameters; no defaults are needed. Even with waveform keywords, the [Ramp] keyword continues to be required so that the IBIS model remains functional in situations that do not support waveform processing. The following discussion lists test details and default conditions.[5]

I-V Tables:

I-V tables for CMOS models:

- typ = typical voltage, typical temp deg C, typical process
- min = minimum voltage, max temp deg C, typical process, minus x%
- max = maximum voltage, min temp deg C, typical process, plus x%

I-V tables for bipolar models:

- typ = typical voltage, typical temp deg C, typical process
- min = minimum voltage, min temp deg C, typical process, minus x%
- max = maximum voltage, max temp deg C, typical process, plus x%

The semiconductor vendor specifies nominal, min, and max temperature. The default range is 50 deg C nom, 0 deg C min, and 100 deg C max temperatures.

The semiconductor vendor (based on numerous fab lots, test chips, and process controls) should statistically determine x%. The value of x need not be published in the IBIS file, and may decrease over time as data on the I/O buffers and silicon process increases.

Voltage Ranges:

Data points for each table must span the voltages as shown in Table 4-10.

Table 4-10. Voltage range span

Table	Low Voltage	High Voltage
[Pulldown]	GND – POWER	POWER + POWER
[Pullup]	GND – POWER	POWER + POWER
[GND Clamp]	GND – POWER	GND + POWER
[POWER Clamp]	POWER	POWER + POWER
[Series Current]	GND - POWER	GND + POWER
[Series MOSFET]	GND	GND + POWER

5 In the discussion of the IBIS Spec, references are often made to temperature (abbreviated as temp). In all cases, temperature refers to junction temperature of the IC die.

As described in the [Pulldown Reference] keyword section, the I-V tables of the [Pullup] and the [POWER Clamp] structures are 'Vcc relative', using the equation: Vtable = Vcc - Voutput.

For example, a model with a 5 V power supply voltage should be characterized between (0 - 5) = -5 V and (5 + 5) = 10 V; and a model with a 3.3 V power supply should be characterized between (0 - 3.3) = -3.3 V and (3.3 + 3.3) = 6.6 V for the [Pulldown] table.

The voltage points must span the range of Vcc to Vcc - 2.2 V when we tabulate output data for ECL type models. This range applies to both the [Pullup] and [Pulldown] tables. Note that this range applies ONLY when characterizing an ECL output.

These voltage ranges must be spanned by the IBIS data. Data derived from lab measurements may not be able to span these ranges as such and so may need to be extrapolated to cover the full range. This data must not be left for the simulator to provide.

4.5.4 Ramp Measurement Methods

Table 4-11 shows the IBIS ramp measurement methods.

Table 4-11. [Ramp] measurement methods

Keyword	Required?	Description
[Ramp]	Yes, except for inputs, terminators, Series and Series_switch model types	Defines the rise and fall times of a buffer. The ramp rate does not include packaging but does include the effects of the C_comp or C_comp_* parameters Sub-Params: dV/dt_r, dV/dt_f, R_load

Usage Rules The rise and fall time is defined as the time it takes the output to go from 20% to 80% of its final value. The ramp rate is defined as: dV/dt = 20% to 80% voltage swing/ Time it takes to swing the above voltage

Other Table 4-11 Notes The ramp rate must be specified as an explicit fraction and it not be reduced. The [Ramp] values can use NA for the min and max values only. The R_load subparameter is optional if the default 50-ohm load is used. The R_load subparameter is required if a non-standard load is used.

Example:

```
[Ramp]
  | variable     typp          min           max
  dV/dt_r       2.20/1.06n    1.92/1.28n    2.49/650p
  dV/dt_f       2.46/1.21n    2.21/1.54n    2.70/770p
R_load = 300ohms
```

Ramp Rates:

The following steps assume that the default load resistance of 50 ohms is used. There may be models that will not drive a load of only 50 ohms into any useful level of dynamics. In these cases, use the semiconductor vendor's suggested (nonreactive) load and add the load subparameter to the [Ramp] specification. The ramp rate does not include packaging but does include the effects of the C_comp parameter; it is the intrinsic output stage rise and fall time-only.

The ramp rates (listed in AC characteristics below) should be derived as follows:

a) If starting with the silicon model, remove all packaging. If starting with a packaged model, perform the measurements as outlined below. Then use whatever techniques are appropriate to derive the actual, unloaded rise and fall times.
b) If: The Model_type is Output, I/O, or 3-state (not open or ECL types); attach a 50 ohm resistor to GND to derive the rising edge ramp. Attach a 50-ohm resistor to POWER to derive the falling edge ramp.
c) If: The Model_type is Output_ECL, I/O_ECL, 3-state_ECL; attach a 50-ohm resistor to the termination voltage (Vterm = VCC - 2 V). Use this load to derive both the rising and falling edges.
d) If: The Model_type is either an Open_sink type or Open_drain type; attach either a 50 ohm resistor or the semiconductor vendor suggested termination resistance to either POWER or the suggested termination voltage. Use this load to derive both the rising and falling edges.
e) If: The Model_type is an Open_source type; attach either a 50-ohm resistor or the semiconductor vendor suggested termination resistance to either GND or the suggested termination voltage. Use this load to derive both the rising and falling edges.
f) Due to the resistor, output swings will not make a full transition as expected. However the pertinent data can still be collected as follows:
   Determine the 20% to 80% voltages of the 50-ohm swing.
   Measure this voltage change as dV.
   Measure the amount of time required to make this swing dt.
g) Post the value as a ratio dV/dt. The simulator extrapolates this value to span the required voltage swing range in the final model.
h) Typ, min, and max must all be posted, and are derived at the same extremes as the I-V tables, which are:

Ramp rates for CMOS models:

typ	typical voltage, typical temp deg C, typical process
min	minimum voltage, max temp deg C, typical process, minus y%
max	maximum voltage, min temp deg C, typical process, plus y%

Ramp rates for bipolar models:
typ typical voltage, typical temp deg C, typical process
min minimum voltage, min temp deg C, typical process, minus y%
max maximum voltage, max temp deg C, typical process, plus y%

where the semiconductor vendor specifies nominal, min, and max temp. The preferred range is 50 deg C nom, 0 deg C min, and 100 deg C max temperatures. Note that the derate factor, y%, may be different than that used for the I-V table data. This factor is similar to the x% factor described above. Just as for I-V tables, temperatures are junction temperatures.

i) During the I-V measurements, the driving waveform should have a rise/fall time fast enough to avoid thermal feedback. The specific choice of sweep time is left to the modeling engineer.

4.5.5 Rising and Falling Waveform Measurement Methods for V-T Data Tables

Table 4-12 shows the rising and falling waveform measurement methods.

Table 4-12. Rising and falling waveform measurements are V-T data tables

Keyword	**Required?**	**Description**
[Rising Waveform], [Falling Waveform]	No	Describes the shape of the rising and falling edge waveforms of a driver. Sub-Params: R_fixture, V_fixture, V_fixture_min, V_fixture_max, C_fixture, L_fixture, R_dut, L_dut, C_dut

Usage Rules Each [Rising Waveform] and [Falling Waveform] keyword introduces a table of voltage versus time points that describe the shape of an output waveform. These voltage versus time points are taken under the conditions specified by the R/L/C/V_fixture and R/L/C_dut subparameters. The table itself consists of one column of time points, then three columns of voltage points in the standard typ, min, and max format. The four entries must be placed on a single line and must be separated by at least one white space. All four columns are required. However, data is only required in the typical column. If minimum or maximum data is not available, use the reserved word NA. The first value in the time column need not be '0'. Time values must increase as one parses down the table. The waveform table can contain a maximum of 1000 data rows. A maximum of 100 waveform tables are allowed per model.
Table 4-12 Notes Note that for backward compatibility, the existing [Ramp] keyword is still required. The data in the waveform table is taken with the effects of the C_comp parameter included
A waveform table must include the entire waveform; that is, the first entry (or entries) in a voltage column must be the DC voltage of the output before switching and the last entry (or entries) of the column must be the final DC value of the output after switching. Each table must contain at least two entries.

Thus, numerical values are required for the first and last entries of any column containing numerical data. a single point in time when the input stimulus is assumed to have initiated a logic transition. All waveform extractions should reference a common input stimulus time in order to provide a sufficiently accurate alignment of waveforms. The first line in each waveform table should be assumed to be the reference point in time corresponding to a logic transition.

For example, assume that some internal rising edge logic transition starts at time = 0. Then a rising edge voltage-time table might be created starting at time zero. The first several table entries might be some lead-in time caused by some undefined internal buffer delay before the voltage actually starts transitioning. The falling edge stimulus (for the purpose of setting reference time for the voltage-time table) should also start at time = 0. And, the falling edge voltage-time table would be created starting at time zero with a possibly different amount of lead-in time caused by a possibly different but corresponding falling edge internal buffer delay. Any actual device differences in internal buffer delay time between rising and falling edges should appear as differing lead-in times between the rising and the falling waveforms in the tables just as any differences in actual device rise and fall times appear as differing voltage-time entries in the tables.

A [Model] specification can contain more than one rising edge or falling edge waveform table. However, each new table must begin with the appropriate keyword and subparameter list as shown below. If more than one rising or falling edge waveform table is present, then the data in each of the respective tables must be time correlated. In other words, the rising (falling) edge data in each of the rising (falling) edge waveform tables must be entered with respect to a common reference point on the input stimulus waveform
The 'fixture' subparameters specify the loading conditions under which the waveform is taken. The R_dut, C_dut, and
L_dut subparameters are analogous to the package parameters R_pkg, C_pkg, and L_pkg and are used if the waveform includes the effects of pin inductance/capacitance.

In IBIS 4.1 the description of the test circuit for measuring waveforms has been expanded and improved. Figure 4-9 shows the [Test Load] keyword test circuit:

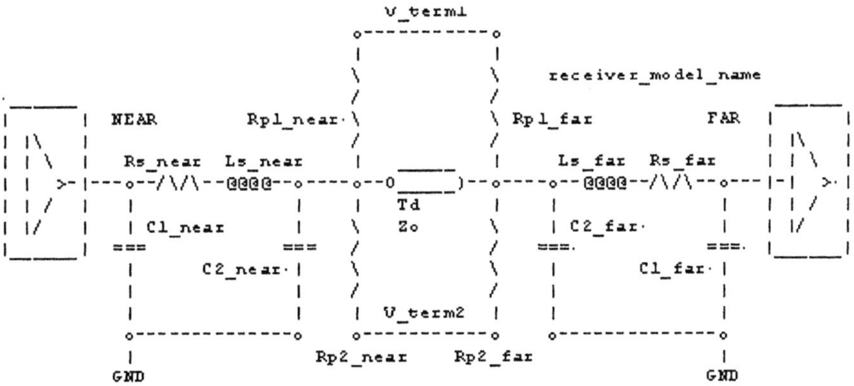

Figure 4-9. [Test Load] keyword circuit from IBIS 4.1

Note: The use of L_dut, R_dut, and C_dut is strongly discouraged in developing Waveform data from simulation models. Some simulators may ignore these parameters because they may introduce numerical time constant artifacts. Only the R_fixture and V_fixture subparameters are required, the rest of the subparameters are optional. If a subparameter is not used, its value defaults to zero. The subparameters must appear in the text after the keyword and before the first row of the waveform table.

V_fixture defines the voltage for typ, min, and max supply conditions. However, when the fixture voltage is related to the power supply voltages, then the subparameters V_fixture_min and V_fixture_max can be used to further specify the fixture voltage for min and max supply voltages.

Note: Test fixtures with R_fixture and V_fixture, V_fixture_min, and V_fixture_max only are strongly encouraged because they provide the BEST set of data needed to produce the best model for simulation. C_fixture and L_fixture can be used to produce waveforms , which describe the typical test case setups for reference.

Note: In most cases, two [Rising Waveform] tables and two [Falling Waveform] tables will be necessary for accurate modeling. An example is shown below:

```
[Rising Waveform]
R_fixture = 50
V_fixture = 0.0
C_fixture = 50p       | These are shown, but are generally not |recommended
L_fixture = 2n
C_dut = 7p
R_dut = 1m
L_dut = 1n

 | Time      V(typ)       V(min)       V(max)
 0.0000s     25.2100mV    15.2200mV    43.5700mV
 0.2000ns    2.3325mV     -8.5090mV    23.4150mV
 0.4000ns    0.1484V      15.9375mV    0.3944V
 0.6000ns    0.7799V      0.2673V      1.3400V
 0.8000ns    1.2960V      0.6042V      1.9490V
 1.0000ns    1.6603V      0.9256V      2.4233V
 1.2000ns    1.9460V      1.2050V      2.8130V
 1.4000ns    2.1285V      1.3725V      3.0095V
 1.6000ns    2.3415V      1.5560V      3.1265V
 1.8000ns    2.5135V      1.7015V      3.1600V
 2.0000ns    2.6460V      1.8085V      3.1695V
 | ...
 10.0000ns   2.7780V      2.3600V      3.1670V
|
```

[Falling Waveform]
R_fixture = 50
V_fixture = 5.5
V_fixture_min = 4.5
V_fixture_max = 5.5

| Time	V(typ)	V(min)	V(max)
0.0000s	5.0000V	4.5000V	5.5000V
0.2000ns	4.7470V	4.4695V	4.8815V
0.4000ns	3.9030V	4.0955V	3.5355V
0.6000ns	2.7313V	3.4533V	1.7770V
0.8000ns	1.8150V	2.8570V	0.8629V
1.0000ns	1.1697V	2.3270V	0.5364V
1.2000ns	0.7539V	1.8470V	0.4524V
1.4000ns	0.5905V	1.5430V	0.4368V
1.6000n	0.4923V	1.2290V	0.4266V
1.8000ns	0.4639V	0.9906V	0.4207V
2.0000ns	0.4489V	.8349V	0.4169V
...	|...	...	...
10.0000ns	0.3950V	.4935V	0.3841V
...	...	...	...

All tables assume that the die capacitance is included. Potential numerical problems associated with processing the data using the effective C_comp (or C_comp_* values as appropriate) for effective die capacitance may be handled differently among simulators.

4.5.6 C_comp Notes

The C_comp and C_comp_* subparameters define die capacitance. These values should not include the capacitance of the package. C_comp and C_comp_* are allowed to use NA for the min and max values only.

Notice that the C_comp parameter of a multi-stage buffer is defined in the top-level model. The value of C_comp therefore includes the total capacitance of the entire buffer, including all of its stages.

Since the rising and falling waveform measurements include the effects of C_comp, each of these waveforms must be generated with the total C_comp present, even if the various stages of the buffer are characterized individually.

4.6 WEB SITES FOR IBIS VISUAL EDITORS AND OTHER TOOLS

As of this writing (2006), there are some IBIS visual editors (GUIs) available from some software companies. Remember to check out the freeware at: http://www.eigroup.org/ibis/tools.htm. For some additional addresses, refer to "Appendix F, List of Websites."

4.7 TDR/TDT - VNA MEASUREMENTS

Time Domain Reflectometry (TDR [124]) and frequency domain Vector Network Analysis (VNA [1]) are the two basic measurement techniques for analyzing and extracting high-speed interconnections.

A TDR instrument sends a very fast step pulse, with 25ps to 35ps rise time to a Device-Under-Test (DUT). The DUT behaves as a transmission-line structure. Reflections are generated in the DUT and sampled by the TDR instrument, which is equivalent to a very fast, very wideband oscilloscope. Based on the reflections, transmission-line impedance (RLGC matrix), the location of faults and discontinuities, time delay (length), and topology can be extracted. Time Domain Transmission (TDT) extends these basic capabilities to include crosstalk, insertion loss, skin effect, and dielectric loss.

A VNA instrument sends out a sweep of sine waves to the DUT. Reflections are again generated and received by the VNA instrument, which has a very narrow band input filter, synced with the transmitted signal. The swept, frequency-dependent S-Parameters can be extracted along with insertion loss and return loss. The two measurements of input reflection coefficient are related by the Fourier transform. Thus, for TDR (time domain):

$$Z_{DUT}\,(input) = Z_o[(1+\rho)/(1-\rho)] \qquad (4\text{-}21)$$

And, VNA (frequency domain)

$$Z_{DUT}(input) = Z_o[(1+S_{11})/(1-S_{11})] \qquad (4\text{-}22)$$

Likewise, when we extract the characteristic impedance, Z_o,, of the transmission line structure inside the DUT we have the relationship:

$$Zo = [(R+j\varpi L)/(G+j\varpi C)]^{1/2} \qquad (4\text{-}23)$$

Figure 4-10 is a transmission line characteristic section, where:

$\varpi =$ $2\pi f$ and f = frequency

R = series resistance of the transmission-line section. Frequency-dependent skin effect adds to R

L = series inductance of the transmission-line section

G = parallel conductance of the transmission-line section. Frequency-dependent dielectric loss adds to G

C = parallel capacitance of the transmission-line section

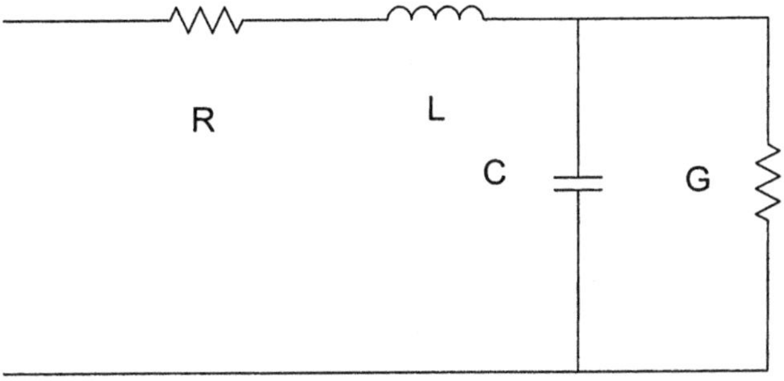

Figure 4-10. Transmission line equivalent circuit / section

For the low-frequency approximation of R=0, G=0 (when $f \leq$ approximately 5 GHz), equation (4-23) reduces to:

$$Zo = \sqrt{L/C} \qquad (4\text{-}24)$$

For further study a "TDR and VNA Measurement Primer" can be found at: http://www.tdasystems.com/library/appnotes/TVMP-0404.pdf

4.8 RLGC MATRIXES

RLGC matrixes in IBIS are used to model pin parasitics, both self and mutual. RLGC matrixes in IBIS are broken into their constituent R, L, G and C matrixes. It is left to the simulator company to recombine them in an appropriate manner. Most of the time the G elements are so small that this matrix is set to zero or approximated to zero. However, when |G| = zero (an ideal open), that value that can cause convergence problems in simulators during matrix math operations.

4.8.1 Types of Matrixes

The arguments for the following matrix types are explained next: Diagonal_matrix, Banded_matrix, Sparse_matrix, and Full_matrix.

The Diagonal_matrix is used to specify uncoupled models. This subparameter is exactly the same as using the Banded_matrix described in the next paragraph with a Bandwidth=0. It has the added benefit of requiring half as many lines, as no instances of the [Row] keyword are required. A Diagonal_matrix is one whose entries are guaranteed to be zero if they are not on the main diagonal.

The Banded_matrix is used to specify the coupling effects up to X pins on either side of a given pin. A Banded_matrix is one whose entries are guaranteed to be zero if they are farther away from the main diagonal than a certain distance, known as the bandwidth. The bandwidth for a Banded_matrix must be specified using the [Bandwidth] keyword. Symmetry is exploited to reduce the amount of data by eliminating any entries below the main diagonal of the matrix.

A *Sparse_matrix* is expected to consist mostly of zero-valued entries. Unlike the Banded_matrix, there is no restriction on where the nonzero entries can occur. Symmetry is exploited to reduce the amount of data by eliminating any entries below the main diagonal of the matrix.

When the *Full_matrix format* is used, the coupling between every pair of elements is specified explicitly. Assume that the matrix has N rows and N columns. The Full_matrix is specified one row at a time, starting with row 1 and continuing down to row N. Each new row is identified with the [Row] keyword. Symmetry is exploited to reduce the amount of data by eliminating any entries below the main diagonal of the matrix.

Tables 4-13 through 4-15 are taken from the IBIS Connector Spec, which contains definitions of various types of matrixes.

4.8.2 Examples of Matrixes

To define and illustrate the appearance of a full, banded and banded symmetric matrix consider a 6 pin C matrix (capacitance):

The main-diagonal elements: C_{11} $C_{22...}$are the self-capacitances of the pins to ground.

The off - main-diagonal elements: C_{12} $C_{13...}$are the mutual capacitances of pins 1 to 2, 1 to 3, …

Table 4-13. Full Matrix

C_{11}	C_{12}	C_{13}	C_{14}	C_{15}	C_{16}
C_{21}	C_{22}	C_{23}	C_{24}	C_{25}	C_{26}
C_{31}	C_{32}	C_{33}	C_{34}	C_{35}	C_{36}
C_{41}	C_{42}	C_{43}	C_{44}	C_{45}	C_{46}
C_{51}	C_{52}	C_{53}	C_{54}	C_{55}	C_{56}
C_{61}	C_{62}	C_{63}	C_{64}	C_{65}	C_{66}

Usually, the subscript index, C_{ij}, is a rough indicator of distance. The greater the ij difference the farther apart the pins are. Mutual parasitics also fall off rapidly with separation. So, many elements can be approximated to zero and eliminated from the matrix, leading to:

Table 4-14. Banded matrix (band of 3)

C_{11}	C_{12}	C_{13}	-	-	-
C_{21}	C_{22}	C_{23}	C_{24}	-	-
C_{31}	C_{32}	C_{33}	C_{34}	C_{35}	-
-	C_{42}	C_{43}	C_{44}	C_{45}	C_{46}
-	-	C_{53}	C_{54}	C_{55}	C_{56}
-	-	-	C_{64}	C_{65}	C_{66}

Most packaging mediums are homogeneous and isotropic within the volume of interest. This means that the capacitance from pin 1 to pin 2 equals the capacitance from pin 2 to pin 1. The rules of matrix algebra say that we can simplify a symmetric matrix by eliminating all the elements below (or above, but not both) the main diagonal, leading to:

Table 4-15. Banded symmetric matrix

C_{11}	C_{12}	C_{13}	-	-	-
-	C_{22}	C_{23}	C_{24}	-	-
-	-	C_{33}	C_{34}	C_{35}	-
-	-	-	C_{44}	C_{45}	C_{46}
-	-	-	-	C_{55}	C_{56}
-	-	-	-	-	C_{66}

Also be aware that an IBIS subcommittee produced an Interconnection Modeling Spec (ICM) that greatly expands package, connector and interconnection modeling capabilities. The ability to handle section, trees, SPICE, S-Parameter, RLGC style matrixes, and touchstone formats, as well as self and mutual pin impedances and couplings, is being added.

Version 1.0 of the specification was ratified and released in September 2003. More information can be found in [56].

4.9 FIELD SOLVER RLGC EXTRACTION FOR ICS

Semiconductor suppliers can do RLGC modeling of the interconnection (as used in IBIS modeling) that the end users of devices cannot do as easily.

Electromagnetic field solver software can extract the electrical pin and package-parasitic parameters of an IC device. The input required is a mechanical model and drawing of the packaged device, its lead frame, and wire bonds. A customer of the semiconductor supplier usually will not receive such proprietary information.

Likewise, the semiconductor supplier can construct dummy packages without the die, but with interconnection and lead frames included, and measure them with TDR – VNA techniques.

Therefore the customer is usually reduced to asking the semiconductor supplier for the data on the electrical pin and package parasitic parameters of the IC device they wish to model. That course of action is recommended.

4.10 WHAT IS MODEL SYNTHESIS?

Analog model synthesis was a feature of the SPICE-like SABER® simulator [103, 104, 105] from Analogy, Inc. Their tool could create a simulation model from a graphical waveform. Normally the waveform is the frequency response of the circuit and devices being modeled. This method is essentially the same as data acquisition and curve fitting when the waveform is generated by measurements. This method is similar to macromodeling when the waveform is generated by simulation of a complex circuit. The end result in either case is an N-Port behavioral model.

Thus, circuit *measurement* can be employed on a physical part, connector, etc., to derive its transfer function. Initially, it might be difficult to model the part. But once the measured transfer function is known, it becomes possible to synthesize its internal circuit. Circuit synthesis has long been employed for analog and RF filter design and many references explain how to do it [139].

4.11 TEST EQUIPMENT PROVIDERS

Test equipment is provided by a variety of manufactures. Competition has forced many changes in this area as with many other high-tech fields. For instance, HP appears to have spun off all their remaining test instruments under the Agilent brand. Agilent appears to have shrunk the product line to include only network analyzers, impedance analyzers, parametric analyzers and fully integrated device modeling systems. Meanwhile, Tektronix has gotten deeply involved in Signal Integrity, and still provides parametric analyzers (fancy curve tracers).

Tektronix:	http://www.tek.com/
LeCroy:	http://www.lecroy.com/
Agilent:	http://eesof.tm.agilent.com/products/85122a-a.html
Instek:	http://www.instek.com/
Fluke:	http://www.fluke.com/us.asp

4.12 SOFTWARE FOR TEST EQUIPMENT CONTROL

Software is available to help users acquire, analyze, and present test measurements. The following terms are commonly used in test equipment product brochures:

- *Acquire:* Software for test equipment control helps make interfacing with and controlling measurement hardware simpler. Software for test equipment control can create virtual instruments.
- *Virtual Instruments:* Consist of an industry-standard computer equipped with software, plug-in boards, and driver software that provide the functionality of traditional instruments with much added flexibility and adaptability.
- *Analyze:* Software for test equipment control has many built-in functions for extracting useful information from data.
- *Present:* Software for test equipment control has a wide selection of visualization features.

One such program is LabVIEW® from National Instruments.

http://www.ni.com/

http://www.ni.com/LabVIEW/

Another program is "85190A IC-CAP® Parameter Extraction and Device Modeling Software:" from Agilent:

http://eesof.tm.agilent.com/products/85190a-b.html

TDA Systems provides software and test equipment for TDR and TDT measurements:

http://www.tdasystems.com/

Several of these programs provide selected capabilities from the following list:

- Simulation measurements from SPICE simulators can be imported and compared directly with lab instrument measurements of the same signals. Time Domain Reflectometry (TDR) measurements can be acquired.
- Temperature, voltage, current, frequency, and resistance can be acquired.
- Full capabilities to extract SPICE, IBIS, Scattering Parameter, and other linear and non-linear models.
- Curve fitting, FFT and frequency analysis, time domain analysis, eye diagrams, probability/statistics, signal processing, DC/RMS levels, and total harmonic distortion can be run and extracted.
- Post-processing for charting, graphing, and manipulating presentations.

4.13 SUMMARY

Direct measurement of model parameters can serve two purposes. If the parameter we are measuring can be derived from modeling, then we can verify the model derivation by measurement. On the other hand, some SPICE model parameters cannot be derived from device modeling. So the only way to get their magnitude is by measuring them in the laboratory.

It is important to verify models. Most published device models (including behavioral models) are derived from a SPICE model, rather than derived from direct measurement. The SPICE model itself may include errors and incorrect assumptions that may not have been caught by measurement. Any original errors in the SPICE model then get propagated into the derived models.

Chapter 5

USING STATISTICAL DATA TO CHARACTERIZE COMPONENT POPULATIONS

In-control semiconductor processes yield quality models

Abstract: Component manufacturing processes produce populations of components that vary in their characteristics. Therefore, the component models must represent that variance. Additionally, the manufacturing process must be in statistical control. Concepts used to examine whether a process is in statistical control include: statistical distributions, Parts-Per-Million (PPM) quality, Six-Sigma Quality, histograms, control charts, special causes, and common causes.

5.1 WHY PROCESS VARIATION IS IMPORTANT

In any manufacturing process, random variation is natural and should be expected. To detect process control problems, suppliers use population statistics and process-monitoring charts. The charts are useful for monitoring processes, controlling processes, and describing model parameter variation. Process variation results in device behavior variation, which should be represented by model variation, and a range on model parameter values. Designing circuits to not be overly sensitive to variation is one of the chief objectives for any engineer.

5.2 ACHIEVING PROCESS CONTROL WITH POPULATION STATISTICS

Histograms and control charts are a large part of statistical process control (SPC). Histograms and control charts present population statistics to help guide the control of a process. SPC involves:

- Monitoring a process by measuring units produced by that process
- Calculating and plotting the population statistics of the measurements using charting tools like histograms and control charts
- Taking action to eliminate the special causes of process variation
- Demonstrating that the process is *in statistical control*

Many manufacturing and tolerance issues are involved in achieving Six-Sigma quality. Components used in circuits (that is, systems of components), and in PC boards undergo further manufacturing steps during which additional issues can affect their performance. Those manufacturing issues are mostly beyond the scope of this book.

The use of components in systems involves the simulation of many models operating together to predict system behavior. Plus components influence each other's operation in a design and interact with bias, temperature, and other variables. These interactions and variables can lead to modeling issues.

After achieving SPC, how do we achieve six-sigma quality? It involves:

- Taking action to eliminate the common causes of process variation
- Making a good match between process centering, capability, and circuit design needs

5.3 BASICS OF POPULATION STATISTICS

Among the first questions that a model user typically asks about populations of components and devices are:

- What is the component's typical behavior?
- What is the component's range of behavior for a population of units produced and delivered over an extended time period?
- Does the model of the component accurately represent the typical behavior and range of variation?

To answer these questions, model makers must gather data about how the population behaves *on average*. Then the model maker uses the data to try to predict how the population will typically behave in the future. When we select a particular unit from the population, we do not know which part of the distribution it came from—unless we measure it directly. We can predict only what the population's behavior will *probably* be.

Does knowing the population's past behavior help predict its future behavior? The answer is yes under normal circumstances and for component processes running *in control.*[1]

In control means that our model of population statistics is representative, repeatable, and well behaved. Unexpected, significant, and random chance can still happen, but is unlikely without raising significant warning flags. However, an in-control process still has random variation.

Usually, in model verification, we take model parameter measurements on a particular unit/device, make predictions about its behavior in a circuit, and then measure its behavior in that circuit. Verification occurs when the predictions match the measurements. This unit-by-unit direct verification is not cost-effective on a day-by-day basis. We use it only rarely–when we want to understand our models, simulations, and process control. Verification is just as effective when a distribution of *predictions* matches a distribution of *measurements.*

5.3.1 Normal Distribution

What kinds of statistical distributions give us the most confidence to predict the future? The usual answer is a Normal distribution of the property being measured. When describing a part and its processes, other kinds of distributions may be just as natural, but they occur far less frequently in nature. The Normal distribution occurs most frequently. Carl Friedrich Gauss (1777-1855) described this distribution mathematically, so it is also called a Gaussian distribution.

Most technical people are familiar with the concept of a Normal distribution. A Normal distribution displays how a population of units exhibits a range and frequency of behavior–when measured for any particular property. This distribution has a mean (or average) value where the peak in the frequency of measured behavior occurs. It then falls off symmetrically as illustrated in the histograms of Figures 5-1 and 5-2.

Histograms are visual representatives of data. One example of a histogram is a bar chart (see Figure 5-1). A histogram is generated by measuring data on some component and model property value for a sample of units and then counting the units at each measurement.

Figure 5-1 shows a process that is well centered between minimum and maximum specification limits and has good guardbands for not exceeding those limits.

---

[1] Also called in statistical process control (SPC).

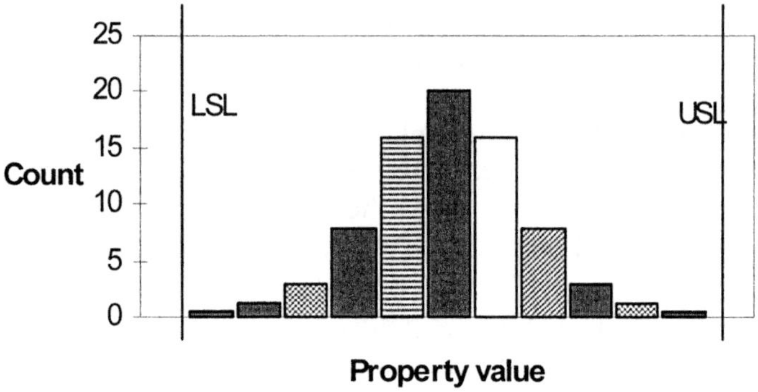

Figure 5-1. A histogram plot of a Normal (Gaussian) population distribution

In Figure 5-1:

- The horizontal axis is the property value scale.
- The vertical axis is the count or frequency-of-occurrence scale.
- The envelope or curve that fits around this accumulation of data approximates a Normal curve.
- LSL = Lower Spec Limit.
- USL =Upper Spec Limit.

Note that the minimum and maximum property values on each of the individual bars correspond to a measurement bin. For example, if the central bar has a middle value of 5.0, the bin might extend from 4.9 to 5.1. All the units that are within these measurement points would be counted in that bin with an average value of 5.0.

Consider why this Normal distribution is so common. For example, when turning a particular shaft diameter on a lathe, the objective is to hit the correct diameter *exactly* every single time. But achieving this objective is probably impossible. *Random variation* makes the process miss the target, causing the results to spread out above and below the target.

In addition to random variation, there can be *special causes* that cause the results to drift in a predictable manner. Example: cutting tool wear and periodic replacement. We say that the process is *in control* if the *only* cause of variation is random. In other words, the variation is caused only by random or statistical variation.

What about the distribution of shaft diameters? Will all production parts work properly in the machine we are building? Are they fit for use? Also, can we say what we will continue to produce? Will our process be repeatable? And, what do these considerations have to do with modeling?

The answers to these questions involve *process control, process improvement,* and *characterization* of the process.

- *Process control* requires that the measurements, testing, procedures, and controls be in place to ensure repeatability. If we switch to another type of steel for our drive shaft, the results may not be predictable from our previous data.
- *Process improvement* requires that we diagnose the causes of random and special variation. Then we work to reduce or eliminate these causes or their effects. If tool wear and replacement are causing our drive shaft diameters to vary too much, we can consider changing the tool steel, shortening the tool replacement cycle, sensing the wear and adjusting the lathe settings, and using soft metal bearings in the machine we are building
- *Characterization* requires that we describe and depict the units that our process produces. The population distribution for our drive shaft may be all that we need. But for modeling and simulation of drive shaft friction and wobble, we may need to select a number of units from our distribution and measure other properties of the shaft. To save ourselves wasted effort, we want to characterize drive shafts that are fit for use and come from a repeatable process.

We have used a simple example of turning shaft diameters on a lathe, but semiconductors are considerably more complex because:

- Semiconductor properties are neither necessarily linearly independent nor linear. For example, selecting a number of units for beta variation for characterization of a model is not the same as selecting for breakdown voltage.
- Semiconductor properties vary in complex ways with variables like bias conditions and temperature.
- Semiconductor models themselves are usually complex.

For example, we can select transistor units based on their beta to measure and characterize the gain-bandwidth curves of the selected units. Should we measure and report separately the typical, minimum, and maximum units? Or should we average the results for all three units and report them as a single curve?

5.3.2 Definitions of Mean and Standard Deviation

What do we mean by a typical, minimum, and maximum unit? To answer this question, we will discuss some important formulas, charts, and curves describing the Normal distribution function. Also, what are the shorthand ways of referring to Normal distributions? The answers to both make use of the *population mean* and the *population standard deviation*.

- *Typical* corresponds to the population mean. A formula for calculating it follows Figure 5-2 in equations (5-2) and (5-3).
- *Standard Deviation*, or sigma (σ), corresponds to the average root mean square of the population spread from its mean. A formula for calculating it follows Figure 5-2 in equations (5-4) and (5-5). The more dispersed a population is, the greater is its standard deviation.
- *Minimum* corresponds to three-sigma deviations[2] *below* the population mean.
- *Maximum* corresponds to three-sigma deviations *above* the population's mean.

The terms just defined are identified in Figure 5-2. The area under the curve corresponds to the percentage of the data in the sample represented. Thus +/-3σ encompasses 99% of the product sample.

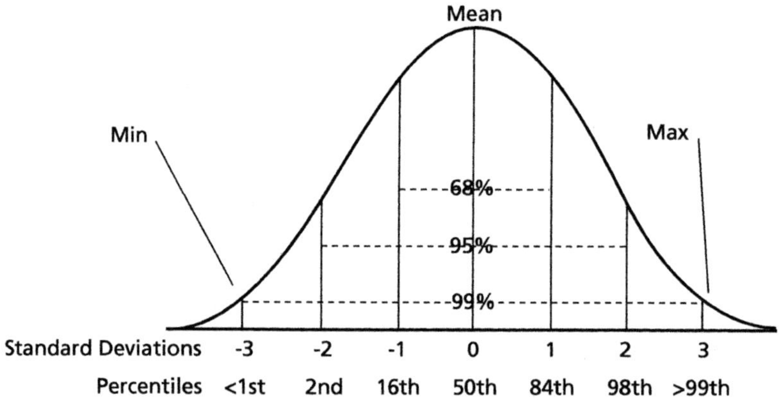

Figure 5-2. A curve of the Normal distribution

Equations (5-1) to (5-5) describe the Gaussian or Normal probability density function. These formulas are used to calculate the mean and spread of a random variable for the case of a discrete variate p with probability function k(p) = kp, or a continuous variate x defined by its probability density function k(x).

Gaussian, or Normal, probability Density Distribution Function:

[2] The convention of min/max corresponding to +/-3σ/-3σ is not a formal one, but just an observation of what seems to be common practice.

$$\phi(x) = \frac{1}{\sigma\sqrt{2\pi}} \ell^{[-\frac{(x-\mu)^2}{2\sigma}]} \tag{5-1}$$

Average or Mean for all p:

$$\mu = \sum_{p=1}^{n} k_p x_p / n \tag{5-2}$$

$$\mu = \int_{-\infty}^{+\infty} x * k(p) dx \tag{5-3}$$

Standard (or rms) Deviation, sigma, from the Mean

$$\sigma = [\sum k_p (x_p - \mu)^2]^{1/2} \tag{5-4}$$

$$\sigma = [\int_{-\infty}^{+\infty} (x-\mu)^2 k(x) dx]^{1/2} \tag{5-5}$$

5.3.3 Histograms Reveal Process Capability

When designing a circuit, we should use parts that are in statistical process control, meet our design needs, and are repeatable. How do we recognize when we might have problems meeting these objectives?

We can use population distributions to see whether a process is predictable and capable of meeting its tolerance limits. Only when the real-world behavior of a device population is predictable and capable of meeting its tolerance limits over time will models of the device's behavior be reliable and predictable.

Let us look at the histograms in Figure 5-3 and what they tell us about issues of quality and process control. Under each histogram is the interpretation of what the shape indicates.

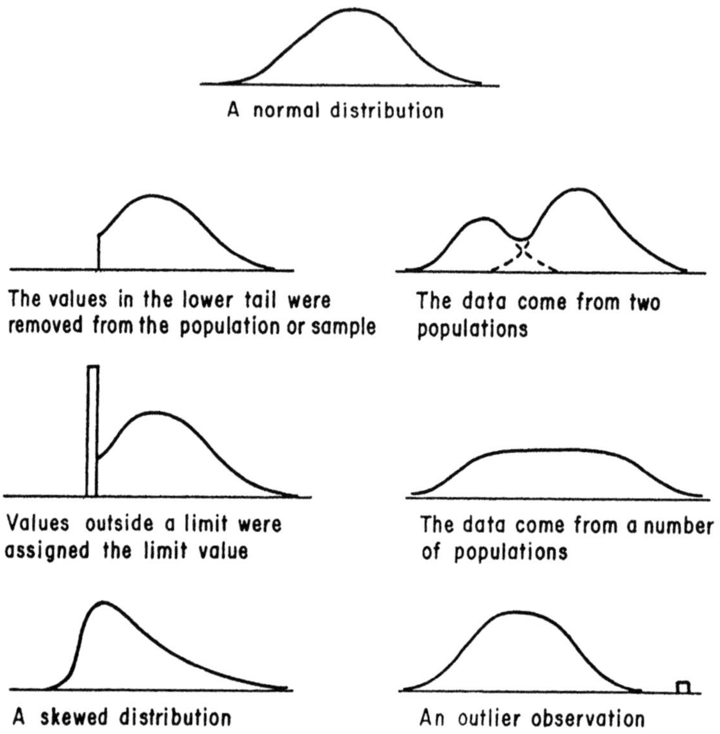

Figure 5-3. Histogram shapes and interpretations [89]

The interpretation of histogram shapes is an important part of quality control and process control. In some references [12, 20, 54, 61, 62, 89, 91, 95], many other shape examples are presented.

Non-Normal distributions can be modeled mathematically, mostly from the point of view of process control, but the modeling gets more difficult. One example of a non-Normal distribution resulting from a special cause is the periodic replacement of dirty particulate filters in semiconductor processing, when the replacement cycle is too long. In this case, the effects of particulate contamination ramp up and then reset periodically to unobservable on a histogram or control chart. This contamination can cause a skewed distribution on some device or model properties.

We next observe how histogram shapes change over time. Changes over time give us more information about that process and its suitability and stability. Figure 5-4 shows an unstable and unpredictable process. Over time, the mean of this unstable process shifts back and forth over a target value (centerline) and its spread increases and decreases. Its process control is poor, so creating a component model based on these results will be difficult and unrewarding.

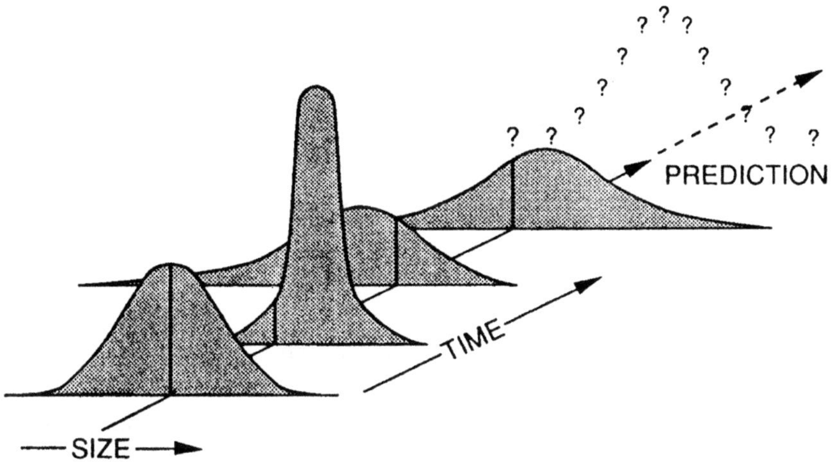

Figure 5-4. Process unstable [89]

The characterization of a *process* is the description of various behaviors over the spread of the process. When we base a model on the process, we want the process to be stable and predictable so the model will be stable and predictable.

As a first step to process improvement, we must diagnose and remove causes of steady process drift, cyclical behavior, bouncing between two extremes, and other special causes. These *special causes* are due to definite, diagnosable causes. The remaining causes of variation are random. They show no detectable patterns and are labeled *common causes*. Common causes are usually harder to diagnose, control, and eliminate.

Figure 5-5 shows the result of some process improvement. The newer, repeatable, stable process is in statistical control. All special causes are removed or mitigated, and future distributions are predictable. Therefore, device models created from component data predict the performance of future parts. All variation is random, as revealed by histogram plots and control-charting tools.

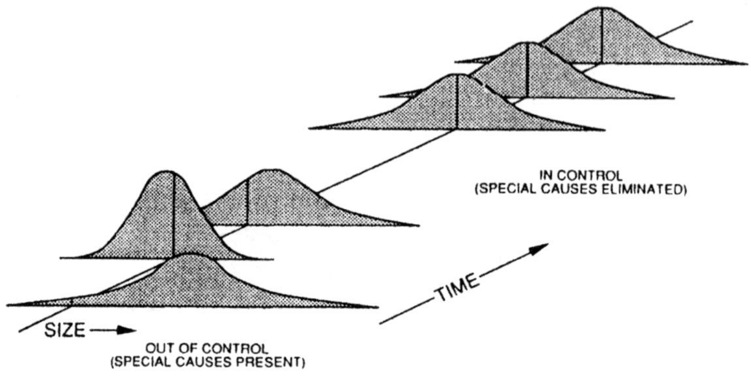

Figure 5-5. Process control [89]

We now ask, does the process meet its tolerance requirements? Is it capable? Perhaps not at first. Reducing random variation due to common causes usually proves more difficult than removing special causes. However, suppose we are smart and persistent enough to reduce random variation too. Then we would see less spread in the process, as shown in Figure 5-6.

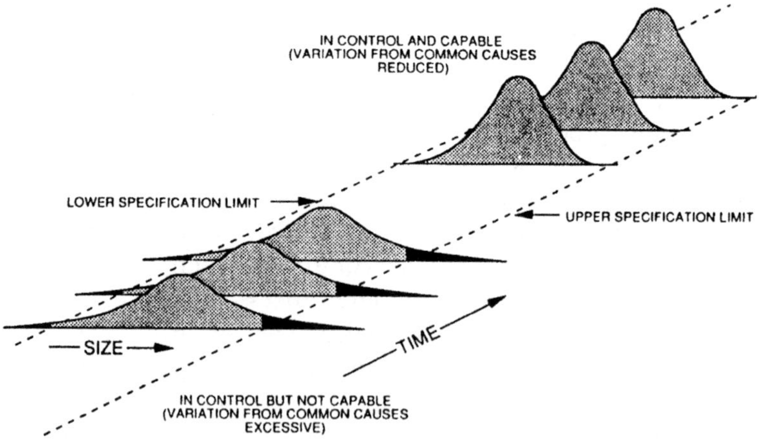

Figure 5-6. Process capability [89]

A process that stays well within the Lower Spec Limit (LSL) and Upper Spec Limit (USL), plus being well centered has benefits. The benefits include reduced sensitivity to component, board, and model variability. Therefore, the demands on model accuracy and detail are reduced. Such a process is called *capable*.

The responsibilities of achieving a good match between process capability and in-circuit insensitivity to variation are shared by the user and supplier. The user *could* just as well design circuits centered on what the supplier can deliver. Otherwise, the user could demand a different, tighter process. If the goal is to produce six-sigma designs, not only capable processes, but also variation tolerant, desensitized design techniques need to be employed.

5.3.4 Control Charts for Process Control and Capability

Control-charting tools are a way of plotting process mean and deviation with respect to time. They help to visualize process control statistics. The USL (Upper Spec Limit) and LSL (Lower Spec Limit) are usually set at +/-3 standard (sigma) deviations. If USL and LSL are set tighter than +/-3 sigma, this is called safety margins, tolerancing, or guardbanding.

For many years, Motorola has been pushing the idea of +/-6 standard (sigma) deviations as part of their six-sigma initiative. Today many companies are advocating this idea as part of their quality improvement programs.

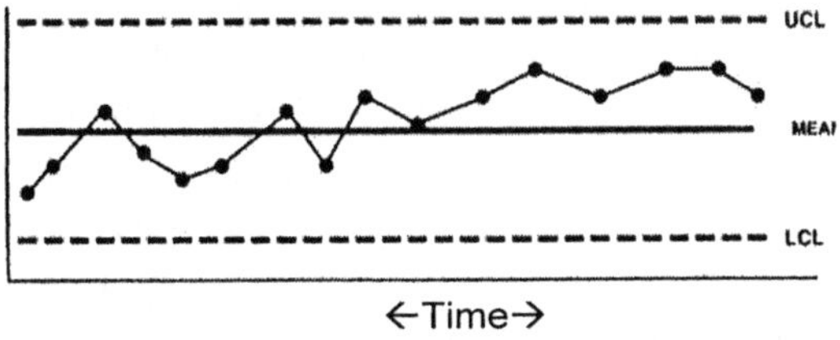

Figure 5-7. A control chart with the process mean plotted [54]

In Figure 5-7 the process mean moves up and down as time progresses. Time is the horizontal axis. Perhaps this random variation is acceptable, but notice the steady upward drift over time. This indicates a special, diagnosable cause. Additional patterns that show up in control-charting tools and their interpretation are presented in [20, 54].

To understand the advantages and burdens of six-sigma design, let us now look at quality and defect levels.

5.3.5 Quality and Defect Levels

In a tolerance window of +/-3 sigma, LSL-to-USL, we have a total of 6 sigmas of deviation. In a tolerance window of +/-6 sigma, LSL-to-USL, we have a total of 12-sigmas of deviation.

In a Normal distribution, an important population statistic is the number of units that fall *outside* an acceptance window that is centered on the mean and is a selected number of standard deviations above and below (+ and -) the mean. When measured as a fraction of a million units, it becomes a popular measure of quality. The measure is called *defect parts per million units* (PPM).

Table 5-1 illustrates the defects in PPM. Note that min/max for +/-3 sigma (first row of data) 2 times 66,810 PPM = 133,620 PPM. Table 5-1 also shows that min/max for +/-6 sigma (last row of data) 2 times 3.4 PPM = 6.8 PPM (or ~ 7 PPM).

Table 5-1. PPM falling outside of sigma deviations from mean

Deviation from Mean in +/− σ	PPM
3.00	66,810
3.50	22,750
4.00	6,210
4.50	1,350
5.00	233
5.50	32
6.00	3.4

The six-sigma quality philosophy says that to achieve high quality, the designer should strive to design a property tolerance window that fits comfortably around +/-6 sigma of the device population's statistics on that property. Likewise for latent defects present in the population. For example, ESD damage from handling should be reduced to below six-sigma.

With six-sigma quality, it becomes unnecessary (even detrimental from the extra handling) to test the parts to see that they pass specification. Six-sigma quality is designed in—not tested in.

The traditional incoming inspection Acceptance Quality Level (AQL) was commonly set at 1% defect level. This equals a 10,000 PPM defect level.

5.4 CHARACTERIZATION FOR SIX-SIGMA QUALITY

This topic discusses the following key points about designing for six-sigma quality:

- Processes that are in control and capable produce quality components and models.

- Processes that are in control and capable enable us to design for six-sigma quality.

However, it is very difficult to demonstrate six-sigma quality in a design. The traditional way to *verify* quality is by *measurement*. To demonstrate six-sigma quality implies taking significant measurement samples on a production run of 20 million units! Today, however, manufacturers and users have short product runs and rapidly evolving technologies and designs. Therefore, *verifying* six-sigma quality with defect levels of a few PPM has to be taken somewhat on faith. To verify six-sigma today requires a new approach. This new approach is one in which simulation of component and product distributions from models are combined with actual measured data to demonstrate the desired quality level.

The smallest sample size required for the simplest of statistical inferences is 25 units. A smaller sample size falls in the realm of pure chance. Today's product runs and testing usually fall far short of the necessary numbers for statistical inferences to be made. But there is a solution. Semiconductor suppliers standardize their *technology processes* and then use TCAD modeling to predict specific device behavior.

Two example technology processes are the 1.3-micron CMOS process and the ABT process. Technology processes have higher unit runs, better data, and more accumulated experience associated with them than individual component catalog entries. Semiconductor houses are mostly fabless. They employ common wafer fabrication houses for device suppliers. The characterized *processes* are fairly well modeled and verified with laboratory measurements. The process models are then used to simulate individual device behavior and the modeling and characterization is done from those simulation results. Far fewer actual measurements of packaged IC devices are done. Usually only enough measurements are taken to verify that no big errors were made. Therefore, the assertion that it is the processes that are verified and individual device parameters are *simulated* is generally true.

The (probable) statistical spread on device parameters is extracted by modeling using TCAD software on a characterized process. This is done for publishing both the data sheet and the model. The future will see even more use of TCAD tools to simulate and predict semiconductor device model parameter spreads. But there is a caution—the TCAD and SPICE parameter information needed to control a process is quite different than what is needed to characterize and model circuit behavior of a product made from that process.

Another component characterization and modeling issue has to do with the many interacting, non-linear, bias, temperature and higher-order effects present in semiconductor behavior. Of course, there will be random variation in a process. But the variation should be small enough so that:

- Higher-order effects do not become a modeling problem.
- Specific modeling assumptions are valid over the range of variation of the process.

For example, suppose we select min-typical-max BJT transistor units (measured or modeled) based on a mid-current beta measurement. We can characterize these units in two interesting ways. In a data sheet, we will publish two sets of curves:

- A beta versus Ic curve for looking at gain versus drive current capability
- A set of gain-bandwidth contours versus Vce and Ic to look at biasing a transistor for adequate RF gain.

Both behaviors are very important in amplifier modeling and simulation. From previous experience, we suspect that higher-order effects will be present in the second curve. To do a complete characterization and modeling of the component, we need a separate curve for the min, the typical, and the max gain-bandwidth units. Generally, such behavioral curves do not scale by a simple multiplication factor. For beta versus Ic, we hope, especially if we normalize the data, that we can manage with a single curve that represents the behavior over a wide range of drive currents and population spread. However, if we let our process spread get too wide, this simplification will not work. Higher order effects occur and the min-typical-max units follow different behavioral curves.

In the 1970s, semiconductor supplier companies attempted to sell everything produced from each chip/die/process line. Wide process variation was not viewed as a problem, even though product from the extremes of the distribution could be viewed as entirely different devices. Device behavior from a very wide process does not scale with a few, simple variables. Semiconductor behavior is non-linear and process variables interact. If one or more of these variables is changed over a wide range, the behavior of the units will follow very different curves. As an example, when manufacturing a discrete BJT transistor, we are controlling the interaction of a matrix of n-by-n non-linear equations, where n is approximately 25. Thus, if you let the range of variation become too large, process control becomes very complex. And, relating a model to a wide process is not very profitable.

If a component's statistical process spread is sufficiently narrow, a min-typical-max simulation model can represent the range of model variation. Thus, good SPC results in accurate modeling. With today's better IC process control, +/-5% variation around typical is more common. Second-order effects are usually not a problem, device models can be scaled and process control with a few, simple variables is often possible.

Another desirable objective is to measure semiconductor properties on the unpackaged device at wafer probe. This is where less cost has been incurred on possible failing units and more control can be exerted. We can reduce the cost of testing by carefully choosing which properties to measure. The DC properties of a transistor are a good choice for some process controls. At wafer probe and previous process steps, there are also other excellent quantities that can be monitored. For example, sheet resistivities through the use of test structures and chips.

The issue for a circuit designer is not just whether device properties can be monitored for good process control—it is whether the semiconductor supplier will translate that knowledge into a complete and useful model for *circuit simulation* use. Once process control is achieved, there is no incentive to go to the extra expense of making a good circuit model unless there is a make-or-lose-the-sale incentive imposed by the user community.

5.4.1 Common Characterization Parameters

Process characterization means describing and depicting the product. The current practice is usually strongly oriented to measurement-based process-oriented verification of TCAD and SPICE parameters. This is followed by SPICE parameter extraction for specific devices. Rules for scaling up and scaling down are applied and individual device-type data sheets are produced.[3]

Technology continues to shrink the size[4] of individual devices and features, while adding more and more gates to a chip. As a result, TCAD and SPICE modeling undergo constant revision. When setting out to create a new process, and making a significant change to some basic variables, we must rely on the existing process models. But results need to be measured periodically to verify modeling assumptions. Models need to be continually adjusted to fit new physical realities.

To incorporate measurements of results into modeling, follow this procedure:

---

3 Scaling does not work very well on the latest, most advanced semiconductor processes. Oxides that an only 3 atoms thick, dopant concentrations that are in the few hundreds of atoms, and other factors ensure that the usual statistical modeling upon which scaling depends will not work any more.

4 The 90 nm process is coming on line and 35 nm is being developed as of this writing.

1. Decide which variables to change. Use Design Of Experiments (DOE) to guide the choices. Avoid changing too many variables at once.
2. Set up the choices, such as mask feature size and doping and diffusion schedules.
3. Implement the changes and verify that the basic targets are met.
4. Repeat the process runs to ensure that the target can be met consistently. Make measurements, gather some initial population statistics, and verify or adjust the models.
5. Explore the design space by deliberately targeting runs for the expected long-term extremes of process control. Take more population statistics and gather more samples.
6. Verify the process spread and shape of the population statistics across +/- 3-sigma for the key parameters by targeting to produce that spread. DOE and the study of a number of process variables are appropriate approaches. At this stage, the composite population statistics are used to target a sample of units equally spread out across the process. This weighted sample can be used for more expensive, labor-intensive testing and as a characterization sample.
7. Measure the characterization units for characteristic curve data for special characterization parameters. Examples are beta versus Ic and gain-bandwidth.
8. Publish the characteristic curves and appropriate population statistics, in a data sheet. Finalize and publish any changes appropriate to the process model.

5.4.2 Economics of Characterization Parameters

Some semiconductor parameters are easier to measure and characterize than others. DC parameters at a specific bias point can be done fairly automatically. High-speed AC and switching parameters require more sophisticated test fixtures and variables sweeping. Therefore, population data gathered on the DC testing side is used to select for process-representative units. These representative units are used to measure the more complex properties. For example, consider some measurements on a discrete BJT.

Population statistics gathered at β_mid_current are used to select for representative units. These same population statistics are used to extract the process mean and sigma. Data sheet min-max are usually set at +/-3 sigma plus some guard band. For the special characterization parameters plots, the results from a few devices, selected to represent the process mean, are put on the data sheet. Special characterization parameters plots include collector characteristic curves, beta versus Ic, gain-bandwidth and more.

An extra special complete job of characterization includes curves for the process mean plus its min-max corners. For an IBIS model, provision is made to extract model parameters at the min-max process corners. In addition, the effects of temperature and voltage are taken into account. Therefore, the slow-weak (slow switching, weak drive) population device is run at low power supply and low temperature for a BJT. The data for this corner is entered in the min column of the IBIS file. The fast-strong (fast switching, strong drive) population device is run at high power supply and high temperature for a BJT. The data for this corner is entered in the max column of the IBIS file.

A composite device representing an average of the data taken from many devices for many parameters would not be a real device. Averaging specific, single properties across a process is OK for population statistics, but usually meaningless for specific unit behavior with respect to bias, frequency, and other conditions.

Additional characterization parameters, usually AC characteristics, can be measured using the weighted sample units. These parameters depend on the types of application the device is designed to perform in. For example: gain-bandwidth for amplifiers and switching rise time for switches. The extra parameters can include junction capacitances, thermal resistance, saturated switching speeds, gain as a neutralized RF amplifier, and matrix model parameters. Because of the time and expense of the testing required, we only occasionally take these measurements.

For certain tests, expendable units may need to be used because the tests are destructive or damaging. For example, additional BJT parameters characterized during the data sheet phase might include BVebo and Safe Operating Area (SOA).

Correlation of measurements between supplier final test and customer incoming inspection–of difficult parameters like switching speed and RF gain for lot acceptance–is an art and science in itself. Today it is unusual to see this kind of effort put into the physical measurement of device parameters for data sheet and model characterization. Most data sheet information is now provided from simulations.

5.5 SIX-SIGMA QUALITY FOR MODELING AND DESIGN

For achieving good designs, good characterization and modeling data are a necessary but not sufficient condition. The component user must also have a knack for designing variation tolerant and desensitized circuits. Less effort put into good design for a component application means that more effort must be put into accurate component characterization, modeling, and testing to achieve acceptable product performance.

To reduce the burden on modeling, effort should include the use of these example design techniques:

- Negative (degenerative) feedback in amplifiers
- Current mirrors in ICs to stabilize against temperature drift
- Matched terminations on transmission lines to stabilize against noise generated timing errors and false switching
- Bypassing of power supplies to avoid ground and power bounce

Also required is a product design process that makes good use of modeling and simulation. Examples include:

- Applying modeling and simulation at the right time and right amount in the development process
- Taking action to avoid or fix potential problems as predicted by simulation. Obviously, if we do not use the modeling and simulation results, why waste time doing them?
- Leaving nothing to chance before prototyping by at some point:
  - Simulating every circuit on a board
  - Simulating for (at minimum) worst-case min-typical-max variations and using more sophisticated statistical design methodologies where warranted

5.6 SUMMARY

Manufacturing processes vary, and they produce device populations that vary in their characteristics. With the aid of statistics, this variation can be studied and understood. Device variation can be accounted for in model parameters measured or simulated over the population of devices produced. Circuits should be simulated with that spread of model parameters representing the spread of devices that will be used in the manufacture of a given product.

To *characterize* a component (device) means to describe and depict it for predicting its performance in a circuit application. To characterize implies that the manufacturing process that produces the component must be repeatable, predictable, and capable of proper operation in its intended use. A model can be no more predicable than the device population it represents.

PART 3: SELECTING COMPONENTS AND THEIR MODELS

Chapter 6

USING SELECTION GUIDES TO COMPARE AND CONTRAST COMPONENTS

Making design tradeoff choices among component properties

Abstract: Selection guides are a much-simplified form of data sheet and present a broad look at making tradeoff comparisons between devices. They also introduce the idea of standardizing component selections, particularly on commodity-type parts. Standardizing component selections makes it much easier to manage commodity-type parts.

6.1 TOOLS FOR MAKING COMPONENT CHOICES

Selection guides are graphical visual aids for making tradeoff comparisons between components. In addition to selection guides, the following tools are available to assist in making component comparisons and design choices:

- Selection guides provide general information about components.
- Data sheets provide more detailed information.
- Simulations provide the most detailed information, particularly about a component's behavior in a proposed circuit.

How does the user select the right devices to meet circuit requirements? The answer depends on many factors, including the exact requirements, availability, cost, size, and data provided by the supplier.

Selection guides simplify and eliminate a lot of the information provided in a data sheet. They simplify the level of detail and present an abstracted measure of *gain, switching speed,* and other performance parameters. Selection guides present only a few variables and a small amount of data at a time. They provide a top-down way of pointing to a device and its data sheet. How the format of any particular selection guide is set up, and which parameters are used, depends on the applications addressed. Finding a device that looks suitable for a circuit often starts with a top-down comparison of diverse semiconductor technologies. For example, a side-by-side comparison between a BJT and a MOSFET is a more simplified comparison of parameters.

Data sheets[1] provide more detailed information than selection guides. They are good for making detailed comparisons and choices between very similar parts. But scanning multiple data sheets while absorbing and understanding their information can be difficult. For example, direct comparison of the gain of a BJT with the gain of a MOSFET uses different parameters and concepts at a detailed level. When comparing dissimilar devices, designers may need to make many choices that are far less detailed.

Simulation. To make an informed choice, we may need even more information than a data sheet can provide. That is where modeling and simulation of a proposed choice gets applied.

Table 6-1 summarizes the types of tools available for comparing and researching components. The table progresses from abstract to detailed.

Table 6-1. Summary of tools for making component choices

Tool	Purpose	Uses
Selection guides (most general information)	Less detailed overview of components. Compares some key parameters of several similar devices. Usually provide by a single supplier for the selection of devices.	Helps the user make quick tradeoffs between many devices.
Data sheets	Specifications created by the supplier after the properties and behavior of a population of devices, from a statistically in-control process, are characterized.	Selection of a suitable device and specification of its functions, properties, and directions on how to use it in a circuit
Simulation results and Application Notes (most detailed)	Simulated and/or measured results of performance of devices in typical circuit applications.	Helps the user with detailed technical results, analysis, and technical advice on the proper application of the device(s).
Purchase specifications (some simple, some detailed)	Supplier-user contract. A purchase spec is usually not needed for a commodity part from a trusted supplier.	Pins down key deliverables on a purchased device. Important for critical applications where the supplier must ensure the deliverables are met.

1 See "Chapter 7, Using Data Sheets to Compare and Contrast Components."

6.2 TEAM MEMBERS USE OF SELECTION GUIDES

Many different design team members in the component, model supply, and user chain can use selection guides. These design team members see the selection process in terms of their own particular concerns. Selection guides, if done well, can then provide a common language between participants with different backgrounds and concerns, including the:

- *Circuit (Logic) Designer:* Selection guides help the user to organize a set of components systematically and to compare and contrast their benefits and shortcomings in a circuit application.

- *Semiconductor Engineer:* A selection guide like Table 6-2 helps a semiconductor device design or application engineer organize information about how different types of devices address particular marketplaces. Table 6-3 type of information combined with Table 6-4 type of information tells users the right kind of device to pick for different kinds of applications. Most suppliers would rather see customers selecting the right kind of product than the wrong kind.

- *Modeling Engineer:* Selection guides help the modeling engineer to organize a set of components systematically and to standardize on a sparse but versatile sub-set of the choices available. There are a limited number of ways to adapt and optimize semiconductor properties to deliver performance. Similar devices usually do not differ much in performance—marketing hype to the contrary. In particular, building a high-quality model library for commodity parts requires a major investment, and resources must be conserved.

- *Marketing and Standardization:* Thinking about standardization helps us to compare and simplify the technology world. For instance, how do 74ACT00, 74HCT00, 74FCT00, and other technologies differ? How does an amplifier whose mid-current beta is characterized at 100 mA differ from one whose mid-current beta is characterized at 120 mA?

  Even though standardization applies best to commodity-type products, it can help the Marketing Department organize its thinking about emerging technologies particularly those in which the application space is just starting to come into focus. Only when the technology is very new and innovative— when people have not yet figured out how they are going to use it—does standardization become a constraint on creative thinking.

- *Component Librarian and EDA Tools:* Standardization helps component librarians store, retrieve, and make models accessible to EDA tool users. The idea of putting devices into categories operates the same way as the Dewey decimal system does for books in libraries. The models can also be attached to other essential EDA information, such as schematic symbol and board footprint and models from different suppliers of the same commercial part number can be given the same internal part number (like a book number).

- *Product Management and Quality:* Categorizing, simplifying, and standardizing device choices helps improve product management and quality. This directly affects model management and quality. Simpler is always better. Complexity creates confusion especially when it is unnecessary.

6.3 SELECTION GUIDE EXAMPLES

In this topic, there are five examples of selection guides. Three examples involve small-signal BJT amplifiers and switches show how understanding properties can help in selecting the best device and type of model for a particular circuit application. The fourth example shows switching speed comparisons between several semiconductor IC technologies. The fifth example shows switching standards for some IC logic families. Additional selection guide information on standard logic ICs can be found in "Appendix K, Logic Selection Guides."

6.3.1 Example 1: BJT Technology

Table 6-2 presents a selection guide of general purpose, low-power BJT transistor amplifiers and saturated switches. Construction and diffusion doping details are shown for different types of applications along with their outstanding electrical performance characteristics. This type of selection guide is helpful to semiconductor device, application, and modeling engineers.

Table 6-2. Small-signal BJT semiconductor device construction

Type	Description
Low Level - Low current - Low Noise	Typically 10x15 mils die with low overall doping. Features rapid low-current β turn-on, high input impedance and low 1/f audio noise. Superseded by JFET technology. High breakdown voltages (BVcbo of 40V, Ic of 50mA). But poor power-handling capacity.
General Purpose	Typically from 10x15 to 250x250 mils die. All-around devices intended to fit most applications moderately well. Drive currents from about 200mA to 50A and BVceo breakdowns from 30V to 200V. Gain bandwidths of about 300MHz at the low-power end to 1MHz at the high-power end.
RF Amplifiers, Oscillators and Mixers	Typically not much larger than 10x15 mils to minimize capacitances. Many construction innovations to optimize gain bandwidths of 900MHz and beyond: Ballasted emitters, steeply graded junctions, graded base, Faraday shields, high emitter periphery to area, and high overall doping to minimize high-current β roll-off. Breakdowns usually below 15V and drives from a few tens to perhaps 200mA at best. A whole variety of RF types (as diverse as most of the categories discussed here put together) were developed by Motorola Semiconductor. But that discussion is too much for the purposes of gaining insight into semiconductor modeling.
Saturated Logic Switches	Typically 10x15 mils to perhaps 22 mils square. Somewhat similar to RF types (for high edge rate). Sometimes gold- (npn) or platinum- (pnp) doped to kill base minority-carrier storage time. Rise and fall times down to 1-2ns and turn-on and turnoff of 15ns on the smaller devices. Breakdown voltages 10-to-15-volts. Higher leakage currents because of the gold and platinum enhanced recombination centers.
High Voltage Video Amplifier - Driver	One example is a 30 mils square die. Optimized for high breakdown voltage and rugged 2nd Breakdown Safe Operating Area capability with several construction features. Including: extra-thick base and collector regions, junction field plate, deep diffusions, triple (collector) diffused, surface-isolation guard / equipotential rings and extra-clean processing. Gain-bandwidth of 80MHz, Ic max of 250mA and 300V BVceo breakdown. Used for CRT video circuit drive.
Switchmode Power Switches	Larger die - 200 mils square and larger. Higher overall doping levels. Low saturation voltages. Characterized and optimized for fast switching speed at high voltages and currents. Rugged Safe Operating Area capability. Breakdown voltages to 800V and switching currents to 20A - not in the same device.
Darlington Amplifiers and Switches	Achieves β multiplication with two transistors on one die in emitter follower cascade. About 30 mils square die with the same general characteristics as general-purpose types. From here to the concept of integrated circuit is not a great distance. Cascaded βs of 1000 to 15,000 depending on the device.

6.3.2 Example 2: Low-Power BJT Saturated Switch

Table 6-3 presents a selection guide of general purpose, low-power BJT transistor saturated switches.

Table 6-3. Switching-current capability versus switching speed and breakdown voltage

Max t_{off}	Max Ic(sat)						
	1200mA	**1000mA**	**750mA**	**500mA**	**300mA**	**200mA**	**150mA**
20ns						MPS2369 15V npn HSS	MPS3640 12V pnp HSS
25ns					PE4313 12V pnp HSS		
30ns				MPS3646 15V npn HSS			
40ns	2N3722 30V npn CD	2N3467 40V pnp CD					
60ns		2N3724 30V npn CD					
200ns						2N3257A 40V pnp GPA/SW	2N3904 40V npn GPA/SW
300ns			2N4400 35V npn GPA/SW	2N4401 40V npn 2N4403 40V pnp GPA/SW			2N3906 40V pnp GPA/SW
500ns	2N3019 80V npn GPA/SW	2N4033 60V pnp GPA/SW					

LEGEND TO TABLE 6-3

- Across the top row, column headings are max Ic(sat)
- Down the first column is max t_{off}
- In the individual cells, voltage quantities are max BVceo
- An example of the data sheet device number is MPS2369
- GPA/SW = General Purpose Amplifier and Switch
- CD = Core Driver
- HSS = High Speed Switch (npn -gold doped, pnp – platinum doped)

6.3.3 Example 3: Low-Power BJT Amplifier

Table 6-4 presents a selection guide for general-purpose, low-power BJT transistor amplifiers. Note that in each cell both npn and pnp 2N-numbers appear. Each npn and pnp pair presents a *complementary pair.* Also note that entries like 2N4400/01-MMBT4401-2N2222A show that the same transistor chip is being used for the whole string of 2N-numbers. The individual data sheets may represent different package and/or breakdown voltage selections of a single chip line. Note that some devices, for example 2N3904, appear in both Tables 6-3 and 6-4.

Table 6-4. Small-signal BJT amplifiers

npn	pnp	Ic (max)	Bvceo (max)	Pc (max)	Package	ft
2N3904 MMBT3904	2N3906 MMBT3906	200mA	40V	0.7W 0.35W	T092 SOT23	350MHz
2N4400/01 MMBT44012 N2222A	2N4402/03 MMBT4403 2N2907	500Ma	40V	1.2W 0.35W 1.2W	T092 SOT23 T018	250MHz
MPSA05/06M MBTA06 2N3020	MPSA55/56 MMBTA56 2N4031	1A	80V	1.5w 0.35w 5.0w	T092 SOT23 T039	100MHz
2N5551 MMBT55512 N3500	2N5401 MMBT5401 2N3495	500mA	150V	1.5W 0.35W 5.0W	T092 SOT23 T039	100MHz

6.3.4 Example 4: Logic Family Switching Speeds

In this topic, we look at a mixture of old, middle-aged, and new semiconductor technologies used for high-speed switching. These families of devices are called glue logic. They glue large, modern ICs (ASIC, FPGAs, large memories, microprocessors, and custom) together by performing level shifting. Table 6-5 compares switching speed between different IC semiconductor technologies.

For a translation and explanation of the acronyms (LS, HC, LVTTL, and other technologies) look up any acronym on:

http://www.acronymfinder.com/

Table 6-5. Switching speed and some logic families

Basic Technology	Type	Tr (ns)	Bandwidth (MHz)	1/2 λ in stripline (inch)
TTL	LS	6	53	16.67
	ALS	3	106	8.33
	FAST	2	159	5.55
CMOS	HC	4	79	11.11
	FACT	2	159	5.55
	ALVT	0.8	398	2.22
ECL	10H	1.0	318	2.78
	100K	0.7	455	1.94
	ECLinPS	0.5	637	1.39
GaAs		0.2	1590	0.56

6.3.5 Example 5: Logic Family Switching Standards

Figure 6-1 shows the voltage levels that some different IC logic families (technologies) operate at, such as input switching thresholds, Vil and Vol.

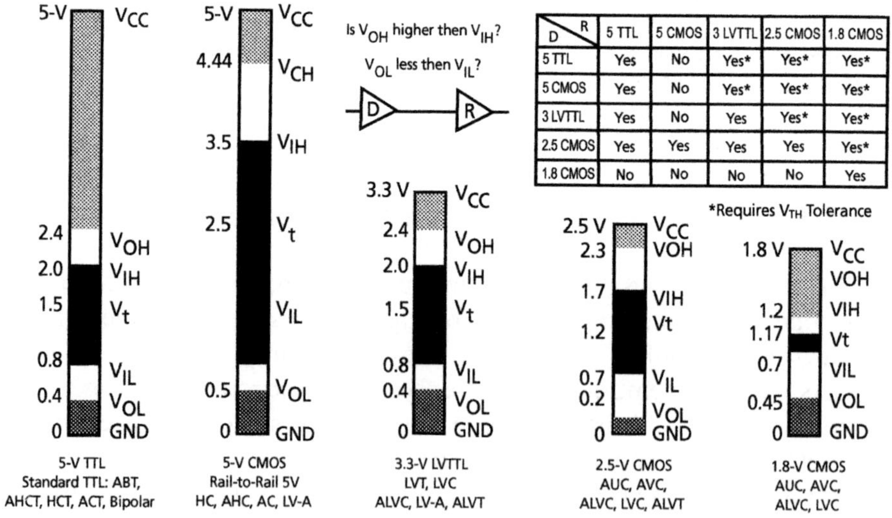

Figure 6-1. Comparison of switching standards for some logic families
Courtesy of TI Semiconductors

6.4 SELECTION GUIDES HELP COMPONENT STANDARDIZATION

Standardizing in commodity-type component areas can improve the selection range and help us get enough models for good part coverage. The chances for volume discounts are also improved.

One benefit of all this classification is a reduction in the number of individual types of parts that must be stocked, modeled, monitored, and generally managed to cover a given range of applications. We can reduce the number of line items from many tens of thousands to a few thousand while still accomplishing the same freedom-of-design choice and coverage.

Tables 6-3 and 6-4 show a set of sparse (but versatile) devices. They can be better managed for full qualification. It is fairly easy to save data and archive this set of devices in a database where they are ready for selection and use in a design. Standardization and re-use of *commodity* components allows leveraging of our investments in support activities.

Support resources must be conserved so that they can be applied to models for new and custom parts and modules. Using a non-standard device that provides no clear advantage over approved, standard parts is a refuge for engineers who substitute minor change for true creativity and innovation. This allows them to put their own unique stamp on a design without investing any real creative effort. It gives the sizzle (appearance) of innovation without the steak (actuality).

We further develop this standardization idea in "Chapter 15, Working with the Model Library."

6.5 SIMULATION AS A SELECTION GUIDE

We have just shown some examples of device-selection guides that use the information presented on data sheets in an abstracted form. Aside from logic levels and switching thresholds, we really did not cover logic devices. For logic ICs, the component selection decision involves setting up the required logic functionality. For circuits with critical timing, the decision involves signal integrity, or other HSDD issues. The final purchase decision might involve modeling and simulation. Leaving aside logic functionality, let's look at HSDD issues. Modeling and simulation of HSDD issues assists in the device selection process.

Before choosing a modeling and simulation method, we need to determine which type of behavior to examine. One chance is to compare typical switching waveforms for some high-speed digital switching technologies.

These waveforms were generated as a selection guide for logic designers at 3Com-CommWorks Division. They graciously allowed the complete results to be posted on the IBIS Committee website under Articles and Training where you can view or download LogicFamilySI.doc (31-Jul-00 09:31 946k)

Results for (5V) ABT, ACT, ACTQ, AHC, AHCT, ALS, AS, HC, HCT, LS, S, VHC, (3.3V) AHC, AHCT, ALVC, ALVCH, ALVT, HSTL, LV, LVC, LVCH, LCX, LVT, LVX, SSTL, (2.5V) ALB, F/FAST, NC, VCX, (2.1V) BCT, BTL/FB, LVT, CBT, CBTLV, GTL/GTLP/GTL+, ECL and LVDS are included.

The following example illustrates how early and repeated simulations can help the design process:

> A device selection and product design engineer asked to simulate a particular net that had a driver and six daisy-chain receivers. The IC component was still in development and board layout had not started. The IC supplier was able to provide a generic, single driver IBIS model. The clock speed on the net was 4MHz. The proposed schematic-driven net was entered in the EDA tool with attached driver model, receiver models, and generic transmission-line sections. Characteristic impedance (Zo), stripline and microstripline configuration, dielectric material and length can also be inputs. From the simulation, big reflection problems were found. It turned out that rise- and fall-time on the driver was about 100ps on this 4MHz net. This was due to die-shrink of the output transistors!

This IC was successfully implemented, but it took close attention to the design and the component starting early in the project.

6.5.1 Simulation of Unterminated and Terminated Nets

Let's look at some 3.3 Volt logic simulations. Figure 6-2 shows the schematic of an unterminated point-to-point net used for the simulation waveform plots that follow. Figure 6-3 shows the same net terminated.

General simulation settings were done at 50 MHz and 1 cycle with the typical data of the IBIS model selected. The transmission line was set at Zo = 60 ohms and 1 ns delay.

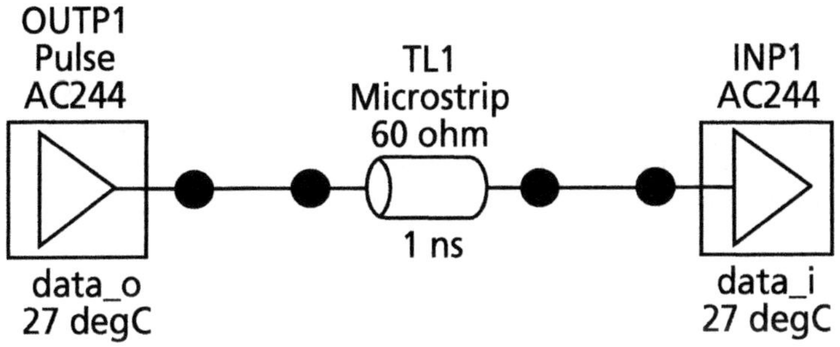

Figure 6-2. Un-terminated simple topology

For terminations, a V-I table (curve) with Rin = R_L = 65 ohms was used. This provides a small, non-zero reflection coefficient at the receiver:

$$\rho = \frac{R_L - Zo}{R_L + Zo} \tag{6-1}$$

Depending on input capacitance of the receiver and the frequency content of the driver, reflection coefficient (matched) is around 0.04.

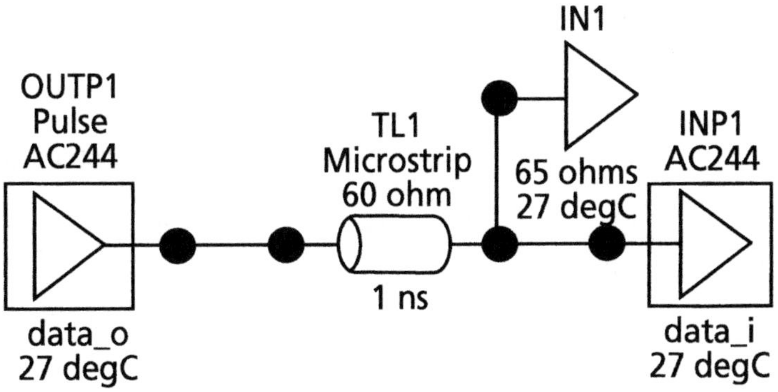

Figure 6-3. Terminated simple topology

6.5.2 Results of Simulating Devices in the Unterminated and Terminated Nets

How does the user select the right semiconductor devices to meet circuit requirements? The answer depends on many factors, including the exact requirements, available selections, cost, size, and data provided by the supplier. If the initial selection does not have the performance required—even after he or she has gotten deeply into the design process—the user might need to revisit the selection process. Figures 6-4 through 6-7 show some results of terminated and unterminated nets employing different logic family technologies with and without clamp diodes on the nets.

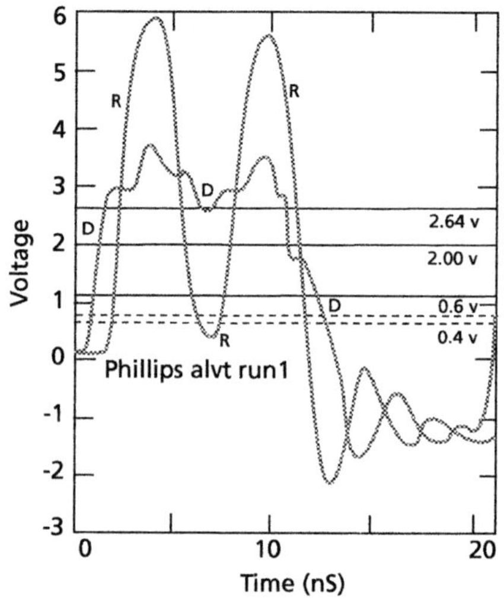

Figure 6-4. Un-terminated Philips ALVT

Phillips advanced low voltage CMOS technology

Driver = 74ALVT162245_OUT

Receiver = 74ALVT162245_IN: run1

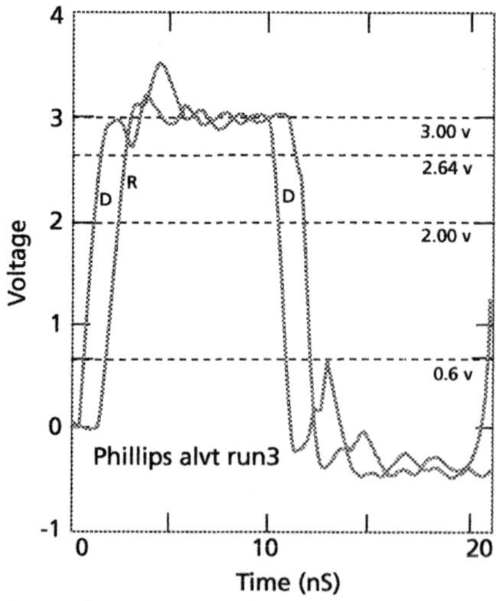

Figure 6-5. Terminated Philips ALVT

Philips low voltage CMOS logic

Driver = 74ALVT162245_OUT

Receiver = 74ALVT162245_IN: run2

The waveforms that are generated in the simulations tell us a lot. Our choices are based on the amount of ringing, overshoot, and other signal integrity issues. The results also depend on the circuit we have placed in our model and whether or not they are terminated. The results also depend on our skills as circuit designers. Termination (transmission-line impedance matching) of the net obviously affects the signal reflections set up. This also depends on the switching speed of the technology selected, the voltage level being switched, the particular implementation of the technology selected including the presence of clamps, output drive strength, and (if fast enough) even package parasitics.

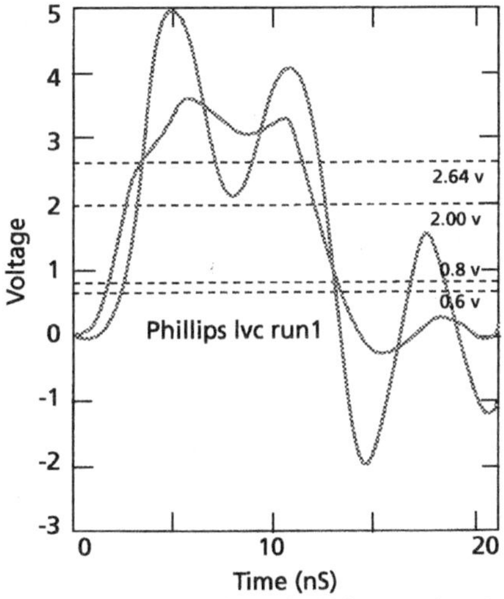

Figure 6-6. Un-terminated Philips LV/LVC

Philips low voltage CMOS logic

Driver = 74LVC16245_OUT

Receiver = 74LVC16245_OUT: run1

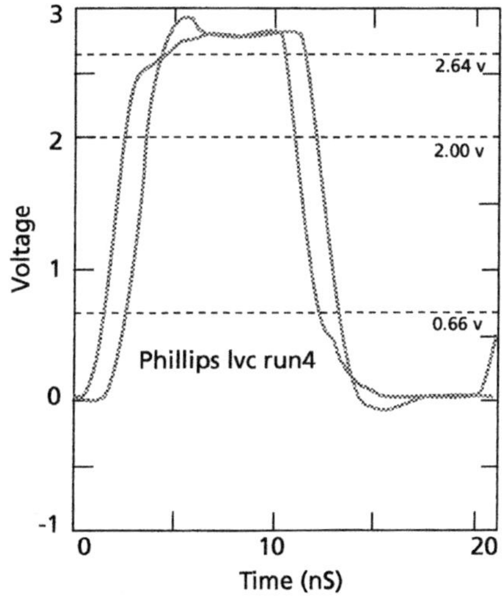

Figure 6-7. Terminated Philips LV/LVC

Philips low voltage CMOS logic

Driver = 74LVC245_OUT

Receiver = 74LVC245_OUT: run3

There are various measures, judgments, and quantifications of high-speed digital-device performance. Particular device selections and how they are used (applied) may be more suitable, or less suitable, for a given application. From Figures 6-4 through 6-7, we can conclude that some technologies appear to be designed to work into matched terminated transmission lines but do not work well otherwise. We can answer design questions with more confidence after a particular device is simulated and modeled. The simulations should be done:

- With or without various circuit design choices, (for example termination or no termination on a particular net), and
- Over a range of variability

It is crucial to make good choices early. This is because about 90% of the eventual cost (and the chances for success or failure) of developing a product is determined by the time 10% of the decisions have been made. We cannot make a good design out of bad choices. Likewise, we can make good device selections, but it is still up to us to properly design and layout the circuits they are used in.

A design task of implementing one FPGA may include multiple semiconductor technologies, signal integrity, power integrity, EMI/EMC, power dissipation, timing and logic function design. A datasheet for a 1000 pin FPGA is overwhelming and the IC may interact strongly with the design of the circuit and the board layout.

6.6 RIGHT-THINKING

The following example from the early days of the semiconductor industry teaches us valuable lessons about adopting "right thinking" about quality:

> Originally, the semiconductor industry focused on marketing. Quality control was half-hearted. Process control was very wide and poor. Beta range on a BJT could easily be 10:1. The tails of such a wide distribution behave significantly differently than the process mean.
>
> Management's strategy was "Find a home for everything." Therefore, the process/product line would be marketed under several different data sheets, which was not only based on assembling the die in different packages.

> Marketing the tails of the distribution, usually at reduced prices, eventually created business. The data sheets for the tails of the distribution were usually sketchy, concentrating on a few low-cost DC test measurements. Expensive testing, such as AC gain-bandwidth, was avoided or minimal.
>
> Producing the tails of the distribution to meet shipments could be chancy. This eventually led to using more than one semiconductor wafer starting-resistivity on a given chip line to meet shipments. This also led to sourcing the "same" device from different chip lines.
>
> Keeping track of which diffusion run should go to which customers was spotty. Quality, product consistency, and model accuracy are impossible in such an environment.

It took the Japanese quality revolution to change how we think about process control–to get product consistency, quality, and model accuracy.

6.7 SUMMARY

The first step to good modeling and simulation is to choose a good component for a particular design. If an unsuitable choice is made, no amount of simulation effort can fix the design. Selection guides help engineers in making good device choices and can also help organize the standardization of device selections for commodity-type parts.

Chapter 7

USING DATA SHEETS TO COMPARE AND CONTRAST COMPONENTS

Making a good match between a component and its in-circuit use is the first, critical modeling step

Abstract: Data sheets are documents created by suppliers for marketing the behavior of their semiconductor devices. The data sheets characterize device behavior over the device's range of operation and population spread. The data curves in the data sheets are similar to behavioral models. By comparing the behavior of various devices, engineers can choose the best device suitable for a particular design. Since the data sheets were developed as marketing tools, engineers must exercise judgment when using them.

7.1 DATA SHEETS AS PRODUCT DESCRIPTIONS

Data sheets are the primary documents for communicating critical technical information to designers. When suppliers market a device, they provide data sheets to convey the technical information. Suppliers may also provide selection guides and models, but data sheets are usually much more available.

The original system of data sheets was issued under the Joint Electron Device Engineering Council (JEDEC), which is a standards committee set up by the Electronic Industries Alliance (EIA).[1] This standards committee had a system of registering devices (1Nxxxx for diodes, 2Nxxxx for transistors, and 3Nxxxx SCRs and other discrete devices).[2] JEDEC was

---

1 Formerly Electronics Industries Association.

2 The history of standards associations in the development of electron devices is given at http://www.jedec.org/Home/about_jedec.cfm.

created in 1960 growing out of standards efforts in vacuum tubes and radio. In 1965 a division addressing ICs and microelectronics was formed.

The standardization and registration of device data sheets (specifications) never really succeeded. Some semblance of order was achieved in the issuance of device numbers. But it pays to make a direct one-to-one comparison of data sheets *and* models from different manufacturers such as the 74LVT245SC (Low Voltage Octal Bidirectional Transceiver with 3-STATE Inputs and Outputs) from Fairchild and the 74LVT245SO from Philips

Data sheets include the following critical information needed for the wise selection and proper operation of a device:

- Electrical behavior—for an IC with many different I/Os (that is, multiple behaviors)
- Electrical behavior under different circuit conditions
- Electrical behavior from different parts of the population distribution
- Pin connections to internal points
- Parasitic elements associated with the internal connections
- Test fixturing used to measure behavior
- Package dimensions, pin locations and pin dimensions
- Thermal elements associated with the dissipation of heat generated in the package
- Safe operation conditions such as maximum power dissipation and temperature rise
- Logic operation (as appropriate) and switching thresholds
- Other important data

Let's examine a sample data sheet for the 2N3904[3] BJT. Figures 7-1 and 7-2 show two pages of the 2N3904's data sheet. For the full data sheet, see "Appendix A: Sample Data Sheets."

[3] The 2N3904 has been manufactured and sold for about 50 years by Fairchild Semiconductor. And, it is still being sold! That's a pretty amazing product lifetime! Therefore, this component has been around long enough to have been thoroughly characterized and modeled. But we should not expect to find and download from the Internet a complete SPICE model (all 404 parameters characterized).

FAIRCHILD
SEMICONDUCTOR ™

2N3904 / MMBT3904 / PZT3904

2N3904 MMBT3904 PZT3904

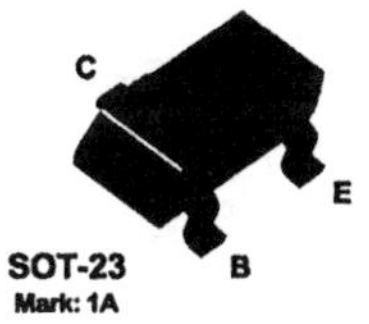

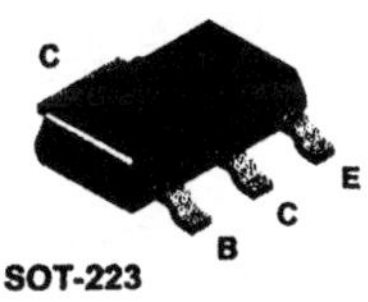

NPN General Purpose Amplifier

This device is designed as a general purpose amplifier and switch. The useful dynamic range extends to 100 mA as a switch and to 100 MHz as an amplifier.

Absolute Maximum Ratings* T_A = 25°C unless otherwise noted

Symbol	Parameter	Value	Units
V_{CEO}	Collector-Emitter Voltage	40	V
V_{CBO}	Collector-Base Voltage	60	V
V_{EBO}	Emitter-Base Voltage	6.0	V
I_C	Collector Current - Continuous	200	mA
T_J, T_{stg}	Operating and Storage Junction Temperature Range	-55 to +150	°C

*These ratings are limiting values above which the serviceability of any semiconductor device may be impaired.

NOTES:
1) These ratings are based on a maximum junction temperature of 150 degrees C.
2) These are steady state limits. The factory should be consulted on applications involving pulsed or low duty cycle operations.

Thermal Characteristics T_A = 25°C unless otherwise noted

Symbol	Characteristic	Max			Units
		2N3904	*MMBT3904	**PZT3904	
P_D	Total Device Dissipation Derate above 25°C	625 5.0	350 2.8	1,000 8.0	mW mW/°C
$R_{\theta JC}$	Thermal Resistance, Junction to Case	83.3			°C/W
$R_{\theta JA}$	Thermal Resistance, Junction to Ambient	200	357	125	°C/W

*Device mounted on FR-4 PCB 1.6" X 1.6" X 0.06."

**Device mounted on FR-4 PCB 36 mm X 18 mm X 1.5 mm; mounting pad for the collector lead min. 6 cm^2.

 2N3904/MMBT3904/PZT3904, Rev A

Figure 7-1. Sample data sheet for 2N3904 page 1
With permission from Fairchild Semiconductor

2N3904 / MMBT3904 / PZT3904

NPN General Purpose Amplifier
(continued)

Electrical Characteristics
T_A = 25°C unless otherwise noted

Symbol	Parameter	Test Conditions	Min	Max	Units
OFF CHARACTERISTICS					
$V_{(BR)CEO}$	Collector-Emitter Breakdown Voltage	I_C = 1.0 mA, I_B = 0	40		V
$V_{(BR)CBO}$	Collector-Base Breakdown Voltage	I_C = 10 μA, I_E = 0	60		V
$V_{(BR)EBO}$	Emitter-Base Breakdown Voltage	I_E = 10 μA, I_C = 0	6.0		V
I_{BL}	Base Cutoff Current	V_{CE} = 30 V, V_{EB} = 3V		50	nA
I_{CEX}	Collector Cutoff Current	V_{CE} = 30 V, V_{EB} = 3V		50	nA
ON CHARACTERISTICS*					
h_{FE}	DC Current Gain	I_C = 0.1 mA, V_{CE} = 1.0 V I_C = 1.0 mA, V_{CE} = 1.0 V I_C = 10 mA, V_{CE} = 1.0 V I_C = 50 mA, V_{CE} = 1.0 V I_C = 100 mA, V_{CE} = 1.0 V	40 70 100 60 30	 300 	
$V_{CE(sat)}$	Collector-Emitter Saturation Voltage	I_C = 10 mA, I_B = 1.0 mA I_C = 50 mA, I_B = 5.0 mA		0.2 0.3	V V
$V_{BE(sat)}$	Base-Emitter Saturation Voltage	I_C = 10 mA, I_B = 1.0 mA I_C = 50 mA, I_B = 5.0 mA	0.65	0.85 0.95	V V
SMALL SIGNAL CHARACTERISTICS					
f_T	Current Gain - Bandwidth Product	I_C = 10 mA, V_{CE} = 20 V, f = 100 MHz	300		MHz
C_{obo}	Output Capacitance	V_{CB} = 5.0 V, I_E = 0, f = 1.0 MHz		4.0	pF
C_{ibo}	Input Capacitance	V_{EB} = 0.5 V, I_C = 0, f = 1.0 MHz		8.0	pF
NF	Noise Figure	I_C = 100 μA, V_{CE} = 5.0 V, R_S =1.0kΩ,f=10 Hz to 15.7kHz		5.0	dB
SWITCHING CHARACTERISTICS					
t_d	Delay Time	V_{CC} = 3.0 V, V_{BE} = 0.5 V,		35	ns
t_r	Rise Time	I_C = 10 mA, I_{B1} = 1.0 mA		35	ns
t_s	Storage Time	V_{CC} = 3.0 V, I_C = 10mA		200	ns
t_f	Fall Time	I_{B1} = I_{B2} = 1.0 mA		50	ns

D

*Pulse Test: Pulse Width ≤ 300 μs, Duty Cycle ≤ 2.0%

Spice Model

NPN (Is=6.734f Xti=3 Eg=1.11 Vaf=74.03 Bf=416.4 Ne=1.259 Ise=6.734 Ikf=66.78m Xtb=1.5 Br=.7371 Nc=2 Isc=0 Ikr=0 Rc=1 Cjc=3.638p Mjc=.3085 Vjc=.75 Fc=.5 Cje=4.493p Mje=.2593 Vje=.75 Tr=239.5n Tf=301.2p Itf=.4 Vtf=4 Xtf=2 Rb=10)

E

Figure 7-2. Sample data sheet for 2N3904 page 2
With permission from Fairchild Semiconductor

On the sample data sheet in Figures 7-1 and 7-2, note the following points:

A) An illustration of three different packages—designated TO92, SOT23, and SOT223— the 2N3904 transistor die is assembled in.
B) Absolute maximum ratings for current, voltage and temperature.
C) Maximum power ratings and thermal characteristics in each of the three packages.
D) DC current gain (also denoted by beta, β, and h_{FE}) and other active state characteristics and parameters. The data sheet gives beta's minimum values over various bias conditions and its min-to-max value at a mid-current bias condition. The min-to-max values on β have been set at 100 to 300. This implies that the process mean is at 200 and the +/-3 sigma limits on the population are 100 and 300 respectively, plus some guard-band. But we do not know whether our associations are true.
E) The SPICE model is shown. That is unusually complete information for a discrete data sheet.

Overall, the parameter information on the sample data sheet is sparse compared to the author's recommendations on process control, characterization, modeling, and substituting DC measurements for AC measurements.

7.2 ARE DATA SHEETS ACCURATE AND COMPLETE?

Since a data sheet helps in selecting the best device for an application, it is important that the data sheet be both accurate and complete.[4] But how complete is the information on the average data sheet?

And how accurately do we need to know the mean and +/-3 sigma values on, for example, β? For example, do we need to know it within?

- An engineering approximation (about +/-10%)?
- Or, to within +/- 0.5%?

Consider the following: If β≥50, we can design a common emitter amplifier with an amplifier voltage gain of 10 by using some degenerative resistance in the emitter, Re. Thus the gain will be set by the ratio of

---

4 The need for performance delivered versus data sheet accuracy applies equally to both models and components.

collector resistance, Rc, and Re. Then Rc/Re = 10 sets the gain. By using degenerative feedback, the amplifier gain becomes independent of β and its variations over a wide range of conditions.

A designer*'s* need for data sheet accuracy relates to three considerations:

- At which stage is the designer working in the design process?
- How tight are the design constraints (at any given point)?
- Was the right component choice made to begin with?

The closer to the delivery date, and the more critical the design, the greater the need for data sheet accuracy. So the sooner the designers get insight into how a device will perform, and the cleverer they are at using non-sensitive circuit design, the less they will be at the mercies of model data errors. Plus at least half of all struggles over model accuracy issues are rooted in a poor choice of device. Bad selections are almost always expensive mistakes and make it difficult to achieve high-yield, high-quality production.

To avoid problems, do *not* choose a device that:

- Is sensitive to variations or stresses in the product's environment.
- Requires testing and rejecting a portion of the population.
- Does not have the capability to run correctly in the product (For example, one that goes into high-current beta rolloff at peak drive demand).

For example, choosing a device that is too fast or too slow for an application results in a situation in which no amount of struggle over modeling accuracy, iterations of simulation runs, or constraints on device specifications results in a good design.

Designers cannot ignore the issue of data sheet and model accuracy. How much care is exercised over the data in an average data sheet? Well, the answer varies—from not much to substantial. But there is a whole host of "let the buyer beware" disclaimers attached to data sheets and models that do not exactly build buyer confidence. The engineer needs to ensure that the supplier delivers what was promised in the data sheet and model.

Data sheets and models have many disclaimers regarding accuracy. To deal with disclaimers versus the actual design requirements, the designer may need to negotiate a *Purchase Spec* or contractual agreement with the supplier. This agreement ensures that the supplier delivers what is advertised in their data sheet and device model.

Sometimes, especially when pushing the state-of-the-art, the designer needs to specify tight device tolerances. Tolerance limits on a data sheet have traditionally been for +/-3 sigma around a process mean. Currently, on digital switching parts, the parameter tolerance windows are actually fairly tight–more like +/-5% variation for a +/-3 sigma spread. Usually, the data sheet commits to a much wider spread than +/-5% for a 3-sigma window. Whenever this kind of padding of spec limits is done, we call it guard-banding.

Minimum (Min) and Maximum (Max) limits on SPICE, IBIS, and other models (when they are included on the data sheet) are usually for +/-3 sigma spreads. For directly translatable parameters, it's not unusual for data sheets and models to disagree. An example of a directly translatable parameter is beta, measured under the same bias conditions on the data sheet and in the SPICE model. The usual explanation for the disagreement given by suppliers is, "Well, we guarantee the *data sheet*."

What guarantees are they talking about? The same disclaimers are used on data sheets and model files. Both disclaimers usually deny that the data bears any resemblance to a real device, living or dead.

By using variation-tolerant design methods, we can deal with the supplier's specsmanship and marketing pitch. Using a simple worst-case simulation works fine on all but the most critical nets. These critical nets may need more sophisticated statistical design methods, desensitization, and robust design methods.

For a +/-6 sigma design goal (everything works without the need for testing), the designer must go outside the limitations of the data sheet and work directly with the supplier. Be aware that the +/-3 sigma process spread must fit well inside a +/-6 sigma design-tolerance window. And when setting out to achieve six-sigma design, designers need a much better grasp of issues, like long process tails and other surprises.

Finally, just as devices perform poorly outside of their limitations and specified test conditions, models perform poorly outside of their assumptions and limitations.

7.3 SELECTING A COMPONENT THAT IS FIT FOR USE

How do we begin selecting a component fit for use in a particular circuit? We now ask:

- Do we use models or do we use data sheets to make the selection?
- What are the criteria to use in making the selection?
- How do we compare components?

7.4 USING DATA SHEETS TO BEGIN THE SELECTION PROCESS

Data sheets, models, and selection guides are tools that can assist in selecting components. For a first pass, we usually start with a data sheet when making a selection. When we compare similar devices, the comparisons are straightforward. When we compare slightly dissimilar devices, the comparisons are less straightforward, but not too difficult. When we compare very dissimilar devices, we must be aware that designing their power supply, bias, coupling circuitry, and loading are quite different, even while comparing similar performance properties. For example, performance of BJT transistors versus MOSFETS as an audio amplifier output.

7.4.1 Comparing and Contrasting Components

Data sheets usually present component behaviors and functions over a range of conditions in an understandable way. A model may give more detailed answers of specific performance under specific conditions. But it is usually harder to grasp the component's overall performance from the model, unless it is simulated under many conditions. However, in making a final choice between components in a circuit, it is common to model and simulate the component behavior.

The criteria to use to make a selection are the in-circuit performance capabilities that are needed. We may need multiple electrical performances, but no single component delivers equally satisfactory results across all requirements. Likewise, several different choices provide a mix of results when trading one set of performances for another.

For instance, we might be interested in the flat mid-current beta range comparisons between several BJT chips that are scaled up in size. In addition, we might be interested in the maximum gain-bandwidth (f_t) comparisons between those same chips. The tradeoff between flat beta range and maximum f_t between chips that are scaled up in size is that while we gain in the former parameter, we usually lose in the latter.

7.4.2 Comparing Dissimilar Devices

Below are three examples of comparing dissimilar devices.

Example 1: We might be interested in the switching speeds and logic thresholds between different logic families that are from different semiconductor technologies. If we compare ABT (a BiCMOS technology) with ALVC (a CMOS technology) we tend to stick to performance parameters that are a bit more abstracted. And we recognize that bias conditions, like power supply voltages, are likely to be different for these two technologies. However, it is a pretty good assumption that similar performance parameters will be given on the data sheets for the dissimilar devices because they will be designed to do the same job. Logic switching per this example.

Example 2: We might be interested in gain-bandwidth and the ability to drive the video plates of a TV tube. We may make the decision to start with a BJT device and want to compare a higher-voltage general-purpose device with a high-voltage video driver device designed specifically to apply to such a circuit. Table 7-1 summarizes different physical designs of BJTs optimized to address different applications.

When we compare apples-to-oranges, we have to pick performance measures that apply to both and recognize that we will indeed need to design their circuitry differently for bias and other characteristics.

Example 3: We might be interested in the performance of semiconductor power amplifiers versus vacuum tube power amplifiers as final output stages in an audio system. Here we would tend to rely on all the tools available for making circuit applications choices: data sheets, selection guides, and modeling and simulation.

7.4.3 Data Sheets Are a Type of Model

A *circuit model* of a device is used in a way that allows the user to numerically compute specific performance under specific sets of conditions. If the user steps the conditions, he or she can then compute (and plot) performance under a range of conditions.

Contrast the circuit model with the data sheet. The behavioral curves are not usually used directly in a circuit simulator program. The performance over a range of bias conditions can be seen in the characteristic curves of the component as part of the information in its data sheet. Thus for instance, whether the device has adequate gain over an expected range of load currents can be predicted. Other data sheet information includes information about population spreads on key parameters, test measurement conditions and circuits, thermal resistance for power handing capability, and package dimensions.

The idea of a data sheet as a type of model is embodied in the term *characterization model.* A characterization model provides *overall* electrical properties and performance at a glance. Thus, the user is aided in making judgments about a device's ability to meet circuit requirements. The circuit model and the characterization model are sometimes combined on a data sheet when plots of model parameter values for particular bias conditions are included.[5]

7.5 CONSTRUCTION CHARACTERISTICS OF AMPLIFIERS AND SWITCHES

Let us look at some discrete bipolar transistors to illustrate how structure and device design influence electrical properties. Suppose we have a group of circuit applications that requires us to make comparisons between how different BJTs are designed. Table 7-1 presents descriptions of some unique construction characteristics for optimizing different BJT transistor capabilities:

Table 7-1. BJT semiconductor device construction

Type of BJT	Unique Construction Characteristics
Darlingtons	Two transistors in one chip in emitter follower cascade configuration. Results in beta multiplication.
Low Noise	Low current device with low overall dopant levels. Results in rapid low current turn on and low 1/f audio noise and high input impedance. Mostly obsoleted by JFET technology.
General Purpose Low/Mid/Hi Power	All-around devices designed to do a wide range of jobs moderately well from about 1 mA to 30 A and about 30 V to 200 V. Power from .35 W to 250 W in packages from SOT23 plastic to T03 metal can. Bandwidths extend to 400MHz at the low power end drop to 1 MHz at the high power end.
High Current	Germanium Mesa process, alloy junction and diffused base pnp parts for rugged current handling.
High Voltage	Optimized for high breakdown voltage, 300V BVceo for instance, with several design tricks. Includes extra thick base and collector regions, field pates to lessen curvature of breakdown fields, surface isolation guard-equipotential rings, triple diffusion, deep diffusion and careful crystal and surface defect-free processing.
High Speed Saturating Switch	Gold (npn), occasionally Platinum (pnp), doped devices to kill base charge storage time. Also optimized for high bandwidth.
Switchmode	Optimized for low saturation voltages and fast switching speeds at high currents and voltages. Rugged SOAs (Safe Operating Area).
RF Types	Many design tricks are employed to optimize gain and bandwidth. Including steeply graded junctions and steeply graded base doping, faraday shields to decrease feedback capacitance, high emitter periphery to area ratios, and ballasted emitters.

[5] An example of this is presented in a "Chapter 19, Deriving an Equation-Based Model from a Macromodel" on using 2 Port Y-Parameters to design an isolation amplifier. Also sometimes the data sheet contains a SPICE or other model as we saw on the 2N3904 data sheet.

Each part listed in column 1 is designed differently to maximize some characteristics and to minimize others. Having an understanding of this will help in choosing the right part for an application. Later, we will focus on the low-power general-purpose type of BJT and take a more detailed look at some data sheet curve comparisons.

7.6 USING BETA TO EXPLAIN DEVICE TRADEOFFS

Let's investigate devices further and compare some characteristics found on data sheets. We look at device properties to make tradeoffs in how the devices satisfy our circuit applications needs.

The five devices in section 7.9 are considered to be small-signal low-to-medium-power devices.[6] These devices were picked as examples because their physical-electrical behavior is well characterized, relates well to device construction, SPICE parameters, and are interrelated in ways that explains the tradeoff and selection process.

The interaction of various structural properties in a simple BJT model involves about twenty-five interacting first-order properties and parameters. Each time you vary one property or parameter as an independent variable, it affects several other properties and parameters.

Beta, the current-gain of a bipolar junction transistor (BJT), is defined thus: Beta (β) = DC current gain = Ic/Ib. β provides us with a standardized measurement and modeling parameter that is well-recognized and long-accepted as of primary importance. β will also provide us with:

- A way of classifying devices designed for different circuit applications
- A BJT SPICE simulation model parameter
- A way of tying circuit performance to device structure and design
- A way of relating quality issues and device defects to processing cleanliness and other issues

Let us look at β as a performance indicator over a range of operation.

β versus collector current. The effect of varying device operating parameters, like current density flowing through the device's active regions, can be observed with β. One reason devices get scaled up is because at high

---

[6] Compared in size and capability (1-amp collector-currents and 40-volt standoff-voltages) with today's deep sub-micron CMOS, these parts would seem to be in the power devices class. But they are only small-signal devices and they illustrate how radically deep sub-micron device power levels have changed our thinking.

current densities β will collapse in a BJT. Three or four phenomena may account for this behavior depending on device particulars:

- Base conductivity modulation
- Decreased emitter junction injection efficiency
- Emitter de-biasing due to base resistance voltage drop
- Collector junction push-out, or, Kirk Effect [64][7]

Figure 7-3, a plot of β versus current, reveals how the current-density tradeoffs behave.

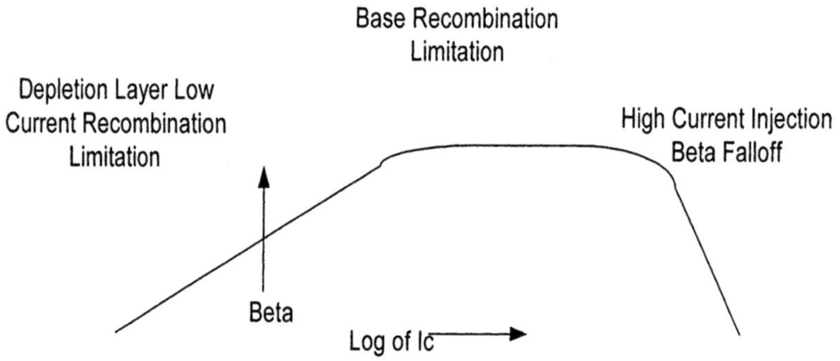

Figure 7-3. Beta versus Ic: The beta turn-on region, mid-current region, and high-current injection falloff regions

There is a turn-on phenomenon in BJTs at low-current (density). In device design tradeoffs what is gained in β holdup at the high-current end is lost at the low-current end. There is a fixed recombination rate in the emitter-base junction that makes for a very inefficient minority carrier injection at very low currents but drops out of significance as the current is increased.

β versus Ic and Vce: BJTs also show Early effect (VAF) on β due to collector voltage increases. The β increases as the effective base region narrows, which is due to the increased collector reversed-bias voltage. Other factors enter in such as doping density, base doping concentration gradient, the targeted absolute value of β, and the narrowness of the base. Figure 7-4, a plot of a family of curves for collector current versus collector voltage for

[7] One consideration the circuit designer has to take into account is the maximum current a chosen device can handle both steady state and as a switch. A switch goes from turn off to full (saturated) turn on.

stepped values of base current, shows this β behavior. Linearity issues, especially for large signal swings, are also illustrated.[8]

Collector characteristics: Figure 7-4 shows a family of curves (collector-current versus collector-emitter-voltage at a set of base-current steps). These curves are labeled to identify several behavioral regions such as collector saturation and voltage breakdown, Early effect, thermal runaway, base punch-through, quasi-saturation [66] and high current beta falloff.[9]

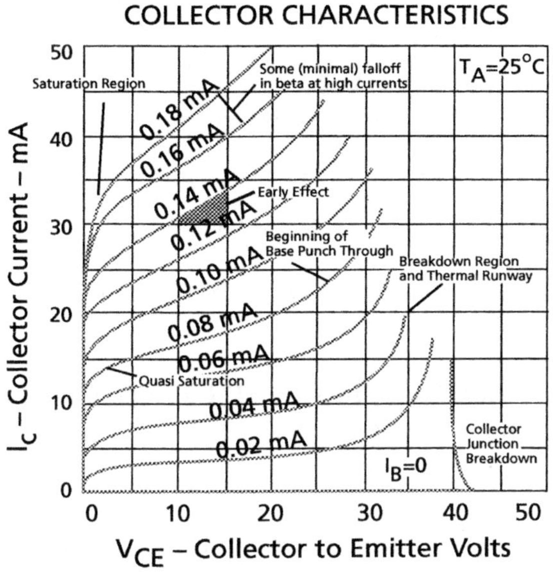

Figure 7-4. Collector characteristics for the 2N3904
Source–Fairchild Semiconductor Discrete Products Databook, c1973

Gain-bandwidth contour plots: Scaling up the size of a BJT carries certain penalties. Larger devices normally have less high-frequency gain-bandwidth capability. Note that f_t = AC-β which is an important SPICE parameter. Figure 7-5 shows plots of contours of constant gain-bandwidth versus collector-current and collector-voltage.

---

8 See "Chapter 4, Measuring Model Parameters in the Laboratory" and "Appendix J: Device Physics" for discussion of beta behavior and formulae that compute their effects.

9 A discussion of saturation is in "Appendix J, Device Physics."

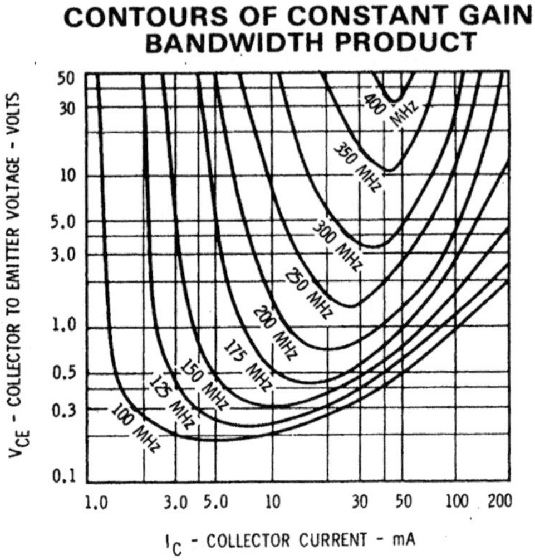

Figure 7-5. Gain-bandwidth (AC β- ft) versus Ic and Vce: 2N3904

7.7 COMPARING FIVE BJTS TO ILLUSTRATE MAKING A SELECTION

Table 7-2 compares five small signal chips that are discrete BJTs. In the same application category, these five examples represent:

- Low, medium and high current (collector currents up to an amp)
- Medium and high voltage application (standoff voltages up to 80 volts)

Table 7-2. Key circuit performance properties

Device	Type	Size (mils)	Bvceo (V)	Ic (max) (mA)	f_t (max) (MHz)	Early effect non-linearity	Pc (max) (W)	Pkg.
2N3904	npn	11x18	40	200	375	Large	0.7	TO92
2N4400	npn	20x20	40	500	350	Large	1.2	TO92
MPSA05	npn	30x30	40	1000	300	Large	1.5	TO92
2N4410	npn	20x20	80	250	250	Moderate	1.2	TO92
2N4402	pnp	22.5x22.5	40	500	300	Large	0.7	TO92

For the 2N4410, a higher starting epi resistivity and thicker die structure is used to achieve a higher breakdown voltage. This results in less Early effect. For the 2N4402, the lower mobility of holes as minority carriers for pnp versus electrons for the 2N4400 npn lowers f_t.

All five devices are "jelly-bean transistors." Jelly-bean is jargon that means a transistor designed to have moderately good performance characteristics in a number of areas without truly optimizing and specializing in one area at the expense of others. The performance areas were gain, high-frequency-gain, switching-speed, standoff-voltage, power-handling, switching-current, high-current-beta-holdup, and noise-figure. All five devices are silicon planar epitaxial process. The five chips are illustrated by:

- A topside chip geometry, including size and some inter-digitated emitter-base topside metal fingers.
- Sets of characteristic curves: Beta versus collector current; collector characteristics, namely Ic versus Vce and Ib, and; contours of constant gain-bandwidth versus Ic and Vce

All the characteristic curves and topside geometries in this section are from the author's [Leventhal] work and appear in the Fairchild Semiconductor Discrete Products Databook, ©1973 [141].

7.7.1 Example 1: Properties of the 2N3904 - 0144 chip

This chip is a npn 11x18 mil jelly bean transistor designed for BVceo = 40 V, Ic max = 200ma. The properties are summarized as follows:

Device	Type	Size (mils)	Bvceo (V)	Ic(max) (mA)	f_t(max) (MHz)	Early effect non-linearity	Pc(max) (W)	Pkg.
2N3904	npn	11x18	40	200	375	Large	0.7	TO92

Figure 7-6 is the topside chip geometry. It shows a modern inter-digitated design meant to maximize emitter periphery to area ratio and minimize base spreading resistance.

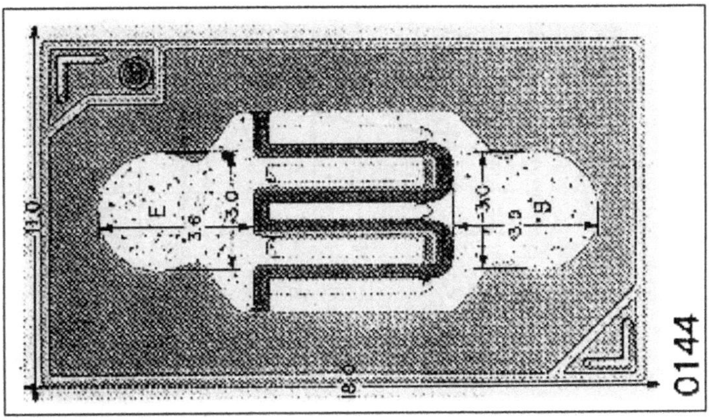

Figure 7-6. Topside geometry of the 2N3904 chip

Figures 7-7 through 7-9: Figure 7-7 shows Beta versus Ic. Figure 7-8 shows collector characteristics. Figure 7-9 shows gain bandwidth. These three figures illustrate important bipolar transistor properties that can be compared from one chip to another.

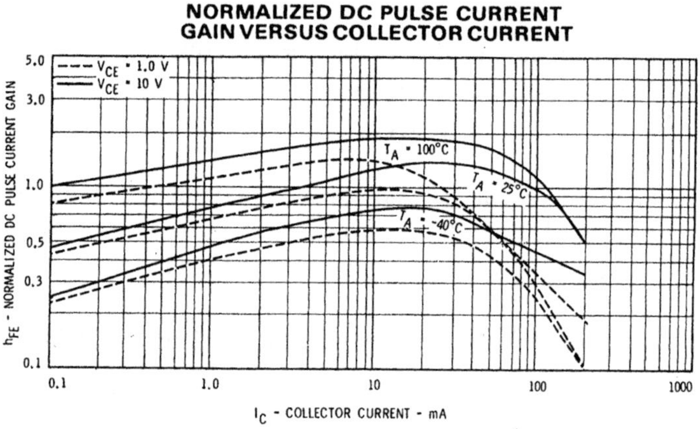

Figure 7-7. Normalized DC beta versus Ic of the 2N3904 low current npn BJT

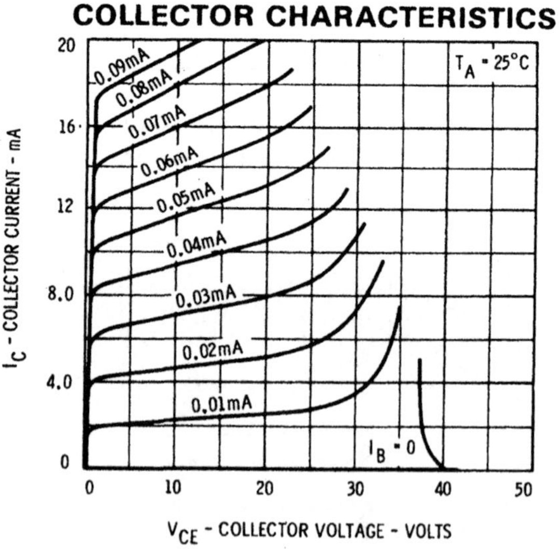

Figure 7-8. Collector characteristics of the 2N3904 low current npn BJT

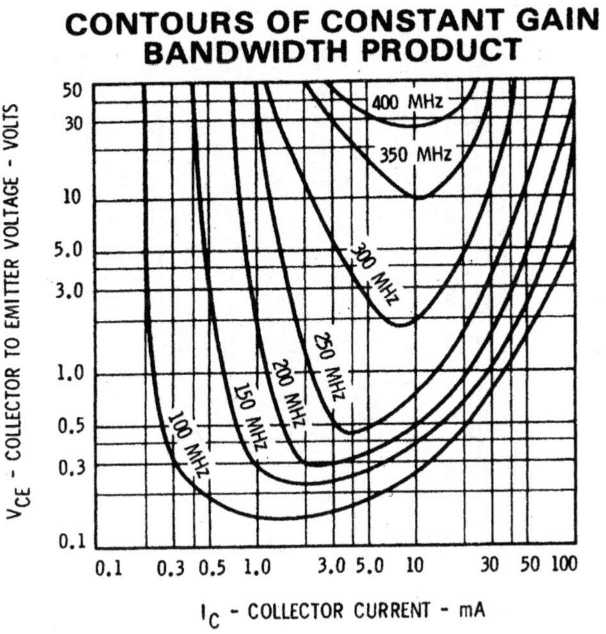

Figure 7-9. Gain-bandwidth contours of the 2N3904 low current npn BJT

7.7.2 Example 2: Properties of the 2N4400/2N2222A - 0145 chip

This chip is a npn 20x20 mil jelly bean transistor designed for BVceo = 40 V, Ic = 500ma. The properties are summarized as follows:

Device	Type	Size (mils)	Bvceo (V)	Ic(max) (mA)	f_t(max) (MHz)	Early effect non-linearity	Pc(max) (W)	Pkg.
2N4400	npn	20x20	40	500	350	Large	1.2	TO92

Figure 7-10 is the topside chip geometry. It shows a modern inter-digitated design meant to maximize emitter periphery to area ratio and minimize base spreading resistance. The base and emitter contact fingers are stubbier to minimize ohmic current drops.

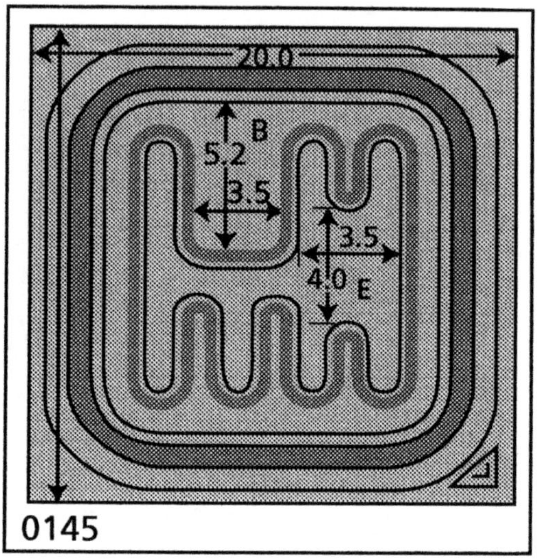

Figure 7-10. Topside geometry of the 2N4400 chip

Figure 7-11 shows Beta versus Ic. Figure 7-12 shows collector characteristics. Figure 7-13 shows gain bandwidth.

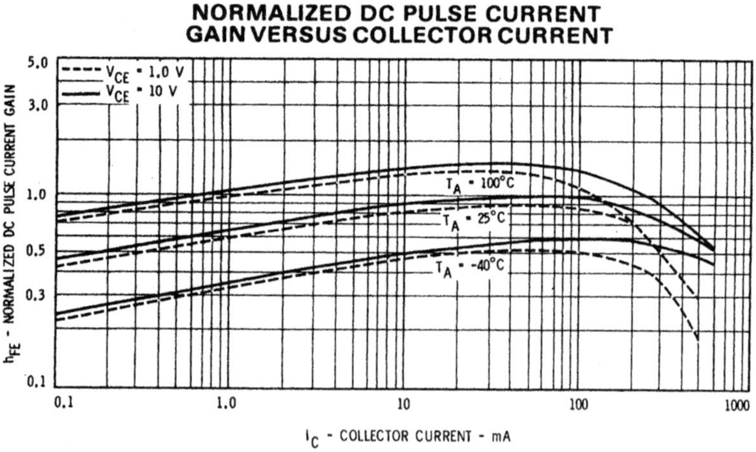

Figure 7-11. Normalized DC beta versus Ic of the 2N4400 medium current npn BJT

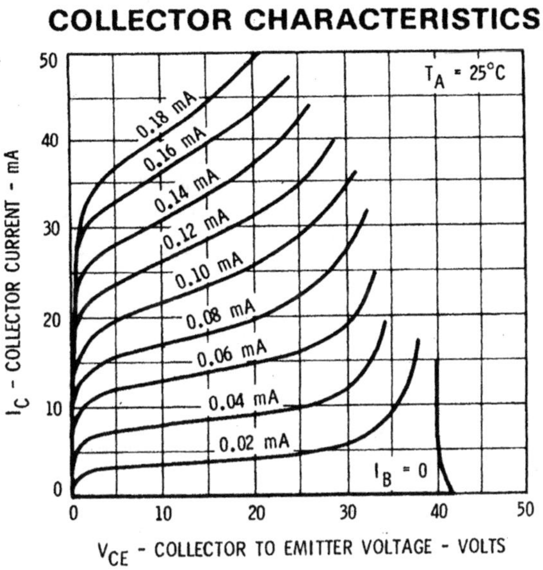

Figure 7-12. Collector characteristics of the 2N4400 medium current npn BJT

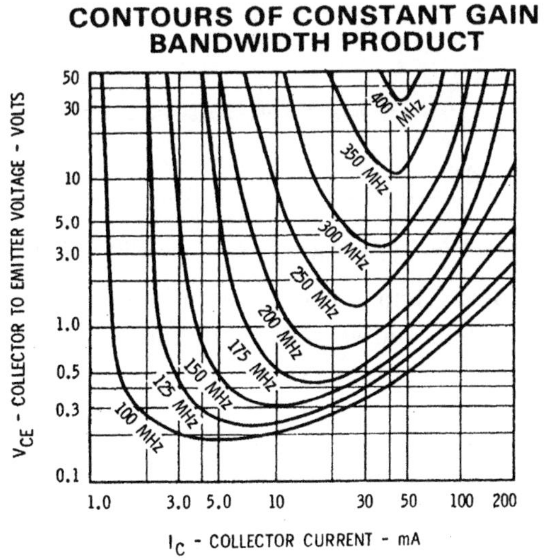

Figure 7-13. Gain-bandwidth contours of the 2N4400 medium current npn BJT

7.7.3 Example 3: Properties of the MPSA05 - 0149 chip

This chip is a npn 30x30 mil jelly bean transistor designed for BVceo = 60 V, Ic = 1000ma. The properties are summarized as follows:

Device	Type	Size (mils)	Bvceo (V)	Ic(max) (mA)	f_t(max) (MHz)	Early effect non-linearity	Pc(max) (W)	Pkg.
MPSA05	npn	30x30	40	1000	300	Large	1.5	TO92

Figure 7-14 is the topside chip geometry.

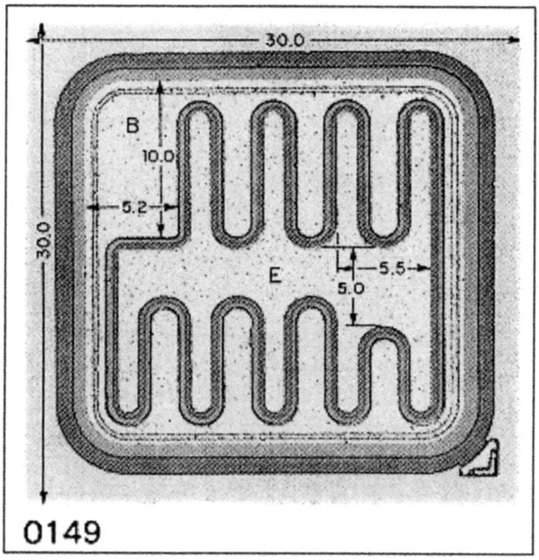

Figure 7-14. Topside geometry of the MPSA05 chip

Figure 7-15 shows Beta versus Ic. Figure 7-16 shows collector characteristics. Figure 7-17 shows gain bandwidth.

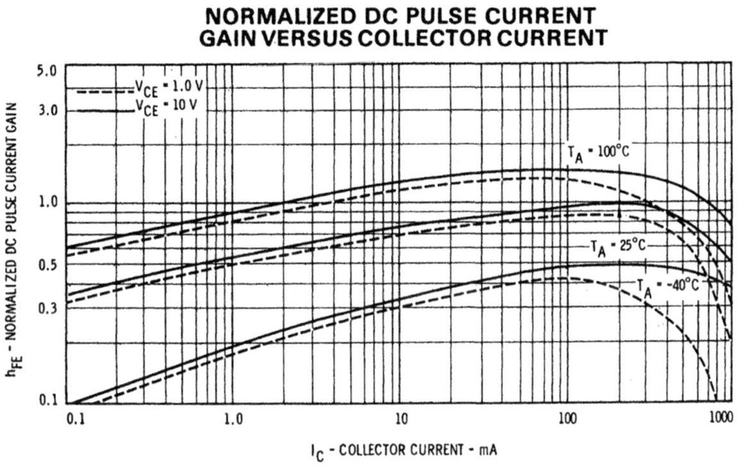

Figure 7-15. Normalized DC beta versus Ic of the MPSA05 higher current npn BJT

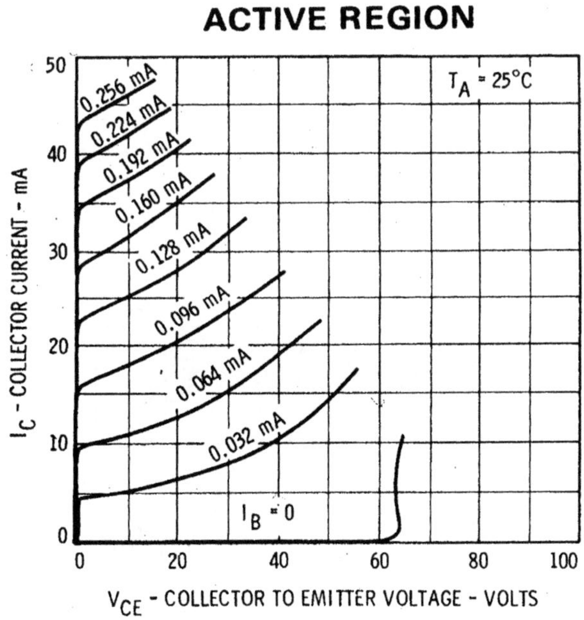

Figure 7-16. Collector characteristics of the MPSA05 higher current npn BJT

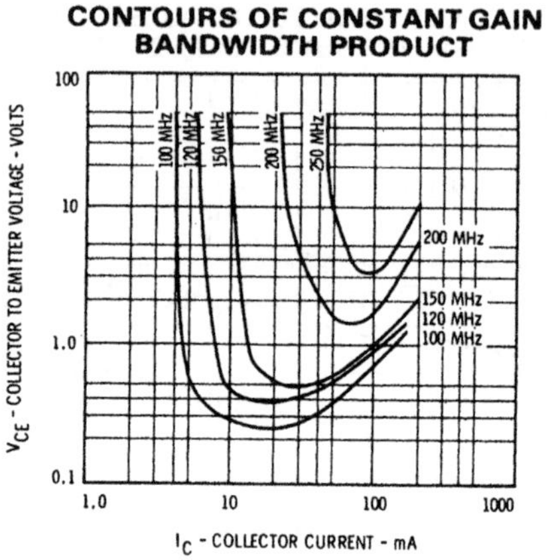

Figure 7-17. Gain-bandwidth contours of the MPSA05 higher current npn BJT

7.7.4 Example 4: Properties of the 2N4410 - 0147 chip

This chip is a npn 20x20 mil jelly bean transistor, but one with a bit higher standoff voltage, designed for BVceo = 80 V, Ic = 250ma.

Device	Type	Size (mils)	Bvceo (V)	Ic(max) (mA)	f_t(max) (MHz)	Early effect non-linearity	Pc(max) (W)	Pkg.
2N4410	npn	20x20	80	250	250	Moderate	1.2	TO92

Figure 7-18 is the topside chip geometry.

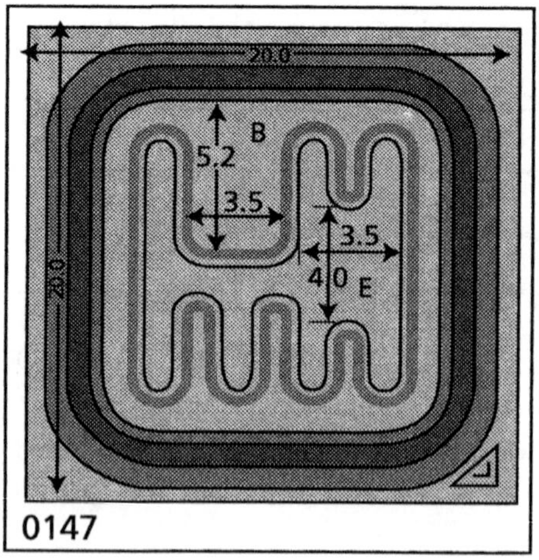

Figure 7-18. Topside geometry of the 2N4410 chip

Figure 7-19 shows Beta versus Ic. Figure 7-20 shows collector characteristics. Figure 7-21 shows gain bandwidth.

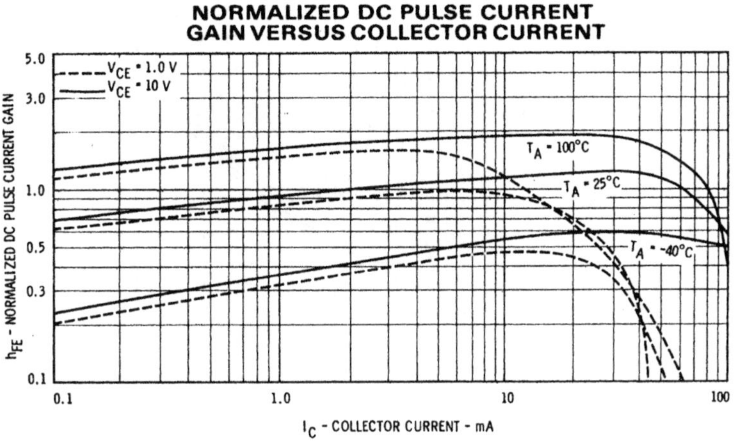

Figure 7-19. Normalized DC beta versus Ic of the 2N4410 medium current – higher breakdown voltage npn BJT

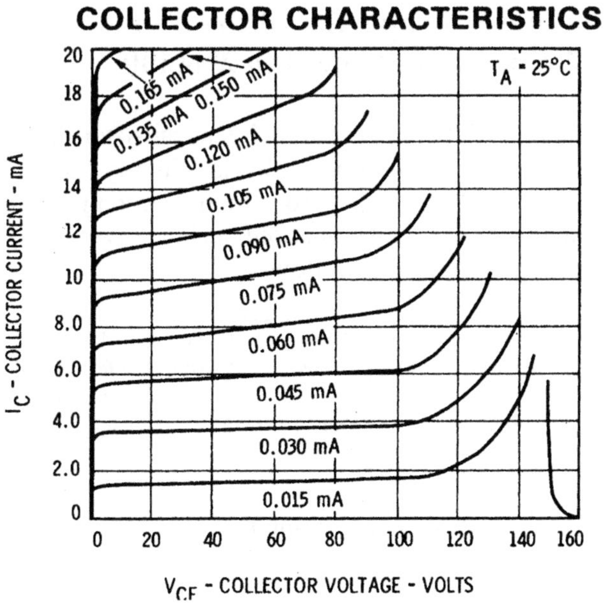

Figure 7-20. Collector characteristics of the 2N4410 medium current – higher breakdown voltage npn BJT

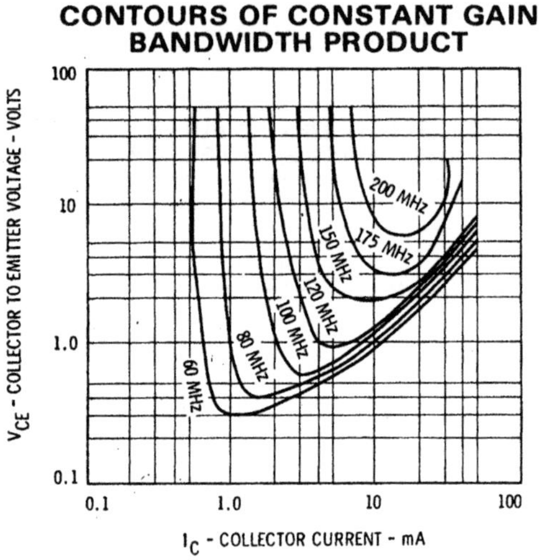

Figure 7-21. Gain-bandwidth contours of the 2N4410 medium current – higher breakdown voltage npn BJT

7.7.5 Example 5: Properties of the 2N4402/2N2905A - 0212 chip: complement to 2N4400

This chip is a pnp 22.5x22.5 mil jelly bean transistor designed for BVceo = 40 V, Ic = 500ma. This pnp chip is slightly larger than its npn complement, the 0145 chip, to help accommodate a surface equipotential/guard ring structure to avoid surface inversion to the die edge. The properties are summarized as follows:

Device	Type	Size (mils)	Bvceo (V)	Ic(max) (mA)	f_t(max) (MHz)	Early effect non-linearity	Pc(max) (W)	Pkg.
2N4402	pnp	22.5x 22.5	40	500	300	Large	0.7	TO92

Figure 7-22 is the topside chip geometry.

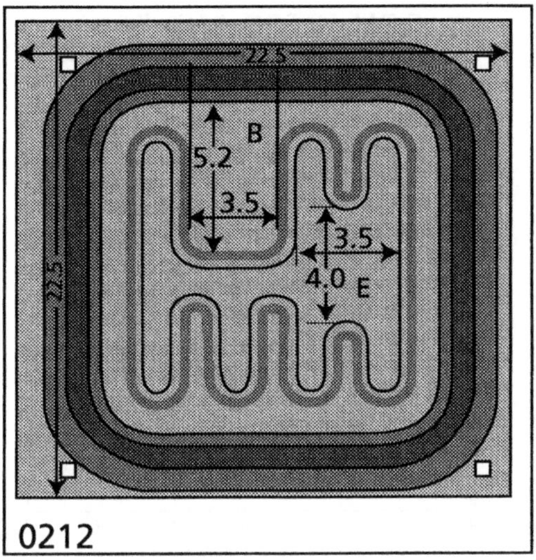

Figure 7-22. Topside geometry of the 2N4402 chip

Figure 7-23 shows Beta versus Ic. Figure 7-24 shows collector characteristics. Figure 7-25 shows gain bandwidth.

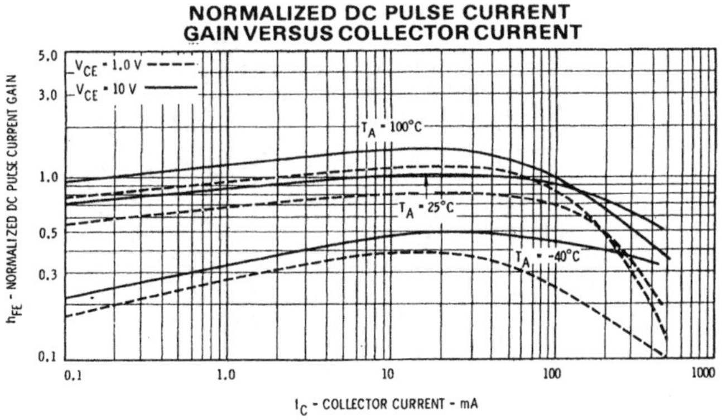

Figure 7-23. Normalized DC beta versus Ic of the 2N4402 medium current pnp BJT compliment to the 2N4400

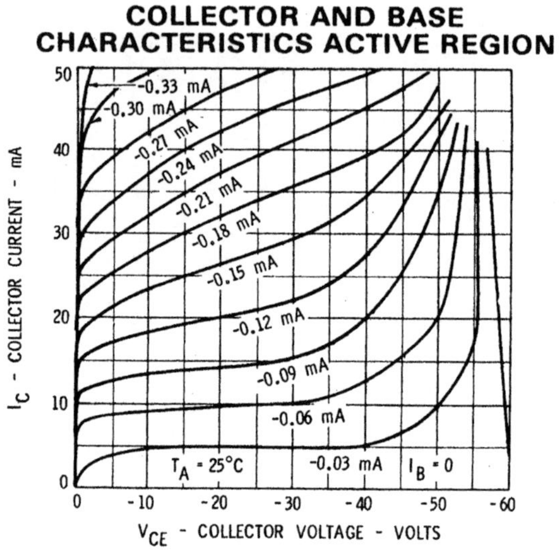

Figure 7-24. Collector characteristics of the 2N4402 medium current pnp BJT compliment to the 2N4400

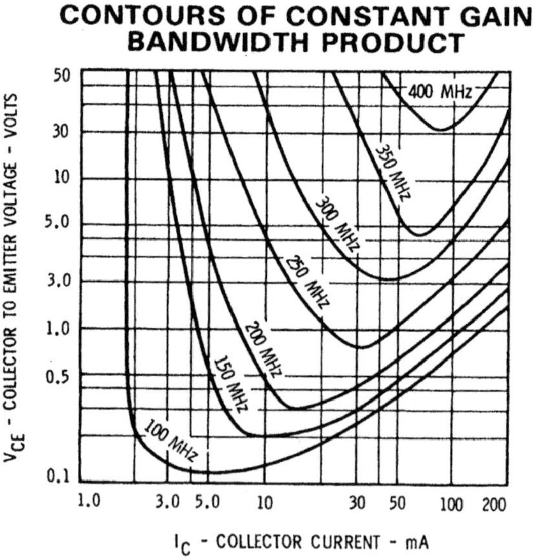

Figure 7-25. Gain-bandwidth contours of the 2N4402 medium current pnp BJT compliment to the 2N4400

7.8 PROCESS FOR MAKING TRADEOFFS

7.8.1 Weighing the Alternatives

To compare the performance of several devices, there are many properties that we could study. In the examples, we have chosen to examine Beta (gain) versus collector current, gain bandwidth, and signal swing capabilities (collector characteristics). We could make tradeoffs related to signal swing capabilities, output drive current, and frequency response. From the five devices shown earlier, we can see that as the devices get larger, they have more signal swing capabilities and output drive current capabilities. But we lose frequency response capabilities. Depending on the needs of the application, we will need to make a choice.

We can refine the analysis by using a load line on the characteristic curves.[10] We know the load impedance, from which we can plot a load line on the collector characteristic curves. We can figure out our quiescent bias point and our min-max current swing. This analysis allows a further refinement of our β versus Ic choice. If frequency response is an issue, we can then look at gain-bandwidth curves.

After we compare the devices, we need to decide which dievice is best for the application. It is likely that we will need to compromise on some properties. We may need to give up performance capabilities in one area to gain capabilities in another. This is called making tradeoffs.

7.8.2 Broadening the Search

What if our selection of candidate devices is not a particularly good one for the application's needs? Perhaps we need to swing only a couple of volts at the load, and we will drive our output as a saturated switch. We notice that the general-purpose BJTs have 40V breakdowns, which we do not need. We also know that switching turn-on and turn-off times need to be kept under control for the application. Switching speeds are not particularly good for general-purpose BJTs. For example, the 2N4401 has a t_{on}, and t_{off} of approximately 300ns. In "Chapter 3, Model Properties Derived from Device Physics Theory," we discussed that some BJTs are specifically designed for saturated switching applications.

We decide to not use our selection of general-purpose BJTs and decide to begin looking at saturated-logic-switch BJTs. This means we have traded-off

---

[10] An example is presented in "Chapter 10, Key Concepts of the IBIS Specification."

one set of performance parameters (that is, structural design) for another set of performance parameters.

Alternatively, perhaps we need a better frequency response than the general-purpose BJTs offered. There are BJTs specifically designed to deliver RF amplifier performance. So we need to think about what tradeoffs to make to get the better f_t performance using devices specifically designed for RF performance.

7.8.3 Relationship of Tradeoffs to Device Design

Not only are β and f_t parameters that are easily related to circuit performance, and SPICE model parameters, β and f_t are also measurement parameters that relate back to the basic physics tradeoffs of the device design. With it, we can see how design choices trade off against each other.

Table 7-3 lists some quantities that can be manipulated in the design of a BJT transistor.

Table 7-3. BJT circuit property tradeoffs versus semiconductor process parameters

To do this...	**Adjust this parameter...**
Maximize γ (emitter injection efficiency)	$Q_E \uparrow$, $Q_B \downarrow$
Maximize α_T (base transfer ratio)	$W_B \downarrow$, $\tau \uparrow$, $a_b \uparrow$
Maximize h_{FE} (DC beta)	$Q_E \uparrow$, $Q_B \downarrow$, $W_B \downarrow$, $\tau \uparrow$, $a_b \uparrow$
Minimize falloff in h_{FE} at low currents	$\tau \uparrow$
Minimize falloff in h_{FE} at high currents	$Q_B \uparrow$, Nac $\uparrow$
Minimize $r_{B'}$ (base spreading resistance)	$Q_B \uparrow$
Maximize BV_{CBO}	Nac $\downarrow$
Maximize BV_{CEO}	$Q_E \downarrow$, $Q_B \uparrow$, $W_B \uparrow$, Nac $\downarrow$, $\tau \downarrow$, $a_b \downarrow$
Maximize radiation resistance	$Q_B \uparrow$, $W_B \downarrow$, $a_b \uparrow$
Maximize f_T (gain bandwidth)	$W_B \downarrow$, $a_b \uparrow$
Minimize Early effect	$Q_B \uparrow$, $a_b \uparrow$
LEGEND Q_E = total emitter dopant Q_B = total base dopant W_B = active base width Nac = dopant concentration in the active collector a_b = base doping gradient τ = minority carrier lifetime in full base $\uparrow$ = signifies increase $\downarrow$ = signifies decrease	

There are many different structure parameters. Despite their complex nature and interactions, a few overriding variables can give insight into the tradeoffs to be made.

Notice that improving one desirable performance property can come at the expense of decreasing another desirable performance property. Performance in a particular circuit application is a compromise and a tradeoff of device design parameters.

7.9 ADDITIONAL CHOICES FOR PICKING A COMPONENT

Component engineer Joe Ziolkowski created the mnemonic ASPOVEPA that summarizes non-technical issues to take into account when choosing a commodity device and model. ASPOVEPA stands for:

A = Available from two or more suppliers?
S = Standardized mature technology and/or industry trend?
P = Performance advantages over existing parts with improved reliability?
O = Ownership cost, recurring and non-recurring?
V = Versatile part, and able to be used in many circuits?
E = Environmentally compatible (lead free)?
P = Production process compatibility?
A = Agency recognition requirements?

This acronym is helpful for remembering that there are other important issues besides electrical performance when selecting a device.

7.10 THOUGHTS ABOUT THE PHYSICAL DESIGN EXAMPLES

Increased die size and region/layer thickness improve device ruggedness. A few of the circuit performance effects that can be seen from the curves of the five example devices presented earlier:

Area Effects:

Die area is increased to provide more current and power handing capability.

- The frequency response ft (gain-bandwidth) goes down. This is largely because of increased die capacitance.
- Current density goes down. Therefore, the low current β-turn-on region extends to a higher total current before the device fully turns on. This is the flip side of extending the transistor's high current β hold-up capabilities.

Doping Density (Conductivity) Effects:

Higher resistivity semiconductor material and lower doping density is used to increase breakdown voltage capability.

- Carrier recombination site density goes down. Therefore, the low current β-turn-on region shrinks to a lower total current when the device fully turns on. The flip side is that emitter emission falls, base conductivity modulation and Kirk effect increase, and the transistor's high current β hold-up capabilities shrink. Early effect [33] can increase.

Increasing Base Width Effects:

- The falloff in beta and gain-bandwidth capabilities is fairly marked as base width increases. This is partially a result of wider, less steeply graded base dopant profile. But also, because a thicker overall vertical structure including a wider base region is used.
- Early effect decreases because the collector depletion region widening that causes it represents a lower percentage of total base width.
- Certain kinds of breakdown (punch-through and corner limited) improve.

And pnp devices possessing lower current carrier mobility show:

- A lower gain-bandwidth.
- A bit quicker high current β collapse.

7.11 SUMMARY

By showing example data sheets, this chapter illustrates making design tradeoffs based on analog, high frequency, and driver impedance behavior. Logic functionality and IC delay times are tradeoff choices one would more commonly expect for logic ICs. But in the case of high-speed devices, their I/O analog behavior is also important. In turn, I/O analog, high frequency, and driver impedance behaviors depend on driver size, structure, and dopant profiles.

Chapter 8

SELECTING THE BEST MODEL FOR A SIMULATION

Comparing and contrasting the advantages and disadvantages of different types of models

Abstract: Several types of models are available for simulating the analog behavior of high-speed designs. Each model type has its strengths and weaknesses, which make it particularly well suited for certain simulation tasks–but not for others. IBIS achieves a good balance between simulation complexity and speed. Therefore, it is the best kind of model to use for the majority of PCB-level simulations. However, recent advances in semiconductor technology require that engineers also know about other types of models. This is especially true when dealing with high-speed digital design and high-frequency issues. When a simulation task requires another type of model, the available model can often be converted to the type needed.

8.1 FROM COMPONENT CHOICE TO MODEL CHOICE

In chapters 6 and 7, we discussed the selection of the *best component* for a design. We now discuss the selection of the best type of *model* to simulate the chosen component's behavior in a circuit. We begin by asking two questions:

- First, is a model–*any* kind of model–available at all?
  - Sometimes a model is available and is exactly what is needed.
  - Sometimes a model is available, but is not the type needed. For example, perhaps we have SPICE, but want IBIS instead.

  - Sometimes no model of any kind is available. In later chapters, we discuss what to do about that.
- Second, what do we want to learn from the simulation?
  - Before choosing a model, it is important to consider the kinds of predictions and answers we want. The more detail, accuracy, and flexibility wanted, the longer it takes the simulation to run.
  - It is also important to consider appropriateness of the simulation results. For example, are we interested in the gain versus frequency of an amplifier circuit, or the noise-margin and timing-margin of a switching network? Also, if we are laying out a PCB on which the component will be assembled, the details of the internal behavior of the IC output driver will often not be of much help.

In chapters 6 and 7, we learned that selecting the best component for the design is not always straightforward. We often need to compromise on desirable features, such as simulation speed versus output drive capability. When selecting the best *model* to simulate the component, we may also need to compromise on simulation capabilities.

8.2 QUESTIONS THAT MODELING AND SIMULATION CAN ANSWER

Typically, when simulating high-speed circuits, we seek answers to the following questions:

a. What signal is transferred from output (generator) to input (load)?
b. What signal is transferred through the load to a succeeding circuit if the load is an amplifier?
c. What signal is transferred to a neighboring circuit that is close enough for electromagnetic coupling (crosstalk) to occur?
d. What signal gets reflected and re-reflected (Signal Integrity) when time delays become significant and mismatching can occur?
e. What signal is induced on power supply lines (Power Integrity) if their source impedance becomes significant?
f. What signal is transferred to distant circuits when wavelengths get short enough for far-field radiation (electromagnetic interference and control-EMI/EMC) to occur? EMI/EMC concerns arise when interference is generated whether the source is conducted, coupled or radiate.
g. What signal is transferred when the behavior of the driver or receiver becomes non-linear? This commonly happens for large signal switching of semiconductors.

To answer those types of questions, we need to consider how the components we choose are connected and operate together. Let's review the definitions of model and system:

- A model is a mathematical representation of a physical thing, and is used to predict its behavior. The mathematical representation can be either a formula, or data to be used in a formula.
- A system is a group of components *operating together*. The common terms for systems of components operating together are *circuits* and *networks (nets)*.

8.3 TYPES OF MODELS

Figure 8-1 is a visual aid presenting the idea that there are many models and many modeling terms. Their uses overlap.

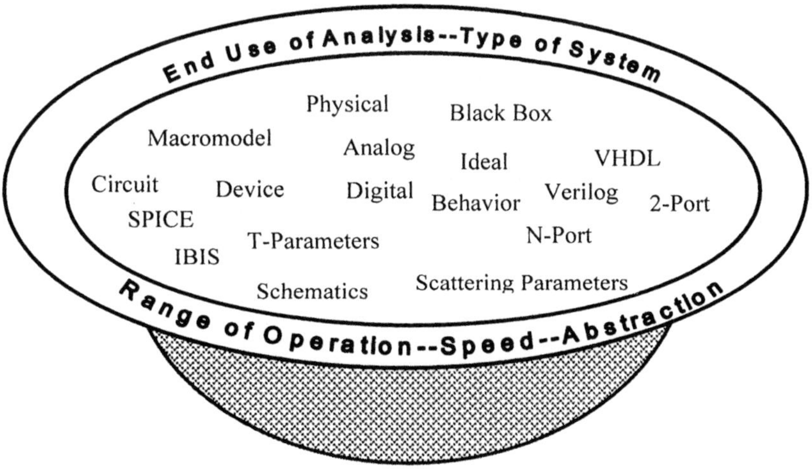

Figure 8-1. The alphabet soup of model types

Models can be classified in several ways, including:

- *Speed* of simulation versus detail of results.
- *Abstraction* versus physical detail level.
- *Analog versus logic operation equation-based models for digital circuits.*
- *Interconnection analysis* in the frequency domain versus the time domain

8.4 USING SYMBOLS AND SCHEMATICS TO REPRESENT MODELS

When people ask to *see* a model, they often mean that they want to see the model's schematic.

SCHEMATIC DEFINITION
A *schematic* is a visual map illustrating the *connection* of models, systems of models, or the internal sub-parts of models into circuits and networks. In a schematic, each model and model element is represented by a symbol. Lines represent wires and show how the elements are connected. Schematics enable one to visualize the models and their interconnections.

Schematics convey one level of information, which is the composition and interconnections of the models, systems of models, and the internal sub parts of models. A second level of information is attached to the schematic symbols. These are the mathematical relations, data and constraints that are used by the simulation engine when running the calculation.

8.4.1 Example of a Simple Schematic

This topic discusses a convention for using schematics to represent models, systems of models, and the internal sub-parts of models. Figure 8-2 shows a very simple schematic consisting of a voltage generator connected across a resistive load.

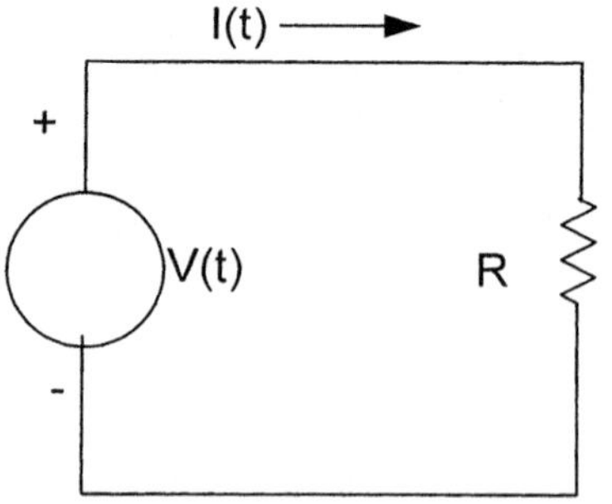

Figure 8-2. A voltage generator (driver or output) and a resistive load (receiver or input)

The generator's symbol is a circle and the resistive load, R, is a zigzag line. A wire is connected from the + output port of the generator to one input port of the resistor. A second wire is connected from the – output port of the generator to the other input port of the resistor forming one continuous circuit loop.

Using a schematic to represent a system of models helps designers solve many issues in the design of high-speed circuits. With the aid of schematics, we can solve for signals around and across the system or circuit. In this simple schematic, we assume the following conditions:

- Time delays of signals through the circuit are negligible
- The generator internal characteristics are negligible
- The load has no associated parasitics or resonances, and neighboring circuits are far away, thus having no influence

The behavior of this simple system is described by: $I(t) = V(t)/R$. This formula is Ohm's Law. It tells us that the current that flows around the circuit is equal to the voltage impressed across the resistor divided by the value of the resistor. Analyzing the behavior of such systems with a computer involves doing the following tasks:

1. Attaching formulas and data to the symbols.
2. Following the conventions for describing the connections to a computer as is done in a SPICE netlist file.
3. Programming the simulation software to follow certain rules for how electrical circuits behave (Kirchhoff's Laws and Ohm's Law).

Specifically, Kirchhoff's Laws state that:

- Voltages around a current loop must sum to zero
- Currents entering and leaving a node (connection of two or more wires) must sum to zero.

Both of Kirchhoff's Laws are a consequence of the conservation of energy.

8.4.2 A More Complex Schematic

Figure 8-3 shows an example of a more complex schematic. Getting any specific information requires zooming in on portions of the schematic. Showing the entire network makes the point that some circuits are very complex. Adding symbols for vias, connector pins, and corners increases the complexity.

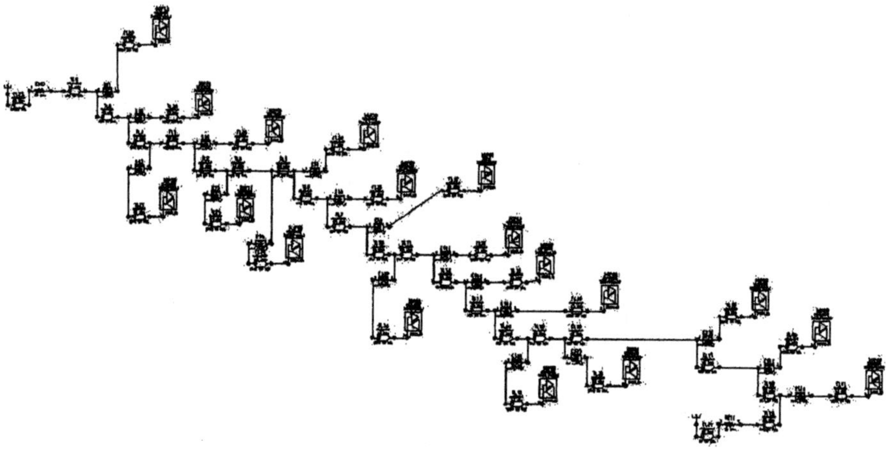

Figure 8-3. An 18-drop daisy-chain backplane network

Additional layers of complexity are associated with high-speed circuit modeling. High-speed circuits are usually characterized by the following conditions:

- Time delays of signals through the circuit are significant.
- Neighboring circuits are close enough to influence each other through coupling and emission phenomena.

Additionally, high-speed circuits may be characterized by the following conditions:

- Generator internal characteristics are significant.
- Associated parasitics or resonances in the load, power supply systems, PWBs, and equipment enclosures are significant.

Simulating for these effects requires additional modeling that takes into account circuit layout, equipment enclosures, and high-frequency effects that change how energy is transmitted down a conductor,[1] power delivery systems, and much more. When modeling the IC, we consider not only the circuit it is part of, but also how the circuit is structured, materials properties of the circuit, neighboring circuits and structures, and non-ideal (zero impedance) power supply systems.

In the case of more complex net connections, more critical timing, and higher switching speeds, it becomes necessary to verify our previous

1 For example, high-frequency skin effect losses.

schematic-driven simulation results with a layout-driven simulation. For added accuracy in modeling the interconnections, the simulator field-solves and models the transmission line sections, vias, pin fields, and connector pins.

8.5 MAJOR TYPES OF MODELS

We now look at the schematics of major types of models to see how they would be represented in a circuit or network. They include:

- *Two-Port Matrix models.*
  Historically, 2-Port matrix models were one of the first model types applied to active semiconductor circuits. They are now mainly confined to passive networks.
- *S-Parameter models.*
  In high-frequency analog RF and microwave circuits, S-Parameter models are applied to linear active semiconductor circuits. In high-speed non-linear digital switching circuits, this model type is confined to passive networks.
- *SPICE models.*
  This model type includes a wide variety of active and passive device models applied to linear and non-linear analog and digital circuits.
- *IBIS models.*
  This model type includes a wide variety of active device models applied to non-linear digital switching circuits.

8.5.1 Two-Port Matrix Models

The black-box N-Port *Matrix Model,* with N from one to a very large number, is among the earliest circuit models. Before semiconductors were invented, 2-Port models were already in use. The schematic in Figure 8-4 shows a 2-Port (input-output) network with input and output voltage and current signals. The box is opaque. That is, we do not know exactly what it contains, and how what it contains is connected. If the internal circuit is known, we can derive its transfer functions and input and output load impedances. We can then model the box with these functions and its external connections, as in *Macromodeling*, and use this model to find the output signals as a function of the input signals.

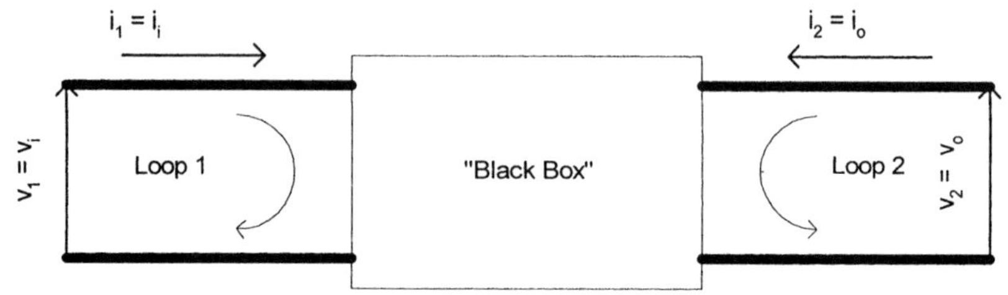

Figure 8-4. Two-Port network

Alternatively, we can measure the output signals as a function of input signals. For example, we can measure currents as a function of voltage as in behavioral modeling, and use this model to find the output signals as a function of the input signals. Further, we can *parameterize* the measured behavior as a set of model elements, one form of which is shown in Figure 8-5, and embed this model in its bias, generator, and load circuitry. The entire circuit can then be solved using Ohm's Law and node and mesh equations. There are only four parameterized elements inside a 2-Port Matrix model. ICs and more complicated networks that use 2-Port models lose some of their correspondence between model elements and device behavior compared to simple single-stage amplifiers and switches.

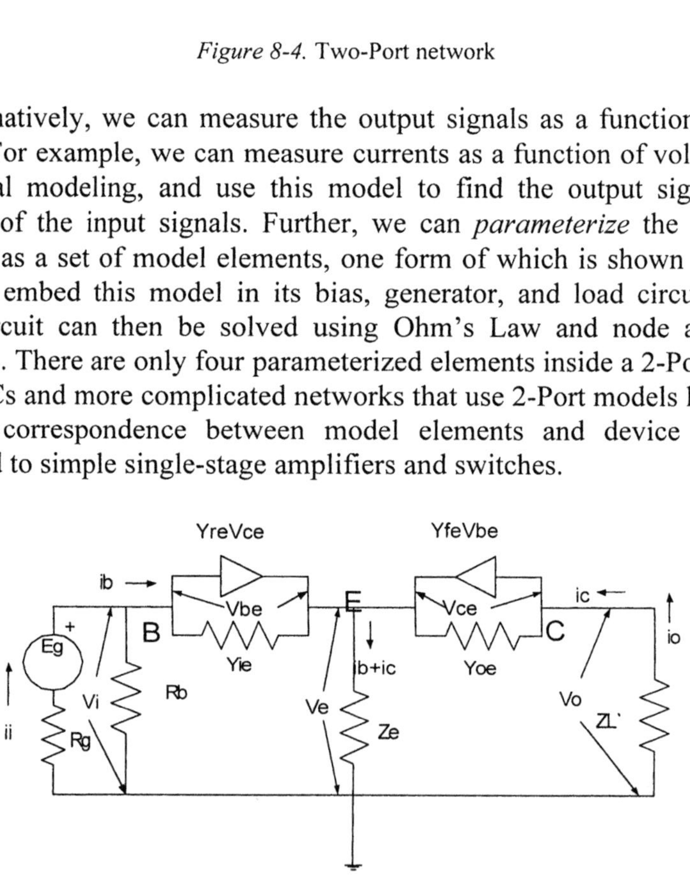

Figure 8-5. Common emitter (CE) amplifier in CE y-parameters with external circuit elements Rg, R_b, Z_e, and Z_L

Matrix models are most often used for linear, small-signal analysis. Table lookups can extend the model for use in non-linear, frequency-dependent, temperature-dependent, and voltage-dependent problems. But limitations are imposed by measurement expense, computer memory expense, and in grasping the key features of the model. Computer memory is getting cheaper fast, but there is a new problem. The new problem is that both 32- and 64-bit calculations are hardware and software dependent.

8.5.2 Scattering-Parameter Models

A modification of the matrix model is the Scattering-Parameter model, shown in Figure 8-6. It was specifically developed to deal with measurement problems imposed by the impracticality of applying ideal measurement shorts and opens. Short circuit and open circuit conditions are applied when measuring transfer and load impedances. At very high frequencies, ideal shorts and opens applied to current and voltage signals become difficult to establish. Instead, transmission line matched generator and load conditions are applied to the model (component). Then transmitted and reflected waves are measured. The main benefit of the Scattering-Parameter model is to extend model parameter measurement validity to very high frequencies. However, the same cautions about changing bias and temperature or driving circuits non-linear apply. As with all matrix models, we have to re-measure the model parameters when we change bias, temperature, or signal swing.

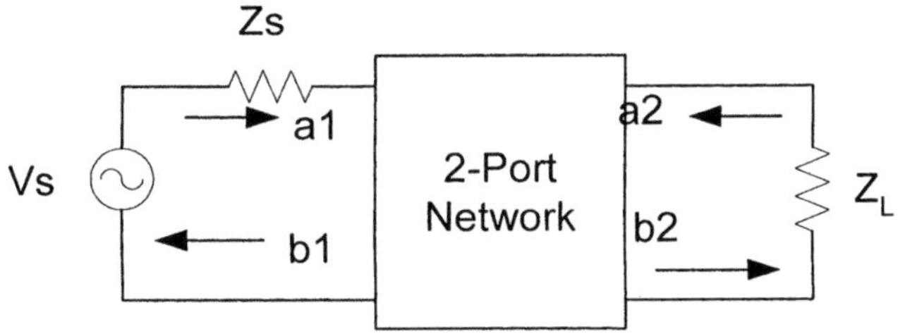

Figure 8-6. A typical 2-Port Scattering-Parameter network

The measurement of the S-Parameters uses the definitions in Table 8-1:

Table 8-1. S-Parameter definitions

Description	Definition	Condition
Input reflection coefficient	$s_{11}=b_1/a_1$	$a_2=0$ (output properly terminated)
Forward transmission coefficient	$s_{21}=b_2/a_1$	$a_2=0$ (output properly terminated)
Reverse transmission coefficient	$s_{12}=b_1/a_2$	$a_1=0$ (input properly terminated)
Output reflection coefficient	$s_{22}=b_2/a_2$	$a_1=0$ (input properly terminated)

where: $Zs = Zo$ sets $a_1 = 0$ and $Z_L = Zo$ sets $a_2 = 0$.

8.5.3 SPICE Models

The difficulty in dealing with changes in bias, temperature, and signal swings in black-box models, plus the need to understand and model the internal, physical workings of transistors, lead to the development of various *Device-Physics Models* as shown in Figure 8-7, for example.

Elements in the model can be modeled with their original device-physics equations. Non-linearities, manufacturing variations, effects of temperature, and voltage can be built into the model. But the common approach is to measure the model parameters under particular conditions and *parameterize* their values in the model. Unless tables of parameter values are used, this approach limits the model's power to handle variation in such parameters.

The circuit diagram shown in Figure 8-7 is simplified compared to real, modern devices. Such BJT models contain more than 40 to 60 model elements. Even with the simplified circuit shown, it would be tedious to solve for its node voltages and mesh currents. The SPICE computer program was developed to deal with the difficulty of solving the circuit equations.

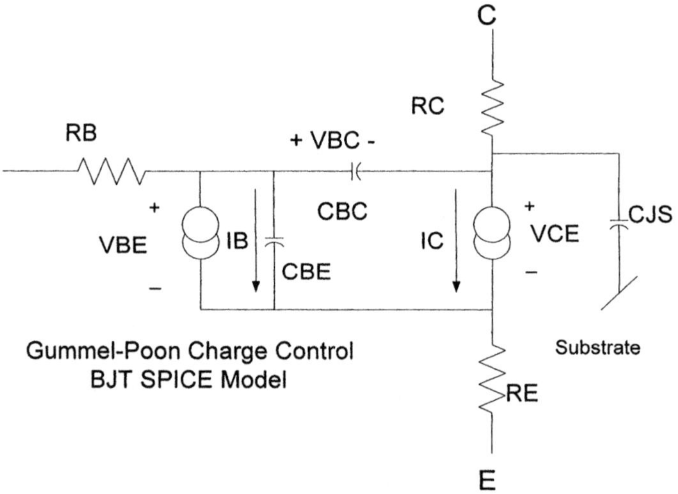

Figure 8-7. BJT equivalent circuit: Source [84].

The SPICE *program* has become synonymous with the SPICE *model*. SPICE is inherently a very good model. But a competitor to a semiconductor supplier can reverse-engineer it to provide information about the inner design of the device.

We can have the following issues with SPICE models:

- The semiconductor supplier may not want to provide this proprietary model.
- The model will run very slowly because the program is solving for all the internal device behavior.
- The internal device behavior information cannot be used and is not wanted at the board level, where we are primarily concerned about signals on the interconnections.
- The non-linear switching behavior is our primary interest and SPICE parameters are not normally measured and calculated under those conditions.

8.5.4 IBIS Models

To resolve SPICE issues, an industry committee developed and adopted IBIS, a new model based on well-established modeling principles. IBIS is an acronym for Input-output Buffer-Information Specification [142, 143]. Inputs (Enable and Signal) are digital, while the Output pad is analog. IBIS is an example of a mixed-signal conceptual model that is usually implemented as a purely analog model.

Measuring the behavior of a black box and creating a model for it is a well established modeling procedure. So is modeling a generator (output) driving a load (input). This is particularly useful when our primary interest is with the signals on the interconnections between the driver and receiver.

The IBIS approach uses the following strategies:

- Data measurements are not "parameterized" as internal model elements.
- Data measurements are swept over the linear and non-linear ranges of switching behavior.
- I-V and V-T data are stored in table format in the model file
- Pin connection information directs a software parser to the correct I-V and V-T tables for a particular I/O.

With this data, we have almost all the information that we need for a basic IBIS model. Adding internal parasitics, switching thresholds, and other information extends the abilities of the model.

This is a simplified explanation of the main structure of the IBIS model. Additional details and refinements extend the IBIS data exchange format capabilities to more complex parts. The IBIS Spec provides a standard for formatting and exchanging model data used in HSDD simulations.[2]

---

[2] Additional IBIS references are [55, 56, 57].

What does the schematic of an IBIS model look like? The Output, or Generator, side is shown in Figure 8-8, and is followed by the Input, or load, side in Figure 8-9.

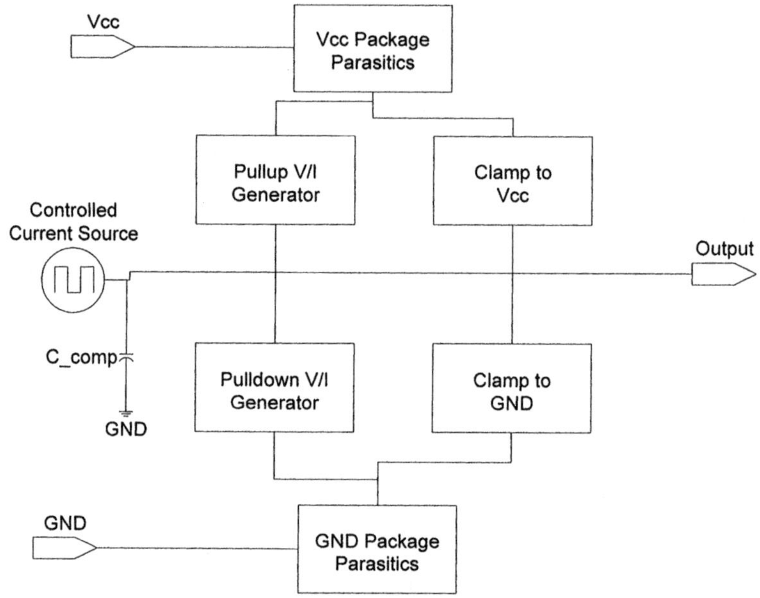

Figure 8-8. IBIS model output (driver) side

The IBIS model has the same limitations as all behavioral models. When bias, temperature, and other conditions change, parameter data for behavioral models have to be re-extracted.

The IBIS Specification now incorporates extensions that allow it to interface with SPICE, VHDL, and S-Parameter hardware description (modeling) languages. A companion spec [56] covers complex connectors and IC packages.

Various proposals [58] are being advanced by both IBIS and the European community to include modeling of the power-supply interconnections for Power Integrity issues. These proposals basically model the in-package and on-die parasitics with an internal I/t curve representing simultaneous switching behavior.

These modeling extensions do not effectively address the issue of EMI/EMC from the IC package. This issue has more recently received some attention in the literature but is still very much in its infancy versus need [13, 119]. Some suggestions have been put forward regarding the issues of noise emissions on power and ground and the conducted emission of core switching noise. These suggestions are discussed in "Chapter 22, Future Trends in Modeling."

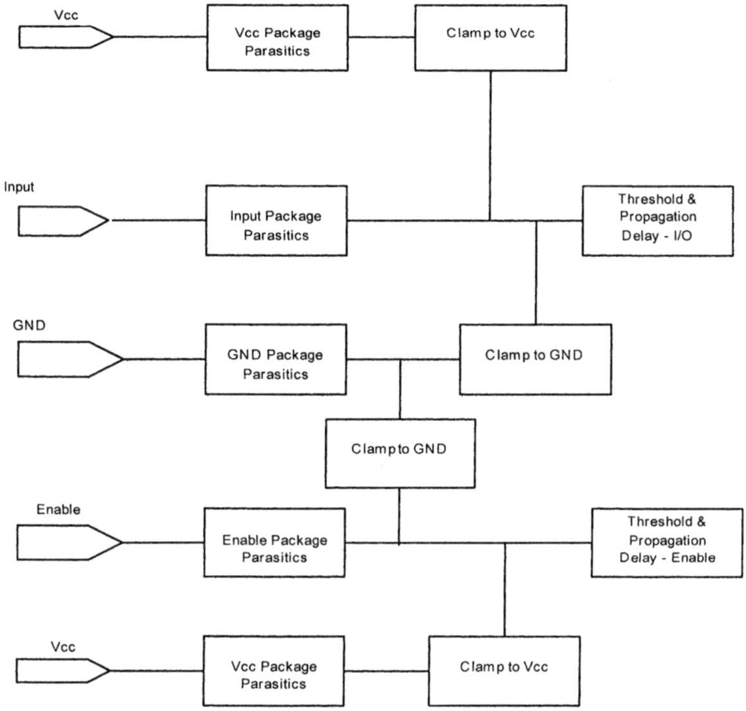

Figure 8-9. IBIS model input (receiver) side

8.6 COMPARE MODELS BY SIMULATION PERFORMANCE

We again list some criteria that can be used to compare and contrast the simulation performance of models, including:

- *Speed* of simulation versus detail of results.
- *Abstraction* versus physical detail level.
- *Analog versus logic operation equation-based models for digital circuits.*
- *Interconnection analysis* in the frequency domain versus the time domain

We are now ready to discuss these topics in greater detail.

8.6.1 Speed of Simulation Versus Detail Level of Results

Table 8-2 compares models by speed. A model can be inserted in a larger system of models for which the overall performance can be calculated.

Table 8-2. Speed of simulation versus detail level in results

Type of Model	Purpose	Simulation Speed	Results
Behavioral models	Usually used when scanning a large design for the first time. At that point, speed is desired over detail.	Fastest ↑	Least detailed ↓
Macromodels	Used early in top-down design approaches when behavioral models are getting decomposed into sub-function blocks		
Ideal-generic models	Used early in the detailed product development when part-specific models have not yet been obtained and only estimates are needed.		
Detailed physical models	Usually used in the verification of a (virtual) physical board design. Accurate and detailed models are particularly important for critical circuits	Slowest	Most detailed

Table Notes:

Behavioral models are made from the measured responses at the input and output ports of a black box containing a circuit. Some formulas and characteristics (actual or implied) of input and output impedance and port-port transfers are derived.

Macromodels are models of a circuit (system) composed of individual parts that have been previously modeled as a whole, unique part or subsystem. Some modeling simplifications and assumptions are usually made at this stage. Equations (actual or implied)[3] can be derived for the following characteristics:

- Input impedance and output impedance
- Port-to-port transfer

The equations are presented as the model for the complete system or subsystem. The input-output impedance and voltage and current transfer ratios of a resistor divider network is a simple case in point.

---

3 SPICE macromodels are actually solved as structured netlists instead of actual transfer functions. This is what is meant by implied equations.

Equation-Based Models are input-output equations (transfer functions) derived directly from a block of circuitry. These equations can include internal capacitance and inductance, voltage and current gain, and other parameters.

AMS Models are analog mixed-signal models. AMS models allow both analog and digital inputs and outputs. The analog section of a model uses equations, while the digital section uses logic (bit/byte) operations. Analog threshold crossings affect digital operations, and digital transitions affect analog operations.

Ideal models are models that are simplified to save simulation time and ignore 2nd, 3rd, and higher order effects. Ideal models also describe ideal components, such as capacitors with no internal inductance that are unrealizable in real life. In spite of this, they can be very useful in analysis. Examples include an ideal current source and an ideal operational amplifier.

Ideal models can be useful as building blocks inside macromodels or in the early stages of a design when ideas are being checked for feasibility. It is interesting to note that models can be at once behavioral, macromodel, ideal, and detailed, as in the case of an OpAmp. It depends on the level of detail and abstraction in the model versus the level of detail and abstraction needed in the analysis task.

Detailed physical models are sometimes not as detailed as expected. Modern, competitive, high-speed digital parts usually have a SPICE model generated by the semiconductor process engineer in charge of their manufacture. That engineer is best situated to know the latest and most detailed component data. However, the engineer in charge typically has no need to know some of the SPICE model parameters used only for circuit simulation and may not include that information in the published model.

In most of the BJT examples in this book, only 24 out of a possible 40 available SPICE parameters are characterized.[4] The remaining parameters are defaulted. Often, only about six SPICE parameters are characterized in a given BJT model.[5]

8.6.2 Abstraction Versus Physical Detail Level

Table 8-3 compares models by abstraction level.

---

4 Exactly how many and what specific BJT parameters are in the model depend on which SPICE version is used.

5 For more detail, see Table J-3 Small Signal General Purpose Amplifiers-& Switches in "Appendix J, Device Physics."

Table 8-3. Abstraction versus physical detail level

Type of Model	Description	Correspondence to Structure of Device Modeled Or Circuit Schematic	Abstraction Level
AMS Model	Linear or Non-Linear equation-based mixed signal model	Structure usually completely obscured, but circuit synthesis can restore it.[6]	From very detailed to very simplified
Behavioral Model	Measured or simulated characteristic curves, data tables, and combinations of these	Not physically modeled from structure. Observed at the terminals of a device.	Completely abstracted, but can be very exact
Macromodel	Functional Model: circuit simplification, circuit build-up, parameterized model methods, and combinations of these	Building-block circuit primitives incorporate simplified functional elements. Equation-based, dependent, controlled sources are used to represent effects that are not modeled by actual components.	Simplified and abstracted
Transistor level circuit model	Primitive Model: devices represented by detailed physical model equations. Schematics aid visualization.	Close to one-to-one correspondence with physical devices and structures	Not abstracted

Table 8-3 Notes:

In *macromodeling,* the necessary equations are extracted from the building blocks available and their topology, or interconnection. In macromodeling, use is made of controlled sources to model non-ideal effects. For example, input leakage current in an op amp. Also in macromodeling, use is made of additional components. For example, in the transistor macromodel to capture higher order effects.

In *behavioral modeling,* after extracting the input-output data (behavior), we go directly to extracting the model properties. We extract these properties without necessarily knowing the internal structure of the black box. Behavioral modeling can apply to low-level primitives or a top-level abstracted, complex hierarchy of interconnections and components.

Transistor modeling is also based on measurements or process simulations. This type of model applies to the lowest level primitives, such as transistors and diodes. Generally, the model is a set of parameters for an

---

[6] Synthesis from equations is difficult. Filter synthesis is one practical application of the procedure.

existing set of equations, such as the SPICE bipolar transistor model discussed previously.

8.6.3 What Does Physical Versus Behavioral Mean?

The concept of modeling abstraction involves going from a detailed to a more general description. This can occur in a continuous series of steps similar to a purification process. At any stage of analysis whether a model is behavioral or physical depends on the user's philosophical point of view.

It is important to know the level of abstraction that is appropriate for both a simulation task, and for modeling the underlying detail or causes. Seeking to understand the design at the deepest level possible builds comfort with the results and subsequent decisions. But the more detail desired, the longer the simulation takes. And what will be done with that detail? Modeling what is going on inside an IC may be comforting to engineers. But the average product design engineer will never make any use of that data. Therefore, the level of modeling and simulation detail chosen should be appropriate to the task.

Models predict behavior, or in other words, performance. All models that describe an object bigger than sub-atomic particles are behavioral in some sense. At the same time, if the variables in the model are perceived as physical variables, the model is also perceived as physical. At almost any level of abstraction, a model can be considered to be both physical and behavioral.

Figure 8-10 shows, in a very generalized way, the process of model abstraction from very detailed to very simplified. The process of circuit model abstraction is presented in more detail in "Chapter 20, The Challenge to IBIS."

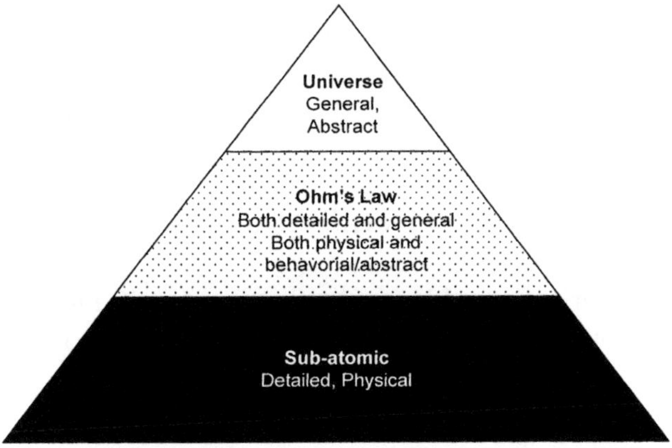

Figure 8-10. Abstraction levels (detailed to general)

Consider the behavioral equation, Ohm's Law. Ohm's Law was derived empirically. In it, the term resistance was originally a constant (fudge factor) that made the equation meaningful. Consider $v = ir$, where:

v = voltage drop across some device
i = current through the device
r = a measured constant of proportionality, resistance

When Ohm's Law was being formulated, resistance was the term chosen to represent the concept of a resistance to current flow when a forcing voltage was applied across a device. v = ir is the behavioral model of the resistor. This behavioral model postulates a physical property that is known as resistance. At the higher-abstraction level, the resistor is a physical thing. If variables (properties) in a formula are physical quantities, the model is said to be a *physical* model.

Resistance is a useful property because, after measuring resistance at one voltage and current pair, we can substitute it in the equation and predict v or i at some other conditions.

As knowledge and experience deepens, we begin to make educated guesses (predictions) about the way that the physical world operates. We then began to talk about first principles in science and engineering. But at root, all our first principles are attempts to explain observation.

Consider now what happens as we seek to deepen our understanding of what resistance is. We develop the equation:

r = (length)/(area)*(conductivity)

Conductivity is thought of as a material property. Then we discover that it may depend on dopant concentration, carrier mobility, temperature, surface roughness, and modulation of channel width by applied voltage.

Now, when we extract the result that (length)/(area)*(conductivity) = r, we say that r represents the *behavior* of underlying physical phenomena. Therefore, the concept of *resistance* is being used in both a physical and behavior sense. At almost any stage of top-down decomposition or bottom-up combination, whether a model is behavioral or physical depends on the user's philosophical point of view.

8.6.4 Level of Physical Detail Modeled

Table 8-4 compares both semiconductor circuit models and device models to the amount of physical construction detail in the models.

Table 8-4. Level of physical detail modeled

Circuit or Device Model	**Description**	**Correspondence to Structure of Device**	**Level of physical abstraction**
Ebers-Moll [34] Gummel-Poon [48][7]	Device Physics Based	One-to-one correspondence. Bipolar Junction Transistor (BJT). Circuit level application.	Lower ↓
Shichman-Hodges [109], Grove-Frohman [42]	Device Physics Based	One-to-one correspondence. MOSFET Transistor (MOSFET). Circuit level application.	
Hybrid-π [83]	Circuit visualization of Gummel-Poon and Ebers-Moll	One-to-one correspondence. Bipolar Junction Transistor (BJT). Circuit level application.	
SPICE, [44, 47] and SPICE derivatives – For example, SABER [104, 105]	Computer implementation of Gummel-Poon, Ebers-Moll, etc.	One-to-one correspondence. Devices include Diodes, BJTs, JFETs, MOSFETs, OpAmps, R-L-C, voltage sources, etc. Modern versions may include many other types of devices. Circuit level application.	
Example: Boyle [19] Operational Amplifier	Macromodel (Also black-box)	Abstracted to and visualized at circuit level: OP1 and OP2 and variants. Circuit level application.	
T-Parameter [36]	Behavioral Model (Also black-box)	3-Terminal circuit visualization of elements. Behavior measured at terminals. Circuit level application	▼
2-Port Parameters: H, Y, Z, G, ABCD [65, 126]	Behavioral Model (Also black-box)	Input-Output Port model measured at terminals with open and short-circuited loads and generators. Circuit visualization of elements. Behavior measured at terminals. Circuit level application	Higher
N-Port	Generalization of 2-Port to N-ports in H, Y, Z, G, ABCD or S Parameters (Also black-box)	Particularly useful for microwave circulators, and cross-coupled package models for IBIS, connectors. Circuit level application.	Higher (Abstract)

7 See also [33, 53, 64, 66, 71, 110, 111].

Table Notes:

Black-box models are usually considered to be behavioral models. In one case, the model data is visualized as lumped elements inside the black box. In that case, it looks more like a macromodel. In the behavioral case, the model data is presented as behavioral tables measured at the ports of the model (IBIS). As presented, both types are totally empirical in nature. That is, they come from taking data measurements at the input and output of the device being characterized.

Both black-box and behavioral model viewpoints normally mask the details of construction and the materials inside the device being measured. Consequently, both models run faster in simulations. But both models lose visibility into internal effects, such as simultaneously changing temperature and voltage. Behaviors at different bias points, frequencies, temperatures, and voltages can be represented by curves, sets of curves, data tables, and so forth. But there is a limit to this approach, which still does not match the flexibility of a model like SPICE that is visualized at the physical level.

Black-box models that have been visualized and parameterized as a set of internal elements can be incorporated in a circuit and analyzed with Ohm's Law and Kirchhoff's Law, leading to closed form equations for entire amplifiers, oscillators, and other types of circuits.[8] There are two common ways of doing this visualization: matrix models and SPICE models. Four-element matrix models are simpler than SPICE and more easily analyzed in closed-form equations.

Parameterized black-box model elements, including S-Parameters, are more commonly measured small signal and linear. They are great for amplifier design in the frequency domain, but not much use in the time domain for non-linear switching behavior. For non-linear large-signal switching behavior in the time domain, it is often more convenient to leave the measurement data in the form of lookup tables like IBIS.

8.6.5 Models Oriented to Interconnection Analysis

Table 8-5 compares black-box models. One example is signal reflection variation with load matching.

---

[8] For more detail on some very useful black-box models, see: "Chapter 4, Measuring Model Properties in the Laboratory," and "Chapter 19, Deriving an Equation-Based Model from a Macromodel."

Table 8-5. Models oriented to interconnection analysis

Black-Box Models	Description	Correspondence to Structure of Device	Mathematical Analysis Domain
Scattering Parameter (S-Parameter)	Behavioral Model	Same as 2-Port and N-Port, but with model measured at terminals with matched loads and generators. No circuit visualization of elements. Instead, reflection and transmission coefficients. Circuit level application.	Frequency
IBIS	Data Exchange Protocol	A combination of elements: Behavioral characteristic curve elements, circuit elements, etc. Device behavior is measured (or simulated from SPICE) at the input and output terminals into known loads. Behavior from input to output of a single device is not modeled.	Time

Table Notes:

S-Parameters were developed for use in RF-Microwave (high-speed analog) applications. Analysis is conducted in the *Frequency Domain.* S-Parameters can be applied to Time Domain problems if transform/inverse-transform methods[9] are applied during simulation. Application to IBIS models requires transforming the I/O output driver signal into the frequency domain for simulation.

IBIS models were developed for use in HSDD applications. Analysis is conducted in the *Time Domain* for large-signal non-linear circuits. IBIS models can be applied to Frequency Domain problems if transform/inverse-transform methods are applied during simulation.

8.6.6 Equation-Based Models for Analog and Digital Circuits

Consider whether a model is analog or digital, particularly at the system level. The next higher level of abstraction takes us to the system, System-On-Chip (SOC), hybrid module and multi-chip module levels. Here the correspondence with the physical structure of a single device has pretty much disappeared. The individual structure and elements of the circuit or

---

[9] Fourier/Inverse-Fourier and Laplace/Inverse-Laplace Transforms are particularly useful tools in this process.

network, including interconnections, are replaced. The entire network is then treated as a single device in an even larger network or system.

At the system level, it often makes sense to abstract much of the analog nature of the circuits to a digital level. For example, a multiplexer or an encoder/decoder might be abstracted from hundreds of transistors to a single digital block, perhaps with a state-dependent delay. At the same time, critical analog functions, such as a low noise input amplifier or a high-power output amplifier might need to still be modeled at the transistor level. An important part of modeling at this level is to make sure all interfaces are correctly modeled: the digital sections load and drive analog sections, and the analog signal changes cause digital data to change state with the expected delay.

Table 8-6 compares system-level analog and digital models.

Table 8-6. Equation-based models for analog and digital circuits

System-Level Models	**Description**	**Correspondence to Structure of a Subsystem**
Analog Examples: SPICE, Verilog-A	Linear and non-linear circuit operation. AMS languages support C-like math functions; Laplace transforms, previous values and slopes can also be used in calculations. SPICE is limited to controlled-source functions.	Represents an analog section as a black box. Equations for sections of the system or network are derived using standard circuit theory. There may be several equations, each representing a functional block, within the analog section
Digital Examples: VHDL, Verilog	Digital circuit operation. Node values and timing delays can have continuous "analog" values. Events are scheduled based on timing delay calculations.	Represents a digital section as a black box. Time delays are derived using standard circuit theory. There may be several equations, each representing a functional block, within the digital section
Mixed-Signal Examples: VHDL-AMS, Verilog-AMS, SABER-MAST	Analog circuit responds to state changes in digital section. Digital section responds to threshold crossings in analog section.	Represents the combination of digital and analog sections within a black block. Analog events can cause digital events to be scheduled, while digital events can cause analog values to change. When a digital event occurs, the analog and digital halves of the model are synchronized in the simulator.

8.6.7 Introducing the Characterization Model

When selecting a part to perform in an application, look at its characteristics before modeling and simulating its performance. There is a connection between modeling and simulation and the performance depicted on a data sheet. The idea of the "Characterization Model" of a device is discussed in "Chapter 7, Using Data Sheets to Compare and Contrast Components."

The *characterization model is another term for a device's data sheet.* The data sheet provides information (as applicable) on a device's logic functionality, power-handling capability, curves of beta versus collector current, gain bandwidth, and more.

We present the characterization model as a new concept. To characterize something means to depict or describe it. The characterization model helps in making good selections of devices to perform in particular designs. A typical use would examine how a device's gain would hold up under maximum current demand.

To make a choice of the best device for an application, it is common to compare data sheets of different devices. Once the choice is made, it is usually followed by a request for the model (SPICE, IBIS, or S-Parameter). The activity of simulating and analyzing the device in the circuit then predicts the performance of everything operating together as a system.

In the low-speed analog and digital circuits of the 1970s, the device-system performance could usually be simulated from circuit topology (schematic) alone. RF was an exception. Today, so many circuits operate so fast, that physical structure and layout on the PCB (and its power planes and enclosure) has to be routinely included in the modeling and simulation.

8.7 ADDITIONAL MODEL COMPARISONS

8.7.1 Behavioral Model Compared to Macromodel

For analog applications, a distinction is usually made between a behavioral model and a macromodel. A macromodel is composed of components and elements. Sometimes these components are blocks of circuitry or function blocks. A macromodel is made by combining models of the individual components or elements. The components in turn can be other macromodels, behavioral models, or a combination of them. We can then say that we can model and simulate the internal behavior of macromodels. In contrast the behavioral model is concerned with the behavior presented to the outside world at the input-output ports of a component.

8.7.2 Digital Compared to Analog in Behavioral Modeling

Digital components have an abstracted model that describes their logic functions. But that does not describe their electrical analog I/O performance. The IBIS behavioral model is used to model the analog behavior of digital parts. However, this too simplistic distinction does not always describe the interactions between digital and analog. First, consider that the analog performance of a digital circuit affects its digital logic performance. Second, logic functions are now used to control analog switching behavior.

A case in point is the bus-hold Intel function. This innovation required an addition to the IBIS Spec to model the new functionality. Another example is programmable drivers, where the presence or absence of a plug-in-board will change the driver, thus the source termination, to a network. It became clear that hardware description and analog macromodeling languages were needed in the IBIS Spec. The IBIS Spec now includes support for various Analog Hardware Description Languages (AHDLs). SPICE itself is one kind of AHDL-like language that is supported in IBIS. However, SPICE is not a true HDL since users cannot edit it.

8.7.3 Behavioral Compared to Physical Modeling Accuracy

As the simulation conditions and the model data approach a true reflection of the actual conditions, the accuracy of a measured behavioral model can approach or even exceed that of a physical SPICE model. This statement is especially true if too little care was exercised in generating the SPICE model. It is also true when the SPICE model is used outside of its assumptions and limitations.

A behavioral model becomes inaccurate when bias conditions, temperature, voltage, and frequency differ significantly from model-measurement conditions for the application being simulated. Many behavioral models do not account for variables like temperature and voltage that change the behavior.

8.7.4 Modeling and Measurement Accuracy Compared

Measurement is usually taken as the most accurate data. But, be careful in such judgments. There is no question that measurement is used to catch modeling and manufacturing mistakes or verify that everything is as expected. But measurement, especially high-speed Signal Integrity and RF/EMI measurements, can be full of mistakes. Performing them is a science in itself. Measurement should verify that the design came out as intended. If environmental factors (like temperature or EMI, or the statistical variations of components and boards) are not appropriately accounted for, it's trusting in luck that the simulation or the measurements will be correct.

8.7.5 IBIS and Other Models in Accounting for Variation

IBIS provides a data exchange format that accounts for some the statistical spread between members of a population of parts with typical, minimum and maximum data. Plus, it provides guidelines for generating the minimum and maximum data. But these guidelines are not requirements. Only typical data is a requirement. However, population spread data in SPICE, S-Parameters, or any other modeling method are strictly an ad hoc addition. There are no standards to set expectations or requirements on such data.

Modeling accuracy and measurement verification is affected by the statistical probability of:

- Outright mistakes in the modeling and/or manufacture of the component or the design.
- Random variation present in all manufacturing processes.

Random variation present in all manufacturing processes also affects whether a given prototype is representative of what performance will be seen in test measurements on any given day or manufacturing over the long run. Unless considerable care and expense is expended we have no way of knowing from what part of a population distribution components came from that went into building the given prototype. Unless we purposely build prototypes from carefully selected components we can only say that what goes into their construction is determined by random chance. Statisticians tell us that a random sample size of less than 25 of anything gives results that match the population distribution only by pure chance.

8.8 RECOMMENDATIONS FOR MODELING

8.8.1 Recommendations for Design Tasks

Only IBIS, SPICE, and Scattering-Parameter models will be rated because widespread technical knowledge, acceptance and availability make them the reasonable choices.

In conclusion, the best choice for the majority (but not all) of HSDD tasks is IBIS. But before dismissing SPICE, S-Parameters, and other models, let's broaden our perspective to include more of the high-technology design process, as shown in Table 8-7.

Table 8-7. Recommendations for design tasks on high-speed boards

Task or Criteria	IBIS	SPICE	S-Parameters
Large signal switching	Excellent	Fair to good	Poor to fair
Linear I/O impedance of drivers and receivers	Fair to good	Excellent	Excellent
Non-linear I/O impedance of drivers and receivers	Excellent	Fair	Poor
Model parameter variability due to process	Good to excellent	Poor to fair	Poor to fair
Model parameter variability due to bias, temperature, etc.	Fair	Excellent	Poor to fair
Effects of non-ideal power-ground paths	Fair	Good to excellent	Good
IC die, package, and pin interconnection parasitics	Good	Good to excellent	Excellent
IC die, package, and pin interconnection crosstalk	Good	Poor to fair	Excellent
Special IC technology	Fair to good	Good	Poor or N/A
Time domain analysis	Excellent	Good	Poor
Frequency domain analysis	Poor or N/A	Good to excellent	Excellent
Wide bandwidth analysis	Good	Good to excellent	Fair to good
Very high frequency analysis	Fair	Fair to good	Excellent
Simulation speed	Excellent	Slow	Slow
Model availability	Fair to good	Fair	Poor to fair

8.8.2 Recommendations for Selected Design Tasks

In making a recommendation for using a particular type of model and type of simulation to use, the authors have to keep in mind that simple answers are the most helpful. To paraphrase: "An analysis should be as possible–but no simpler." In the case of serial data busses operating at several GHz/GBITs, simplistic answers do not work.[10] Simulating serial data bus at GHz/GBIT data rates is a good example that one model type and one modeling method are no longer sufficient for handling today's design challenges. Table 8-8 shows the authors' recommendations of the type of model used to accomplish various design tasks.

---

[10] For more information, see "Transform Methods for Gigabit Serial Nets" (page 237) in this chapter and "SPICE-based Macromodeling Templates by Cadence" (page 602) in "Chapter 20, The Challenge to IBIS."

Table 8-8. Recommendations for accomplishing selected design tasks

To Perform This Task	Use This Recommended Model or Method	Or This Secondary Choice
Design devices	TCAD Software	Device physics models
Select devices	Selection guides and data sheets	Design re-use
Analog	Behavioral models and transform methods	Macromodels
Macromodeling: analysis of behavioral tendencies	2-Port	Various simplified generic forms: For example, a voltage amplifier, etc.
Control system simulation	Behavioral models and transform methods	Macromodels
Analog simulation: amplifiers, oscillators, etc	SPICE	2-Port. Also, various simplified generic forms : For example, a voltage amplifier, etc.
High frequency analog simulation: RF and microwave	Scattering-Parameters	SPICE
Signal Integrity simulation	IBIS	SPICE
Serial data bus at GHz/GBIT data rates	Special case requiring a combination of models and methods	SPICE, IBIS, S-Parameters and Transform methods
EMI simulation: transmitter-receiver portion	Scattering-Parameters	SPICE, IBIS
Power, thermal and reliability simulations	SPICE + thermal resistance	2-Port + thermal resistance
Digital system simulation	Verilog and VHDL	–

8.8.3 Summary of Model Recommendations

SPICE, IBIS, and S-Parameters[11] are the most important types of models in today's circuit simulators.[12] Most simply summarized:

- *SPICE* can be very accurate (if care is taken), very detailed (in predicting non-ideal and subtle effects), and very flexible in being able to represent what happens when temperature and bias conditions change. A SPICE model file is compact. A SPICE model is used by designers of semiconductor devices, semiconductor process engineers, and is the most readily available type of model. Other models are most often simulated and extracted from the original SPICE model.

[11] S-Parameters are a form of 2-Port Matrix model.

[12] There are other types of models, but they play almost no role in user interaction with today's simulators.

The negatives are that: There are no standards for SPICE, there are many different varieties of SPICE and they are often not fully compatible. SPICE can be reverse engineered, difficult to obtain and encrypted when obtained. Designers of semiconductor devices and semiconductor process engineers do not share the same concerns as circuit design engineers and the SPICE model they produce often do not pay correct attention to characterizing SPICE parameters used in circuit design but not in process control.

SPICE models can contain over 40 circuit nodes compared to an average of 4 for an IBIS or S-Parameter model and they simply take too long to simulate for most HSDD tasks.[13]

- An *IBIS* model can be very accurate, but only under bias, load and temperature conditions close to the conditions under which it was measured or extracted. IBIS makes use of data lookup tables for key parameters. These data-lookup tables deal with non-linear switching efficiently, but make for large model files. IBIS simulates very fast. IBIS strikes an excellent balance between speed and accuracy on HSDD simulation tasks where interconnections dominate. IBIS data exchange follows an industry-backed standard and IBIS models are very portable between simulators.

  The downside of the IBIS canned approach to a system of templates (Keywords) to represent I/O is slowness to respond to rapidly evolving I/O. Modern I/O designed for dynamic clamping, flexibly programmed drive strength and other features need more of the output circuitry included in the model than the traditional IBIS approach..

  Consequently, SPICE is now used in over 20% of HSDD modeling, especially gigabit SERDES® links. The IBIS Committee is in the process of evolving the IBIS Specification[14] to take fuller advantage of external circuit calls, macromodeling, and equation based modeling to respond to these issues.

- *S-Parameter* models do an excellent job of handling high-frequency related effects of modeling and simulation. Particularly, these effects include loss and dispersion on interconnects for nets with signal frequencies into the 10s of gigahertz. Thus, S-Parameters are familiar in their application to passive interconnections, but unfamiliar in general,

---

[13] Simulation run time increases as the cube of the number of nodes.
[14] Power Integrity and SSN/SSO issues are also receiving attention.

and specifically to active I/O devices. S-Parameters are data-table based, inherently small signal linear, and frequency domain models. But S-Parameters probably have a growing role to play since the design issues of HSDD and microwave signaling are tending to converge.

8.9 CONVERTING A MODEL TO ANOTHER TYPE OF MODEL

Got the wrong type of model? Not to worry. Many models can be converted to other types. Here is how:

8.9.1 Two-Port Conversions for Z, Y, H, ABCD, and S

Two-port matrix model conversions are commonly used. The matrix conversion formulas can be found in many standard texts on circuit theory or matrix math. N-element square matrices are fairly simple mathematical constructs. Generally, they are an arrangement of n equations in independent variables used, for example, to solve for the currents through a circuit given the voltages across its elements.

For a black-box input-to-output model of a semiconductor device, the N-Port models simplify down to two equations in two independent variables, or, a 4-element, 2-Port matrix.

Conversions Between the Matrix Model Z, Y, H, ABCD, and S

Table 8-9 shows the matrix formulas for converting the model parameters of Z, Y, H, and ABCD.

Table 8-9. Two-Port matrix conversions

	Given			
Find	**Z**	**Y**	**H**	**ABCD**
Z	Z_{11} Z_{12} Z_{21} Z_{22}	$Y_{22}/\|Y\|$ $-Y_{12}/\|Y\|$ $-Y_{21}/\|Y\|$ $Y_{11}/\|Y\|$	$\|H\|/H_{22}$ H_{12}/H_{22} $-H_{21}/H_{22}$ $1/H_{22}$	A/C Δ_8/C $1/C$ D/C
Y	$Z_{22}/\|Z\|$ $-Z_{12}/\|Z\|$ $-Z_{21}/\|Z\|$ $Z_{11}/\|Z\|$	Y_{11} Y_{12} Y_{21} Y_{22}	$1/H_{11}$ $-H_{12}/H_{11}$ H_{21}/H_{11} $\|H\|/H_{11}$	D/B $-\Delta_8/B$ $-1/B$ A/B
H	$\|Z\|/Z_{22}$ Z_{12}/Z_{22} $-Z_{21}/Z_{22}$ $1/Z_{22}$	$1/Y_{11}$ $-Y_{12}/Y_{11}$ Y_{21}/Y_{11} $\|Y\|/Y_{11}$	H_{11} H_{12} H_{21} H_{22}	B/D Δ_8/D $-1/D$ C/D
ABCD	Z_{11}/Z_{21} $\|Z\|/Z_{21}$ $1/Z_{21}$ Z_{22}/Z_{21}	$-Y_{22}/Y_{21}$ $-1/Y_{21}$ $-\|Y\|/Y_{21}$ $-Y_{11}/Y_{21}$	$-\|H\|/H_{21}$ $-H_{11}/H_{21}$ $-H_{22}/H_{21}$ $-1/H_{21}$	A B C D

where:
$|Z| = Z_{11}Z_{22} - Z_{21}Z_{12}$
$|Y| = Y_{11}Y_{22} - Y_{21}Y_{12}$
$|H| = H_{11}H_{22} - H_{21}H_{12}$
$\Delta 8 = AD - BC$

Scattering-Parameters are not measured with AC shorts and opens on the black box model ports, but rather with transmitted and reflected waves. The 2-Port S-Parameter matrix follows.

S_{11}	S_{12}
S_{21}	S_{22}

Therefore, equation sets (8-1) to (8-7) derive the conversions between S-Parameters and Z, Y, H, and ABCD Parameters.

Equation set (8-1) derives S-Parameters from z-parameters. The lower case z is used to indicate normalized values; thus $z_{11} = Z_{11}/Zo$ and $\Delta 1 =(z_{11}+1)(z_{22}+1)-z_{12}z_{21}$

$$S_{11} = \frac{(z_{11}-1)(z_{22}+1)-z_{12}z_{21}}{\Delta 1}$$
$$S_{12} = \frac{2z_{12}}{\Delta 1} \qquad (8\text{-}1)$$
$$S_{21} = \frac{2z_{21}}{\Delta 1}$$
$$S_{22} = \frac{(z_{11}+1)(z_{22}-1)-z_{12}z_{21}}{\Delta 1}$$

Equation set (8-2) derives S-Parameters from y-parameters. The lower case y is used to indicate normalized values; thus $y_{11} = Y_{11}Zo$ and $\Delta 2 =(y_{11}+1)(y_{22}+1)-y_{12}y_{21}$

$$S_{11} = -\frac{(y_{11}-1)(y_{22}+1)-y_{12}y_{21}}{\Delta 2}$$
$$S_{12} = -\frac{2y_{12}}{\Delta 2} \qquad (8\text{-}2)$$
$$S_{21} = -\frac{2y_{21}}{\Delta 2}$$
$$S_{22} = -\frac{(y_{11}+1)(y_{22}-1)-y_{12}y_{21}}{\Delta 2}$$

Equation set (8-3) derives S-Parameters from h-parameters. The lower case h is used to indicate normalized values; thus $h_{11} = H_{11}/Zo$, $h_{12} = H_{12}$, $h_{21} = H_{21}$, $h_{22} = H_{22}Zo$ and $\Delta 3 = (h_{11}+1)(h_{22}+1)-h_{12}h_{21}$

$$\begin{aligned} S_{11} &= -\frac{(h_{11}-1)(h_{22}+1)-h_{12}h_{21}}{\Delta 3} \\ S_{12} &= -\frac{2h_{12}}{\Delta 3} \\ S_{21} &= -\frac{2h_{21}}{\Delta 3} \\ S_{22} &= -\frac{(h_{11}+1)(h_{22}-1)-h_{12}h_{21}}{\Delta 3} \end{aligned} \tag{8-3}$$

Equation set (8-4) derives S-Parameters from abcd-parameters. The lower case abcd is used to indicate normalized values; thus $a = A$, $b = B/Zo$, $c = CZo$, $d = D$, and $\Delta 4 = a + b + c + d$

$$\begin{aligned} S_{11} &= \frac{a+b-c-d}{\Delta 4} \\ S_{12} &= \frac{2(ad-bc)}{\Delta 4} \\ S_{21} &= 2/\Delta 4 \\ S_{22} &= \frac{-a+b-c+d}{\Delta 4} \end{aligned} \tag{8-4}$$

Equation set (8-5) derives abcd-parameters from S-Parameters. The lower case abcd is used to indicate normalized values; thus $a = A$, $b = B/Zo$, $c = CZo$, and $d = D$.

$$\begin{aligned} a &= \frac{(S_{11}+1)(S_{22}-1)+S_{12}S_{21}}{2S_{21}} \\ b &= \frac{(S_{11}+1)(S_{22}+1)-S_{12}S_{21}}{2S_{21}} \\ c &= \frac{(S_{11}-1)(S_{22}-1)-S_{12}S_{21}}{2S_{21}} \\ d &= \frac{(S_{11}-1)(S_{22}+1)+S_{12}S_{21}}{2S_{21}} \end{aligned} \tag{8-5}$$

Equation set (8-6) derives z-parameters from S-Parameters. The lower case z is used to indicated normalized values; thus $z_{11} = Z_{11}/Zo$ and $\Delta 5 =(S_{11}-1)(S_{22}-1)-S_{12}S_{21}$

$$
\begin{aligned}
z_{11} &= -\frac{(S_{11}+1)(S_{22}-1)-S_{12}S_{21}}{\Delta 5} \\
z_{12} &= {}^{2S_{12}}\!/_{\Delta 5} \\
z_{21} &= {}^{2S_{21}}\!/_{\Delta 5} \\
z_{22} &= -\frac{(S_{11}-1)(S_{22}+1)-S_{12}S_{21}}{\Delta 5}
\end{aligned}
\qquad (8\text{-}6)
$$

Equation set (8-7) derives y-parameters from S-Parameters. The lower case y is used to indicated normalized values; thus $y_{11} = Y_{11}Zo$, and $\Delta 6 = (S_{11}+1)(S_{22}+1)-S_{12}S_{21}$

$$
\begin{aligned}
y_{11} &= \frac{(S_{11}-1)(S_{22}+1)+S_{12}S_{21}}{\Delta 6} \\
y_{12} &= -{}^{2S_{12}}\!/_{\Delta 6} \\
y_{21} &= -{}^{2S_{21}}\!/_{\Delta 6} \\
y_{22} &= \frac{(S_{11}+1)(S_{22}-1)+S_{12}S_{21}}{\Delta 6}
\end{aligned}
\qquad (8\text{-}7)
$$

Equation set (8-8) derives h-parameters from S-Parameters. The lower case h is used to indicated normalized values; thus $h_{11} = H_{11}/Zo$, and $\Delta 7 =(S_{11}-1)(S_{22}+1)+S_{12}S_{21}$

$$
\begin{aligned}
h_{11} &= \frac{(S_{11}+1)(S_{22}+1)-S_{12}S_{21}}{\Delta 7} \\
h_{12} &= {}^{2S_{12}}\!/_{\Delta 7} \\
h_{21} &= -{}^{2S_{21}}\!/_{\Delta 7} \\
h_{22} &= -\frac{(S_{11}-1)(S_{22}-1)-S_{12}S_{21}}{\Delta 7}
\end{aligned}
\qquad (8\text{-}8)
$$

8.9.2 Conversions Between BJT Configurations

The three circuit configurations, in which a BJT transistor can be used, are Common Emitter (CE), Common Base (CB), and Common Collector (CC). Table 8-10 is an example of measuring a set of 2-Port admittance parameters in the CE configuration and computing the values that would be measured in the CB and CC configurations.

Table 8-10. CE-CB-CC conversion formulae

CE Measured	CB Computed from	CC Computed from
y_{ie}	$y_{ib}=y_{ie}+y_{fe}+y_{re}+y_{oe}$	$y_{ic}=y_{ie}$
y_{fe}	$y_{fb}=-(y_{fe}+y_{oe})$	$y_{fc}=-(y_{ie}+y_{fe})$
y_{re}	$y_{rb}=-(y_{re}+y_{oe})$	$y_{rc}=-(y_{ie}+y_{re})$
y_{oe}	$y_{ob}=y_{oe}$	$y_{oc}=y_{ie}+y_{fe}+y_{re}+y_{oe}$

We could also present other conversion tables to show how useful these models are in fast computers. However, it's best to have the level of detail remain in and be handled by computer programs.

8.9.3 SPICE to IBIS Conversion

IBIS Golden Parser

The IBIS Golden Parser checks for correct syntax and the presence of required model elements. It is a free software program available from the IBIS Open Forum. Each version of the Golden Parser is backward compatible with earlier versions; that is, ibischk4 can check version 1 files. The most recent version should always be used since it will incorporate all the latest parser bug fixes.

The Parser is available at: http://www.eigroup.org/ibis/tools.htm and is maintained by the IBIS Committee. The website also provides other useful links and free tools. For instance, a SPICE to IBIS conversion utility from North Carolina State University (s2ibis3) can be used to create IBIS models from SPICE. There are also downloadable IBIS tools from EDA providers that make it easier to validate IBIS models, including graphically viewing IBIS tables.

The Golden Parser should be run during file construction on a check-as-you-build basis. Before running the Golden Parser for the first time, it is necessary to download and install the program. Be sure to choose the correct IBIS version (DOS or UNIX) for your platform. The program is run from the

DOS prompt by typing: `ibis_chk <filename>`. The most trouble-free approach is to run a copy of the parser in the same subdirectory as the IBIS model file.

Determine how many unique buffer designs and how many package types are to be tried in the part to be modeled. Each unique buffer and package combination requires a different [Component] model. Find out whether the device will be used in a mixed power-supply environment. For example, the ESD diodes may be referenced to a different supply than the buffers.

In a large device of several hundred pins or more, the construction of a complete IBIS model file can be a lot of work. Consider consulting with the end user to find out if some pins are not going to be simulated and can be ignored or given a default model. For example, static control pins rarely get simulated. According to the IBIS Quality Checklist, a complete IBIS file should have a model for every pin, but this is not required for Signal Integrity simulations early in a design development.

"Chapter 13, Using EDA Tools to Create and Validate IBIS Models from SPICE," shows the use of one commercially available EDA tool for deriving an IBIS model from a SPICE model. The example in this chapter uses Model Integrity® from Cadence Design Systems. Competing EDA tools are available.

Overview of the SPICE Simulation Method

SPICE models are usually considered to be proprietary and are not commonly available to customers from semiconductor suppliers. That is one reason the IBIS Spec was created! To get a SPICE model, especially with process distribution data, a customer usually needs to be doing considerable business with the supplier and sign a Non-Disclosure Agreement. If a SPICE model is obtained, the customer may need to generate the IBIS behavioral tables. If the SPICE model came without min-typ-max data, the customer needs to do the following tasks:

1. Put in estimates of the +/- variation from the typical or nominal values of the SPICE model elements and simulate under those conditions to get IBIS min-typ-max results. In the IBIS file, the IBIS format arranges the results data typ-min-max from left-to-right.
2. Simulate the I-V and V-T tables.
3. Consult with the semiconductor supplier to get a sense of how much their process varies.
4. Most often assume +/- 10% on the current for the I-V MIN-MAX tables and +/- 25% on the time for the V-T MIN-MAX tables.

Today, simulating complex I/O, such as a driver with pre-emphasis, requires capabilities beyond those of traditional IBIS models. Encrypted SPICE models are being supplied under NDAs. But this is not an optimum solution as is discussed in "Chapter 20, The Challenge to IBIS." For one, simulation times with SPICE are much longer.

I-V Tables

To start, use typical process parameters. To get pullup and pulldown tables for a buffer:

1. Set the control and enable inputs to turn on and turn off each pullup and pulldown device as required to go to output logic high and low.
2. Set the SPICE simulator for a DC sweep analysis. This is done by connecting a voltage source to the output node, sweeping it over –Vcc to +2Vcc and measuring the current into and out of the output node. Output clamp diodes, when present, must be disconnected in the SPICE model to obtain the pullup (pulldown) I-V tables. Next, reconnect the diodes and re-simulate to get the total turn off (turn on) I-V tables. After subtracting the pullup (pulldown) portion, the ground (power) clamp I-V current portion is left for inclusion in the IBIS model. Remember that the I-V behavioral table datums are stored (if present) in for portions – pullup, ground clamp, pulldown, and power clamp. These datums are re-added appropriately by the simulator depending on which part of the –Vcc to +2Vcc range the voltage is at.
3. Most of the time simulate the min-max tables as well as typ. Let us say that the min voltage range is the same as the typ voltage range minus 5 %. Then, adjust the sweep voltage range by the same amount when the simulation is run.

V-T Tables

To start, use typical process parameters. To get rise and fall waveforms for an output buffer:

1. Set the control and enable inputs to turn on and off each pullup and pulldown device as required to go to output logic high and low. Refer to the IBIS Spec for specifying the output load or test fixture to drive.
2. Set the SPICE simulator for a transient analysis. Connect a voltage supply to the output node of Vcc. Remember to reconnect any output clamp diodes in the SPICE model if they were disconnected to obtain the pullup and pulldown tables.

3. Suppose this is a single discrete transistor being modeled. Then, drive the SPICE model input with a pulse waveform having a rise and fall time at least as fast (but, not a lot faster) as the fastest expected rise and fall time of the device being simulated. Also, the amplitude of the input drive should be enough to drive from hard on to hard off and vice versa. Most of the time the output buffer of an IC is being modeled. Then, use the inherent internal process rise and fall time for the driver as provided by the semiconductor supplier.
4. The simulated rise and fall time is supposed to be the rise and fall of the buffer itself. So, remove internal pin and package parasitics from the SPICE model before running the simulation. And, remember to use the V-T data at the internal die connection point node in the IBIS file.
5. Remember to connect the load to the proper rail when running the simulations. Suppose the standard 50 Ω load is used. On pullup connect it to the Vcc rail and on pulldown connect it to ground. For open pullup or pulldown devices and ECL use the load and voltage used by the semiconductor supplier when they specified propagation delays.
6. Remember that the rise and fall slew rate interval for IBIS is 20 % to 80 % of the normal full output swing, if ramp (slew) rate information will be used instead of full V-T data.
7. Remember that serious ringing and non-monotonicities tend to invalidate simulations because they are hard to interpret, repeat, reproduce and verify with bench measurements. Clean output waveforms are needed for reliable model data.

Usually simulate min-max tables as well as typ. Let us say that the min voltage range is the same as the typ voltage range minus 5%. Then, adjust the sweep voltage range by the same amount when running the simulation.

8.10 TRANSFORM MODELS FOR SYSTEMS

In solving today's extremely high speed switching networks, transform methods are a very cutting edge method.

Transform methods often allow analysis to be easily switched between time and frequency domains. State machines for logic analysis can be represented using transform methods. Analysis of mixed analog digital systems with sampled data is also facilitated.

8.10.1 Domain Transformations Assist Problem Solving

The boundary between analog and digital is becoming increasingly blurred. For example a recent article discussed the use of spread-spectrum methods to suppress (double-sideband suppressed carrier, Phase-Lock-Loop) the carrier of a high-speed digital clock! This is definitely an analog technique being used in a digital system. Another reason for covering something about system-level modeling subjects in a discussion of semiconductor modeling is that the future holds more System-On-Chip (SOC) designing, where such modeling will be employed.

The design of serial busses operating at 10 GHz frequency rates and above requires a combination of models and modeling methods. In the macromodeling template approach, the template amounts to a structured SPICE-like netlist that can program-call various types of models as best fits the simulation task–SPICE, IBIS, convolved S-Parameter impulse response, or AMS-equation-based type modeling. For more information about the macromodeling template approach, see the "SPICE-based Macromodeling Templates by Cadence" topic in "Chapter 20, The Challenge to IBIS."

Because of these reasons and that transform methods are now being used in high-speed serial data busses, it is important to briefly review them. By transforming (mapping) a problem into a new mathematical domain the mathematics and solution of problem can often be made simpler. The inverse transform, back into the original domain, is usually performed to yield the final answer.

The transform method most immediately familiar to engineers is the Fourier Transform and its inverse between the Time-Domain (physical space-time) and the Frequency-Domain (a new, mathematical domain). We can visualize a single sine wave frequency and think that we are conceptualizing a physical entity. But that is a misconception. The single sine wave frequency is still a time-domain signal. The point is that the frequency domain is a mathematical transformation, or mapping, from the time domain.

System-level transfer function modeling is not a new subject. For example, this method has been used for a long time in the design and development of analog hybrid OpAmps. Such models in conjunction with Bode Gain-Phase graphs are used for designing and optimizing feedback loop compensation circuits. For a given OpAmp chip, the objective is to achieve the maximum gain-bandwidth for an unconditionally stable hybrid thick-film OpAmp, when the bare chip itself is potentially unstable. Transfer-function modeling is used in control system design and has been in use in top-down filter synthesis and design since very early in the 20th century.

Transfer functions can be developed *bottom-up*. That is, from a circuit analysis of the detailed network. But system level transfer function modeling is the usual starting point for analog *top-down* design [81]. Once the transfer function is derived from system constraints, there are various was to physically implement it, such as the Cauer realization for a filter. Also, once the transfer function is derived, it can be *decomposed.*[15] The overall system function-block is usually referred to as a *macromodel.* This block can often be split down into simpler, more physically realizable blocks, such as into an amplifier gain block (function) followed by an interstage filter block (function).

The decomposition of an analog system transfer function and its implementation in hardware is explained well in [81]. In addition to analog systems, there are mixed analog-digital systems employing sampled data.

8.10.2 Transforms in Mixed Technology – Mixed Signal Systems

Consider a gurney carrying a patient into a Magnetic Resonance Imaging (MRI) scanning machine. The gurney is physically moved by a set of electromechanical servomotors driving appropriate mechanical gear trains. There may be hydraulics involved, particularly if a certain amount of inertia in table motions have to be taken into account. The motors will be driven by some kind of low frequency analog electrical power circuit with amplifiers, and filtering. Feedback regarding positioning in the MRI machine will involve a laser-optic, high frequency RF detection link. Data from the positioning link will be converted to digital for processing in a digital circuit controlled by a software program. Control signals from the digital computer will be fed-back to the servo loop to move the patient back and forth into perfect alignment for their test scan. A human operator observing through a TV display will probably intervene in the process.

Long before the first piece of hardware or software is designed and built, this entire system can be, and usually is, modeled and simulated as a virtual design in a computer. Top-down system level design is rapidly advancing just as systems are rapidly becoming more complex

The system design is decomposed down into working on specifically analog, digital, software, communications and communications interface portions. For a digital portion, the IBIS model addresses the analog behavior resulting from fast switching digital signals. VHDL-Verilog (Verilog is an alternative to VHDL) addresses the timing analysis of the digital circuits.

---

[15] For instance, the Heavyside Expansion Theorem method.

8.10.3 Transform Methods for Gigabit Serial Nets

Transform methods convert a mathematical problem from one frame of reference (such as space-time for distributed networks) into another (such as frequency) where the problem is often easier to solve. This process is usually followed by an inverse transformation of the solution back into the original frame of reference. For example, a Laplace transformation of a differential equation, derived from circuit analysis, reduces a time-domain calculus problem into an algebraic (s-domain[16]) problem where it is easier to solve.

A lumped-element circuit equation in the time domain usually results in an ordinary differential equation in time (variable t, as in dV/dt). A distributed-element circuit, typically a transmission line, equation in space and time usually results in a partial differential equation. Examples are variables x and t, as in $\partial^2 V(x, t)/\partial x^2$ in the transmission line equation. Using Laplace transforms of the partial differential equation, for a transmission line in the time domain converts it into the s-domain. There it becomes an ordinary differential equation with respect to x [145]. Thus, formulations of transmission line equations, lossless or lossy, can be readily solved using transform methods.

Above 1 GBIT, S-Parameters (a frequency domain model) provide the best representation of (frequency-dependent) lossy, dispersive interconnections and their impedance discontinuities. Full-Wave solvers are often required in deriving and extracting accurate S-Parameter models for lossy, dispersive interconnections, even for Signal Integrity design.

The task in HSDD, however, is to compute the transient, time-domain response of signals using the frequency-domain S-Parameters. Transformation can be applied to these S-Parameters models if they are linear to bring them into the time-domain. If the models are non-linear, one method is to use the Impulse Response of the network, which is convolved, with the specific spectrum of the signals (direct and reflected) to derive the results.

Other mathematical methods, such as the Harmonic Balance and Envelope Detection methods, are more commonly used in RF, microwave, and EMI/EMC design. In those circuits, the frequency response of signals, wavelengths, and resonance modes are more informative. Because Signal Integrity, Power Integrity, and EMI/EMC are challenges for HSDD, a designer of GBIT serial SERDES nets has to be informed of time-domain, frequency-domain, and transform methods [4].

---

[16] S-domain is distinctly different than S-Parameter. See Laplace transforms in Table 8-11.

Table 8-11 summarizes the various transform models and methods.

Table 8-11. Transform methods

Transform Name	Definition of Transform Type	Transforms/Inverse-Transforms (variable range)
Fourier	Frequency Domain	$F(\omega) = \Phi[f(t)] = \int f(t)e^{-j\omega t}dt$ $(t = -\infty \rightarrow +\infty)$
		$F(t) = \Phi^{-1}[F(\omega)] = 1/2\pi \int F(\omega)e^{j\omega t}d\omega$ $(\omega = -\infty \rightarrow +\infty)$
	Comment: Expresses a complicated time-domain function as a frequency spectrum with magnitudes and phase.	
Laplace	Complex Frequency Domain	(s-plane, $s = \sigma + j\varpi$) $F(s) = L[f(t)] = \int f(t)e^{-st}dt$ $t = 0 \rightarrow \infty$ $f(t) = \left(\frac{1}{2\pi j}\right)\oint F(s)\ell^{st}ds = L^{-1}F(s)$
	Comment: Converts differentiation and integration into algebraic operations. Models for R, L, C and G map directly into the s-plane. Very useful for continuous transfer functions and closed loop system stability analysis.	
Z (a bi-linear transformation from/to the s-plane)	Complex Frequency Domain for time Sampled Data Signals	(z-plane, $z = e^{Ts}$) $F(z) = Z[f(t)] = \Sigma f(nT)z^{-n}$, $n = 0 \rightarrow \infty$
	Comment: Additional uses for continuous and sampled data transfer functions and closed loop system stability analysis. Nyquist plots are used.	
W (a bi-linear transformation from/to the z-plane)	Complex Frequency Domain for time Sampled Data Signals	For communications (w-plane, $w = (z-1)/(z+1)$) $z = (1+w)/(1-w)$ For control systems (w-plane, $w = (2/T)(z-1)/(z+1)$) $z = (1+(wT/2))/(1-(wT/2))$
	Comment: Enables and eases use of Bode diagram and Nichols chart for continuous and sampled data closed loop system stability analysis.	

Now that the authors have introduced transform methods as relevant to HSDD, we will say a few more things about transform methods.

A mathematical frame of reference is called a domain, as in space-time domain and frequency domain. A mixed domain uses more than one frame of reference simultaneously. Mixed domain techniques for modeling and simulation are becoming popular in EDA tools.[17]

[17] The Advanced Design System (ADS2004A)–E8885A Convolution Simulator–from Agilent EEsof EDA is one such tool.

High-speed switching is a transient, non-linear, large-signal problem and it is best handled in the time domain. Simultaneously, high frequency effects such as conductor and dielectric losses, and dispersion, are best modeled in the frequency domain. This is because they are frequency dependent in ways that often involve non-linearities. Losses, dispersion, and decrease of dielectric constant increase with frequency. S-Parameters are the method of choice for such frequency dependent models. The S-Parameters for lossy networks become complex variables, with attenuation and phase information. In other words, the variable contains a real number term and imaginary number term, $\alpha + j\beta$.

Also, to *convolve* something literally means to *fold* it. In this case, the mathematics folds, or maps, the impulse response of the S-Parameter models out of the frequency domain and into the time domain. The convolved impulse response is the transfer function of the network in the time domain. When the convolved function is multiplied by the input waveform function, the output response is produced [4, 146, 149]. Initially, before reflections occur, the switching signal out of the driver sent towards the receiver is the input signal.

With data rates well above one GBIT, the entire signal path from driver output pad to receiver input pad must be simulated to get waveforms, timing delays, and timing margins correct. Both pads are inside their respective IC packages.

To show what can be accomplished with today's sophisticated EM simulators and S-Parameter model extractors, consider a 25 GHz connector. Figure 8-11 shows a partial assembly view of a proof-of-concept 25 GHz connector. The connector was modeled with Sonnet Software's CST Microwave Suites® 3D modeler. This software can extract a full set of S-Parameters: S_{11}, S_{22}, S_{12}, and S_{21}. There are two center pins, each surrounded by a ring of return signal pins, in a side-by-side arrangement. The pin diameters are 2 mils. The impedance of this arrangement was 51.34 Ohms and showed little variation, plotted on a Smith Chart, over the frequency range of interest. Figure 8-12 shows one result, the $|S_{11}|$ magnitude in dB, extracted by the modeler from this arrangement.

The S-Parameter model for this 25 GHz connector can be used in a time domain simulation of a multi-gigabit switching network by employing transform and convolution methods.

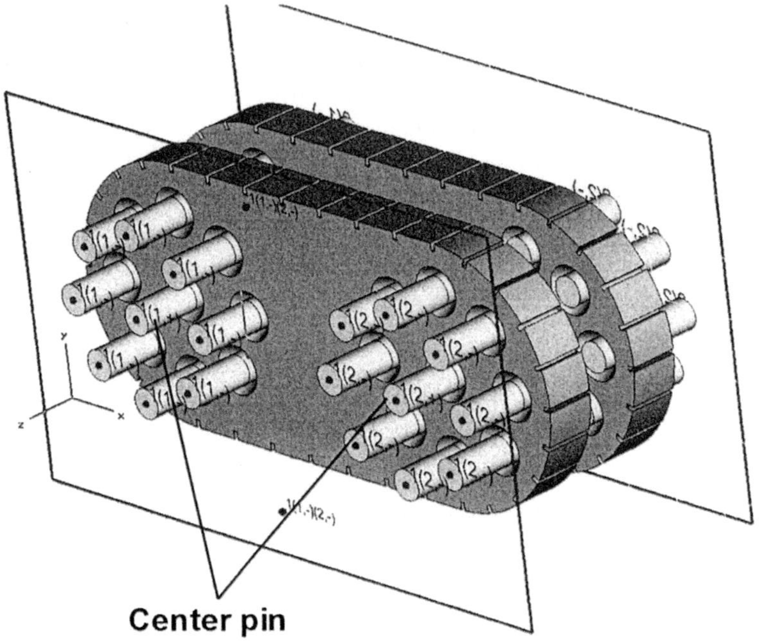

Figure 8-11. Test connector for 25 GHz signal.
Used with permission of Cinch Connector

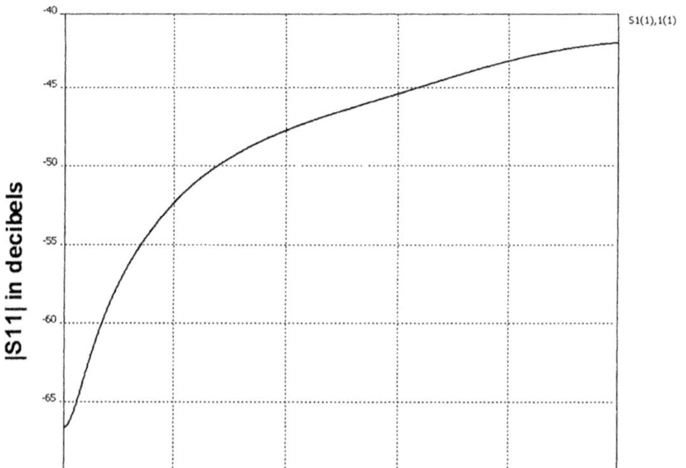

Figure 8-12. $|S_{11}|$ magnitude in dB for the Cinch 25 GHz connector.
Used with permission of Cinch Connector

8.11 SUMMARY

After selecting the best component to use in a design, the next step is to choose the best model to analyze how that device will behave in the circuit. There are several types of models to choose from. Each model type has certain strengths and weaknesses for the necessary analysis tasks. The major types are SPICE, IBIS, and S-Parameters models. SPICE models are very complex and good for tracking physical effects. IBIS models are good for PCB-level high-speed switching on interconnections, but do not track bias changes. S-Parameter models handle high-frequency effects well, but do not model time-domain switching well. Mixed-domain simulations, involving time, frequency, and Laplace mean a combination of models will be used. Mixed analog-digital simulations will involve W- and Z-Transforms.

Chapter 9

MODELING AND SIMULATION IN THE DESIGN PROCESS FLOW

Modeling accuracy requirements increase as the design is fine-tuned

Abstract: It is important to start circuit analysis and simulations early in the design process. Early simulation results can guide strategic choices and lead to the best, most cost-effective design. Early on, accuracy should not yet be a major concern, so engineers can use generic, default, and adapted models. Later, as the design develops, more accurate models may be needed and there is less freedom to make circuit changes. Also, the need increases for more complete coverage of components with their specific models. During design and development, obtaining more and better models is an ongoing process. Engineers and other users need to understand where to find, fix, create, and purchase models in order to have the appropriate models available when needed.

9.1 SIMULATION IN THE DESIGN PROCESS

Whenever we plan to run simulations on any kind of active circuit, we will need to use models of semiconductors. Very early in design development, even before we possess much information, we will need to decide where to get our models. We will need to follow a procedure for making those decisions—from both a systematic point of view and on an individual model basis. The company's model acquisition procedures should allow us to proceed in an orderly fashion. We may also occasionally have a need for emergency acquisition of models. The company's procedures should help us understand the available model sources and to learn which modeling sources to trust. From our OEM point of view, the less need there is to fix models and the more we can trust them, the more we can focus on our main task—to design and develop products.

Let's assume that we have an EDA tool for the physical design and layout of a circuit, and that we work in a company that has a CAD (Computer-Aided-Design) department. Let's also assume that our PCB designs have started switching faster, and will now need to be simulated in order to get them to work properly and within our project schedule. To do so, we will need to purchase EDA simulation tool licenses. Two important criteria for the purchase of the simulation tool licenses are that the tool works with our board layout tool and includes a basic library of good, working IBIS models.

The next step is to understand the most productive way to apply the simulation tool to our product development flow.

9.2 A TYPICAL DESIGN FLOW

Figures 9-1 and 9-2 show a typical design process flow, separated by phases. The phases in the design flow, in order of occurrence, are:

- *Phase A:* Preliminary Analysis–system level timing analysis and generation of timing and noise margin constraints.
- *Phase B:* Model Acquisition, BOM and Netlist generation–runs in parallel with schematic capture and logic design.
- *Phase C:* Pre-Layout Simulation and Constraint Management–Signal Integrity simulation of network topologies. Generation of layout and routing constraints.
- *Phase D:* Layout Simulations and Board Design–Signal Integrity, Power Integrity and Near-Field EMI simulation of board, stackup, routed and placed nets and verification of timing and noise margin performance. Preliminary design for meeting Regulatory requirements.
- *Phase E:* Documentation, Prototype and Verification–verification of simulation results with hardware laboratory measurements. Preparation of data for manufacturing and design reuse.

Figure 9-1 shows the start of the design process in Phase A.

Phase A, Preliminary Analysis is when major architectural and strategic decisions are made. The design engineer, using some generic, preliminary, and default models, can simulate prototypical network topologies from a sketch. Early simulations can give an indication of which nets will be critical, and therefore where to focus our energy to obtain accurate models.

Logic timing and noise margin constraints (whether estimated or simulated) are critical inputs for Signal Integrity and Power Integrity. Answering the question *will the logic work* sets how much noise and delay are tolerable.

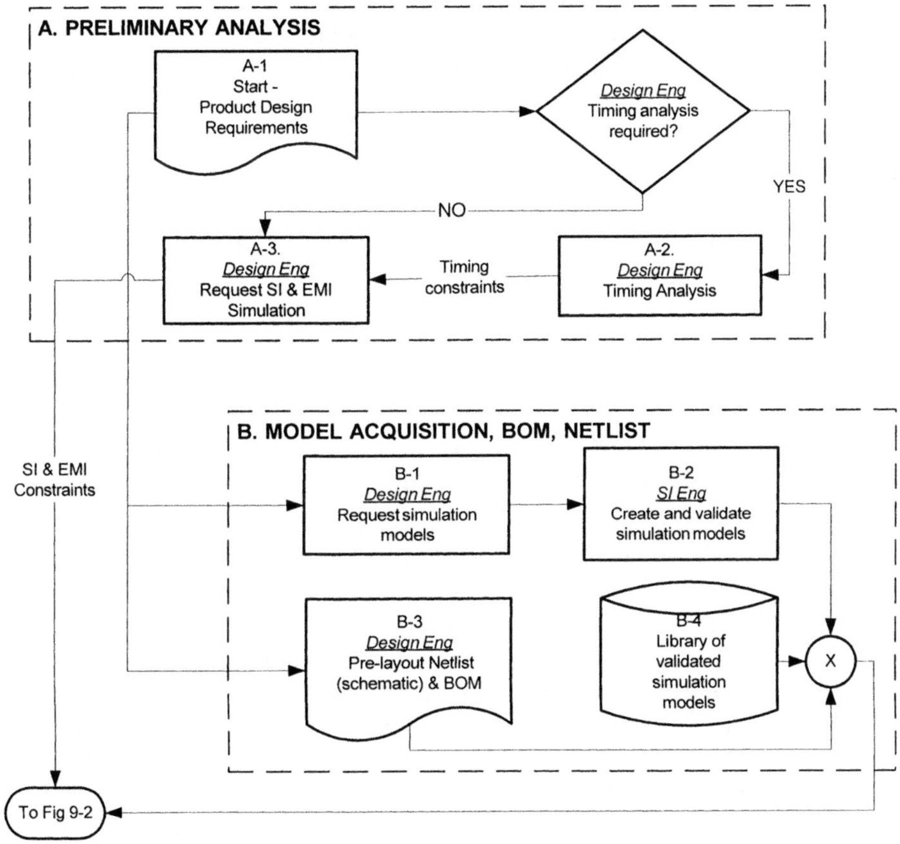

Figure 9-1. Flowchart of early phases of product development, modeling, and simulation

Strategic thinking must take place early in the design process. For example, we need to think about how to configure a clock tree, how much fan-out, loading, propagation delay and other factors to accept. These design choices (and others) affect the accuracy and detail level needed in the models. Simulations can answer questions, such as:

- Will the fastest and slowest drivers work in a heavily loaded network?
- Will the fastest and slowest drivers work on a network with many reflections?
- Is the data for fastest and slowest (typ-min-max) parts even included in the model file?

- Can the typ-min-max process corners on a generic default model be used to answer this kind of question?
- Must we get a more accurate model from the supplier?

Phase B, Model Acquisition, BOM, Netlist. When the system engineer is making strategic design choices, the engineer is often considering alternatives. Early simulation can be a great aid in making those decisions. Since time is usually critical, approximate answers will be adequate.

Quick acquisition of many types of models is important. Not quite as important is whether the models are exact matches to proposed components. For example, even if we plan to use a TI part, a Fairchild GTLP1634 IBIS model could be used to represent the component. What we want to know early on is whether the circuit will be critical. If necessary, we may need to take appropriate actions that can include acquiring a more accurate model, changing a board layout, or changing logic architecture.

Figure 9-2 shows what happens next. That includes Phase C: Pre-Layout Simulation and Constraint Management, Phase D: Layout Simulations and Board Design, and Phase E: Documentation, Prototype, and Verification.

Phase C: Pre-Layout Simulation and Constraint Management. Prior to board layout, we can run many simulations as soon as a preliminary BOM, netlist, and a set of models are available. These simulations should be combined with statistical design techniques to get a distribution of likely results. How many terminations, where they should be placed, and other design constraints (target Zo, stub lengths, bypass capacitor sizing, and placement) can be formulated and passed to the PCB board designer in the CAD group. The Signal Integrity engineer works out the issues about termination, stub length, fan-out, timing, and modeling issues with the logic engineer.

Phase D, Layout Simulations and Board Design, is when the modeling and simulation are refined by the layout and structure of the board. Now the PCB Designer joins the design team.

Phase E, Documentation, Prototype, and Verification. Finally, when the virtual design is completed, it must be prepared for sending to a board fabrication house, prototyped, and tested.

When the Gerber plots and related PCB fabrication drawings are sent to a board manufacturer, we may learn that the manufacturer is unable to meet the requirements. The board manufacturer may inform us there is no way they can deliver the target Zo given the stackup, etch width, and spacing set forth in the Gerber plots.[1] This can happen if we do not accurately model the variation in dielectric constant due to manufacturing issues for FR4 in our simulations.

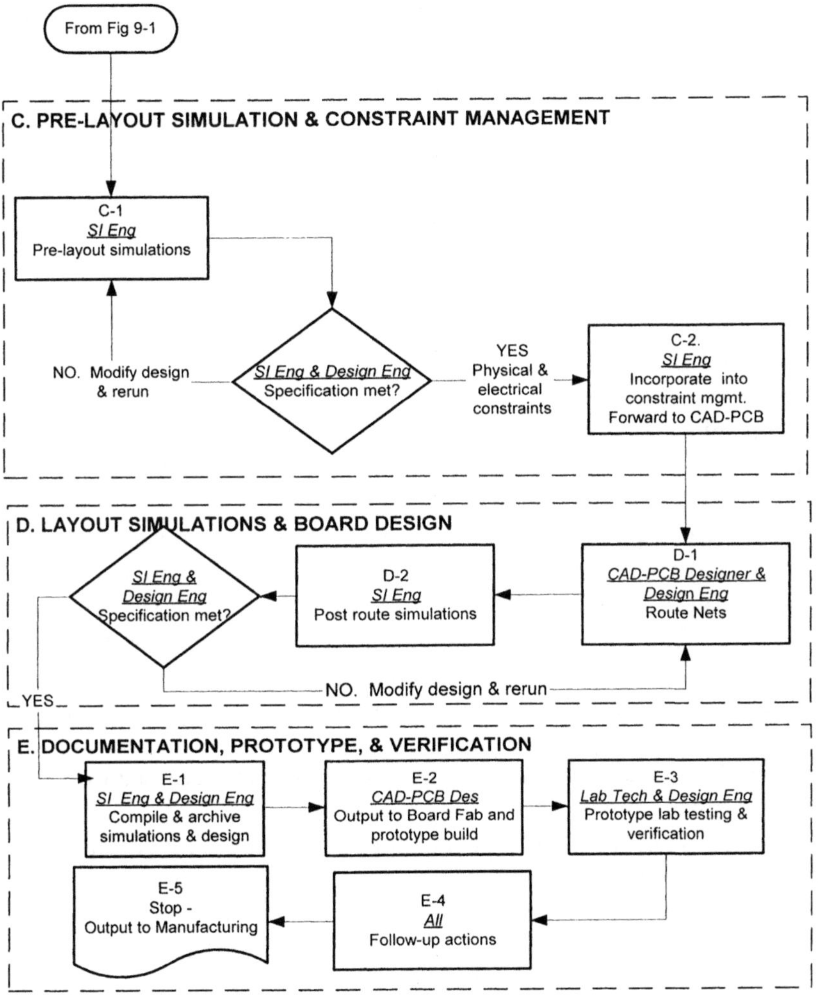

Figure 9-2. Flowchart for pre- and post-layout simulations phases of product development, modeling, and simulation

1 Not all board simulators properly account for effects like a non-uniform dielectric constant across the stackup cross-section due to manufacturing variances.

9.3 STRATEGY OF MODELING AND SIMULATION IN DESIGN

Strategic choices, such as whether to use a parallel or a serial bus architecture, are made early in a design development. Super accurate models are not needed to guide such choices. Even design rules can suffice to guide such choices. Critical nets are usually simulated early in the design cycle. Non-critical nets are usually simulated just before sending the CAD layout to a board fabricator. As the design progresses to building a prototype, the need for more accurate models, and better model coverage usually increases. This is particularly true for densely populated boards, higher switching speeds, and more complex circuit topologies. Fine-tuning of the design, and competing design needs occur on boards late in the design cycle. Many cycles of "model, simulate pre-layout, route, simulate post-layout, and adjust and re-iterate" usually occur in complex designs.

The most critical nets should be routed first. This is when the most freedom to place and route them exists. Gigabit serial data bus nets fall into this category. This means some early simulation with highly accurate models has to be done. Such simulations are normally very costly in computer run time. The reuse of prior successful layouts and routings of such nets in a new design is a particularly good idea. Highly accurate models for use in simulations on cutting edge designs[2] are sometimes needed very early in a design cycle. This early critical modeling need is in contrast to the rule of approximate models being good enough at first.

All nets on a board should be simulated at an appropriate time to avoid surprises on a prototype. The time and expense would be too high if long running highly accurate transistor level simulations were used on many of these non-critical nets.

During most of the design cycle, high-accuracy simulations would be inappropriate. Nets will often get re-routed because of crosstalk issues, accommodating other nets, and many other reasons. IBIS model simulations offer a good balance between accuracy and speed and are recommended for the majority of nets of moderate to low criticality.

---

[2] For example, nets operating at multi-gigabit rates using 90 nm technologies.

9.4 ACQUIRING IBIS MODELS: AN OVERVIEW

9.4.1 IBIS as the Model of Choice

IBIS models offer the following advantages:

- A good tradeoff between simulation details versus speed.
- Portability between simulation platforms.
- Good or excellent availability compared to other model types.
- Ability to represent most available semiconductor technologies.
- Ability to support and interface with the major hardware description languages (HDLs), such as SPICE, VHDL, and Scattering-Parameters for special simulation tasks.
- Certain built-in advantages, such as modeling of an entire IC component and its process corners on a consistent basis.

Therefore, our main focus is on IBIS.

Where and when to get models is affected by the project schedule, the challenge of the design task, and the cost of the source. Early in a project, there is more flexibility in making design decisions and changing logic architecture. Approximate simulation answers usually provide enough information to make such decisions. After a board has been laid out and complexity has been added, more accurate answers are needed, particularly if we have little design safety margin.

9.4.2 Where to Get IBIS Models

A product design company should never be entirely at the mercy of outside sources for models. In the final analysis, design engineers must be masters of their own fates in performing modeling and simulation. If necessary, they should be able to generate their own IBIS models. However, in-house model generation is costly and drains resources from board simulation. Having a trusted outside source for models is a good alternative.

Figure 9-3 lists ten alternative sources for acquiring IBIS models. For details regarding this list, see "Chapter 14, Sources of IBIS Models."

a) Company library
b) EDA provider tool library
c) Downloaded from the semiconductor supplier's website
d) Requested directly from the semiconductor supplier's representative and emailed in response
e) Model library purchased from a third-party modeling service
f) Adapted and inferred from an existing model of similar parts
g) Individual model purchased from a third-party modeling service
h) SPICE-to-IBIS conversion (if the semiconductor supplier sends us the SPICE model)
i) Developed by the semiconductor supplier (in parallel with the development of the IC itself)
j) A request to the SI-LIST email reflector for help in pointing to a possible source for the model

Figure 9-3. Ten sources of IBIS Models

9.4.3 Modeling and Simulation in the Design Development Flow

If a model arrives too late to help us in our design decisions, it will not be beneficial—even if it is the best, most accurate model available. Let us take a look at a typical design process flow and discuss how it impacts our modeling acquisition.

Example 1: Early strategic choices

We have already mentioned the need to make strategic choices as switching speeds continue to increase. For example, whether to stay with parallel bus architecture as opposed to switching to serial bus architecture. Timing margins erode on parallel busses as clock speed increases.

Example 2: Verifying all nets as a design progresses to prototyping

This example concerns a crosstalk problem that arose on a set of bus nets that should not have been a problem. These nets were routed together (parallel), and were among the last routed on a very densely populated board. The CAD designer used the board space that remained. Therefore, the set of nets were routed in parallel all over the board, jumped from layer-to-layer, and were much longer than desired as a result. Crosstalk then became a problem. Complexity was added as a result of the board layout process. Simulations to reveal these surprises must often wait until late in the product design cycle.

Example 3: Design rules can save simulation time and avoid "death by analysis"

In this example, a set of differential, point-to-point, terminated busses were to be routed. These nets would have enough prime board real estate space available to follow the design rules developed to prevent crosstalk. The design rules dictated a set of spacings and maximum parallel lengths dependent on edge rates. The design rules were followed, the board worked properly with respect to crosstalk, and all without ever directly modeling and simulating the nets.

Example 4: Catching a potential problem early

A tentative net topology was set up in Cadence's EDA tool SigExplorer® starting with a sketch. Tentative etch lengths, Zo, and net connections were included in the topology. The circuit was a clock distribution net with a driver at one end and a single, serial, transmission line with 5 receivers spaced along it. The semiconductor supplier provided a preliminary driver output cell IBIS model. The supplier said that the preliminary model was essentially what the final model would be like. The IC was in development for our application.

The clock net ran at 4.5MHz. This is not a particularly demanding application. However, simulations immediately revealed very sharp reflection spikes at the clock edges. The edge rate was about 100ps!

The knee frequency, F_{knee}, is the frequency at which the spectrum of a pulse starts to fall off rapidly (-20dB/decade). F_{knee} is given by:

$$F_{knee} = 0.5 / t_r = .5 / .1 * 10^{-9} = 500MHz \qquad (9\text{-}1)$$

Therefore, significant frequency components existed out to 500MHz! Obviously, this part is switching much too fast to be of practical use in this network. To prevent the net from behaving as a transmission line, time delays on the transmission line have to be less than about 1/6 the rise time. With a rise time of 100ps, and propagation velocities of about 140ps/inch on stripline in FR4, this translates to a maximum length of about 0.12 inches. This is not very practical for a 4.5MHz clock. Terminating this net to eliminate reflections would be difficult considering the frequency-dependent loading of the receiver input capacitances.

The best solution to the reflection problem is to not launch such high edge rates on the net. This issue of excessive edge rate versus application need is becoming very common as geometries shrink to 90nM and below, and switching becomes very fast.

Remedies could include running the driver output through a low pass filter. A better solution would be edge rate limiting circuitry, such as driver scheduling, on the driver output. Dynamic clamping might also be a good idea. Such remedies need to be worked out with the part supplier, and will probably be a cost increase. But without them, the logic will probably not work properly. Whatever the remedies, the essential point of this example is that early simulations alert us to potential problems.

9.4.4 Modeling Needs Change During Design

Pre-layout simulations get refined as the board stackup,[3] realizable Zo, vias and their placement, and other place and route choices get defined. Critical nets[4] are laid out first. As the design proceeds it gets harder and harder for the PCB designer to meet all the design constraints set up on all the nets because of physical and manufacturing constraints. Trial placement and routing of nets must be re-simulated and verified for performance. During this phase, we discover all the layer changes (added vias and discontinuities) the PCB designer had to make. For all these reasons, we find ourselves re-thinking our design constraints and the model accuracies they are based on.

As the board layout progresses, rigidity sets in regarding design choices, and the need for model accuracy usually increases. Non-critical nets (and bus networks) are routed towards the end of the layout process. They get any left over board real estate. As a result, they are often routed from one layer to another many times. These nets often end up much longer than originally planned. Therefore, they can end up being a problem. Nets we thought we would not need to simulate (nets we did not even seek models for) now need to be modeled and simulated.

How do we remember to simulate everything, and simulate it for min-typ-max? We need to build that into our design procedures.

9.4.5 What To Do When a Model Does Not Work

Lack of model availability is a common problem in simulating a design. It is, perhaps, the biggest single barrier to design simulation and analysis. Given a random choice of components with no preconditions, only about 35% of the components come with good, usable models available early in a design cycle.[5] To find, make, or repair models for the remaining parts, we will need to put in substantial effort.

---

[3] The board cross-section, its layers, layer thickness, dielectric constant, etch width, etc.

[4] Previously identified in timing and pre-layout Signal Integrity simulations.

[5] Usable means that any mistakes in the models will be small and can be quickly fixed.

Historically, about 70% of IBIS models do not work as first received. At first, they are neither ready for use nor trustworthy. Before using these models, they must be fixed. Some model files will never work. SPICE models, despite having a 40-year head start on IBIS, are not much better in quality.

The only way this situation will change is when an industry pattern develops showing that the availability of a good simulation model has a major influence on deciding whether a component will be purchased.

9.4.6 Potential Costs of Obtaining Models

There is no fixed formula for computing the costs versus benefits of fixing and generating IBIS model files, but here are some factors to consider before deciding whether to buy a model versus make or fix it.

Three-fourths of model files have potential problems when first received. So the potential for problems is high. The biggest problem is not having any model file available at all. The following list presents some of the considerations that impact the cost of obtaining and fixing models.

- The complexity of problems encountered with the model file.
- The resources for fixing problems with a model file. Resources include knowledge, process and procedures, and time available.
- The urgency of the need in fixing a problem and lead-time in the product development schedule.
- The criticality of the circuit that needs to be modeled and simulated.
- The track record of the model supplier. For comments, see "Chapter 12, Fixing Errors and Omissions in IBIS Models."

We will probably need to fix and generate IBIS models on an ongoing basis. In that case, we should strongly consider setting up processes, procedures, and resources to do so. Among the options to consider, both in individual situations and as management preferences, are:

- Doing the work ourselves
- Relying on our CAE tool provider's library
- Relying strongly on our semiconductor supplier's abilities
- Hiring a third-party modeling service company

Or some combination of the above.

9.4.7 Developing One's Own Models

When a supplier-provided model is unavailable,[6] users may need to create their own model for a device. But there are many perils and pitfalls when creating models. The benefit of developing a model is that we can proceed even if we are unable to get a model elsewhere. In practice, the lack of any readily available model occurs in a significant amount of the nets and components that need simulation.

For simple components (like resistors, capacitors, and inductors), it is quite easy for users to make their own appropriate models. At one end of the technology spectrum, old technologies (developed before IBIS models were adopted) may no longer have any modeling resources committed to them. At the other end of the spectrum, ICs using new process technology and custom designs almost never have models available early on.

Supplier guardbanding and disclaimers regarding parameter limits and model accuracy add to the uncertainty of model data. However, suppliers still work from some knowledge of the device they are modeling. When users try to create a model, they are even further removed from that device knowledge.

When trying to characterize a component, users have a big disadvantage. The users will not get a clear picture of the variations in model behavior that will be encountered in production. The users will be able to obtain only a small product sample, made over a short time span—not the full range of variation that will occur over time.

When suppliers are not forthcoming with population statistics on their models, the user's ability to independently obtain data is made considerably harder. In the earlier days of 1% AQL acceptance (10,000 PPM defect rates) quality (for long-running commodity parts), a few suppliers knew their population distributions. In many cases, they actually did not understand their population spreads that well. This was because of poor quality control, ongoing dynamic process changes, and lack of market insistence on good data. With today's intense competition to shrink feature size, it is again perhaps questionable as to how well suppliers understand and model their processes.

IC suppliers see only a snapshot of product upon which to characterize a component model. Especially with regard to high quality programs like six-sigma[7] or even verifying a three-sigma spread. In today's rapid technology evolution and short product runs, large product runs are largely a thing of the past. Rapid time-to-market and pressure to increase production make it very

---

6 Many users are learning to request models from their suppliers as part of doing business.

7 To *verify* a six-sigma spread requires statistical sampling on a production run of 20 million units!

difficult to characterize products. How can a user devise a six-sigma (6.8PPM defect rates) strategy for designs, when a run of 20 million units is needed to certify a six-sigma population distribution?

9.4.8 Costs of Finding, Fixing, and Making Models

The costs of finding, fixing, making, or buying a model can be quite steep. A company's accounting department can price out the ten options for acquiring models. These options are listed in Figure 9-4.

What are some typical costs for models? First, let's consider the EDA software tool provider whose tools we have purchased. Let's assume we could not find the model we need in the library provided with the simulator. By checking with some EDA providers about additional models, we learn the following alternatives are available:

- Some provide only a time and materials consulting engagement.
- Some say they will provide models free if we buy their software.
- Some want $1200 or more for each model.

Let us say we have about 1000 simple, glue-logic parts we need models for. None of the above cost alternatives look terribly attractive unless the payoff benefits are high. In addition, a single, good, detailed, verified IBIS model from a reputable third-party model provider could cost $40,000 for a large pin count, complex, fast, custom IC.[8]

When a model is not available from the semiconductor supplier or EDA tool vendor, or cannot be adapted from an existing model, the model must either be purchased or made by the OEM's staff. Whether the OEM's staff makes the model, or adapts it from an existing model, the cost can be significant. Either way, in the authors' experience, the model maker must be highly skilled and motivated.[9] Whether the work is done by a staff Signal Integrity Engineer or a Product Design Engineer, it will be non-routine work of some urgency to require interrupting simulation and design tasks. Also, the staff must own expensive software licenses to do the work. Even more expensive are the possibilities of "lost opportunity" costs when the model-maker needs to be doing other design work. The authors estimate the average

[8] $40,000 may seem like a lot of money to spend on model acquisition, but in the overall costs of building a group of critical circuits without simulating them, it can seem insignificant. It is a decision that requires careful consideration.

[9] Model making is a technically very sophisticated activity. Model making must be motivated by a passion to achieve accurate, useful circuit simulations for challenging design problems.

resources needed for modeling as half the time of the available staff involved in modeling and simulation.

At 3Com, the staff costs of acquiring, fixing, and making models averaged about $150,000 a year to support the equivalent of a dozen large server boards developed per year.

The payoff for the investment was realized because we went from an average 3.5 prototype cycles per design to an average 1.1 prototype cycles per design. Not having to scrap and rework prototypes meant a savings of about $1,500,000 per year. These savings in prototype scrap and rework were realized by having first prototypes whose Signal Integrity performance was "right by design." The savings in avoiding project delays and missed market opportunities were about ten times as high as the prototyping savings.

The point of mentioning the savings for doing modeling and simulation is to provide a balanced view of the costs of doing modeling and simulation. Providing models for simulation can involve a big investment. A very important part of realizing the return on that investment is where, when and how simulation is applied in the design process flow.

9.4.9 Knowledge for Acquiring or Making Models

Table 9-1 summarizes the knowledge needed to find, fix, make, and buy an IBIS model.

Table 9-1. What design engineers need to know

To learn...	And to answer this question...	See Chapter...
What is needed in the model file	What are the key parts of an IBIS model?	10, Key Concepts of the IBIS Specification
	How does the IBIS model work?	11, Using IBIS Models in What-If Simulations
How to acquire models (See Table Note 1)	Will it work, and if it does not work, can we fix it?	12, Fixing Errors and Omissions in IBIS Models
	If we cannot find an IBIS model, can we make one from a SPICE model from the supplier?	13, Using EDA Tools to Create and Validate IBIS Models from SPICE
	What are the sources of IBIS models and can we leverage models we do have to work in place of ones we do not have?	14, Sources of IBIS Models
	As we proceed to debug, fix and improve the model how do we control that process?	15, Working With the Model Library

Table 9-1. (cont.)

To learn...	And to answer this question...	See Chapter...
How and what to verify (See Table Note 2)	How do we verify an IBIS model against a previously verified SPICE model?	16, Methodology For Verifying Models
	How do we verify an IBIS model against hardware?	17, Verifying Model Accuracy by Using Laboratory Measurements
	How much accuracy verification do we need versus design margin?	18, Balancing Accuracy Against Practicality When Correlating Simulation Results
	An example of a new idea (theory) about modeling that was carried through to verification with hardware measurements.	19, Deriving an Equation-Based Model From a Macromodel

Table Notes:
1. Unless the company has highly competent sources to do modeling work, engineers may need to do the work themselves. The payoffs in profits and success due to modeling and simulation are too great to pass up.
2. Verification (whether against a previously verified model or directly against hardware) teaches how far to trust a simulation. But we need to be practical in our accuracy goals and be able to trade them off against design cleverness.

9.5 SUMMARY

The process for finding, making, or buying an IBIS model starts with the Product Design Engineer's selection of components and the types of simulations to be run. An important part of that decision is the availability and suitability of the components and models already in the company's component database. If the engineer can use something that is already in the company's database, the engineer will save a lot of money and inconvenience.

When is the best time to begin modeling and simulation? The answer is the earlier the better–and continually as the design heads towards completion and becomes more constrained. Early simulations can be approximate and address only critical circuits. Later simulations must be more accurate and must cover all the nets completely, leaving nothing to chance prior to physical hardware prototyping.

PART 4: ABOUT THE IBIS MODEL

Chapter 10

KEY CONCEPTS OF THE IBIS SPECIFICATION

The IBIS Specification establishes strict rules for model data exchange in order to ensure that model data files are software tool neutral

Abstract: The IBIS Specification is capable of modeling most high-speed digital semiconductor technologies. The IBIS data file makes use of behavioral models and does not contain any internal proprietary models. The IBIS key concepts include the lexicon, syntax, data exchange format, measurement methods, and design applications of the IBIS behavioral model. The full IBIS Specification is large, detailed, and complex. But the key concepts are easier to understand than the entire specification.

10.1 INTRODUCTION

10.1.1 What is IBIS?

IBIS concepts may be confusing to newcomers to modeling and simulation. This is because IBIS concepts include many features and separate functions. Some features may be used at one time, but not others. The features include:

- *Industry standard specification*[1]
  The IBIS Specification establishes the rules for creating model *data files*. The rules are standards established by the IBIS Open Forum Committee and are approved by the members following EIA/ANSI committee

[1] The specification number is ANSI/EIA 656-A.

procedures and then vote on the changes. The rules are a voluntary industry standard established by consensus, not a government regulatory mandate. The rules specify the model elements to include for each device.

- *Data exchange format*
  The data exchange format is established by the IBIS Specification. The rules for data format, syntax, and the order in which model elements can appear are quite strict. These rules enable portability of models between platforms and simulators.

- *Model data files*
  Data files describe models and must conform to the IBIS Specification rules. EDA simulator providers decide which models and aspects of IBIS they will implement.

- *Procedures for checking the quality of IBIS model data files*
  A parser software utility is available to check the data content and format in a data file to ensure that the data file follows the standard. A quality checklist is also provided.

- *Circuit model (schematic)*
  IBIS provides a *schematic* that helps users to *visualize* how different elements in the model are connected in a circuit. The schematic shows both an output side composed of several elements and an input side also composed of several elements. The IBIS circuit model assumes that a voltage generator (output driver side) sends a signal to a load (input receiver side).

- *Modeling methodology*
  There is a basic modeling *methodology* that EDA simulators follow when using IBIS circuit models. Here is how a simulator works with IBIS data files:
  - An IBIS output driver sends a voltage wave, provided by the model file's [Ramp] or V-T data lookup tables, down a network[2] to an IBIS input receiver.
  - Signal reflections occur due to high-speed effects on the network.
  - The reflections continue and may settle out to a steady state, given sufficient time before the next switching event.

---

[2] The network can be as complex as necessary and the input and output characteristics of driver and receivers can be non-linear.

– The reflections at each IBIS Input/Output are calculated[3] with the use of the model file's I-V data lookup tables.

10.1.2 Why was IBIS Developed?

By 1994, modeling and simulation had evolved with two popular types of models, which are:

a) The SPICE model for most modeling tasks, and
b) The S-Parameter model for the microwave circuit niche.

Edge rates for most high-speed digital switching circuits were not high enough to involve S-Parameter modeling. S-Parameter modeling was not friendly to large-signal, non-linear, time-domain switching simulation.

SPICE had a long and successful history, so why was IBIS developed? IBIS was developed because SPICE has several drawbacks. One flaw in SPICE is that it can be reverse-engineered. Consequently, because SPICE models can reveal proprietary information, suppliers do not like to provide them.

Another flaw is that SPICE simulations take too long, because it is solving for the internal behavior of the model. What usually dominates board network performance is the behavior of the interconnections between devices. Simulating all the nets on a large server board could easily require well over 500,000 complex simulations. This becomes an impossible simulation task using SPICE models. A PCB circuit designer would not normally use most of the data available in internal SPICE model results anyway. Solving a SPICE circuit requires solving all the internal nodes of the SPICE models. Simulation run times increase as the cube of the number of circuit nodes– and a SPICE model can contain over 50 nodes. The workload of simulating a board using the IBIS model is manageable because IBIS models have only three to six internal nodes.

Additionally, SPICE models did not normally provide for:

- Min-typical-max process corner data.
- Modeling multi-pin ICs, with a mix of many different models and pin parasitics on a consistent basis.
- Standardized data elements portable between all the different simulation platforms available to run a given type of model.

---

3 The bounce diagram method [Bergeron, 14] is used when calculating reflections.

- Portability between platforms and simulators. Even today, there are many different flavors of SPICE in use and data is still often not portable between them.

10.2 IBIS SPECIFICATION

The IBIS Specification combines many different modeling elements under syntax and formatting rules so that model data files can be easily exchanged between device suppliers, simulation software providers, and product designers. Because there is an IBIS model standard, concepts like the quality of a particular model, judged against the standard, can be applied. IBIS is the first model supported by an effective and functioning industry standards committee.

IBIS Specification Version 2.1 contained most of the IBIS key concepts. Later versions have added capabilities that are mostly in specialized areas. *IBIS* is an acronym for *Input-output Buffer Information Specification.* IBIS was first ratified as an industry standard in 1994 (version 1.0). IBIS is a relatively new modeling method compared to SPICE, which has been around since the mid 1950s. It was developed to facilitate the modeling of the analog behavior of high-speed digital circuits and to facilitate data exchange. The IBIS Specification is a work in progress: New versions incorporating improvements get ratified and issued from time-to-time (current version 4.1, ratified 2/2004).

The full IBIS Specification is a substantial amount of reading material for users to absorb at once. The authors recommend using the specification like an engineering handbook: that is, look up and apply the details as needed.

The IBIS Specification is a set of rules for entering the appropriate device data in a file that EDA tools and users can reference. IBIS provides for some very detailed information exchange about models. However, not all features are implemented in all EDA simulators at any one time. And not all features are needed in any one model data file at any one time.

Version 4.1 of the IBIS Specification is 142 pages of closely spaced intensely technical detail. The entire specification can be viewed or downloaded at http://eda.org/pub/ibis/ver4.1/ver4_1.doc For a more complete understanding of IBIS, the authors suggest reading the documentation listed in Tables 10-10 and 10-11.

10.2.1 Types of Data Elements in IBIS

There are three main data elements (plus many secondary elements) in the IBIS Specification. They are:

- Specification requirements
- Specification format and syntax, because IBIS is a data exchange protocol
- Specification parameter-measurement techniques

The IBIS Specification is like a template, although an inefficient one. Once a model file is created, checking it with the IBIS Golden Parser is similar to running it through a spell-checker. Creating the model file and getting all the information correct and in the right places is usually a complex job. Commercial EDA tools that help with the model creation process are available.

Two types of data elements make up the bulk of the data entries of an IBIS data file.

- An I-V lookup table – given a voltage (and a process corner typ or min or max), the lookup table returns a current. Ohm's Law is implicitly "solved," but since the resistance is non-linear the data is stored as an I-V lookup table. Both input and output impedances are needed.
- A Ramp V-T lookup table – given a time (and a process corner typ or min or max), the lookup table returns a voltage. This is no different than describing the time domain output waveform of a generator.

10.2.2 Sample Page of IBIS

Here are a few sample pages of a single IBIS keyword from the IBIS Specification:

```
| Keyword:  [Pin Mapping]
| Required:  No
| Description: Used to indicate the power and/or ground buses to which a given driver,
| receiver or terminator is connected.
| Sub-Params:  pulldown_ref, pullup_ref, gnd_clamp_ref, power_clamp_ref,
|           ext_ref
| Usage Rules:  The [Pin Mapping] keyword names the connections between POWER
|           and/or GND pins and buffer and/or terminator voltage supply
|           references using unique bus labels. All buses with identical
|           labels are assumed to be connected with an ideal short. Each
|           label must be associated with at least one pin whose
|           model_name is POWER or GND. Bus labels must not exceed 15
|           characters.
|
|           Each line must contain either three, five or six entries.
|           Use the reserved word NC where an entry is required but a bus
|           connection is not made (see below).
```

The first column contains a pin name. Each pin name must match one of the pin names declared in the [Pin] section of the [Component].

For buffers and terminators, the remaining columns correspond to the voltage supply references for the named pin. Each [Model] supply reference is connected to a particular bus through a bus label in the corresponding column.

The second column, pulldown_ref, designates the ground bus connections for the buffer or termination associated with that pin. The bus named under pulldown_ref is associated with the [Pulldown] I-V table for non-ECL [Model]s. This is also the bus associated with the [GND Clamp] I-V table and the [Rgnd] model unless overridden by a label in the gnd_clamp_ref column.

The third column, pullup_ref, designates the power bus connection for the buffer or termination. The bus named under pullup_ref is associated with the [Pullup] table for non-ECL [Model]s (for ECL models, this bus is associated with the [Pulldown] table). This is also the bus label associated with the [POWER Clamp] I-V table and the [Rpower] model unless overridden by a label in the power_clamp_ref column.

The fourth and fifth columns, gnd_clamp_ref and power_clamp_ref, contain entries, if needed, to specify additional ground bus and power bus connections for clamps. Finally, the sixth column, ext_ref, contains entries to specify external reference supply bus connections.

The usage of the columns changes for GND and POWER pins. For GND pins, the pulldown_ref column contains the name of the bus to which the pin connects; the pullup_ref column in this case must contain the reserved word NC. Similarly, for POWER (including external reference) pins, the pullup_ref column contains the name of the bus to which the pin connects; the pulldown_ref column in this case must contain the reserved word NC.

If the [Pin Mapping] keyword is present, then the bus connections for EVERY pin listed under the [Pin] keyword must be given.

If a pin has no connection, then both the pulldown_ref and pullup_ref subparameters for it will be NC.

The column length limits are:

[Pin Mapping] 5 characters max

```
|               pulldown_ref     15 characters max
|               pullup_ref       15 characters max
|               gnd_clamp_ref    15 characters max
|               power_clamp_ref  15 characters max
|               ext_ref          15 characters max
|
|          For compatibility with models developed under previous IBIS
|          versions, [Pin Mapping] lines which contain ext_ref column
|          entries must also explicitly include entries for the
|          pulldown_ref, pullup_ref, gnd_clamp_ref and power_clamp_ref
|          columns. These entries can be NC, as explained above.
|
|          When six columns of data are specified, the headings
|          gnd_clamp_ref, power_clamp_ref and ext_ref must be used on
|          the line containing the [Pin Mapping] keyword. Otherwise,
|          these headings can be omitted.
||------------------------------------------------------------------------
[Pin Mapping] pulldown_ref pullup_ref gnd_clamp_ref power_clamp_ref ext_ref
|
1          GNDBUS1   PWRBUS1   | Signal pins and their associated
2          GNDBUS2   PWRBUS2   | ground, power and external
|                              | reference connections
3          GNDBUS1   PWRBUS1    GNDCLMP     PWRCLAMP
4          GNDBUS2   PWRBUS2    GNDCLMP     PWRCLAMP
5          GNDBUS2   PWRBUS2    NC          PWRCLAMP REFBUS1
6          GNDBUS2   PWRBUS2    GNDCLMP     NC
7          GNDBUS2   PWRBUS2    GNDCLMP     NC       REFBUS2
|                    | Some possible clamping
|                    | connections are shown above
| .                  | for illustration purposes
| .
11         GNDBUS1   NC        | One set of ground connections.
12         GNDBUS1   NC        | NC indicates no connection to
13         GNDBUS1   NC        | power bus.
| .
21         GNDBUS2   NC        | Second set of ground connections
22         GNDBUS2   NC
23         GNDBUS2   NC
| .
31         NC        PWRBUS1   | One set of power connections.
32         NC        PWRBUS1   | NC indicates no connection to
33         NC        PWRBUS1   | ground bus.
| .
41         NC        PWRBUS2   | Second set of power connections
42         NC        PWRBUS2
43         NC        PWRBUS2
| .
51         GNDCLMP   NC        | Additional power connections
52         NC        PWRCLMP   | for clamps
|
```

```
| .
71          NC       REFBUS1   | External reference connections
72          NC       REFBUS2
|
| The following [Pin] list corresponds to the [Pin Mapping] shown above.
|
[Pin] signal_name model_name R_pin L_pin C_pin
|
1   OUT1        output_buffer1     | Output buffers
2   OUT2        output_buffer2     |
3   IO3         io_buffer1         | Input/output buffers
4   IO4         io_buffer2         |
5   SPECIAL1    ref_buffer1        | Buffers with POWER CLAMP but no
6   SPECIAL2    io_buffer_term1    | GND CLAMP I-V tables; two use
7   SPECIAL3    ref_buffer2        | external reference voltages
11  VSS1        GND
12  VSS1        GND
13  VSS1        GND
21  VSS2        GND
22  VSS2        GND
23  VSS2        GND
31  VCC1        POWER
32  VCC1        POWER
33  VCC1        POWER
41  VCC2        POWER
42  VCC2        POWER
43  VCC2        POWER
51  VSSCLAMP    GND                | Power connections for clamps
52  VCCCLAMP    POWER              |
71  V_EXTREF1   POWER              | External reference voltage pins
72  V_EXTREF2   POWER              |
|=========================================================================
```

10.2.3 Summary of Key Sections of the IBIS Specification

IBIS and its parser software recognize a system of keywords, model parameters, and subparameters. Simulator programs then use some or all of this data to perform analyses.

Keywords are written between square brackets, [Keyword]. Example: [IBIS Ver]. Getting the syntax and the file format correct lets simulators find the right data for a given task. The IBIS Specification itself runs all the data elements and *more* together in one file. The *more* can consist of measurement techniques, technical notes, and advice to the model supplier. The specification was written, amended, and added to by a large and diverse committee in over a ten-year period. The specification's organization has been recently improved for readability, but it is still not totally organized and compartmentalized.

The order in which the keywords appear follows a prescribed order and grouping. *In presenting and explaining these keywords, we do not cover most of them in detail.* For that, there is the IBIS Specification itself, plus some additional documentation found on the included CD.

10.2.3.1 Header Section Rules

The Header section of the data file contains some important information, specifically the keyword [IBIS Ver], which tells the simulator or parser which version of the IBIS Specification to use.

A typical data file entry starts like this:

```
[IBIS Ver]        2.1
```

Most of the Header keywords call out information for the user, not the simulator. They are: [Comment Char], [File Name], [Date], [Source], [Notes], [Disclaimer], and [Copyright].

The Header section information is important. It is usually worthwhile to open the data file and scan through this information. For instance, is this the latest file revision?

10.2.3.2 Component Section Rules

The IBIS data file contains from one to many components. In IBIS usage, a component is a particular IC chip, in a particular package, measured at particular power supply, temperature, and load condition.

A typical data file entry starts like this:

```
[Component]       Virtex-II
```

An IBIS data file can contain several [Component] sections. For example, if we place the same die in a different package (BGA versus flip-chip for example) or measure it at a different supply voltage (3.3V versus 1.8V for example), we have to add a new component section. Each new combination gets its own [Component] section.

The Component section has many important keywords grouped under it. Not all these keywords are used in any one component and there are precedence and override rules associated with them. Some important keywords that are always there are [Manufacturer], [Package], and [Pin]. Global values for pin parasitics (R_pkg, L_pkg, C_pkg) for a particular package can be entered under the [Package] keyword. The information under [Pin] is covered in the next topic.

Additional keywords grouped under [Package] are [Package Model], [Alternate Package Models], [End Alternate Package Models], [Pin Mapping], [Diff Pin], [Series Pin Mapping], [Series Switch Groups], [Node Declarations], [End Node Declarations], [Circuit Call], and [End Circuit Call].

10.2.3.3 Pin and Pin Parasitic Section Rules

Under each Component is a list of pins. All pins must be listed (more about FPGAs later)–even No-Connects (NC).

A typical [Pin] data file entry starts like this:

```
[Pin]   signal_name   model_name   R_pin  L_pin  C_pin
1       LVTTL2F        LVTTL2F
```

The [Pin] list tells the simulator what model_name to look for in the IBIS data file in order to get the modeling information for that pin. Nearly as important as the pin list is signal_name, which is the logic function signal name. It is often the only reliable link in debugging a pin list when the pin numbers on a board layout and the pin numbers in the IBIS data file do not match.

The R_pin, L_pin, and C_pin subparameters describe the parasitics due to the internal connections between the I/O die pad and the package external pin of the I/O models on an individual pin basis. The bondwire from the die pad to the package pin is usually physically too small, and too short electrically, to act as transmission line structure. But the values of C_pin, R_pin, and L_pin can be large enough to act as a filter to V-T waves entering or leaving an I/O. For GBIT high-speed serial I/Os, the bondwires also acts like a transmission line.

The IBIS data file can assign parasitic values globally to a particular component or individually to each pin. The IBIS package model can also assign pin-to-pin mutual coupling parasitics. A separate spec, the IBIS Interconnect Modeling Spec, can also model the bondwire as a distributed network, a transmission line, and add additional modeling sophistication.

The IBIS Keywords [Package], [Pin], [Package Model], [Pin Mapping], and [Diff Pin] are concerned with the internal connections and model cell assignments of a device. These Keywords allow for increasing levels of detail regarding connections and cell types.

[Package]:
Allows for a default set of parasitics (R_pkg, C_pkg and L_pkg) to be globally assigned to the device's pin-to-semiconductor die connections.

[Pin]:
Allows *all* the pins (actually I/O cells on the semiconductor die) to be assigned their proper model type. [Pin] also allows pin-to-semiconductor die connection parasitics (R_pin, C_pin and L_pin) to be individually called out by pin. R_pin, C_pin, and L_pin will override the values in R_pkg, C_pkg, and L_pkg. When the R_pin, C_pin, and L_pin properties are measured, they obviously are for a specific semiconductor die + package combination. Depending on how the device was set up to be measured, test socket or PWB, the user should be aware that a given set of pin parasitics are specific to a given test-socket/padstack/via combination also.

There are some cases where bare die behavior (I-V and V-T curves) can be inserted in different package types and the effects of the pin parasitics added in making characterization information more "portable." Certainly, "what-if" simulations can be run, varying pin parasitics data.

[Package Model]:
Allows the creation of a separate data file (optional) *packagemodel*.pkg in which additional detail can be supplied. This data file must be in the same directory as the *ibismodel*.ibs file, or the [Package Model] Keyword must be included in the .ibs file. This model allows for the description of cross coupling between pin connections, and between pin connections and package body. This is done through a set of R, L, and C matrices that describe the self and mutual values of the connections. IBIS does not officially support a G (conductance) matrix. Most often the values of G are extremely small and are approximated to zero.

In a large package model of N pins (N-by-N matrix), these pin parasitic matrices can be very large. Their combination into an RLGC matrix and solution can be computer-intensive. Therefore, IBIS is set up to handle sparse and banded matrices where coupling between non-adjacent and shielded pins is insignificant. To handle multi-pin connectors, some simulator companies adapted the Package Model. Also, the RLGC matrix can handle the situation of a set of cross-coupled transmission lines (etch) or cable wires.

[Pin Mapping]:
Allows a set of I/O cells to be associated with particular ground and power pins (busses). Today, it is often the practice in large, high-speed digital packages to provide multiple power and ground pins that serve particular sets of I/O cells. Also, multiple pullup and pulldown rail voltages can be assigned to those busses.

[Diff Pin]:
Allows for pairs of I/O pins to be set up as differential pairs. Offsets in driver launch time delay or input threshold voltage (mutually exclusive) can be specified.

The IBIS Specification clearly states that all pin information must be given. Since spelling out *all* the pin information can be labor intensive and tedious on a large unprogrammed FPGA, following the IBIS Specification's rules can be counterproductive. When the FPGA is being co-developed with a board, the IBIS file from the semiconductor supplier lists all the model types. But the pin lists are incomplete. Instead, once the FPGA is programmed, the semiconductor supplier usually has a software utility for generating a pin list for a specific programmed device.

All the pins must be specified in translating an IBIS model to a file that is to operate with a particular software company's simulator. If they are not specified, a parser or simulator error is generated. If we are in a hurry to check out a few I/Os for a device, we can temporarily make the unused pins NC = No Connect.

In some data files, the [Model Selector] keyword, a refinement on [Model], appears at the end of the Package and Pin section.

10.2.3.4 Model and Model_Type Section Rules

After the simulator picks up the model_name subparameter information in the pin list, it looks for the [Model] Keyword in the sections of the data file that appear after the pin list.

When it finds a data file entry like:

```
[Model]       LVTTL2F
```

the necessary connection between pin and model data is made. The sub-parameters that are part of the [Model] keyword and grouped under it are:

- Model_type, Polarity, Enable, Vinl, and Vinh
- C_comp, C_comp_pullup, C_comp_pulldown, C_comp_power_clamp, and C_comp_gnd_clamp

- Vmeas, Cref, Rref, and Vref
- Rref_diff, and Cref_diff.

In addition, the following keywords, many with their own subparameters, are grouped under the [Model] keyword:

- [Model Spec]
- [Receiver Thresholds]
- [Add Submodel], and [Driver Schedule],
- [Temperature Range], [Voltage Range], [Pullup Reference], [Pulldown Reference], [POWER Clamp Reference], [GND Clamp Reference], and [External Reference]
- [TTgnd], and [TTpower]

Model_type tells the simulator how the I/O port behaves at an abstract level of understanding. Typical data *must* be supplied for all model I/Os and their I-V and V-T table data. min and max data is optional. Model_type is a required subparameter of the [Model] keyword. The three main *Model_types* are:

- *Input:*
  An Input model functions only as a receiver. Vinl and Vinh (input threshold levels) must be defined. Power and/or Ground Clamp I-V curves must be defined if they are supplied in the device.

- *Output:*
  An Output model functions only as a Driver. Voh and Vol (output high and low limits) are not part of the IBIS Spec. However, most simulator companies put such information to good use. Power and/or Ground Clamp I-V curves must be defined if they are supplied in the device. Pullup and Pulldown I-V curves, as present in the device, must be supplied. This model always sources and/or sinks current and cannot be disabled. Buffer switching speed information in the form of output rise and fall ramp rates or V-T rise and fall curves must be supplied. If V-T curves are supplied they supersede the ramp rates.

- *I/O:*
  This is a type of model where the pin is connected to device cells that can function either as a driver (output) or a receiver (input) depending on the enabling logic. During simulation the EDA tool GUI usually provides a means of designating the active driver on a net. The default is usually 3-

state if a driver is not designated as the active driver. The default is usually input if an I/O is not designated as the active driver.

Other *Model_type* options follow:

- *3-state:*
  This is a type of output driver model. It indicates that an output can be disabled. That is, put into a high impedance state.

- *Open-drain:*
  This is a type of output driver model with an open pullup side. This name is retained for backward compatibility.

- *I/O_open_drain:*
  This device indicates a combination of I/O and Open_drain behavior.

- *Open_sink:*
  This is a type of output driver model with an open pulldown side. The user supplies a pullup resistor and power rail connection.

- *I/O_open_sink:*
  This device indicates a combination of I/O and Open_sink behavior.

- *Open_source:*
  This is a type of output driver model with an open pulldown side. The user supplies a pulldown resistor and/or ground/pulldown rail.

- *I/O_open_source:*
  This device indicates a combination of I/O and Open_source behavior.

Similar to the above, we have:

- *Input_ECL/PECL*
- *Output_ECL/PECL*
- *I/O_ECL/PECL*
- *3-state_ECL/PECL*

Note: ECL is an acronym for "Emitter Coupled Logic." PECL is an acronym for "Positive Emitter Coupled Logic." These types of technology follow some different conventions for pulldown than the previous types.

- *Terminator:*
  This model is an input-only device that can have an analog loading effect on the circuit being simulated but has no digital logic threshold. Examples are resistors, diodes, and capacitors.

Notice that logic function terms like AND, NAND, and NOR are not used in IBIS.

Typical entries under the [Model] keyword appear as follows:

```
Model_type      I/O
Polarity        Non-Inverting
Enable          Active-Low
Vinh = 2.0V
Vinl = 0.8V
Vmeas = 1.40V
Cref =  0pF
Rref =   1.00M
Vref =  0.000V
C_comp     10.00pF      10.00pF      0.00pF
|
|
[Temperature Range]       25.00       100.00       0.000
[Voltage Range]       3.30V      3.00V      3.60V
```

10.2.3.5 I-V Data Section Rules

In the I-V section of the data file are the keywords [Pulldown], [Pullup], [GND Clamp], and [POWER Clamp] as present in the device and the model.

A typical I-V data table entry starts like this:

```
[Pulldown]
| voltage     I(typ)          I(min)          I(max)
|
  -3.30      -5.48mA         -3.66mA         -6.34mA
  -0.70      -5.48mA
```

Subparameters and keywords grouped with the I-V curves can include [Rgnd], [Rpower], [Rac], and [Cac], [On], and [Off], [R Series], [L Series], [Rl Series], [C Series], [Lc Series], [Rc Series], [Series Current], and [Series MOSFET]. Other terms and ideas used in the IBIS Specification for the different types of current-voltage tables included:

- *Range of I-V Data:* The I-V curve data is presented as a table of current measurements taken at a series of output voltages that covers a range of

–Vcc to +2Vcc. The current sourced or sunk by the device is measured with currents into the device being positive by convention.

- *Measurement Conditions:* The data is measured by placing a stepped and swept voltage source on the output and measuring (simulating) the current into or out of the pin. The device is allowed to settle to semi-quiescent conditions. Allowing a buffer to reach quiescent conditions is a lot easier to measure, but this can be misleading in modeling dynamic switching behavior. Adding pin parasitic information better models the high-speed and high-current switching behavior. For further details, refer to the IBIS Spec, especially the data derivation section.

- *Simulation From SPICE:* IBIS data is most likely to be simulated from a SPICE deck rather than being measured. The same technique of hooking up a (virtual) constant voltage source, sweeping it, and measuring output current is used. Often the SPICE model assumes zero diode resistance on clamps. It is not uncommon to see clamp currents of 10^{20} amps in the data tables as a consequence. But, this is not real data and needs to be corrected when found.

- The Four I-V Curves *[Pullup], [Power Clamp], [Pulldown], and [Ground Clamp]:* The current at an output is the total of the pullup and pulldown and the clamp curves when present. But these curves are stored in four separate data tables (not counting min-max conditions) in the model. IBIS includes an explanation of measurement techniques for separating out these individual elements. When simulating from a SPICE deck, the same effect is achieved by disabling and disconnecting the various elements in turn in the simulation. Table 10-1 lists the four IBIS I-V data tables.

Table 10-1. The four I-V data tables

Keyword	**Description**
[Pulldown]	Low-state driver I-V table data.
[Pullup]	High-state driver I-V table data.
[GND Clamp]	I-V table data of ESD diode structure.
[POWER Clamp]	I-V table data of ESD diode structure.

When a simulator is run, the simulation tool's task is to recombine this I-V data in a manner relevant to the I/O's region of operation. All drivers (Outputs and I/Os) have some type of voltage-time (V-T) tables requiring at least [Ramp] data. The V-T data is much more relevant at today's switching speeds.

- *The IBIS Voltage Convention:* The voltage in the table (ECL and PECL parts can be exceptions) for pullup and power clamp curves is:

$$V_{table} = V_{cc} - V_{output} \tag{10-1}$$

Therefore, the voltage across the pullup/clamp circuit is referenced to Vcc and not to ground—by convention. At the same time, the pulldown curves are normally referenced to ground. The range –Vcc to +2Vcc is chosen because those are the limits of the voltage swings that can be seen on a transmission line with high reflection (high mismatch) coefficients when switching a driver from Vcc high state to 0V low state.

10.2.3.6 V-T Data Section Rules

The V-T table data section begins with the [Ramp] keyword and its subparameters of dV/dt_r, dV/dt_f, and R_load. The [Rising Waveform], and [Falling Waveform] keywords, when present in the model, along with their subparameters: R_fixture, V_fixture, V_fixture_min, V_fixture_max, C_fixture, L_fixture, R_dut, L_dut, and C_dut appear in this section. The [Ramp] data is always required because not all simulators implement V-T data tables.

A typical V-T data table entry starts like this:

```
[Ramp]
| variable      typ            min            max
  dV/dt_r       0.22/0.66n     0.16/0.82n     0.30/0.58n
  dV/dt_f       0.26/0.95n     0.19/1.16n     0.33/0.82n
R_load = 50.00
|
[Rising Waveform]
R_fixture = 50.00
V_fixture = 0.000
| time          V(typ)         V(min)         V(max)
|
  0.000S        0.000V         0.53uV         0.000V
  0.20nS        0.000V         0.55uV         -0.46uV
  0.40nS       -4.58uV
```

Other terms and ideas used in the IBIS Specification for the voltage versus time switching behavior of the model types attached to the I/O pins are:

- *Ramp:* The simple expedients of measuring rise and fall slew rates of an output driver is the easiest way of describing its voltage versus time behavior when it switches. The ramps described by dV/dt can be a good approximation to the rising and falling edge waveforms, but often they are not.

- *Ramp, Slew Rate, and Rise and Fall Time:* Very often the driver switching behavior is described in terms of its rise and fall times. The various definitions and terms used to describe rise and fall times can be confused, because there is no single set definition outside of the IBIS Spec. Also be aware that:

  - *Minimum Rise and Fall Time* corresponds to maximum Rise and Fall Slew rate (or, alternatively edge rate), which corresponds to Fastest Switching. An IBIS data file arranges its data columns typ-*min*-max left-to-right instead of *min*-typ-max, which most simulators and data books follow. Formatting can get confused in the translation from databook → IBIS data file → simulator file → simulator user interface.

Typ-Fast-Slow Rise and Fall Time: Users need to keep a few things straight about switching speed. Namely:

- *Typical Rise Time* corresponds to typical (typ) Rise Time (t_r) and typical (typ) Slew Rate (dV/dt_r) and, for parts normally held low, typical drive high/turn-on time. Similar definitions hold for the fall time/turn-off time.

- *Fast Rise Time* corresponds to minimum (*min*) Rise Time (t_r) and maximum (max) Slew Rate (dV/dt_r) and, for parts normally held low, minimum drive high/turn on time. Similar definitions hold for the fall time/turn off time. Note that holding a logic part low usually means that it is turned on and drawing current!

- *Slow Rise Time* corresponds to maximum (max) Rise Time (t_r) and minimum (*min*) Slew Rate (dV/dt_r) and, for parts normally held low, typical drive-high/turn-on time. Similar definitions hold for the fall time/turn-off time.

Rise/Fall-Time Definitions: Data sheets and IBIS define Rise/Fall Time differently. Details are covered in "Chapter 17, Verifying Model Accuracy by Using Laboratory Measurements." The IBIS definition of Rise Time (and similarly Slew Rate) is the time to go from 20% to 80% of the total transition, when the output is loaded by R_load to Vcc for falling transition.

R_load defaults to 50 ohms if it is not defined. ECL model types use R_load to Vcc - 2 V for both edges.

V-T Data Tables: V-T data tables are the voltage vs. time of the rising and falling waveforms of the output of a driver into a specified test fixture. They take a little more labor to measure and then enter the data than the simple expedient of measuring the slopes of the rising and falling waveforms (dV/dt) and entering that data.

10.2.3.7 Additional Keywords and Subparameters Rules

Many additional keywords and subparameters can be used in a data file. To mention only the major sections, they can include [Test Data] (Golden Waveform data), [External Model] (other hardware description models supported), [Submodel], [External Circuit], [Define Package Model], each with its own grouped keywords and subparameters.

10.2.3.8 [End] Keyword Section Rules

Do not include more than one naked [End] in a data file. If you want to indicate the end of a section, hide it behind a comment character "|."

For example:

```
|  [End LVTTL2F]
```

10.2.4 Additional IBIS Options and Rules

10.2.4.1 Package and Pin Parasitic Option Rules

Describing pin and package parasitics can start out very simply or progress to a fairly sophisticated system just within the IBIS Specification itself – never mind anything as complex as the IBIS Interconnect Modeling Specification. IBIS can assign R-L-C parasitic values to pins in the following manner:

- Globally to pins: Using R-L-C subparameters to the [Package] keyword
- Individually to pins: Using R-L-C subparameters to the [Pin] keyword
- Individually to pins plus mutual coupling between pins: Using R-L-C subparameters to the [Package Model] keyword

Therefore, some additional information is provided in Table 10-2.

Table 10-2. IBIS package and pin properties

Keyword	Description	Required?
[Package Model]	Name of the package model, if supplied. Can be supplied in a separate .pkg data file if in the same directory as the .ibs data file. Package models allow for pin-pin coupling parasitics. Overrides defaults in [Package]	Optional
[Define Package Model]	Required if [Package Model] used	IBIS-Yes 3Com-Yes
[Manufacturer]	Manufacturer of the parts that use this package Required (IBIS and 3Com) if [Define Package Model] used	
[OEM]	Manufacturer of the parts that use this package Required (IBIS and 3Com) if [Define Package Model] used	
[Description]	Human readable package description Required (IBIS and 3Com) if [Define Package Model] used	
[Number of Pins]	Tells parser how many pins to expect Required (IBIS and 3Com) if [Define Package Model] used	
[Pin Numbers]	The ordered arrangement of the pin numbers Required (IBIS and 3Com) if [Define Package Model] used	
[Model Data]	Begins RLGC matrices which overrides [Pin] and [Package] parasitic element defaults Model I/O cell assignments (types) are still controlled by [Pin]	See Table Note D at the end of this table
[Resistance Matrix] [Inductance Matrix] [Capacitance Matrix]	A 1x1 matrix is self-resistance of a single line, a 2x2 matrix is self and mutual resistances of 2 coupled lines, etc. The a_{11}, etc., elements are self. The a_{13}, etc., elements are mutual. The R, L and C matrices (G elements are usually zero or neglected) form the basis of coupled line (or connector or package) structures and all transmission lines.	
[Row]	Used to denote the start of a new row in a matrix when the rows are more than 80 characters long.	
[Bandwidth]	The distance, in number of matrix elements on either side of the main diagonal, that is, band beyond which the matrix elements are guaranteed to be zero. That is, no significant coupling Yes, if Banded_matrix matrices used.	
Banded_matrix	Subparameters of R, L and C matrices. A type of matrix. Used when a full matrix for a large BGA, etc., array would be unnecessarily huge.	
Sparse_matrix	Subparameters of R, L and C matrices. A type of matrix. Used when a full matrix for a large BGA, etc., array would be unnecessarily huge.	

Table 10-2. (cont.)

Keyword	Description	Required?
Full_matrix	Subparameters of R, L and C matrices. A type of matrix. Used when a full matrix is used.	
[End Model Data]		
[End Package Model]		
[Pin Mapping]	Used to tell which power/ground busses a particular driver, receiver or terminator cell, or their elements, are connected to.	Optional
pulldown_ref pullup_ref gnd_clamp_ref power_clamp_ref	Subparameter of [Pin Mapping]. Each pulldown bus has a unique name. To meet minimum ground /power bounce, devices are being supplied with several such pins. Connections are designed to minimize ground/power impedance and xtalk. All entries with identical labels are assumed to be connected together. Each unique I/O pin must be connected to at least one pin whose model name is POWER or GND. Otherwise they are all NC.	Optional
[Rac]	Resistance of internal RC shunt termination If present in device.	IBIS-Yes 3Com-Yes
[Cac]	Capacitance of internal RC shunt termination If present in device.	IBIS-Yes 3Com-Yes
[Rgnd]	Internal package resistance of ground pin if and only if [Model_type] is Terminator If present in device	IBIS-Yes 3Com-Yes
[Rpower]	Internal package resistance of power pin if and only if [Model_type] is Terminator and present in device.	IBIS-Yes 3Com-Yes
[Diff Pin]	Associates differential pins and their threshold voltages and timing offsets. If present in device. See also Polarity in [Model].	IBIS-Yes 3Com-Yes
inv_pin	Subparameter of [Diff Pin]. The inverting pin of the pair. If [Diff Pin] used.	IBIS-Yes 3Com-Yes
Vdiff	Subparameter of [Diff Pin]. Specified output or input threshold differential voltage. IBIS says required if [Diff Pin] used. But, 3Com says optional	IBIS-Yes 3Com-Opt
tdelay_typ tdelay_min tdelay_max	Subparameter of [Diff Pin]. Launch delay of the non-inverting pin relative to the inverting pin – typical. IBIS says required if [Diff Pin] used. But, 3Com says optional	
Table Note D: There is little reason to use the [Package Model] keyword unless intending to include RLGC matrices. The G (conductance) matrix is almost always zero. The R matrix is often zero, and when included, is almost always the self-resistance of the pin. The C and L matrices contain the mutual coupling elements that are the chief motivation for using the [Package Model].		

10.2.4.2 Model_Type Subparameters

Special usage rules apply to the subparameters in Table 10-3. Some definitions are included for clarification:

Table 10-3. IBIS Model_type subparameters

Subparameter	Description
Input	These model types must have Vinl and Vinh defined. If they are not defined, the parser issues a warning and the default values of Vinl = 0.8 V and Vinh = 2.0 V are assumed.
I/O	
I/O_open_drain	
I/O_open_sink	
I/O_open_source	
Input_ECL	These model types must have Vinl and Vinh defined. If they are not defined, the parser issues a warning and the default values of Vinl = -1.475 V and Vinh = -1.165 V are assumed.
I/O_ECL	
Terminator	This model type is an input-only device that can have analog loading effects on the circuit being simulated but has no digital logic thresholds. Examples of Terminators are: capacitors, termination diodes, and pull-up resistors.
Output	This model type indicates that an output always sources and/or sinks current and cannot be disabled.
3-state	This model type indicates that an output can be disabled, that is put into a high impedance state.
Open_sink	These model types indicate that the output has an OPEN side (do not use the [Pullup] keyword, or if it must be used, set I = 0 mA for all voltages specified) and the output SINKS current. Open_drain model type is retained for backward compatibility.
Open_drain	
Open_source	This model type indicates that the output has an OPEN side (do not use the [Pulldown] keyword, or if it must be used, set I = 0 mA for all voltages specified) and the output SOURCES current.
Input_ECL	These model types specify that the model represents an ECL type logic that follows different conventions for the [Pulldown] keyword.
Output_ECL	
I/O_ECL	
Series	This model type is for series models that can be described by [R Series], [L Series],[Rl Series], [C Series], [Lc Series], [Rc Series], [Series Current] and [Series MOSFET] keywords.
Series_switch	This model type is for series switch models that can be described by [On], [Off], [R Series], [L Series], [Rl Series], [C Series], [Lc Series], [Rc Series], [Series Current] and [Series MOSFET] keywords.
Additional Notes: A Terminator model uses one or more of the [Rgnd], [Rpower], [Rac], and [Cac]. However, some models may have only a subset of these keywords. For example, an input structure normally only needs the [Voltage Range], [GND Clamp], and possibly the [POWER Clamp] keywords. If one or more of [Rgnd], [Rpower], [Rac], and [Cac] keywords are used, then the Model_type must be Terminator.	

10.2.4.3 External Models in IBIS

The IBIS Committee has expanded the IBIS Specification to accommodate submodels, external models, and hardware description languages. Some of today's I/O buffer circuitry has become quite complex. This is in response to user needs and increased switching speeds. The Committee realized that not all expansions of IBIS modeling capability are efficiently handled by Keyword use and expansion. Therefore, they have accommodated to supporting SPICE, S-Parameter, AMS, and Macro models. A complete discussion of these IBIS features is found in "Chapter 20, The Challenge to IBIS."

10.2.4.4 Additional Data Used in IBIS Simulators

The following data is sometimes available and used in simulating IBIS models. None of it is an IBIS Specification requirement:

Vol	Used to automate flight time measurements. Not an IBIS Specification requirement. Output low state voltage - high limit of population.
Voh	Used to automate flight time measurements. Not an IBIS Specification requirement. Output high state voltage - low limit of population.
Technology	Not an IBIS Specification requirement. Examples: GTLP, BTL, and CMOS.

10.3 SAMPLE IBIS DATA FILE

We sometimes need to construct an IBIS model data file by cutting, pasting, and editing from pre-existing, incomplete, but properly constructed IBIS data files. This often arises when working with an FPGA or ASIC. The simulation and release of a board often runs far ahead of when a final IBIS model can be released jointly by the semiconductor supplier and user.

The following pages show a shortened IBIS data file with a few pins called out and one I/O cell model. It is common for the data file of a large IC to run to hundreds of pages and include much numerical data. To keep the sample short, most of the mid-range data in the I-V and V-T tables is omitted. The sample data file is separated into the following sections:

- Header
- Component
- Pins and Pin Parasitics
- Model and Model Types
- I-V Data
- V-T Data
- End Keyword

10.3.1 Header Sample

The Header section (shown below) provides general information and is not directly used by the simulator. For details on each keyword, see Table 10-4.

```
|************************************************************************
|
|                              ACME, Inc.
|
|                    IBIS Models for Virtual device
|
|************************************************************************
[IBIS ver]        2.1
[File name]       abcdefg.ibs
[File Rev]        3333.3
[Date]          2/22/1905
[Source]        Hspice simulations (18-07: TT)
[Notes]         ACME IBIS model for abcdefg I/O.

        abcdefg LVDS drivers require the use of external
        resistors to provide the correct voltage levels.
        Please read ACME application note XXXXX "Abcdefg LVDS
        Drivers& Receivers:  Interface Guidelines" for details

       COMMENTS: Models include process corners for MIN and MAX.

[Disclaimer]       The  data  in  this  model  is  derived  from  SPICE
simulations  using  modeling  information  extracted  from  the  target
process.  While  a  great  deal  of  care  has  been  taken  to  provide
information that is accurate, this model is considered preliminary as
it has not been verified by actual silicon measurement. Treat the data
in  this  model  as  preliminary  until  actual  silicon  verification  is
performed.

[Copyright]      Copyright 2002, ACME Inc., All rights reserved
```

Table 10-4. Header keywords

Keyword	Description	Required?
[IBIS Ver]	Version of IBIS Specification used in data file. Must be the first keyword in any IBIS data file. Comment lines can precede. Informs parser and software of the IBIS template version. Thus, what to check for and use. **Comment:** Do not confuse with [File Rev]	IBIS-Yes 3Com-Yes
[Comment Char]	For defining a new comment character **Comment:** The default IBIS comment character is a "|" (pipe)	Optional
[File Name]	Name of the model data file. 8 ASCII lowercase characters max. All lower case, ending in .ibs extension for model data files, .pkg extension for separate package model data files, and .ebd for electrical board description files. 8 characters max plus extension. Must match the actual data file name **EXACTLY**. Include [File Rev], [Date], [Source], [Notes], and | comments after the comment character, |. Especially, comments on what was done to kluge together a model.	IBIS-Yes 3Com-Yes
[File Rev]	Revision number of the model data file. **Comment:** Revision numbers must have meaning.	IBIS-Yes 3Com-Yes
[Date]	Date created or last revised.	Optional
[Source]	Originator of IBIS data file.	Optional
[Notes]	Explanations and comments.	Optional
[Disclaimer]	Usually "- - for modeling only - - not guaranteed - - "	Optional
[Copyright]	Example: " - - all rights reserved - - "	Optional

10.3.2 Component Sample

The Component section (shown below) provides information on specific, marketed IC-die-package-voltage-temperature combination of a device. For details on each keyword, see Table 10-5.

```
|                    Component abcdefg
|*************************************************************************
[Component]         ABCDEFG
[Manufacturer]      ACME Inc.
[Package]
| variable          typ              min              max
R_pkg               0.258            0.1870           0.3280
L_pkg               5.1nH            2.4000nH         7.8000nH
C_pkg               1.1pF            0.8000pF         1.4000pF
|
|*************************************************************************
```

Table 10-5. Component keywords

Keyword	Description	Required?
[Component]	Name of the component modeled. Each section begins with a new [Component] keyword if the .ibs data file contains data for more than one component. **Comment:** Beginning of the IBIS description. The sub-skeleton of the [Component] keyword is the essence of simplicity. That is, unless you have a specialized part that requires specialized keywords to describe it. Examples are ICs that include features like [Series Pin Mapping] as in analog switches, and [Driver Schedule] as in the sequencing of a variable-output-drive part.	IBIS-Yes 3Com-Yes
[Manufacturer]	Maker of the component **Comment:** Second source model data is usually different from the first source model data.	IBIS-Yes 3Com-Yes
[Package]	Default R_pkg, L_pkg, and C_pkg parasitics applied globally to all component pins. Typical values must be specified. If min and max are missing they must be noted with "NA."	IBIS-Yes 3Com-Yes

10.3.3 Pins and Pin Parasitics Sample

The Pins and Pin Parasitics section (shown below) provides the complete list of pins in a component and what they are connected to internally. For details on each keyword, see Table 10-6.

```
|
[Pin]   signal_name        model_name       R_pin      L_pin      C_pin
1       XYZ+                 XYZ
2       PGND                GROUND
3       VCCO                POWER
```

Table 10-6. Pins and Pin Parasitics keywords

Keyword	Description	Required?
[Pin]	Associates the component's various I/O models to its external pins and signal names. Parameter values here override defaults in [Package]. All pins are supposed to be specified with a [Model] name. R_pin, C_pin, and L_pin can be particularized to a pin and are often optionally supplied. Pins can be specified with POWER, GND or NC (no connect). **Comment:** [Pin] a list of pins and what they're connected to in the way of I/O models, including power, ground, and NC – no connect pins.	IBIS-Yes 3Com-Yes
signal_name	Subparameters of [Pin] from same in data book. Often missing.	IBIS-Yes 3Com-Opt
model_name	Subparameters of [Pin] name of I/O type for that pin Often missing, but essential.	IBIS-Yes 3Com-Yes
R_pin L_pin C_pin	Subparameters of [Pin] Rpin, R_pin, R_pkg, Rpkg, Rdut and R_dut are all equivalent. Likewise, C_pin, etc., L_pin, etc. But, Rpin, etc., allows for parasitic values specific to a pin and overrides Rpkg and Rdut when present. Often missing, but important. See Table Note A at the end of this table.	IBIS-Yes 3Com-Yes
Table Note A: IBIS data files store their current-voltage (I-V) and voltage-time (V-T) behavioral table data for the ICs in the form of look-up tables from which a simulator retrieves the necessary data. The simulators usually interpolate as necessary between data points. The common method of solution is to use the reflection diagram method (Bergeron) at the output and input of a driver-receiver connected by a length of transmission line. If there are bias and terminating elements on a line these are handled as SPICE elements and solved as lumped circuits at the points of connection.		

10.3.4 Model and Model_Type Sample

The Model and Model Types section (shown below) provides information on one particular model (from, perhaps, many in an IBIS data file) attached to one or more pins in a component. For details on each keyword, see Table 10-7.

```
|*****************************************************************************
|                              Model XYZ
|*****************************************************************************
|
[Model]             XYZ
Model_type          I/O
Polarity            Non-Inverting
Enable              Active-Low
Vinh   = 1.4V
Vinl   = 1.0V
Vmeas=    1.2000V
Cref =  10.0000pF
Rref =    1.0000M
Vref =  0.000V
C_comp              5.0000pF              4.0000pF              6.0000pF
|
|
[Temperature Range]     25.0000              100.0000              0.000
[Voltage Range]        2.5000V              2.3750V              2.6250V
```

Table 10-7. Model and Model_Type keywords

Keyword	Description	Required?
[Model]	Begins behavioral description of one of the component's I/O cells.	IBIS-Yes 3Com-Yes
	Syntax: 20 characters max. Each model type must begin with the keyword [Model]. The model name must match the one that is listed under the [Pin] keyword. A .ibs data file must contain enough [Model] keywords to cover all of the model names specified under the [Pin] keyword, except for those model names that use reserved words (POWER, GND and NC). Model names with reserved words are an exception and they do not have to have a corresponding [Model] keyword.	

Table 10-7. (cont)

Keyword	**Description**	**Required?**
Model_type	Subparameter of [Model] Some I-V and/or V-T tables may not apply to certain I/O types. Must be one of the following: Input, Output, I/O, 3-state, Open_drain, I/O_open_drain, Open_sink, I/O_open_sink, Open_source, I/O_open_source, Input_ECL, Output_ECL, I/O_ECL, or Terminator.	IBIS-Yes 3Com-Yes
Polarity	Subparameter of [Model]—inverting or non-inverting.	Optional
Enable	Subparameter of [Model]— Active High or Active Low.	Optional
Vinh	Subparameter of [Model] – input threshold - high limit of population (for Input and I/O). **Comment**: Used to automate flight time measurements. **Syntax:** See Note B at the end of the table.	IBIS-Yes 3Com-Yes
Vinl	Subparameter of [Model] – input threshold - low limit of population (for Input and I/O). **Comment:** Used to automate flight time measurements **Syntax:** See Table Note B.	IBIS-Yes 3Com-Yes
Vmeas	When switching test circuit data is supplied, buffer delay can be recalculated for different loading and can be subtracted from total delay to get wire delay for timing simulators. Testload for dV/dt measurements. See Figures 10-19 and 10-20. **Syntax:** See Table Note B.	Optional
Cref		Optional
Rref		Optional–same as R_load
Vref		Optional
C_comp	Subparameter of [Model] - Die capacitance of the I/O cell. Should not include package capacitance. **Syntax:** See Table Note B.	IBIS-Yes 3Com-Yes
[Temperature Range]	Actual die temperature range in measurements. Default 50/0/100 C° **Comments:** Provide if different than default values **Syntax:** See Table Note B.	IBIS-Yes 3com-Yes
[Voltage Range]	Power supply voltage with tolerance. Range over which model operates. **Syntax:** See Table Note B.	IBIS-Yes 3Com-Yes
Table Note B: Typical must be specified. If data for *min* and max columns is not available that must be noted with NA. Consists of 3 columns (typ-min-max) of 9 characters max each. On each line at least one space must separate each column entry from its neighbor		

10.3.5 I-V Data Sample

The I-V Data section (shown below) provides data for all of the I-V curves associated with a particular model. For details on each keyword, see Table 10-8.

```
[Pulldown]
| voltage      I(typ)              I(min)              I(max)
|
  -2.5000      -88.8600mA          -60.2580mA          -0.1064A
  -2.3000      -88.8600mA          -60.2580mA          -0.1064A
  -2.1000      -88.8600mA          -60.2580mA          -0.1064A
      •
      •
      •
[Pullup]
| voltage      I(typ)              I(min)              I(max)
|
  -2.5000        0.2955A             0.2311A             0.3407A
  -2.3000        0.2697A             0.2106A             0.3114A
  -2.1000        0.2437A             0.1899A             0.2816A
      •
      •
      •
[GND_clamp]
| voltage      I(typ)              I(min)              I(max)
|
  -2.5000       -1.8030A            -1.7360A            -1.8280A
  -2.4000       -1.6860A            -1.6300A            -1.7060A
  -2.3000       -1.5680A            -1.5230A            -1.5850A
      •
      •
      •
[POWER_clamp]
| voltage      I(typ)              I(min)              I(max)
|
  -2.5000        7.3920nA           67.8700nA           67.5200nA
  -2.4000        6.9270nA           63.8500nA           64.0300nA
  -2.3000        6.4030nA           60.4200nA           60.7700nA
      •
      •
      •
```

Table 10-8. I-V Data keywords

Keyword	**Description**	**Required?**
[Pulldown]	Low state driver I-V table data. **BY DEFINITION** measured over –Vcc to + 2Vcc range. Referenced to ground, therefore the same as output I-V table as in data book format.	IBIS-Yes 3Com-Yes Yes, if present in device - must include at least typical data. If min and max missing they must be NA
[Pullup]	High state driver I-V table data. **BY DEFINITION** measured over –Vcc to + 2Vcc range. V-data referenced to Vcc power supply, not ground.	
[GND Clamp]	I-V table data of ESD diode structure. **BY DEFINITION** measured over –Vcc to + 2Vcc range. Can be present on inputs and/or outputs.	
[POWER Clamp]	I-V table data of ESD diode structure. **BY DEFINITION** measured over –Vcc to + 2Vcc range. Can be present on inputs and/or outputs. V-data referenced to Vcc power supply, not ground.	

10.3.6 V-T Data Sample

The V-T Data section provides voltage versus time switching data for a particular model. For details on each keyword, see Table 10-9.

```
[Ramp]
| variable        typ                    min                    max
dV/dt_r        1.4070/0.3280n         1.3026/0.4908n         1.4946/0.2610n
dV/dt_f        1.4472/0.3288n         1.3535/0.4905n         1.5291/0.2665n
R_load = 0.1940k
|
[Rising Waveform]
R_fixture = 0.1940k
V_fixture = 1.2000
V_fixture_min = 1.1000
V_fixture_max = 1.3000
| time            V(typ)                 V(min)                 V(max)
|
  0.000S          41.3000mV              53.1100mV              37.3300mV
1.000e-10S        41.2200mV              53.0700mV              37.2850mV
   0.2000ns       41.5600mV              53.1100mV              38.1400mV
   0.3000ns       42.6700mV              53.3867mV              40.3600mV
   0.4000ns       43.9800mV              53.6900mV              37.9200mV
       •
       •
       •

[Falling Waveform]
R_fixture = 0.1940k
V_fixture = 1.2000
V_fixture_min = 1.1000
V_fixture_max = 1.3000
| time            V(typ)                 V(min)                 V(max)
|
  0.000S           2.4210V                2.2680V                2.5580V
1.000e-10S         2.4210V                2.2680V                2.5580V
   0.2000ns        2.4210V                2.2680V                2.5580V
       •
       •
       •
|
| End [Model] XYZ
|
[End]
```

Table 10-9. V-T Data keywords

Keyword	**Description**	**Required?**
[Ramp]	Slew Rate of output driver. IBIS format is data columns left to right – typ-min-max – usually leading to confusion. **Includes** effect of C_comp. **BY DEFINITION:** Measured over 20% to 80% of full 0% to 100% swing of output.	IBIS-Yes 3Com-Yes
dV/dt_r	Rise Slew: measured with R_load to Vcc when driving high. Subparameter of [Ramp] 20%-80% range set to avoid non-linearities (ambiguities) made worse by -Vcc to +2Vcc range.	Yes, if [Ramp] present
dV/dt_f	Fall Slew: measured with R_load to GND when driving low. Subparameter of [Ramp] 20%-80% range set to avoid non-linearities (ambiguities) made worse by -Vcc to +2Vcc range.	Yes, if [Ramp] present
R_load	Test load for measuring [Ramp] parameters.	Yes, if not 50 Ω
[Rising Waveform]	V-T table data of rising edge. Supersedes dV/dt_r when present. Will be measured over 0% -100% of the output swing. Better than [Ramp] data because non-linearity detail provided. See Table Note C.	Optional
[Falling Waveform]	V-T table data of falling edge. Supersedes dV/dt_f when present. Will be measured over 0% -100% of the output swing. Better than [Ramp] data because non-linearity detail provided. See Table Note C.	Optional
R_fixture	Test fixture data for measuring [Rising/ Falling Waveform]	IBIS-Yes 3Com-Yes
C_fixture, L_fixture	Testload for Rise/Fall Waveform data measurements. See Figures 10-19 and 10-20.	Optional
V_fixture		IBIS-Yes 3Com-Yes
V_fixture_min, V_fixture_max		Optional
R_dut, C_dut, L_dut,		Optional

Table Note C: Inclusion of V-T curves rather than just simple slew rate / ramp rate data is becoming more common. In lower frequency and simple topologies (point-point for example) V-T curves do not add a lot of accuracy. But, in topologies that tend towards a lot of high frequency reflections (multi-drop busses with significant stub lengths) they can be a key differentiator in getting an accurate simulation. This is particularly seen in bus drivers that have been designed to avoid such reflections by having soft turnon and turnoff (example GTLP). The behavior of such devices in complex topologies cannot be modeled by simple slew rate information.

10.4 PARSING AND CHECKING IBIS DATA FILES

10.4.1 Checking IBIS Data Files for Syntax and Lexical Errors

> It is extremely important to check all IBIS data files with the IBIS Golden Parser at two critical times: when the data file has been created initially and when a new model is received.

After we construct an IBIS data file, it is essential to check it for mistakes. Most IBIS data files are too long and detailed to check manually. An IBIS data file of medium complexity can have ASCII text that is over 80 pages long. The IBIS Committee provides free, downloadable software to help check IBIS data files. The IBIS Golden Parser can be found at: http://www.eigroup.org/ibis/tools.htm

Follow the instructions about platform. Download and use the latest version, which is backward compatible with previous versions of IBIS. The easiest way to use it is to:

- Place the ibisfile.ibs and the parser ibischk4.exe in the same subdirectory.
- At a command line prompt, enter:
  ```
  ibschk4 ibisfile.ibs > ibisfile_parser_results.txt
  ```
- Check the results for errors and warnings.
  Errors will prevent the data file from working in some simulators.
  Warnings may indicate a problem that prevents getting good results.

10.4.2 Running the Golden Parser on a Sample

In the earlier topic "Sample IBIS Data File," we examined an IBIS data file for a FPGA component. It was created by cutting and pasting from a master IBIS file provided by FPGA supplier. It is normal practice for suppliers to provide a master file for FPGA where the user often finishes the IBIS data file construction. This is because until the user programs the FPGA, there is no way for the supplier to know exactly what to provide.

The sample file was checked three times with the IBIS Golden Parser. The initial IBIS data file was constructed by using Microsoft Word instead of an ASCII text editor. On the first run, there were many mistakes. After switching to an ASCII text editor, the file was re-checked with the Golden Parser. On the second run, the file still did not pass and the user received this error message.

```
IBISCHK4 V4.0.1
Checking abcdefg.ibs for IBIS 2.1 Compatibility...
ERROR - Component 'virtual-E': Model 'GROUND' for Pin '3' not
defined.
Errors  : 1
File Failed
```

The data file was then edited again. On pin 3, the word GROUND was changed to GND. On the third run, there were no error messages. The following message indicated success.

```
IBISCHK4 V4.0.1
Checking abcdefg.ibs for IBIS 2.1 Compatibility...
Errors  : 0
File Passed
```

We can see that the syntax rules are applied very strictly. This is done to maintain software tool neutrality and preserve IBIS data file portability.

10.4.3 Viewing I-V and V-T Tables Graphically

Acknowledgement:
The I-V curves in Figures 10-1 through 10-6 were generated using Cadence Design System's SpecctraQuest®.

A plot of some typical I-V and V-T tables as curves helps to visualize the concepts we have just discussed.

10.4.3.1 I-V Curves

Figures 10-1 through 10-4 show how some of the I-V tables look when plotted as curves and viewed with a Graphical User Interface (GUI).

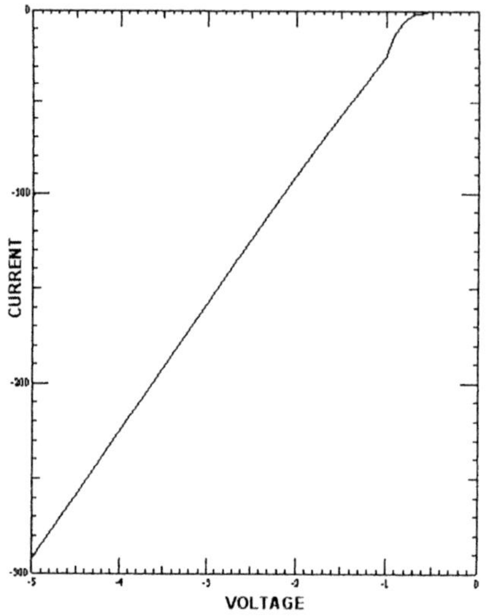

Figure 10-1. CDSDefault I/O GND clamp I-V curve

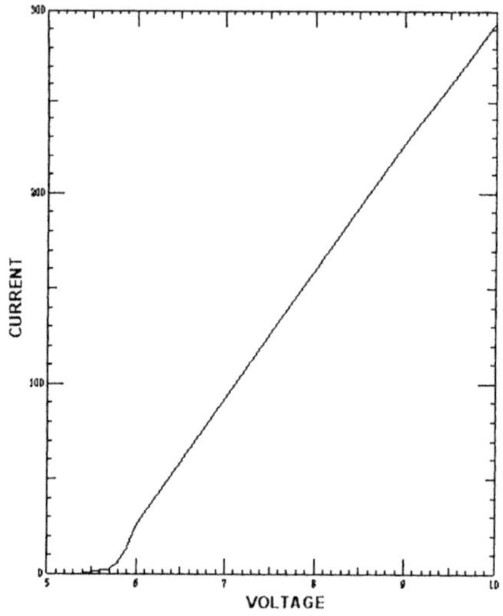

Figure 10-2. CDSDefault I/O Power clamp I-V curve

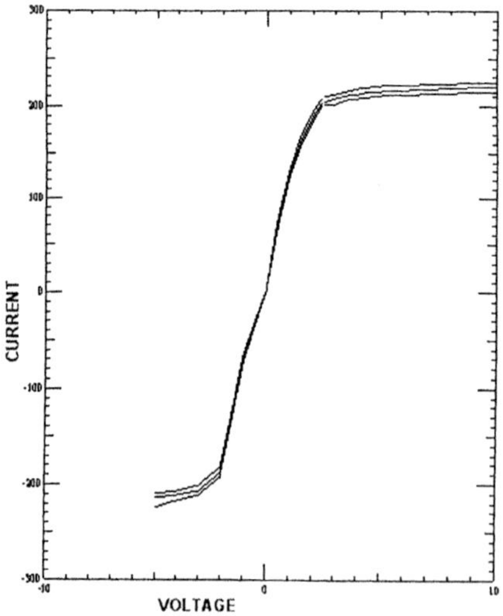

Figure 10-3. CDSDefault I/O Pulldown I-V curve

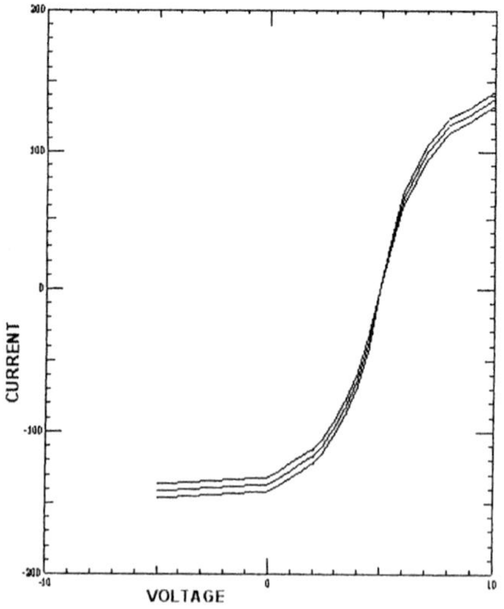

Figure 10-4. CDSDefault I/O Pullup I-V curve

10.4.3.2 V-T Curves

Figures 10-5 and 10-6 show how some of the V-T tables look when plotted as curves and viewed with a Graphical User Interface (GUI).

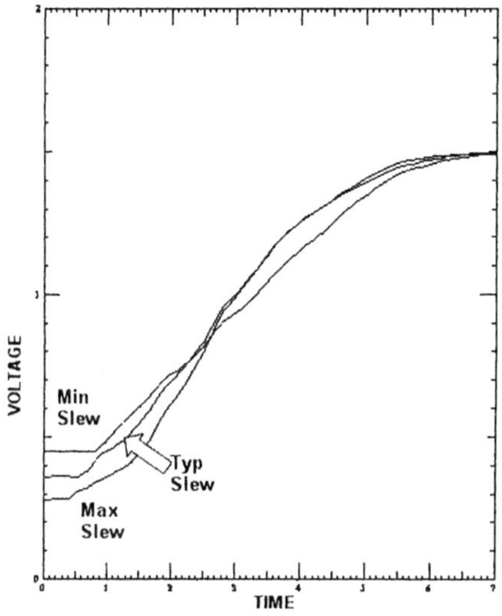

Figure 10-5. Rising waveform V-T curve

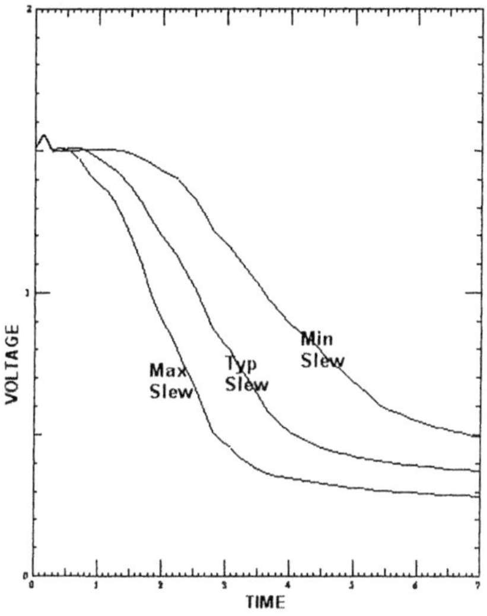

Figure 10-6. Falling waveform V-T curve

10.5 SCHEMATIC OF A BASIC IBIS MODEL

A schematic is a tool for visualizing a circuit model. Figure 10-7 shows the output side of a basic IBIS I/O cell, and Figure 10-8 shows the input side of a basic IBIS I/O cell. Schematics for all other IBIS I/O cells build on these two schematics.

The schematics in Figures 10-7 and 10-8 show what elements might be connected in a specific device. In a specific device, some elements may be missing and others may be added. For instance, either or both of the clamp diodes may be missing on a given output or input. Or, as in the case of an open pullup, the pullup element may be missing. But the elements are connected between the I/O pin (Output) and the power (Vcc) and ground pin (GND) as shown, when present in the device.

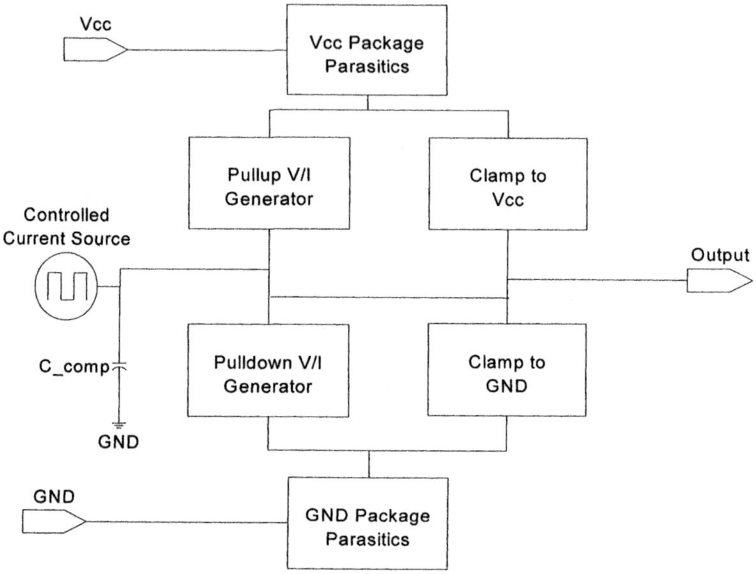

Figure 10-7. Basic IBIS model output (driver) side schematic

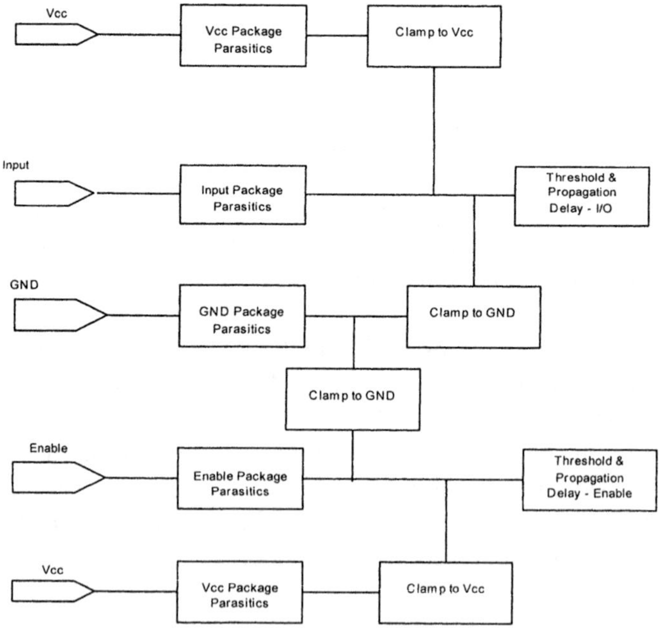

Figure 10-8. Basic IBIS model input (receiver) side schematic

What is implied when we say an element is connected between two points? In the case of the I-V data lookup tables, when a given voltage is applied across an output, a particular current results. An approximate answer is that the element acts like a dynamic non-linear resistor. The non-linear resistor describes the behavior of the pullup/pulldown/power-clamp/ground-clamp part of a digital output switch.

The Vcc and GND package parasitics shown in Figures 10-7 and 10-8 can be more technically involved than the schematics might indicate. They point out that significant parasitic elements can exist in the power and ground path inside an IC package at today's edge rates. Most IBIS data files treat the power and ground supply paths as ideal, leading to ideal (or negligibly differing from ideal) Vcc and zero node voltages. That is not the case for many real devices.

Consider how those package parasitics might be measured and modeled. The IBIS Specification says that the I-V curves are to be measured from the die pad. But in most cases, a packaged device, with parasitics lumped in with the I-V characteristics, is the device upon which the data is actually measured. Not a bare die.

We *also* know that most IBIS models are simulated from SPICE and that package parasitics are not included in the I-V characteristics. This is like the bare die situation we desire. However, if we need to do Power Integrity simulations, we are still in trouble. We need the package parasitics in the power and ground path and it is unlikely that circuit designers will be able to get that information from semiconductor suppliers.

As of this writing, the IBIS Committee is discussing modeling the power and ground parasitics in IBIS data files to enhance IBIS model capabilities. The parasitics associated with the output and input pins are straightforward and handled fairly well in IBIS. They can be visualized as part of the output and input pin connectors. Even mutual parasitic coupling between pins can be modeled. IBIS can include simple modeling of power and ground parasitics, but a more sophisticated approach is likely to be incorporated when the specification's next release is issued. The power and ground modeling proposed in ICEM 62014-3 with clarifications and improvements will probably be adopted.

10.5.1 Circuit Elements in an IBIS Data File

IBIS is not a physical model in the same sense as SPICE. Instead, IBIS is a data file with some circuit element data in it. What kinds of circuit elements are found in an IBIS data file and how are those elements connected in a circuit?

IBIS has two independent halves to it: an output side and an input side. IBIS also provides that a given cell connected to an I/O pin can be enabled either as an Output or an Input, depending on what is present in the physical device, pin use, and user instructions. IBIS can be visualized as a schematic, as shown in Figures 10-7 and 10-8; SPICE can also be visualized as a schematic. The difference is that most of the sub-elements of the IBIS model are visualized only as "black boxes." The exceptions are elements like pin parasitics, which are visualized as discrete Rs, Ls, and Cs.

For the most part, IBIS sub-elements are data, whereas SPICE and matrix model sub-elements are parameterized Rs, Ls, Cs, and dependent generators. Device physics equations and/or measured data may underlie the parameterized SPICE elements. A few IBIS data elements are single-point data at a given process corner. For example, Vih (input threshold voltage – high) is a single data point quantity that is measured at the typical process corner. But the other data elements are look-up tables of data. For example, the ground clamp I-V "curve" measured at the typical process corner.

IBIS look-up data-table sub-elements are referred to as behavioral models because they measure (and store as data) behavior at the pins of a device. Most often, this data is generated by simulation from a semiconductor company's proprietary SPICE model. It is occasionally generated by direct measurement. All of the matrix models are generated by measurement of behavior at the pins of a device. But the matrix model data gets *parameterized* and stored as circuit elements of the model. The behavioral parts of the IBIS model are its I-V data tables (for its dynamic, non-linear impedance behavior) and its V-T data tables (for its dynamic, non-linear switching signal behavior).

Strictly speaking, timing information is *not* included in IBIS. Setup and hold data and t_{co} (time-to-clock-out) are specifically not included. We would expect to find timing data in VHDL models. Signal Integrity simulation with IBIS models is commonly used for timing *verification* of targeted time-of-flight or time delay on the interconnections.

The previous topic discussed the basic IBIS I/O cell and its elements. In addition to the basic elements, an IBIS data file is also capable of containing many enhancements. For example, driver scheduling and programmable output strengths. For adding more advanced capabilities, IBIS provides a system of adding new keywords. Once the new keywords are approved, they are fairly standard and inflexible, but they ensure model data portability.

10.6 HOW IBIS CIRCUIT MODELING METHODOLOGY IS USED

The following key concepts explain how an IBIS data file is traditionally[4] used in a simulation:

- *IBIS is a behavioral model.* The behavior at its input and output pins is described, not what happens between input and output pins.
- *IBIS driver is used as an output generator signal* applied to an interconnection circuit, which transmits signal to a load.
- IBIS assumes that an *"event generator"* tells the output generator/driver to switch and send its signal down a transmission line to a load/receiver.
- IBIS helps simulate circuits where there is sufficient *time delay* for reflections to be set up by mismatches to the transmission line impedance.
- IBIS assumes that the *signal into a receiver is the end of the cause-effect chain.* That is, IBIS does not address what happens inside the receiver and thereafter.

10.6.1 Use of Driver I-V and V-T Curves to Design Terminations

The material in this topic is adapted from a presentation at Cadence's user group meeting, CDNLive! 2005, and is used with permission [135]. The topic explains the basic concepts and operation of the IBIS behavioral model.

10.6.1.1 Sequence of Events When a Driver Switches

When a driver I/O buffer switches, it sends a rising or falling signal wave edge out of its output port onto a circuit. After the switching event is over, the V-T characteristics of the driver do not enter into circuit behavior until another switch event occurs. The driver spends most of its time in the steady-state condition, where subsequent behavior is determined by its I-V curve.

The circuit characteristics determine what happens to the signal launched out of the driver. The circuit characteristics are its connections, time delay, driver and receiver impedances, passive structures, and (for transmission line

---

[4] Modern advanced I/O, such as SERDES, requires the simulation of some internal I/O behavior. See "Chapter 20, The Challenge to IBIS" for more details.

behavior) the impedance and loss characteristics of the transmission line sections.

The circuit behaves as a transmission line if round trip signal propagation time delay is two or three times longer than the rising and falling edge time delay. Mismatch signal reflections at the receiver sends a return signal back towards the driver, creating noise on the signal. Re-reflections will continue adding noise.

There are two ways to manage mismatch reflections:

- By avoiding impedance transformations.
- By matching driver and receiver impedances to the transmission line's impedance (Zo).

Receiver impedances are usually much larger than the transmission line impedance, calling for a parallel termination for matching. This has the drawback of shifting DC voltage levels on the line. Therefore, input threshold switching gets shifted. Serial termination at the driver (Rseries), whose impedance is usually less than that of the transmission line, is usually favored.

The actual voltage signal launched onto a transmission line depends on the load impedance seen directly across the driver output port and the driver's internal impedance. The instantaneous circuit formed between the supply voltage output and the return signal path is a series divider string. The string is composed of the driver's internal impedance and the load impedance seen directly across the driver output port. The assumption is that the rest of the circuit is too far away in terms of time delay (transmission line behavior) to be "seen" by events at the driver.

10.6.1.2 Driver Curves, Load Lines and Operating Point

Knowledge of the driver's impedance characteristics allows a designer to appropriately select series termination schemes and driver strengths. Information about a driver's impedance characteristics (Zout) and driver strength are contained in its I-V curves.

Let's look at some I-V curves for a 1.8V HSTL CMOS driver. Figure 10-9 reviews the normal operating regions of a CMOS gate.

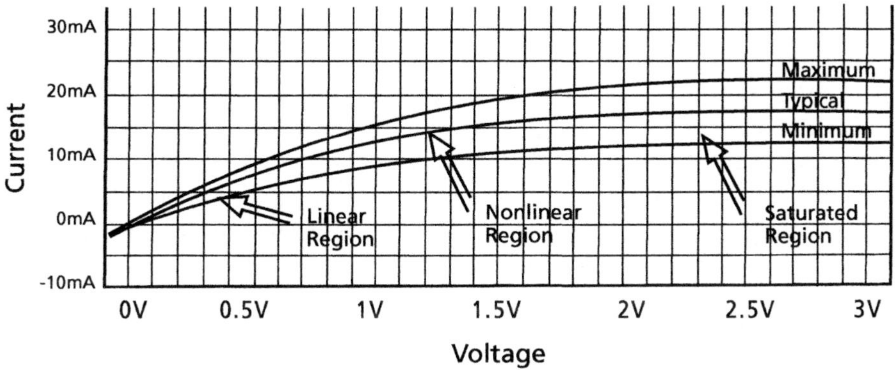

Figure 10-9. 1.8V HSTL CMOS I-V curves in the normal operating region [135]

If we know Zout, we can compute the step voltage launched on the line and we can select Rseries so that:

$$Zout + Rseries = Zo \tag{10-2}$$

Zout can be determined using load line analysis[5] on the I-V curves to determine the operating point. Figure 10-10 shows a typical load line plot.

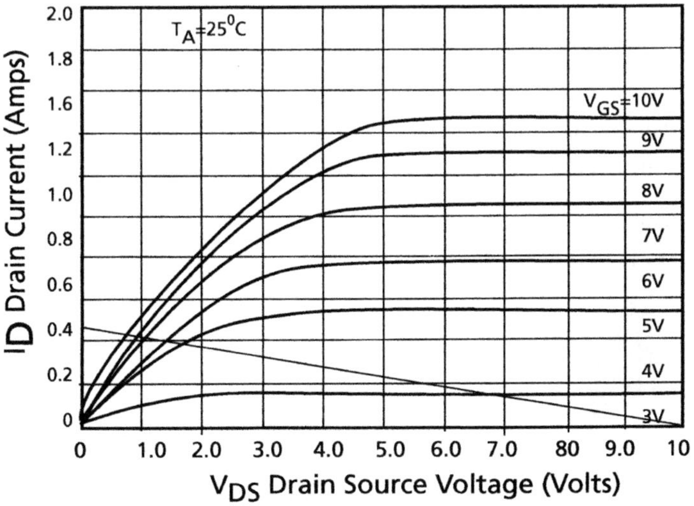

Figure 10-10. Typical load line plot on 1.8V HSTL CMOS I-V [135]

---

[5] Another term is "load line analysis for amplifier circuits." Both terms are equally valid for digital circuit analysis. Digital circuits just operate with V_{GS} saturated.

Figure 10-11 shows a set of 1.8V HSTL CMOS Pulldown driver curves, where V_{GS} = VDDQ and the load is 50 ohms. 1.8V HSTL CMOS uses reflected wave switching. Figure 10-11 is a plot with the min-typ-max data displayed when $I_D = 0$, $V_{GS} = 1.8V$ and when $V_{GS} = 0$, $I_D = 36$ ma.

The load line intersects the typical I-V curve at 23 mA and 0.64V. The output voltage switches down to 0.64 volts when this buffer switches, and it overdrives current into a 50-ohm line in the nominal case.

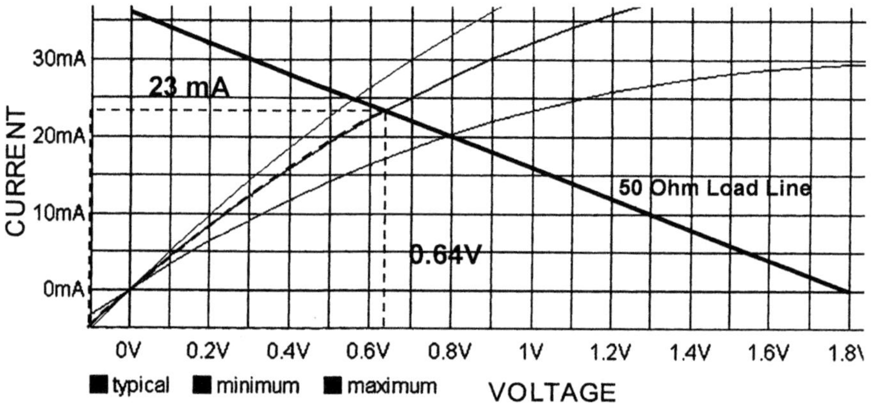

Figure 10-11. 50-ohm load line plot on 1.8V HSTL CMOS I-V with min-typ-max curves displayed [135]

Figure 10-12 shows the topology of a network using this HSTL CMOS driver. The simulation result is shown in Figure 10-13.

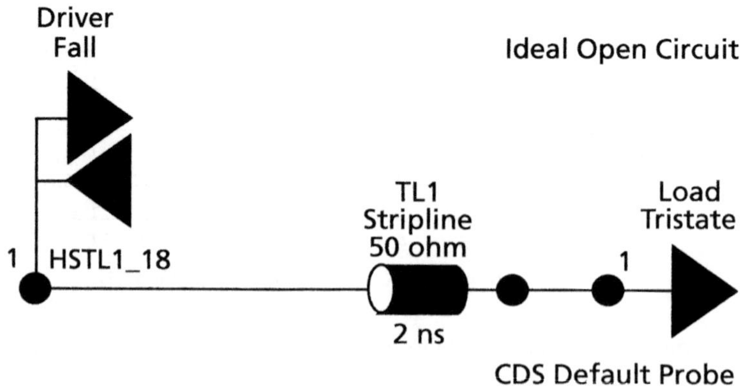

Figure 10-12. 50-ohm open line topology simulated [135]

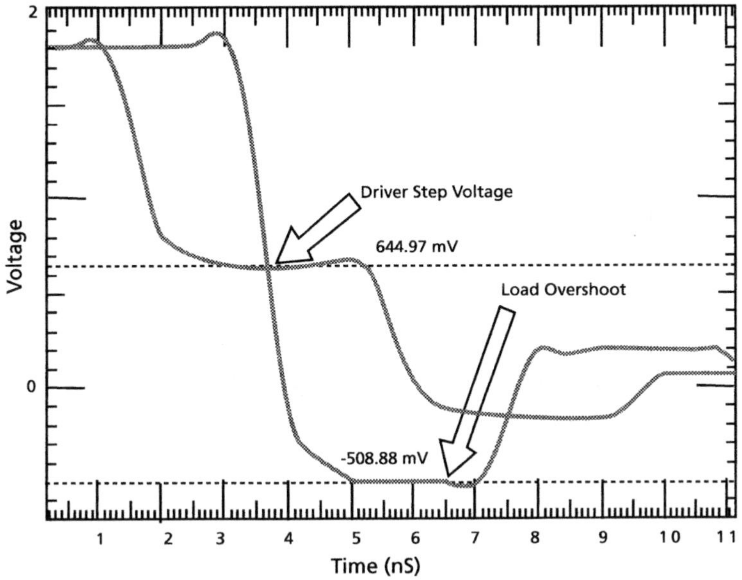

Figure 10-13. 50-ohm open line simulation results [135]

The ideal step voltage for reflective wave switching is VDDQ/2. Therefore the voltage, E, launched on the line should be E = VDDQ/2. The current launched on the R = 50-ohm line should be:

$$I = E/50 = VDDQ/2/R = 1.8/100 = 18\ mA \tag{10-3}$$

for the ideal reflection behavior at the receiver. The Zout of the driver at this operating point is:

$$Zout = I/E = 0.64/23 = 28\ ohms \tag{10-4}$$

We need to adjust the output strength and effective output impedance of this buffer.

10.6.1.3 Driver Strengths and Termination Schemes

We want to adjust the effective output impedance of this buffer so that E = 0.9V and I = 18 mA, then output resistance looks like 50 ohms.

$$Zout + Rseries = 50 \tag{10-5}$$

Since Zout = 28 ohms, Rseries = 22 ohms. Figure 10-14 shows the new network topology. The simulation result is shown in Figure 10-15.

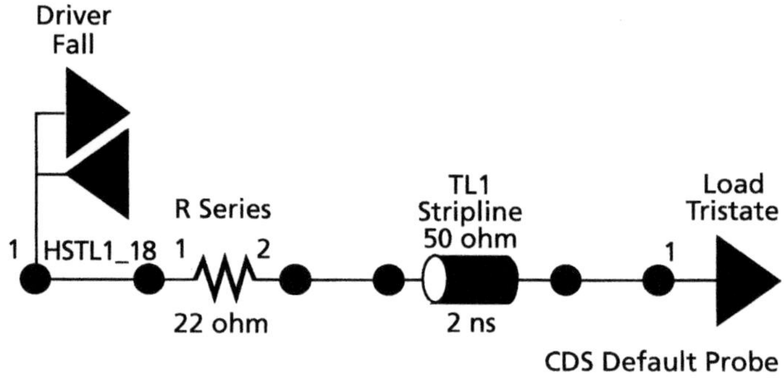

Figure 10-14. 50-ohm open line 22-ohm series terminated topology [135]

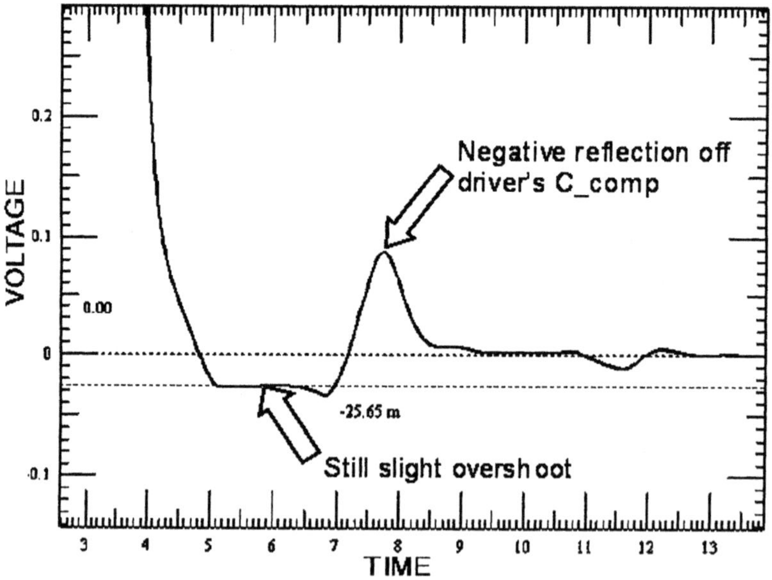

Figure 10-15. 50-ohm open line 22-ohm series terminated simulation results [135]

We still have a slight overshoot of 25mV below ground and a small amount of negative reflection off of the driver's C_comp. What is the problem?

10.6.1.4 Series Termination and Its Effect on the Operating Point

A series-terminating resistor changes the transistor's operating point. The load line will now be 50 + 22 = 72 ohms to VDDQ. If the I-V curve is not linear between the old and new operating points, Zout will change. Figure 10-16 shows the shift in load line in the nonlinear characteristic region.

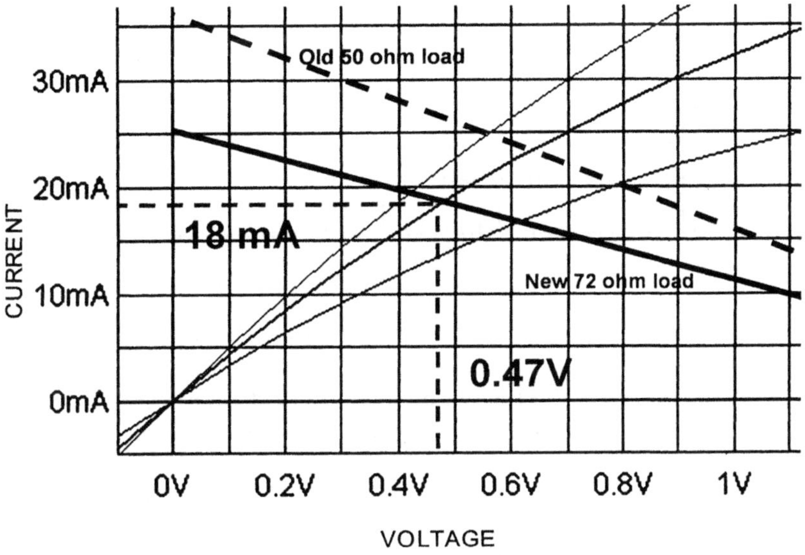

Figure 10-16. 72-ohm load line plot on 1.8V HSTL CMOS I-V with min-typ-max curves displayed [135]

The new Zout = 0.47V/18mA = 26.1 ohms. The Rseries should be adjusted to 23.9 ohms. The results of simulating with this new value of Rseries are shown in Figure 10-17.

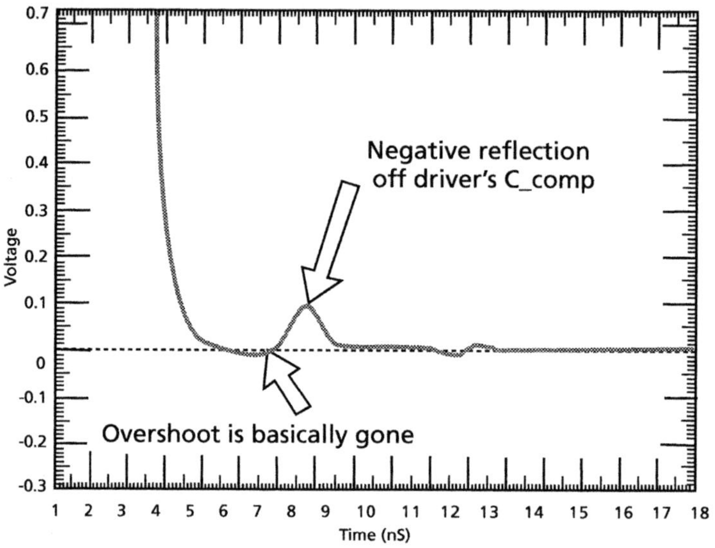

Figure 10-17. Adjusted 23.9-ohm series terminated simulation results [135]

Figure 10-18 shows the results of selecting a standard resistor value of 24 ohms and simulating for Pullup and Pulldown behavior at the receiver.

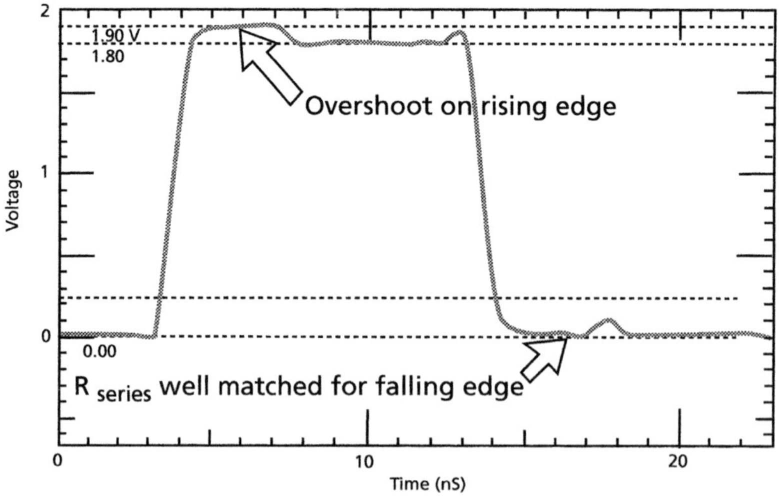

Figure 10-18. Standard 24-ohm series terminated simulation results, rising and falling edges [135]

This is a HSTL Class 1 buffer. We would expect it to be designed for point-to-point operation without termination. That requires Zout effective =

50 ohms. Zout effective ranges from 21 to 26 ohms requiring a series resistor to avoid overshoot. Pullup and Pulldown are not well matched: 21 versus 26 ohms. This can either be a buffer design issue or a modeling issue.

10.7 IBIS TEST CIRCUITS

The following schematics, Figures 10-19 and 10-20, are for test circuits in the sample IBIS data file.

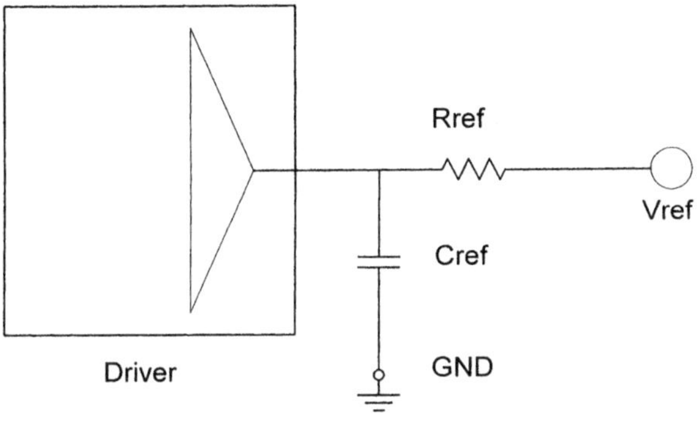

Figure 10-19. Testload for dV/dt measurements.

When this switching test fixture data is supplied, buffer delay for different loading conditions can be simulated. For details, see the subparameters associated with Ramp, Rising, and Falling Waveforms under Model_type.

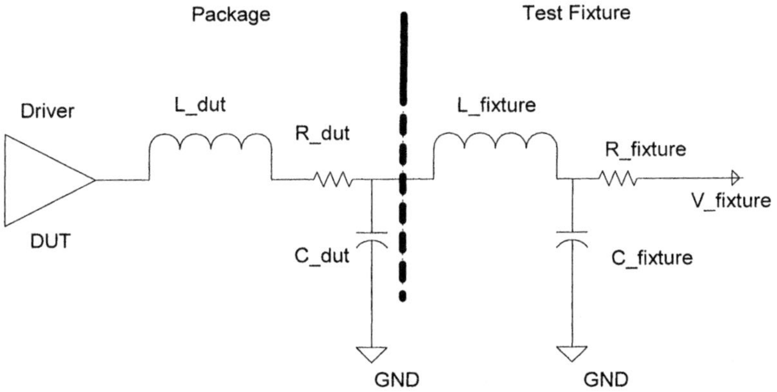

Figure 10-20. Testload for rise/fall waveform data measurements

10.8 ISO 9000 PROCESS DOCUMENTATION FOR IBIS MODELS

10.8.1 ISO 9000 Process Documentation

ISO 9000 is a standard regulating company internal quality systems. Certification, performed by Regulatory agencies, allows a company to sell electronic equipment to the European marketplace. One of the chief concerns of the regulations is the control of EMI emissions and interference with other electronic equipment. ISO 9000 specifically addresses a company's quality systems. Model integrity and the effectiveness of simulations in ensuring the design behavior of equipment gets involved in the quality systems. Therefore, the process of obtaining, validating, verifying, and using models gets involved in the quality system.

The quality system, particularly for processes and procedures, should be documented under ISO 9000. The genesis of all documentation and certification is the answer to two questions:

- Can we describe the company's processes and procedures?
- Can we demonstrate that we are following the documentation?

ISO 9000 regulations and documentation can be very helpful in promoting quality, reliability, and process improvement. But, if not done properly, ISO 9000 processes can be very constraining, bureaucratic, and difficult to meet. Much of design is situational, requires the good judgment of the designers, and cannot be spelled out in a regulatory document. The trick is to strike a balance with the hierarchy of ISO 9000 documentation and how it is enforced.

10.8.2 Hierarchy of ISO 9000 Documentation

The hierarchy of ISO 9000 documentation, listed from most important, includes:

- *Process and procedures* documentation (most important). It is what the regulators follow in their certification visits. We should expect the regulators to be strict on how the process is documented and followed. Therefore, when writing ISO 9000 process and procedure documentation, we need to avoid unnecessary detail.

- *Specifications* documentation (next most important in the examination and enforcement process). Their interpretation is loosely enforced and may not even be involved in the regulator's visit. All the same, we need to differentiate between how-to explanations and hard specification requirements. If the company writes how-to explanations, they should be written as guidelines–not process or procedures.

- *Design guidelines, explanations, and how-to advice.* These documents are rarely involved in an ISO 9000 certification. When we create documentation, the detailed explanations, advice, and what-ifs go here. This documentation should not be given an ISO 9000 document number and control sheet.

10.8.3 ISO 9000 IBIS Documentation

Tables 10-10 and 10-11 list available ISO 9000 [59] IBIS process documentation.

Table 10-10. IBIS documentation (See Appendix E)

Document	Description	Based On IBIS Version
IBIS_Syntax.doc	Details of allowed character size of data fields, format and other syntax issues separated out of the full IBIS Specification for simplification. http://www.eda.org/pub/ibis/training/3com-docs/	3.2
Internal IBIS Spec.doc	Details what the data fields are, which are required (as set by OEM purchase spec) or optional, and how the data is measured separated out of the full IBIS Specification for simplification. http://www.eda.org/pub/ibis/training/3com-docs/	3.2
Simulation Model Request Form.doc	A check off list of model data used by the product design engineer to tell the modeling and simulation engineer what data is truly required for a particular model. http://www.eda.org/pub/ibis/training/3com-docs/	3.2
Tree Diagram 4_1.txt	Similar to a table of contents. Organizes the keywords and their subparameters. http://eda.org/pub/ibis/ver4.0/tree4_0.txt	4.1
IBIS ver4_1.doc	Latest ratified IBIS Specification as of 12/26/04 http://www.eigroup.org/ibis/specs.htm	4.1
IBIS ICM ver 1_0.doc	Latest ratified IBIS Interconnect Modeling Specification as of 12/26/04 http://www.eigroup.org/ibis/specs.htm	ICM 1.0

Table 10-11. ISO 9000 IBIS model process documentation. (See Appendix E.)

Number	Title	Comments
F0673	Simulation Request Form	Lists the nets the designer wants simulated, at what frequencies, worst case, and any other special conditions. Lists the kind of reports (data) needed and any terminations to be optimized.
E0171	Simulation Model Creation and Updating Process	This is the process that will be followed in creating or updating a simulation model when it must be generated by the Signal Integrity support staff. It is mainly of interest to the Signal Integrity support staff.
E0172	Simulation Model Verification Process	This is the process followed to verify the accuracy of a simulation model.
E0167[6]	Simulation Model Procurement Process	These are the procurement steps needed to insure that a simulation model is *available* when it is needed to run simulations on a design.

10.8.4 3Com's Design Guideline Documentation

Acknowledgement
In appreciation to 3Com for developing ISO 9000 processes for handling IBIS models and for making the documentation available to the general public on the IBIS Committee website at http://www.eda.org/pub/ibis/training/3com-docs/

Affecting the ISO 9000 documentation is a strategy for dealing with a few primary specification issues:

- Not every parameter available in the IBIS Specification is needed for every device. At the same time, many parameters are optional as written in IBIS. This includes minimum and maximum values, V-T tables, and much more, but these may be too important to remain optional. Therefore 3Com provided the Product Design Engineer with a means to ask for specifics in the model.
- 3Com provided the Product Design Engineer with a means to ask for specifics in the simulations to be run by the Signal Integrity Engineer. In addition, the Product Design Engineer is required to communicate certain information—like the clock speeds for the various nets plus the power supply nets and their voltages.

---

6 E0167 is proprietary and is not publicly available. Listing it in the table highlights the need for developing such a procedure, which can involve acquisition decisions and getting expenditure approvals.

The IBIS Committee's website has a lot of material on modeling, timing, and much more [142, 143, 144]. It includes some ISO 9000 IBIS documents. For example, while working under ISO 9000, 3Com engineers captured lessons learned, design guidelines, tips, and explanations in the following documents:

- *IBIS – How To Use.doc*
  Explanations and examples of IBIS model simulations as applied to choosing which optional model elements to add to your IBIS model request
- *LogicFamilySI.doc*
  Shows a few sample simulation waveforms for comparison purposes from as many different logic families as possible.
- *SignalIntegrity.doc*
  More discussions of design methodology, keywords, and their physical explanation. Plus EMI control rules, interpreting simulation results, test measurements, and technical references.

10.8.5 IBIS Committee

Intel and a few other companies originated the IBIS model. By May 1993, 20 participating companies had formed themselves into the voluntary *IBIS Open Forum Committee* under the auspices of the Electronic Industries Association (EIA).

To ratify changes and set future directions, the IBIS Open Forum communicates via e-mail, teleconferencing, and in person. Teleconferencing occurs every three weeks on average and in-person summit meetings occur a few times a year. Many communications are handled via the IBIS email reflector. For help or to subscribe/unsubscribe, email majordomo@eda.org with the appropriate command messages in the body:

```
Commands:
help
subscribe    ibis       <optional e-mail address, if different>
subscribe    ibis-users <optional e-mail address, if different>
unsubscribe ibis        <optional e-mail address, if different>
unsubscribe ibis-users <optional e-mail address, if different>
```

- To send a request, email: ibis-request@eda.org
- To view or download the IBIS Spec, find links to model suppliers, get training materials, and much more, go to the IBIS website at: http://www.eigroup.org/ibis/

- To suggest changes and improvements, submit a BIRD (Buffer Issue Resolution Document). Go to the archive of BIRDS. At the bottom is a link for instructions on submitting one.
- To view past discussions and email threads, go to: http://www.eda.org/pub/ibis/users_archive/

10.9 SUMMARY

The IBIS model data file is the most practical model to use for most high-speed circuit simulation. It offers the best compromise between simulation speed and model complexity. That balance is accomplished by ignoring the internal behavior of drivers and receivers and by modeling just the behavior at the terminals.

Although the key concepts are simple, the details of IBIS are very complex. This makes it extremely important to have good processes and good software tools for making and validating IBIS data model files. Two important tools for validation are the IBIS Committee's Golden Parser and Quality Checklist.

Chapter 11

USING IBIS MODELS IN WHAT-IF SIMULATIONS

What happens when we change model parameters? What can we learn from virtual experiments?

Abstract: To understand how a model behaves in a design, engineers should experiment with the model and observe the simulation results. This chapter describes seven virtual experiments in which parameter values are changed and the effects of the changes are simulated. The simulated results are compared to circuit theory to ensure that they are consistent.

Acknowledgement:
All simulations and screen shots in this chapter were generated with Cadence Design Systems, Inc., EDA tools—particularly their SigNoise and SigExplorer tools, version 13.0.

11.1 A NEW METHOD OF DESIGN AND DEVELOPMENT

Traditional circuit design and development relied on building a hardware prototype to perform debug of possible design problems. Some limited use was made of modeling and simulation. Today, extensive use is made of modeling and simulation, with the objective of having the first prototype work correctly. This is also called *right by design.*

11.2 VIRTUAL EXPERIMENTS

By performing virtual experiments, in other words simulating "what-if" questions with the aid of EDA software tools, engineers gain experience and confidence with models and simulation. In these experiments, we apply the ideas of virtual design to the semiconductor models themselves. *We can do what-ifs with models that we cannot do with the physical parts.* This chapter presents several virtual experiments. Some of the experiments are very simple. Often the simpler an experiment, the better the understanding gained. We use seven experiments and study the effects of:

1. Output Drive Capability
2. C-comp Loading
3. Edge Rate and Frequency Content of a Driver on Reflections
4. Using V-T Data Versus a Ramp
5. Parasitics and Packaging
6. Environmental and Population Variables
7. Verification of Vol from Model Simulation Against the Data Sheet

We first discuss our virtual experimental techniques and the model used for our experiments.

11.3 VIRTUAL EXPERIMENT TECHNIQUES

A what-if question is a good way to learn more about a model and a simulator. For example, we may ask, "What if we change parameter x by amount y? What can we learn from virtual experiments?" To answer these questions, we perform one or more simulations.[1] When we run a simulation, we should also be able to explain the simulation results in terms of electric circuit theory.

When we experiment to carry out the what-if simulations with models, we should use the following techniques:

---

[1] Tip: There are two ways to make parameter changes: via ASCII text editors or through the tool's GUI. Some EDA tools are very flexible in letting us modify model parameters. Whenever changing an IBIS model file for what-if purposes, be sure to do it on a local copy, not the company's library file copy. Also be sure to follow similar procedures when making changes through the tool GUI.

- Study one issue at a time. In the physical world, changing one parameter may cause other parameters to change. But the beauty of a model and a simulator is that it is easy to control one factor at a time.
- Eliminate the complication of signal reflections by using perfectly terminated transmission lines. Many of the examples in this chapter make use of perfectly terminated lines, or lines with small, frequency-independent reflections to simplify an investigation.
- Change several parameters at a time plus study the effects of reflections. This last technique approximates looking at the effects of process and environmental variables. Process and environmental variables are not directly accessible for modification in IBIS, but are changed by simulating from the typ-min-max corners in the IBIS model file.

Key issues in any circuit design often include: signal source drive capability, loading, frequency-related effects, parasitics effects, process variability, and the effects of voltage and temperature variation.

11.4 PROPAGATION DELAY IN HIGH-SPEED NETS

The behavior of a high-speed network is often dominated by propagation delay and impedance mismatch reflections on a transmission line. During this chapter's experiments, reflections can make it difficult to separate out different driver behaviors. We start by eliminating those reflections, which simplifies the simulations.

The experiments in this chapter follow the roadmap below:

1. *We use a perfectly matched, pure resistive load.* There will be no reflections. Output voltage waveform behavior is directly dependent on the output impedance of the driver and the transmission line impedance.
2. *We use a slightly mismatched, pure resistive load.* There will be a small reflection that is independent of the frequency content of the driver switching waveform. This adds small amounts of reflection effects back into the simulation.
3. *We use a slightly mismatched, resistive load in parallel with some receiver input capacitance.* There will be more complex reflections that depend on the frequency content of the driver switching waveform and are more realistic.

After we learn about the driver behavior, we add the reflections back in to make the simulations more realistic. When we add back in the effects of driver C_comp, and driver and receiver clamping, behavior gets still more complex.

11.5 WHY WE USE THE IBIS MODEL

Throughout this chapter, we use the IBIS model because that is what we want to learn about. IBIS models provide a good balance between detail and simulation speed. Simulators that use the IBIS model load the following kinds of information:

- The IBIS data file I/O buffer information including I-V, V-T, parasitics, switching thresholds, and other data.
- Transmission-line models, either extracted from PCB layout or as prototypes from schematic layout.
- SPICE models for termination resistors, capacitors, and protection diodes.

After loading and extracting the necessary models and data, the EDA tools proceed to run the simulations.

At a single point (node) on a circuit, all HSDD simulators are basically SPICE simulators solving Ohm's Law in the Time Domain. The simulator solves for currents and voltages through and across SPICE elements. IBIS model currents and voltages are found in the look-up tables of the model file. Currents are summed at nodes and analysis proceeds.

At this point in our reasoning we ask, "How is any of this reasoning different than solving an Ohm's Law circuit?" Answer: It is not different. But then a difference does happen with high-speed signals. When signal time delay across the network is long enough compared to signal transition times, reflections can occur. The delay sets up the potential for reflections due to impedance mismatches. Then the interconnecting network acts as a transmission line. The transmission line becomes significant, even dominant, in what happens to the signal.

How are the transmission line sections handled in the analysis (simulation)? The most common method used to solve the transmission line behavior is the bounce (Bergeron) diagram method [14]. Signals are propagated from driver to receiver. When reflections occur, the magnitude of the reflected current and voltage signals are computed and propagated back and forth on the transmission line. That is, reflected signals are bounced back and forth on the transmission line until they settle out to the steady state or until another switching event takes place. V-T and often I-T waveforms are computed at appropriate points (such as driver and receiver) by summing the reflected waves at those points.

To visualize the IBIS driver-switching event, we should think about the IBIS dV/dt data. Figure 11-1 shows a current source charging a capacitor. The voltage is a ramp with a slope of i/C as in: i = C(dV/dt). This C is not identical with C_comp.

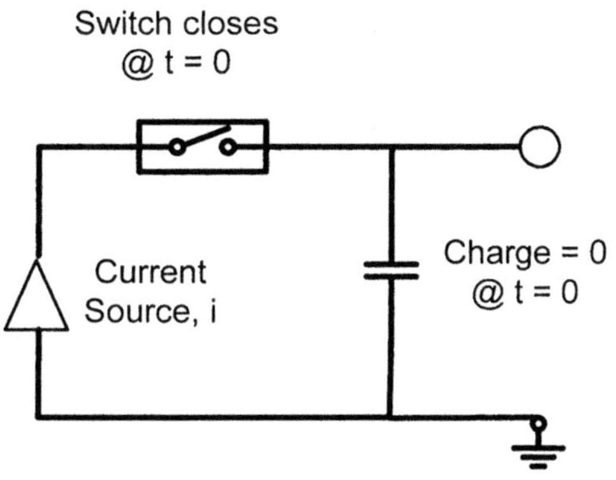

Figure 11-1. Envisioning the IBIS internal generator

Actually, the ramp rate is limited by all the internal driver linear and non-linear small-signal and large-signal characteristics, including real and effective reactive elements—not just slew-rate rise-time-limited. Internal Miller Capacitance [87] also plays a role. From a simulator point of view, the ramp slew-rate limit is a behavioral model of an internal rise-time-limited component. In any event, something (a state event) switches the output, which then ramps the voltage up depending on this internal drive limit and the loading seen at the driver.

Several other elements in an IBIS data file permit a more complete Signal Integrity simulation of a complete IC package. Additional elements of information can be included in the model file: model I/O cell connections, and power and ground assignments to groups of I/O cells. We can define parasitics for each pin and possible internal coupling between pins, either on a generic or an individual basis. Thus, power and ground pins that are separated out and grouped or bussed together can be modeled. We can also define pins that act as differential pairs.

11.6 DATA USED IN EXPERIMENTS

Figure 11-2 presents an example set of IBIS I-V data for use in the virtual experiments. The data from the Cadence Design Systems, Inc., IBIS model CDSDefaultIO.dml file was copied to use as an example.

```
(PullDown
    (Reference Voltage
     (minimum 0 )
     (typical 0 )
     (maximum 0 ) )
 (VICurve "-5.0 -215ma    -210ma    -225ma
          -4.0 -212.0ma -207.0ma -217.0ma
          -3.0 -207.0ma -202.0ma -212.0ma
          -2.0 -188ma    -183ma    -193ma
          -1.0  -70ma     -65ma     -75ma
           0.0    0ma      0ma       0ma
           0.5   70ma     65ma      75ma
           1.0  127ma     122ma     132ma
           1.5  164ma     159ma     169ma
           2.0  188ma     183ma     193ma
           2.5  203ma     200ma     208ma
           3.0  207ma     202ma     212ma
           3.5  210ma     205ma     215ma
           3.0  207ma     202ma     212ma
           3.5  210ma     205ma     215ma
           4.0  212ma     207ma     217ma
           4.5  214ma     209ma     219ma
           5.0  215ma     210ma     220ma
          10.0  220ma     215ma     225ma" ) )
||
  (PullUp
    (Reference Voltage
     (minimum 5 )
     (typical 5 )
     (maximum 5 ) )
    (VICurve "10    -142ma    -137ma    -147ma
           5    -137ma    -132ma    -142ma
           4.5  -133ma    -128ma    -138ma
           4    -128ma    -123ma    -133ma
           3.5  -123ma    -118ma    -128ma
           3    -118ma    -113ma    -123ma
           2.5  -110ma    -105ma    -115ma
           2      -98ma     -93ma   -103ma
           1.5    -83ma     -78ma    -88ma
           1      -64ma     -59ma    -69ma
           0.5    -38ma     -33ma    -43ma
           0       0         0         0
```

Figure 11-2. Cadence CDSDefaultIO.dml file (1 of 2)

```
        -1     64ma      59ma      69ma
        -2     98ma      93ma     103ma
        -3    118ma     113ma     123ma
        -3    118ma     113ma     123ma
```

```
        -1     64ma      59ma      69ma
        -2     98ma      93ma     103ma
        -3    118ma     113ma     123ma
        -4    126ma     121ma     131ma
        -5    137ma     132ma     142ma" ) )
||

 (GroundClamp
   (Reference Voltage
    (minimum 0 )
    (typical 0 )
    (maximum 0 ) )
   (VICurve "0 0 0 0
        -0.1 0 0 0
        -0.4 -0.1ma -0.1ma -0.1ma
        -0.5 -0.5ma -0.5ma -0.5ma
        -0.6 -1.2ma -1.2ma -1.2ma
        -0.7 -2.4ma -2.4ma -2.4ma
        -0.8 -6ma -6ma -6ma
        -0.9 -13ma -13ma -13ma
        -1.0 -25ma -25ma -25ma
        -5.0 -293ma -293ma -293ma" ) )
||
  (PowerClamp
   (Reference Voltage
    (minimum 5 )
    (typical 5 )
    (maximum 5 ) )
   (VICurve "0.0 0 0 0
         -0.1 0 0 0
         -0.4 0.1ma 0.1ma 0.1ma
         -0.5 0.6ma 0.6ma 0.6ma
         -0.6 1.2ma 1.2ma 1.2ma
         -0.7 2.4ma 2.4ma 2.4ma
         -0.8 6ma 6ma 6ma
         -0.9 13ma 13ma 13ma
         -1.0 25ma 25ma 25ma
         -5.0 293
```

Figure 11-2. Cadence CDSDefaultIO.dml file (2 of 2)

Note that the clamp curve currents in Figure 11-2 for typ-min-max are all equal at a given voltage. This means the data is entered in the file, but the process corners for min-max are not represented. In the Cadence Design Systems' Inc. GUI, this data can be plotted as curves.

11.7 EXPERIMENT 1: OUTPUT DRIVE CAPABITY VERSUS LOAD

The first experiment is concerned with the question: *What happens at a driver and a receiver due to the internal drive capability and the loading on the driver?*

All simulation results waveform plots use the following key:

- D = driver output.
- R = receiver input.

11.7.1 Driver Output Impedance and Matching

We now take a look at how the internal drive limit and the loading interact with the transmission line impedance, Zo.[2] The output impedance of a driver is significant in two ways.

- First, output impedance is of interest in terms of how it matches the Zo of the transmission line to which the driver is connected.
- Second, it is of interest when understanding what the magnitude of the wave it launches down the transmission line will be.

IBIS does not include any parameters for driver (or receiver) impedance directly. But such information is implicit in a device's I-V curves. We need to think of the Zout of the driver chip itself as the slope of those curves:

$$Zout = \Delta V / \Delta I \tag{11-1}$$

The V/I curves can be quite non-linear. But, in the active, unclamped midrange they can usually be approximated by a straight line calculation of:

$$Zout = |(V_2 - V_1)/(I_2 - I_1)| \tag{11-2}$$

with the absolute value signs, | |, indicating that Zout will be real, and non-regenerative, for most any situation encountered in Signal Integrity analysis.

---

[2] Load line analysis with driver-characteristic curves is a more detailed way to match the driver to the transistor line impedance. This method is presented in "Chapter 10, Key Concepts of the IBIS Specification."

Zout is usually quite low and less than the Zo of the transmission line. Thus, it's quite common to see a series-terminating resistor matching a driver to a transmission line where the object is to have:

$$Zo = Ztermination + Zout \qquad (11\text{-}3)$$

When a driver sees the reflected energy wave returning from a receiver (or impedance discontinuity), it can be either in an on or off state.

- If on, Zout is given by equation (11-2).
- If off, Zout goes to a high impedance state.

It is somewhat more likely to parallel terminate if the round trip reflection propagation time delay is large enough (greater than the pulse width) for the driver to have turned off.

The effects on matching Zo of C_comp and pin parasitics (C_pin, R_pin and L_pin) are added in the same manner as discussed under Receivers in the next topic. Figure 11-3 shows how this topology is connected.

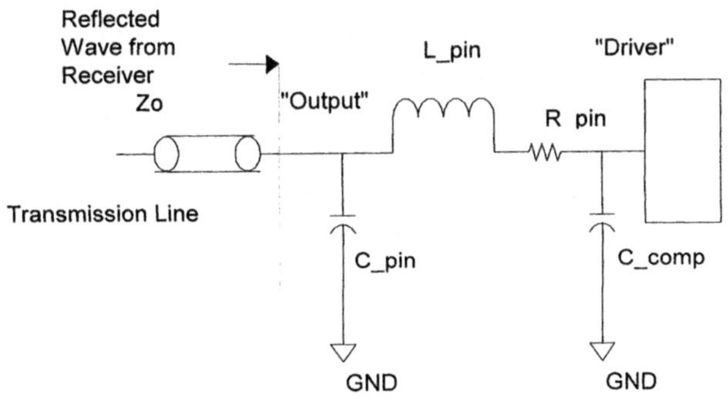

Figure 11-3. Driver output parasitics

As mentioned above, Zout is also of interest when analyzing its effects related to a low impedance and/or heavily loaded line. So far in our discussion, we have been assuming that our driver is a constant voltage generator. A constant voltage generator assumes a zero or insignificant internal generator impedance, Zg.

The condition that Zg be insignificant is not always met. Since Zg = Zout, we can look to see whether it is insignificant. The fraction of the driver output voltage swing launched down the line will be given by:

$$Vout' = (Vout)(Zo')/(Zo' + Zout') \qquad (11\text{-}4)$$

Where Zout' = driver impedance plus internal parasitics, and Zo' = net transmission line impedance presented to the driver. The impedance presented to the driver includes the effects of lumped distributed loads and parallel termination at the driver. Likewise, the voltage sent out of a series terminator at a driver will be reduced because the terminator will act like an added internal generator resistor.

11.7.2 Receiver Input Impedance and Matching

In the discussions of impedance and reflection coefficient, we alternate between talking about I-V curves and impedance. These terms are synonymous as used here. Impedance and transmission lines refer to circuitry, while waveforms refer to signal generation and content. Signal generation and content are not generally discussed until we get to ramp rates and V-T curves. The response of a circuit to a digital signal involves impedance, parasitics, and impedance matching.

Almost all receivers have very high input impedance. The input impedance changes when clamping diodes are present and the input voltage overshoot turns them on. Once the clamping diodes turn on, they ideally present a short circuit to the clamping rail, whether it is Vcc or ground.

Within the active range, the input impedance, Zin, consists primarily of the loading capacitance that the device presents to the circuit (as a first approximation). This capacitance usually consists of C_comp from input pin to ground. Pin and package parasitics can add a couple of pF to C_comp.

Matching the receiver input impedance, Zin, to the transmission line impedance, Zo, often consists of putting a shunt termination equal to the transmission line impedance in parallel with the receiver. More exactly, for a matched load:

$$Zo = Zload = (Ztermination)(Zin)/(Ztermination + Zin) \qquad (11\text{-}5)$$

Where Zload represents the total adjusted input impedance. The degree to which the input impedance (with or without matching termination) does not satisfy this equation is the degree to which energy from the incident switching wave is reflected back towards the source. We can analyze the effect mathematically by looking at *reflection coefficient*, ρ:

$$\rho = (Zload - Zo)/(Zload + Zo) \qquad (11\text{-}6)$$

ρ can vary from +1 to –1, or +100% to –100%. The + sign means that the reflected wave is the same polarity as the incident wave, and the – sign means that the reflected wave is of the opposite polarity from the incident wave.

When looking at input impedance, the next level of detail is to add in the effects of pin parasitics per the circuit of Figure 11-4.

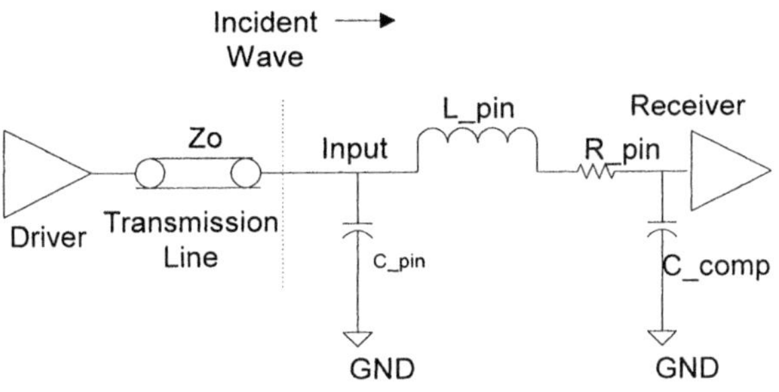

Figure 11-4. Receiver input parasitics

When computing the input impedance to the receiver, we consider the effects of C_pin, R_pin, L_pin, and C_comp in the topology shown. This will become the new Zin in equation (11-5).

An even more detailed look at input impedance would replace the single line pin parasitic model consisting of C_pin, R_pin and L_pin in Figure 11-4 with an N-by-N RLGC matrix set that would include the effects of mutual coupling and conduction between pins and between pin and package.

So far we have discussed the effects of loading and matching at the driver and receiver. We have ignored the true nature of the driver generator impedance other than to mention that the I-V data characterizes it. We now begin to investigate what happens with driver output impedance (sometimes called driver strength) by the simple step of scaling the I-V data. That is, we will use the original model presented earlier in "Data used in Experiments" and create a clone of the model, where the current magnitudes are multiplied by 4 (¼ the output impedance, 4 times the strength). Using the original, and X4 model, we will see what happens with simulation.

11.7.3 Scaling the I-V Data and Simulating the Results

The data for the original CDSDefaultIO.dml model has an output impedance in the range of 20 ohms on the pullup side and 10 ohms on the pulldown side.

We will first drive lines perfectly terminated in their characteristic impedances (no reflections from the receiver) of $Zo_1 = 72$ ohms and $Zo_2 = 18$ ohms and look at the results. Next, we will multiply the currents at a given voltage times 4 (makes the driver output impedance ¼ as large as originally) and again simulate and look at the results.

In Figure 11-5 the first pulse train is the driver at 50 MHz. The 3nS delayed (t_{pd} of transmission line= 3nS) pulse train is the receiver. Next we will drive the 18-ohm line with this model. There are no reflections, as expected, and Vout is about 4.13 volts.

Figure 11-6 shows that Zout is effectively about 18 ohms because Zout ≈ Zload gives approximately half the supply voltage out. There are no reflections, as expected, and Vout is about 2.1 volts. Next, we'll drive the 72-ohm line with the 4X scaled up driver.

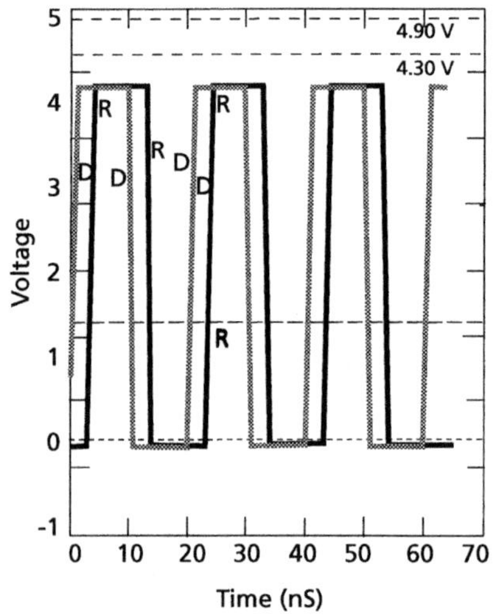

Figure 11-5. CDSDefault model driving 72 ohms
Vout ≈ 4.13V. Driver Zout estimated at about 18 ohms. Receiver matched to the transmission line.

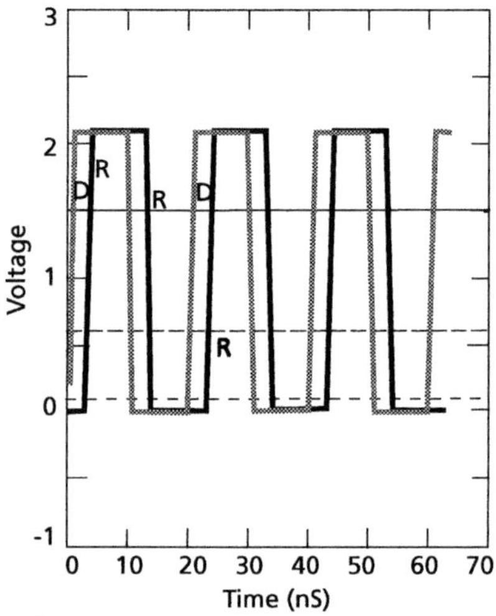

Figure 11-6. CDSDefault model driving 18 ohms
Vout ≈ 2.1V V ≈ ½ our original Vout.

Figure 11-7 shows that Vout is close to the full output capability of the driver ideal internal voltage generator, or 5 volts. Here, our Zout is about 4.5 ohms or only about 6.25% of the 72-ohm load on the driver.

Finally, we'll drive the 18-ohm line with our 4X scaled up driver. Figure 11-8 shows that driving a line with ¼ the 18 Ohms Zo of our first simulation, with a driver having 4 times the drive capability, gives the exact same output swing of 4.13 volts. Just like Ohm's Law claims!

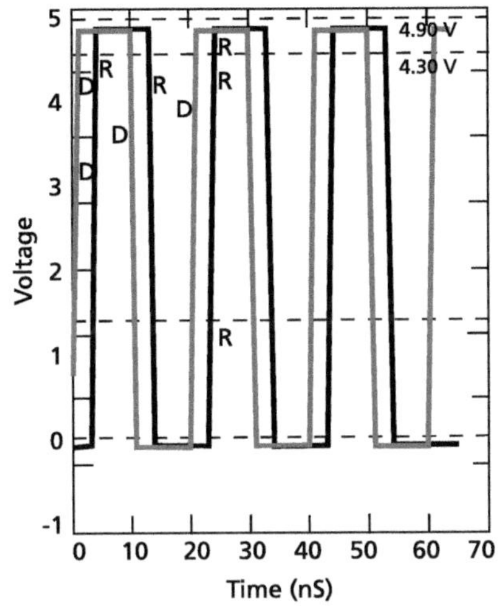

Figure 11-7. 4X I-V model driving 72 ohms
Vout ≈ 4.8V

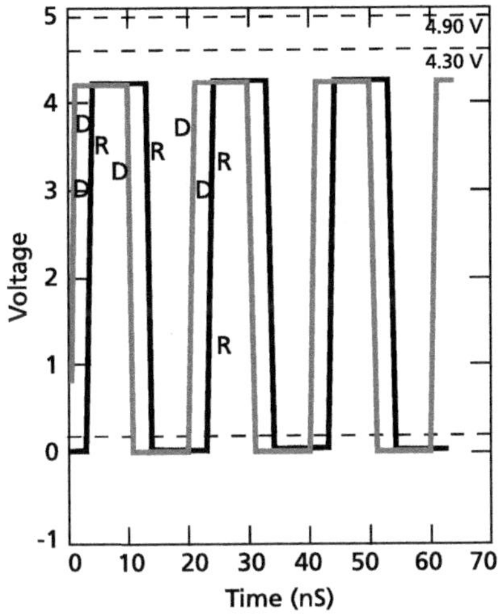

Figure 11-8. 4X I-V model driving 18 ohms
Vout ≈ 4.13V

Summarizing these results in Table 11-1 shows that, in the absence of reflections, the voltage out of a driver exactly follows Ohm's Law – as expected. No great surprise. But when learning how to manipulate the model, specifically its output impedance, we have picked up new understanding and skills.

Table 11-1. Results of driver impedance versus load impedance

Driver Model	Vsupply (V)	Vout (V)	
		Zo_1=72Ω	Zo_2=18Ω
CDSDefault Model, Zout≈18Ω	5.0	4.13	2.1
4XCDSDefault Model, Zout≈4.5Ω	5.0	4.8	4.13

In the real world, we hardly ever have a perfectly matched termination at the receiver. The capacitive loading of the receiver C_comp will almost surely set up some reflections of the energy at the highest frequencies present in our driver output waveforms. This leads us to the next set of experiments that correspond to a more real-world situation.

11.8 EXPERIMENT 2: C_COMP LOADING

Experiment #2 is concerned with the question: *What are the effects of C_comp loading on the driver?*

11.8.1 Capacitive Loading Effect On Reflections

When reviewing the effects that C_comp has on signals, we learn that C_comp affects the following electrical behavior:

- The matching impedance presented to a transmission line that can lead to signal reflections.
- The capacitive loading and rolloff filtering of high frequencies that can sometimes be seen in a simulation.
- The setting of ramp rate out of a driver.

This last bullet needs to be thoroughly understood. It is important to know that the effect of C_comp on dV/dt or V-T *is already included* in the dV/dt and V-T data. Wrong answers will be calculated if a simulator uses the C_comp data as an additional load on the output and adjusts the initial ramp rate accordingly. This will be double-counting the effect of C_comp. On the receiver side, the problem of double-counting C_comp's effect does not arise.

The actual ramp rate will also be affected by the stray parasitic capacitance due to the bondwire from die to pad. Driver Vout is defined as measured at the *die pad.* Usually this effect is negligible.

The actual ramp rate will also be affected by the external capacitive loading on the output of the driver. This is primarily due to the capacitive loading of the transmission line and the lumped loading of any termination circuitry placed on the driver.

In the previous topic, "Scaling the I-V and Simulating the Results," we used a 72-ohm load to ideally terminate a 72-ohm line. We used the CDSDefaultIO model as a receiver, reduced its I-V curves to a single characteristic of a 72-ohm resistor, and reduced its C_comp to zero. Thus we constructed the 72-ohm load. The same approach can be used to construct an open-circuit-receiver load and then to add back in various values of C_comp. This creates an idealized receiver model without clamps, in which we can study the effects of C_comp alone.

In our default model t_r(IBIS) = 0.8nS, a critical frequency of interest is 1/edge-rate, which is 1.25 GHz. Another critical frequency of interest is F_{knee} = 0.5/edge-rate [60], which is 625 MHz. Most of the digital pulse energy is concentrated below this frequency. Reactive impedance of a capacitor is:

$$Xc = 1/j2\pi fC \qquad (11\text{-}7)$$

At the two frequencies and picking three values of C_comp, we get the magnitude of Xc as shown in Table 11-2:

Table 11-2. Some capacitive loads versus frequency

Frequency (MHz)	|Xc| (ohms)		
	2.5pF	**5 pF**	**10 pF**
625	101.92	50.96	25.48
1250	50.96	25.48	12.73

Reflection Coefficient, ρ, is given by:

$$\rho = (ZL - Zo)/(ZL + Zo) \quad (11\text{-}8)$$

On a 72-ohm line, we get the reflection coefficients shown in Table 11-3:

Table 11-3. Reflection coefficients for the above loads

Frequency (MHz)	ρ		
	2.5 pF	**5 pF**	**10 pF**
625	.172	-.171	-.477
1250	-.171	-.477	-.7

The following three figures show the results of simulating a pure capacitive load (no clamping) of:

- 2.5 pF (Fast Simulation) (see Figure 11-9)
- 5 pF (Typical Simulation) (see Figure 11-10)
- 10 pF (Slow Simulation) (see Figure 11-11)

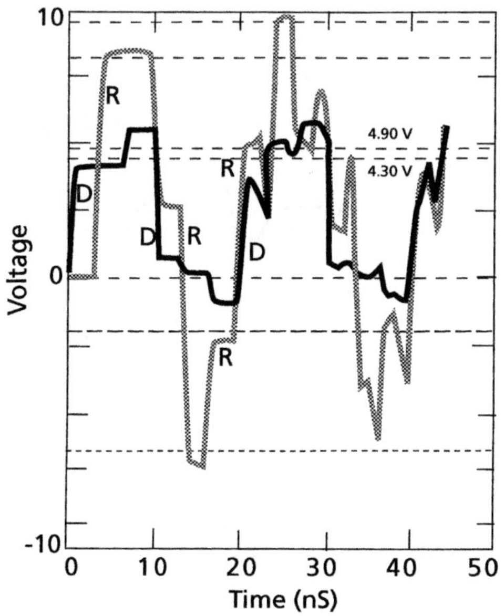

Figure 11-9. Ideal open circuit receiver with C_comp = Cin = 2.5pF, fast driver

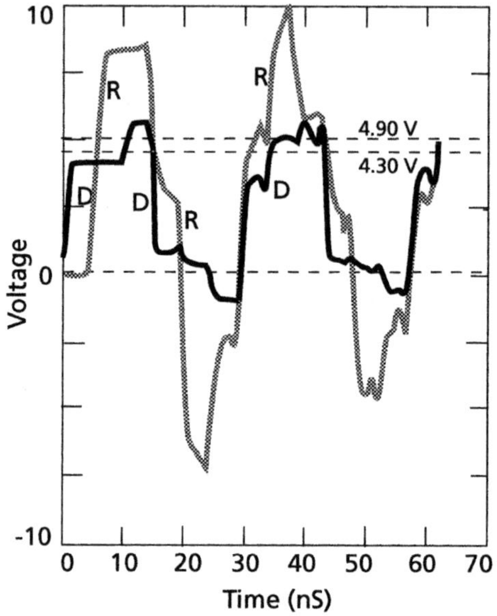

Figure 11-10. Ideal open circuit receiver with C_comp = Cin = 5pf, typical driver

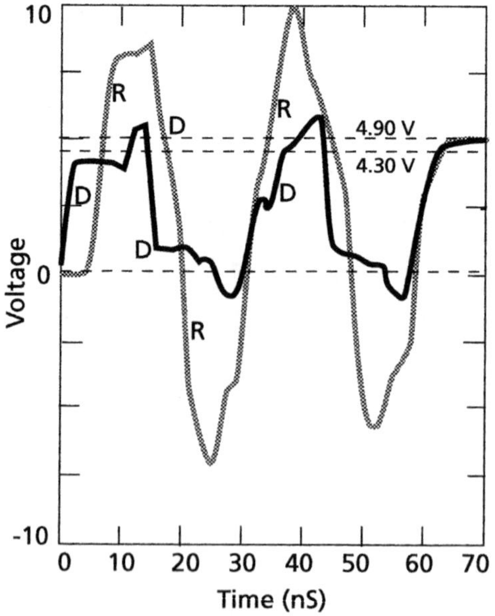

Figure 11-11. Ideal open circuit receiver with C_comp = Cin = 10pf, slow driver

Figure 11-12 shows an additional non-ideal capacitive load simulation with a standard default CMOS Input (6 pF) at 50 MHz.

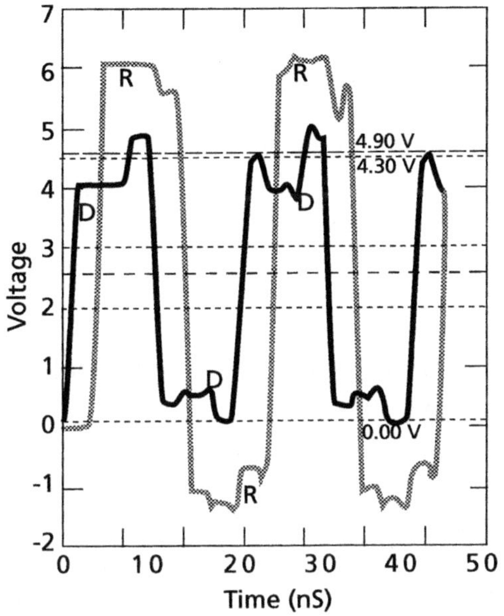

Figure 11-12. Default CMOS input receiver with C_comp = Cin = 6 pf and clamping, slow driver

In Figures 11-9 to 11-11, we see the combined effects of the reflection coefficient, frequency content of the edge rate and frequency rolloff of the input capacitance. Plus in Figure 11-12, we see the add-in effects of clamping.

Scaling the size of C_comp in an IBIS model and re-simulating is just a matter of editing the IBIS file. This is good for doing what-ifs regarding C-comp's effects. But in a real device, C_comp will be interdependent with slew rate. So from this manipulation, we can only draw general conclusions.

By making the area of a driver larger, the current drive capability can be increased. This also increases C_comp. So for drivers made with a given technology, we end with some intrinsic figures of merit that can be difficult to improve on.

11.8.2 Is C_comp Being Double-Counted?

The simulation of an IBIS driver model should include the effect of C_comp in the slew rates and/or V-T curves. Varying C_comp is supposed to have no effect on driver rise and fall as opposed to adding a variable capacitance across the output.

One way to test this is to reduce C_comp to nearly zero and then raise it to a very large value, say 50 pF. Nearly zero, but ≠ 0, avoids computation problems that might arise if the value ends up in a denominator in the computation process. If the buffer rise time changes as C_comp is varied, then C_comp may be getting double counted in the simulation. In that case, we suggest simulating the rise and fall time of the IBIS test fixture with the correct values of C_comp entered in min-typ-max. Then, we can observer whether the IBIS slew rate and/or V-T curves can be reproduced. We should use the IBIS values, as the data sheet values are usually guard-banded.

11.9 ALL-IMPORTANT ZO: ALGORITHMS AND FIELD SOLVERS

An accurate value of Zo is an all-important quantity in modeling and simulating high-speed circuits. Field solution results have been shown to correlate within one or two percentage points to the measured results.[3]

Using algorithms (or formulas)[4] often results in answers that are off by +/-5% and even as much as much as +/-10%, depending on which algorithm is used. It's better to use an EM field solver simulator for extracting Zo, propagation velocity, coupling capacitance, and other transmission line behavior. Besides using an EM field solver, we need to correctly model the board cross-section, etch characteristics, and dielectric characteristics. Across the cross-section, the dielectric will often be non-uniform, especially near the etch profile. This variation can cause a very significant difference in results.

To save on computing time and power, most simulators use a 2½ D (dimension) field solver. In this approach, the field is solved in two dimensions in an x-y plane perpendicular to the etch. The velocity of propagation along the etch is then used in the z-direction. This limits the solver's accuracy at etch transitions, corners, vias, and split planes, and other complex structures. But this is not of practical concern until we begin to switch at edge rates in the 100 pS range.

---

[3] See "Appendix H, Production Realities" for the difficulties in controlling actual Zo in a production environment.

[4] The formulas that were developed to solve PCB transmission line structures stem mostly from empirical curve fitting research that started in the early 1950s. This was before the availability of sufficient computing power to go the field solution route. The chief problem with their accuracy is that they only apply to situations that closely match those in which the curves were generated

11.10 EXPERIMENT 3: EDGE RATE OF A DRIVER AND REFLECTIONS

Experiment #3 is concerned with the question: *What are the effects of V-T switching waveform frequency content on the simulation response?*

11.10.1 Ramp Data: Slew Rate

It's important to review the definitions and terms used to describe switching speed for the following reasons:

- The definitions have changed depending on usage. For example the definition of Rise Time dV range. And Minimum Rise Time corresponds to Maximum Rise Slew rate (or, alternatively edge rate) column, which corresponds to Fastest Switching and Minimum Fall Time—and so forth.
- Instead of min-typ-max, an IBIS file arranges its data columns typ-min-max, left-to-right. The latter format is what most simulators, and data books follow. Therefore, formatting can get confused in the translation from data book→IBIS file→simulator file→simulator user interface.

The user should understand the basic switching speed terms and definitions and remember to keep a few things straight:

- Switching speed:

  Fast Rise Time corresponds to minimum (min) Rise Time (t_r) and maximum (max) Slew Rate (dV/dt_r). For components normally held low, fast rise time also corresponds to minimum turn-on time when the output is driving high. Similar definitions hold for the fall time and turn off time. Note that holding a logic part low usually means that it is turned on and drawing current!

  Typical Rise Time corresponds to typical (typ) Rise Time (t_r) and typical (typ) Slew Rate (dV/dt_r) and, for parts normally held low, typical drive high or turn-on time. Similar definitions hold for the fall time or turn-off time.

  Slow Rise Time corresponds to maximum (max) Rise Time (t_r) and minimum (min) Slew Rate (dV/dt_r) and, for parts normally held low, typical drive-high/turn-on time. Similar definitions hold for the fall time/turn-off time.

- Rise and Fall Definitions:

  The historical definition of Rise Time (and similarly Slew Rate) was the time to go from 10% to 90% of the rail-to-rail (usually 0 to Vcc) swing at the output of a device—similarly for Fall Time. This was the convention followed by most of the electronics industry and data books in particular. In the 1980s, this definition began to change when the long "tails" of CMOS rising and falling waveforms were observed. Then people began to migrate to a definition of Rise Time (and similarly Slew Rate) was the time to go from 20% to 80% of the rail-to-rail (usually 0 to Vcc) swing at the output of a device—similarly for Fall Time.

  IBIS definition of Rise Time is the time to go from 20% to 80% of the total transition when R_load to Vcc loads the output for falling transition. R_load defaults to 50 ohms if it is not defined. ECL model types use R_load to Vcc - 2 V for both edges. Properly explained, IBIS does not talk about slew rate. In particular, the two values of [Ramp] (dV and dt) are *NEVER* divided, but are simply measured and entered into the model. Too many model makers use "slew rate" thinking, and normalize to 1 nsec—this way of thinking is an error. The IBIS keyword [Ramp] has a very specific definition that must be followed.

  Rising ramp rate data is obtained when the buffer changes state so that the output switches from low to high. A load resistance is placed from the output to ground. Falling ramp data is captured with the load resistor tied to Vcc.

11.10.2 Scaling Ramp Rate and Observing the Results

We next look at the effect of ramp rate on some simulations. To do this, we do *not* make use of Slow-Typical-Fast or min-typ-max. This is because the usual practice is to simultaneously vary C_comp, driver output impedance, and ramp rate.

Instead, the typ value of dt is edited in the CDSDefaultIO model to take on the values of 0.9nS, 0.6nS and 0.48nS in turn. Also, we set up a 79-ohm terminating resistance (10% higher than a perfect match) on a 72-ohm line. This is so we can set up some small reflections (ρ = -.046) and see what happens:

Figure 11-13 shows driver and receiver waveforms with a slew rate of dV/dt = 3/0.9nS. Figure 11-14 shows driver and receiver waveforms with a slew rate of dV/dt = 3/0.6nS.

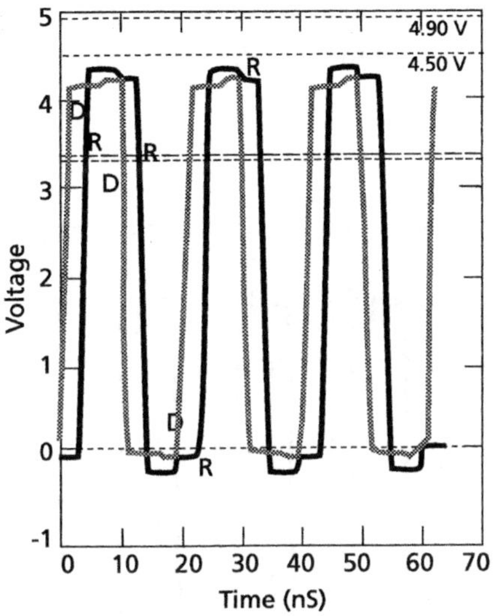

Figure 11-13. Reflections @ dV/dt = 3/0.9ns

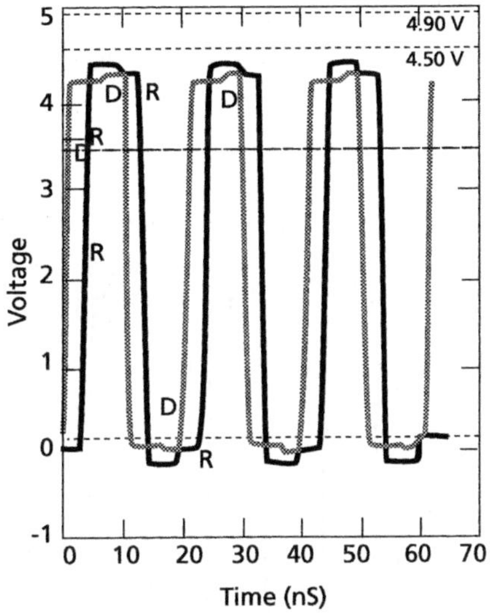

Figure 11-14. Reflections @ dV/dt = 3/0.6ns

Figure 11-15 shows driver and receiver waveforms with a slew rate of dV/dt = 3/0.48ns.

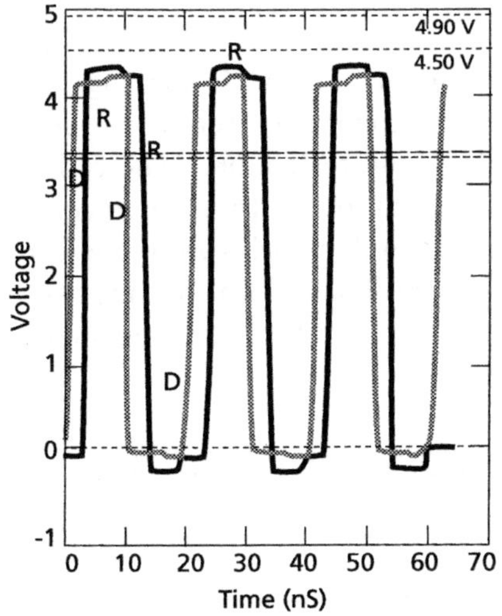

Figure 11-15. Reflections @ dV/dt = 3/0.48ns

The waveforms in Figures 11-13 through 11-15 look nearly identical–dull. Actually, with a pure resistive mismatch, clock rate and edge rate changes do not affect reflection coefficient at all. So we really do not see much change beyond an almost undetectable (at these scales) change in edge rate.

Figures 11-16 and 11-17 show what happens when we add a shunt parasitic capacitance (frequency variable element) of 5pf across the 79-ohm load.

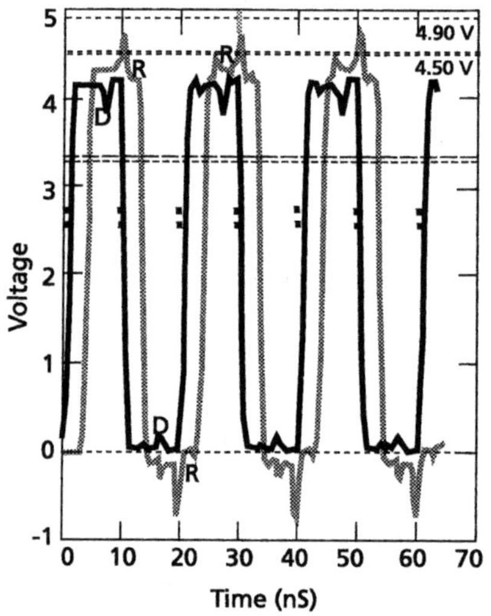

Figure 11-16. Reflections @ dV/dt = 3/0.9ns

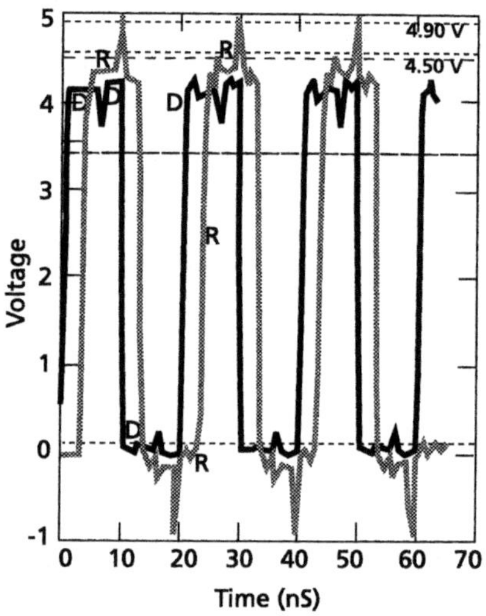

Figure 11-17. Reflections @ dV/dt = 3/0.48ns

The frequency-dependent mismatch loading of the shunt capacitance interacts with edge-rate-dependent frequency content of the signal to produce a spectrum of unique reflections. The reflections and re-reflections combined with time-of-flight from driver to receiver produce a pattern of noise on the signal waveform that varies from one cycle to the next.

Now we can see that the interaction of edge rate frequency content and reactive load element does substantially change the results.

11.11 EXPERIMENT 4: USING V-T DATA VERSUS A RAMP

Experiment #4 is concerned with the question: *What are the effects of using a V-T data table versus a simple Ramp?*

11.11.1 Why Use a V-T Data Table

V-T data are the voltage versus time of the rising and falling waveforms of the output of a driver into a specified test fixture. They take more labor to measure and to enter the data than the simple expedient of measuring the slopes of the rising and falling waveforms (dV/dt) and entering that data.

Simplifying assumptions are made when the dV/dt of the rising and falling waveforms of a driver are used in simulations. The assumption is made that the waveforms are linear (straight lines) from 0% to 100% of the output swing. When they reach the rails, the straight-line curves then make sharp turns.

The simplifying assumptions are just not true for most real devices. Real devices usually show rounding where the rising and falling waveforms are close to the rails and, sometimes, other non-linearities.

The effect of sharp corner V-T curves versus round corner V-T curves is to inject more high-frequency edge-rate related energy into the simulations than is sometimes warranted. This can result in overly conservative designs and problems in correlating simulated to measured results. Some simulators recognize this and make adjustments in how they use ramp data. But exactly what is done is usually proprietary. In the next two topics, we examine these issues.

11.11.2 Examples of V-T Curves

This topic discusses two sets of V-T curves. The first set is purposely linear with sharp corners on the rising and falling waveforms in contrast to the rounded corners of the real V-T curve model. This is done to mask out any rounding algorithms the simulator may be using on ramp data and compare the sharp V-T corner effects directly.

Linearized V-T Curves: Figures 11-18 and 11-19 show the idealized and linearized rising edge characteristics.

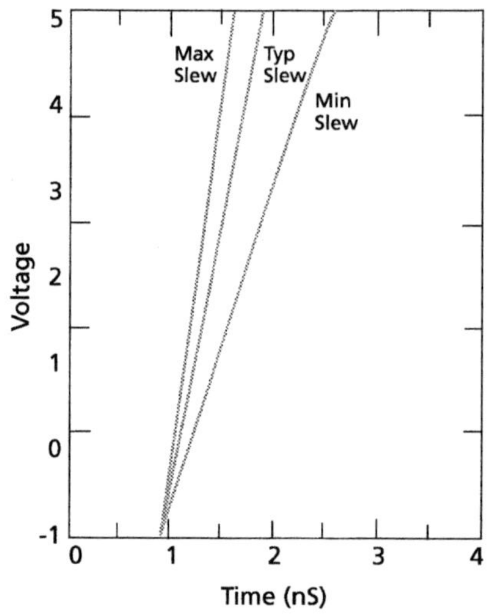

Figure 11-18. Linearized (idealized) rise waveforms

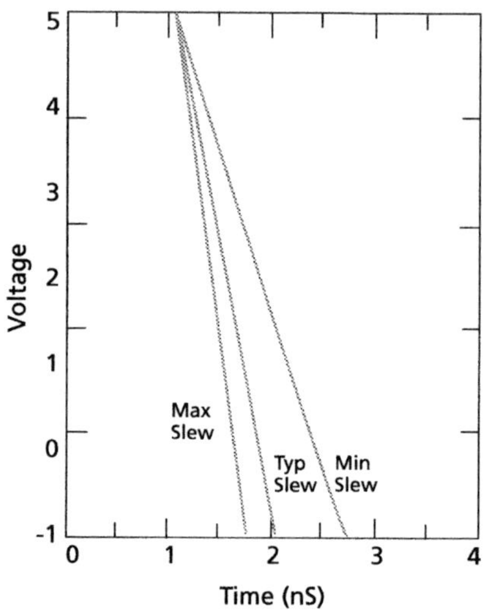

Figure 11-19. Linearized (idealized) fall waveforms

Real V-T Curves: Figures 11-20 and 11-21 show characteristics that are non-linear and similar to those seen in many actual devices.

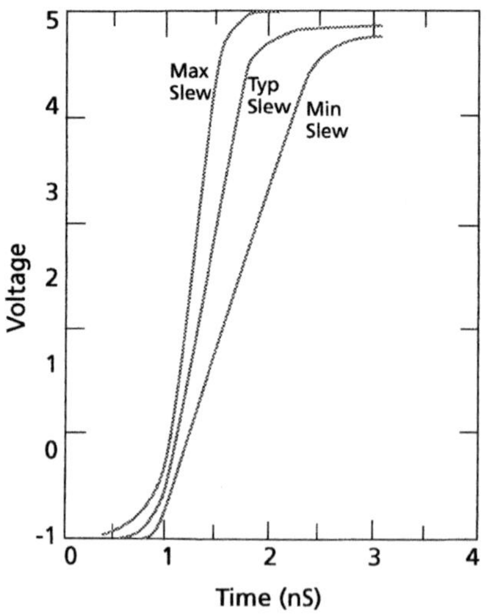

Figure 11-20. Non-linear (real) rise waveforms

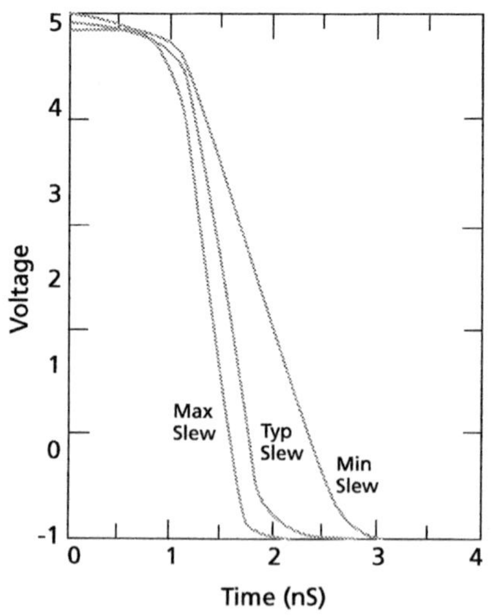

Figure 11-21. Non-linear (real) fall waveforms

11.11.3 Example of Linear and Non-Linear V-T Curves

The pairs of simulation results (ideal-driver and real-driver) are from driving the topology of Figure 11-22. The ideal-driver and real-driver are constructed with a CMOS default I/O model that first incorporates the ideal V-T curves and then uses the real curves, illustrated in Figures 11-18 through 11-21. The pairs of simulation results contrast the results of Slow simulation, Typical simulation, and Fast simulation in turn.

Differences can be seen in the results. These differences would be significant for verification studies where it is desirable for measured and simulated results to agree within a couple of percent. The best approach for measuring the differences in the simulation results is to examine peak-to-peak overshoot-to-undershoot for the two cases.

Unless we are designing right up to the edge on noise margin, these differences are small for the frequencies and circuit conditions given. These differences should not enter into design decisions. It is much better to make design choices, such as adding termination, shielding and using slower edge rates that result in 50% and 100% improvements in noise margin.

Although important for correlation, a few percentage points difference due to V-T curve non-linearity is the wrong approach for design. However, in the next topic (V-T Curves: a Real Example), the results from a controlled ramp rate-soft turnon and turnoff component on a heavily loaded distributed bus can be substantial. In that example, network complexity and switching waveform frequency content interact.

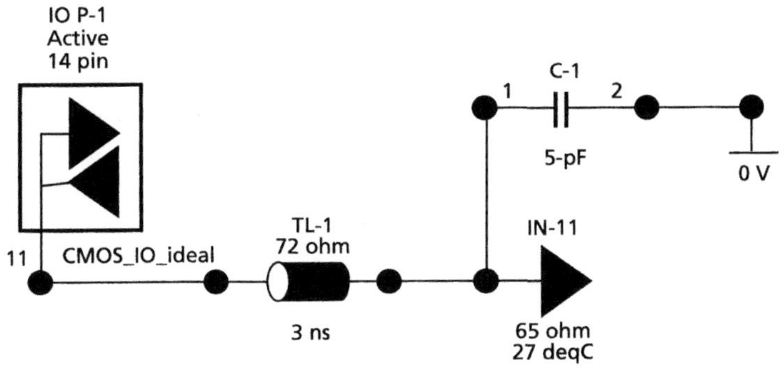

Figure 11-22. Topology for contrasting linear and non-linear V-T curves

Figures 11-23 to 11-28 show simulations with the slow-typical-fast corners combined with ideal and real waveform edge frequency content.

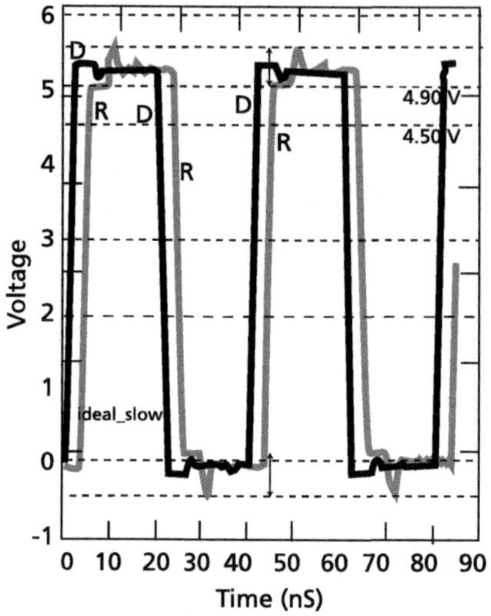

Figure 11-23. Ideal-slow driver

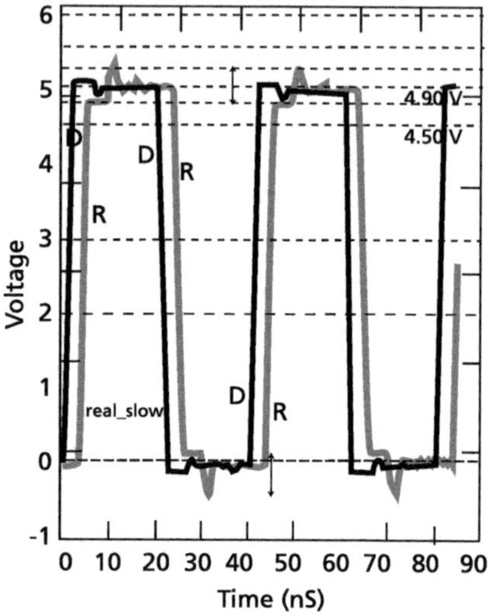

Figure 11-24. Real-slow driver

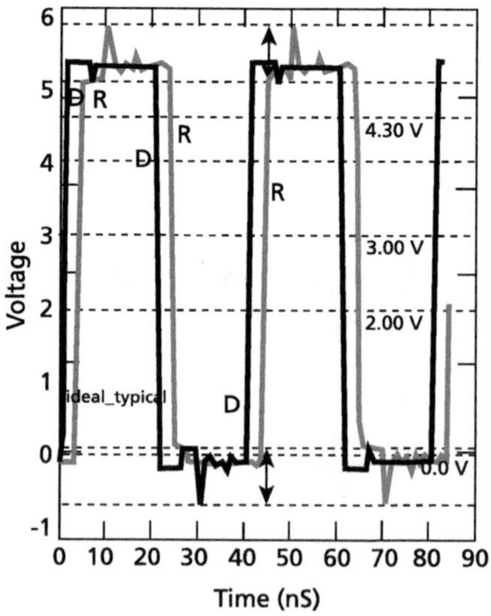

Figure 11-25. Ideal-typical driver

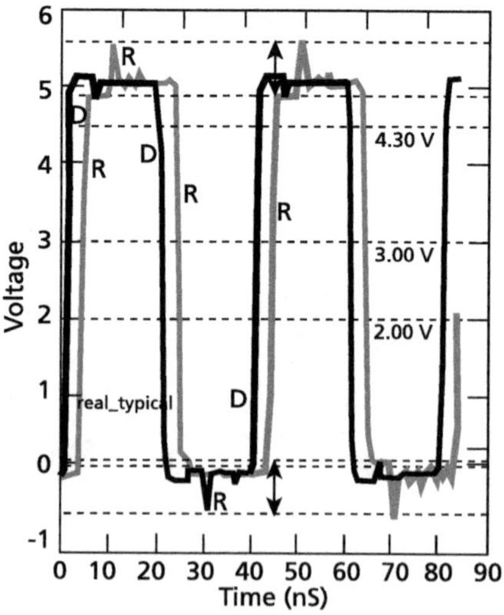

Figure 11-26. Real-typical driver

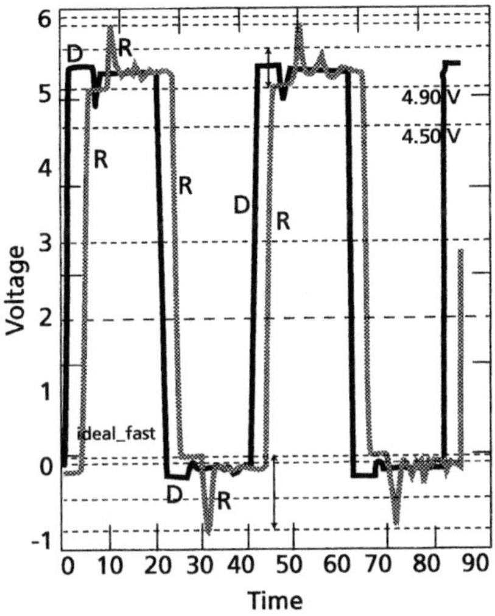

Figure 11-27. Ideal-fast driver

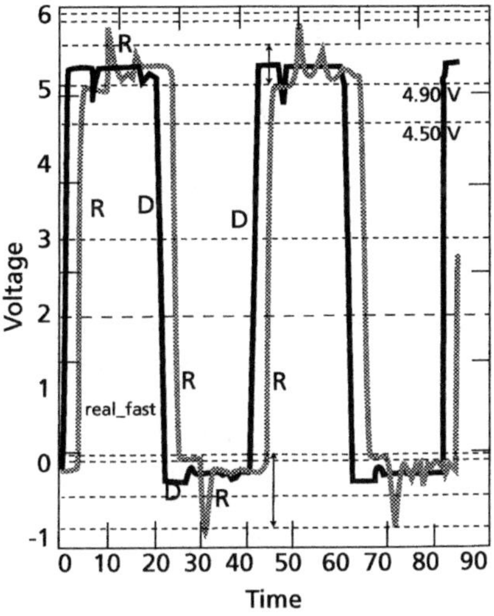

Figure 11-28. Real-fast driver

Summary for this Example

Table 11-4 shows the effect on reflection noise of including (or not including) the extra high frequency content modeled by sharp V-T waveform corners.

Table 11-4. Overshoot, undershoot, and V-T curve corner frequency

Process Corner	Peak-to-Peak Noise (mV)			
	Ideal Driver (sharp V-T corners)		Real Driver (round V-T corners)	
	Driving High	Driving Low	Driving High	Driving Low
Slow	495.58	557.52	471.70	528.30
Typical	743.36	743.36	698.11	679.25
Fast	814.16	943.40	735.85	886.79

The real IBIS V-T curves result in an average 5% less maximum overshoot and undershoot across the board than the ideal IBIS V-T curves for this relatively well-behaved example. These effects are not important in this simple network. But the frequency content effect can be really pronounced, such as in the case of complex backplane busses that set up multiple reflections. One technology developed to work well in such complex networks is edge-controlled GTLP devices. These devices use active feedback to achieve soft turn-on and turn-off and edge rate control. Their reflections are much easier to tame.

11.11.4 V-T Curves: a Real Example

In the late 1990s at 3Com, a multi-drop TDM backplane bus with variable loading was studied to come up with recommendations for the design engineers.

Some of the givens were:

Clock frequency	8 MHz
Backplane	957 mils pitch between connectors, Zo = 63 ohms, 80 mils separation to closest aggressor nets on same layer.
Nets	56 TDM lines with 2 to 18 boards (variable) teed into each net at the connector in daisy chain fashion. But, removing or inserting a board did not interrupt the backplane bus continuity.
Daughter cards	Each daughter card was a bi-directional I/O. That is each could be driver or receiver as determined by switching logic.
Termination	Termination at the ends of the backplane bus would be OK; termination on each daughter card was less desirable.
To be designed	Termination details, stub lengths and technology choices.

Some of the results were:

- Stub lengths were optimized at 1000 mils.
- Technology choice was open-drain GTLP I/Os.
- Termination was one 33-ohm pullup resistor at each end of the nearly 18-inch long backplane bus.

One of the most interesting results was the completely different conclusion about the GTLP technology choice, which was opposite to our first impressions. In between, we spent considerable time optimizing BTLP technology (which was recommended at the time) and investigating several others. What happened to change our mind?

The first time we simulated the GTLP model, we used IBIS version 1.1 models, which did not come with the V-T curves feature. Figure 11-29 shows the composite signal seen on the backplane bus due to one I/O driver output, and seventeen I/O receiver inputs. Here is what that simulation looks like in the final, optimized[5] topology.

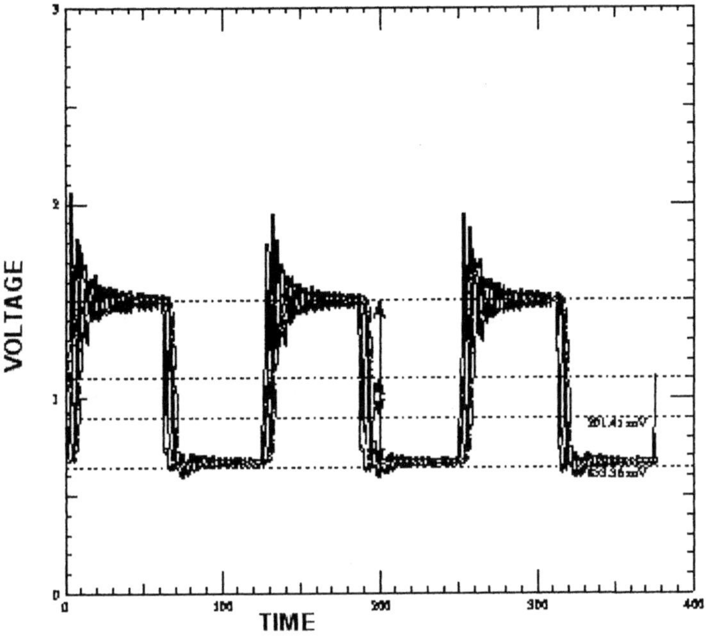

Figure 11-29. TDM-bus envelope of responses: 18 boards

But this component has an active feedback (Gunning Transistor Logic), controlled ramp rate, and soft turnon and turnoff. Figures 11-30 and 11-31 show the properly modeled V-T curves from the IBIS version 2.1 models.

---

[5] The simulator was used to optimize R-C terminators at the I/Os and the value of the pullup resistors at the ends of the backplane bus.

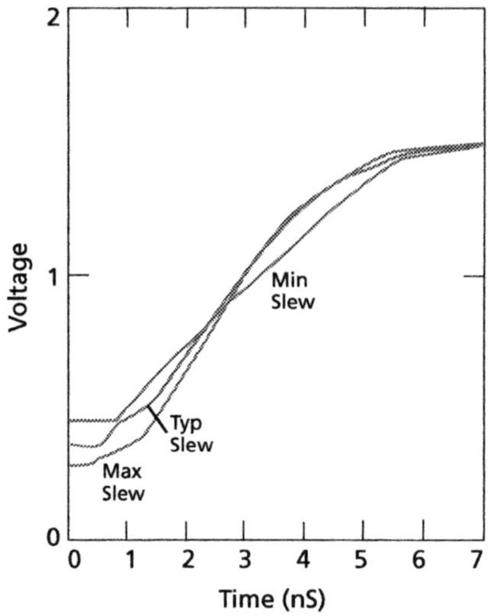

Figure 11-30. GTLP rising V-T waveform

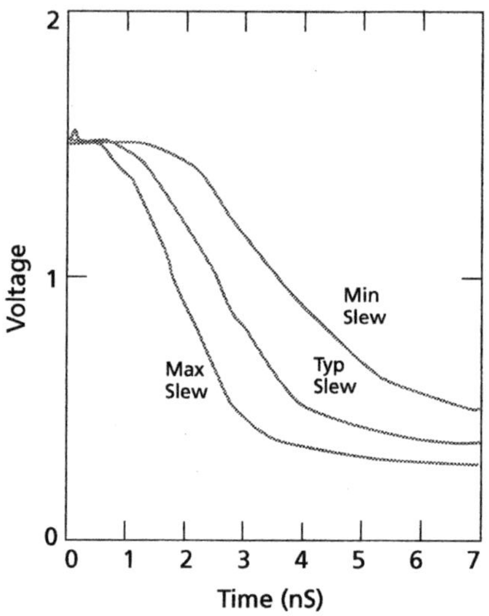

Figure 11-31. GTLP falling V-T waveform

Figure 11-32 shows the results of using the corrected model in the simulation.

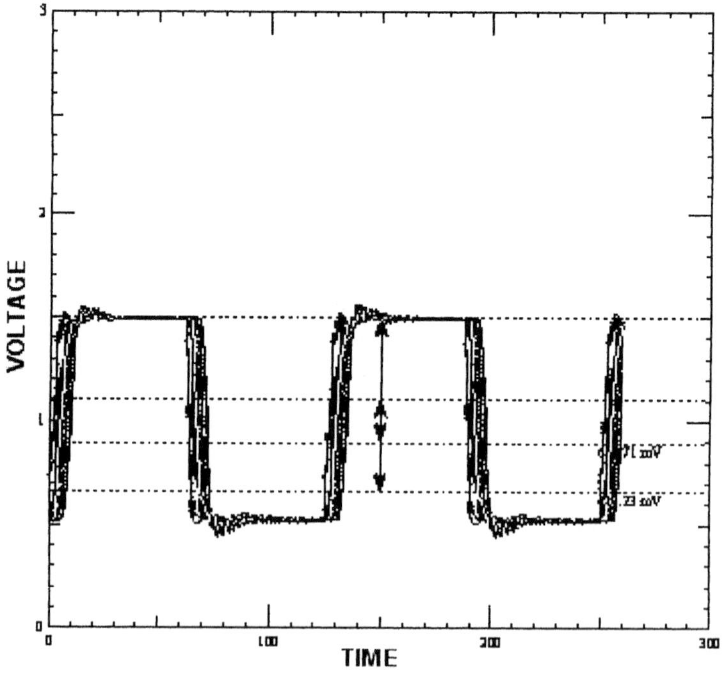

Figure 11-32. TDM-bus envelope of responses: 18 Boards

Some simulation results on first-switch, noise margin, and other Signal Integrity measurements are listed in report format (Figure 11-33).

```
################################################################################
# DF/SigNoise 13.0
# (c) Copyright 1997 Cadence Design Systems, Inc.
#
# Report: Standard Reflection Summary Sorted By Worst Settle Delay
#       Mon Jan 25 16:18:16 1999
###################################################################################
CASE 2: 1000 mil stubs, 63-ohm system, no RC termination,
8 MHz, 33-ohm pullups, 18 boards
load5.top
**************************************************************************************
Delays (ns), Distortion (mV), (Typical FTSMode)
**************************************************************************************
XNet Drvr    Rcvr    NMHigh NMLow OShootHigh OShootLow SwitchRise SwitchFall SettleRise SettleFall Monotonic
------- ----------- ---------- ------ ----- ---------- --------- ---------- ---------- ---------- ---------- ---------
1 A MD7 A IOP-37 31 A IOP-54 1 373.7 344.9 1542   452.5  10.12  10.06  10.98  10.82  PASS
1 A MD7 A IOP-37 31 A IOP-53 1 376.4 347.9 1543   453.2  9.757  9.708  10.61  10.46  PASS
1 A MD7 A IOP-37 31 A IOP-52 1 382.6 357.3 1539   466.7  9.358  9.311  10.24  10.08  PASS
1 A MD7 A IOP-37 31 A IOP-51 1 372.3 343.6 1543   458.4  8.975  8.928  9.846  9.687  PASS
1 A MD7 A IOP-37 31 A IOP-50 1 381.5 354.5 1548   455.8  8.552  8.501  9.481  9.289  PASS
1 A MD7 A IOP-37 31 A IOP-49 1 382   336.4 1560   439    8.12   8.088  9.06   8.863  PASS
1 A MD7 A IOP-37 31 A IOP-48 1 385   323.1 1544   451.8  7.748  7.685  8.646  8.449  PASS
1 A MD7 A IOP-37 31 A IOP-47 1 385.3 303.9 1540   459.6  7.367  7.293  8.267  8.045  PASS
1 A MD7 A IOP-37 31 A IOP-46 1 381.4 305.5 1547   456    6.915  6.861  7.818  7.6    PASS
1 A MD7 A IOP-37 31 A IOP-45 1 387.3 294.8 1545   447.1  6.534  6.456  7.368  7.162  PASS
1 A MD7 A IOP-37 31 A IOP-44 1 369.2 288.3 1547   450    6.106  6.023  6.972  6.735  PASS
1 A MD7 A IOP-37 31 A IOP-43 1 369.3 350.8 1526   475.5  5.642  5.567  6.537  6.295  PASS
1 A MD7 A IOP-37 31 A IOP-42 1 356.4 268.3 1538   458.4  5.241  5.165  6.066  5.84   PASS
1 A MD7 A IOP-37 31 A IOP-41 1 358.4 344.3 1531   472.6  4.825  4.741  5.695  5.431  PASS
1 A MD7 A IOP-37 31 A IOP-40 1 356.3 331.6 1521   478.7  4.396  4.321  5.254  4.987  PASS
1 A MD7 A IOP-37 31 A IOP-39 1 356.4 336.8 1518   488.5  3.952  3.826  4.823  4.53   PASS
1 A MD7 A IOP-37 31 A IOP-38 1 349.7 327.2 1515   490.8  3.51   3.39   4.383  3.992  PASS
------------------------------------------------------------------------------------------------------------------
************************************************************************
Simulation Preferences
---------------------------------------------------------------------
    Variable              Value
 ----------------------- --------------
 Pulse Clock Frequency    8MHz
 Pulse Duty Cycle         0.5
 Pulse Step Offset        0ns
 Pulse Cycle Count        2
```

Figure 11-33. Report: Standard Reflection Summary Sorted By Worst Settle Delay

So with a more complex topology, a more sophisticated and correct model of the V-T behavior makes a big difference in results. In this complex topology, all the T-junctions cause reflection noise. The T-junctions are where the daughter boards tap into the bus. The stub lengths and the distributed, lumped-loaded, and heavily loaded effects on the line also affect the simulation. Eliminating as much high frequency content as possible from the driver waveform makes a big difference. This has the effect of making the driver waveform more trapezoidal – even sinusoidal – and much easier to tame reflections on the net.

11.12 EXPERIMENT 5: PARASITICS AND PACKAGING EFFECTS

Experiment #5 is concerned with the question: *What are the effects of pin and package parasitics?*

11.12.1 Pins, Connections, and Model Assignments

IBIS provides syntax for identifying which model types are connected to which pins. Digital ICs are multi-pin devices containing more than one input and/or one output almost by definition. The types of models attached to each pin, and the pins where the rails are connected to, need to be defined.

The IBIS Keywords [Package], [Pin], [Package Model], [Pin Mapping], and [Diff Pin] are all concerned in whole or in part with the internal connections and model cell assignments of a device. These Keywords allow for increasing levels of detail regarding connections and cell types.

Individual IC components can contain from eight pins to many hundreds of pins. A packaged semiconductor device containing 1000 pins is not uncommon.

11.12.2 Package Parasitics

Package parasitics refers to the global assignment of pin parasitics. C_pkg does represent the total stray capacitive coupling of a pin to ground through the package. C_pkg does not represent the stray capacitive coupling of the package itself to ground.

11.12.3 Pin Parasitics

A manufacturer most often estimates package and pin parasitics (R-L-C). A few manufacturers characterize their bare die and add in parasitics. Some vendors, especially modeling services, de-embed the die behavior and the parasitics from a characterization of packaged devices.

The parameter G (GMATRIX), which is due to conductivity losses, is usually non-existent in most IBIS models. GMATRIX is normally so small it is most often approximated to zero. It represents leakage conductance. The only place G usually appears is in very high frequency simulations of lossy interconnection beyond 5GHz. The value of G is then used in calculating the high frequency bandwidth limiting effects of loss tangent and frequency dispersion on the PC board itself. The dimensions of ICs, connectors and discrete components are usually too small to involve loss tangent and frequency dispersion all the way into the 50^{+} GHz range.

11.12.4 Examples of Pin Parasitics

Following is some text copied from an IBIS file giving the default pin (package) parasitics with some fairly typical values:

|	typ	min	max
R_pkg	0.071	0.063	0.077
L_pkg	4.129nH	3.799nH	4.627nH
C_pkg	0.923pF	0.803pF	1.143pF

We use these values in the next topic to take a look at the effects of pin parasitics on simulation results.

11.12.5 Simulation of Package Parasitics

In this topic, we look at how the magnitudes of R-L-C in pin parasitics change the results of simulation. We again use the expedient of editing the typ(ical) values for these parameters and avoid changing several other properties at the same time. We will increase and decrease (together) pin parasitic R-L-C at both receiving and driving ends of a transmission line. We should probably change edge rate because L and C are reactive elements – maximum edge rate with maximum R-L-C values, and minimum edge rate with minimum R-L-C values.

The topology is a point-to-point driver-receiver with a default CMOS IO cell driving a 3nS 72-ohm characteristic impedance line into a pure resistive load. The load impedance is 79 ohms to set up a small reflection.

Figures 11-34 through 11–36 show driver and receiver waveforms at slew rates of 3.0/0.9, 3.0/0.6, and 3.0/0.48 V/ns respectively.

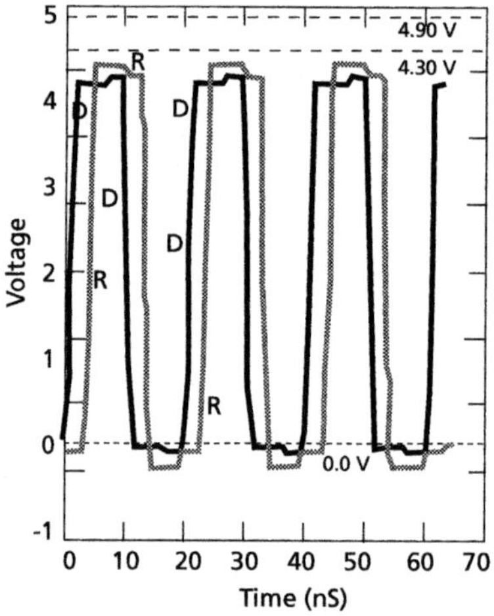

Figure 11-34. Reflections @ dV/dt = 3/0.9ns minimum pin parasitics R, L and C

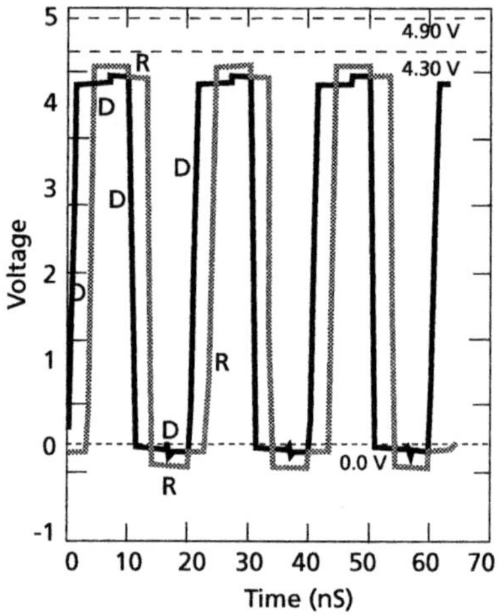

Figure 11-35. Reflections @ dV/dt = 3/0.6 nS. Typical pin parasitics R, L and C

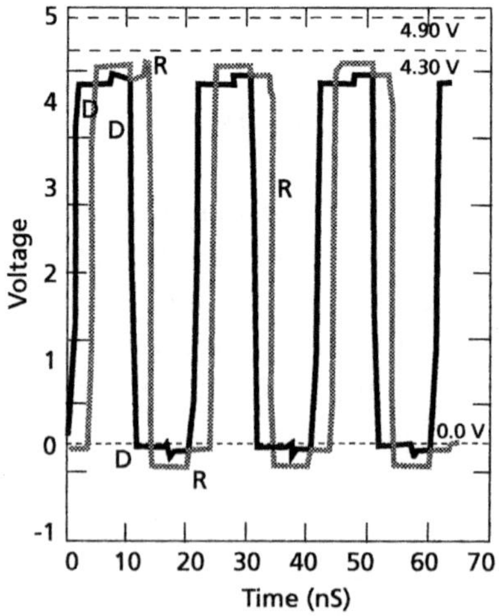

Figure 11-36. Reflections @ dV/dt = 3/0.48 ns maximum pin parasitics R, L and C

The effect of package parasitics and ramp rate and +/-10% changes in their values worst case can easily be seen in the above simulations. But in this example, at the edge rates and RLC magnitudes relative to Zo, they are not a major source of Signal Integrity problems. The actual values varied in these simulations are shown in Figure 11-37:

|	typ	min	max
R_pkg	0.071	0.063	0.077
L_pkg	4.1nH	3.8nH	4.6nH
C_pkg	0.9pF	0.8pF	1.1pF
|	typ	min	max
dV	3.0	3.0	3.0
dt	.6nS	.9nS	.48nS

Figure 11-37. Package parasitics

11.13 EXPERIMENT 6: ENVIRONMENTAL AND POPULATION VARIABLES

Experiment #6 is concerned with the question: *Can the effects of environmental variables be simulated?*

The answer to the question in single word is *no*. The min-typ-max values on the I-V and V-T curves, parasitics and other parameters already have embedded in them worst-case conditions. These conditions include variations in voltage, temperature, and process. The voltage, temperature, and other variables are not directly accessible. Some simulators may add features and "wrappers" that make their capabilities look different. But IBIS is not a physical model like SPICE, where temperature, voltage, and other variables are directly accessible.

IBIS is a behavioral model. Vcc can be varied by +/-5% if the data is in the model, but not by much more. We cannot, for instance, simulate a 5 V model of a part at 3.3 V and get meaningful results. The I-V curves and other data will default back to the 5 V values. To simulate at 3.3 V, we have to generate a 3.3 V model.

11.13.1 Voltages, Temperatures, and Population Spreads

With regard to the IBIS model, power-supply voltage and ambient temperature are not independent, accessible variables. A place that we will see voltage and temperature variables is in the IBIS model properties, such as Vol, Voh, Vil, Vih, the I-V curves, and slew rates. It is important to recognize that in most instances these parameter values represent statistical data of the population distribution run at a combination of variables that will result in worst-case results.

The min and max I-V curves, for instance, are supposed to incorporate the corners of the process population distribution, as well as the effects of temperature and voltage. The [Temperature Range] and [Voltage Range] keywords give the range of those variables, as measured at the die for which the data in the IBIS file is measured.

While min and max data may appear and may have been measured with worst case population units, plus temperature conditions, and supply voltage, there is no direct, unit-by-unit connection between setting those parameters in the simulator and the results predicted. All such effects are subsumed under the min-typ-max (sometimes called fast-typical-slow by the simulator software) data given in the model as typical behavior and worst-case behavior. Whatever conditions of high-voltage, low-temperature, process-corner result in those worst-case responses – they are subsumed under the min-typ-max data. There is no way to de-embed the effect, for example, of the temperature variable and extend the range or precision of its predictions.

11.13.2 Example: Drive Capability – Worst-Case

Figures 11-38 through 11-40 show the simulated results of driving a perfectly terminated Zo = 72 ohm transmission line using Fast-Typical-Slow versions of the default model. For a line with no reflections such as this we can understand Fast-Typical-Slow to mean min-typ-max. We correspondingly get Vout = 4.05 V, 4.13 V and 4.20 V respectively. These differences are due to minimum currents being about 5 ma less and maximum currents being about 5 ma more than the typical case. The min-typ-max values are supposed to represent a worst-case combination of varying temperature, supply voltage and population sample.

This set of curves reflects a case where the driver Zout is small with respect to the line impedance, Zo. About 18 ohms versus 72 ohms, but not an insignificant percentage of Zo. For more discussion on this, see the earlier topic, "Scaling the I-V and Simulating the Results."

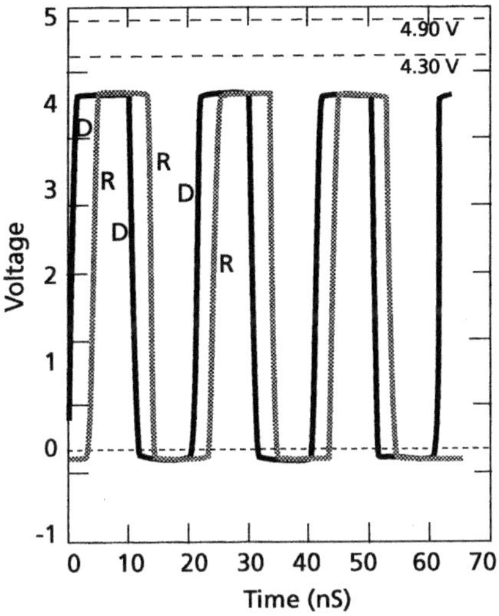

Figure 11-38. Slow default model driving 72 ohms – Vout ≈ 4.05V

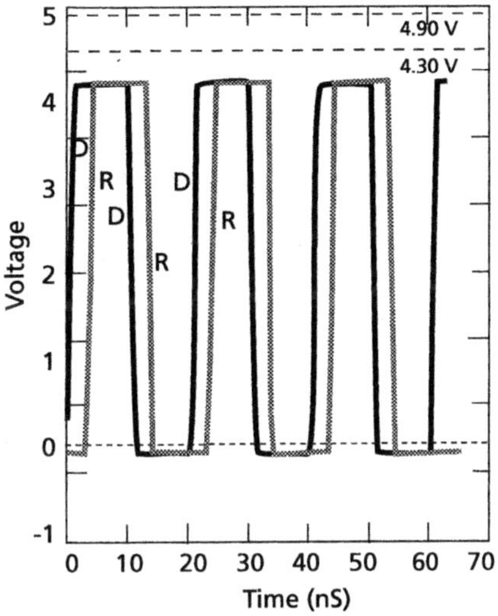

Figure 11-39. Typical default model driving 72 ohms – Vout ≈ 4.13 V

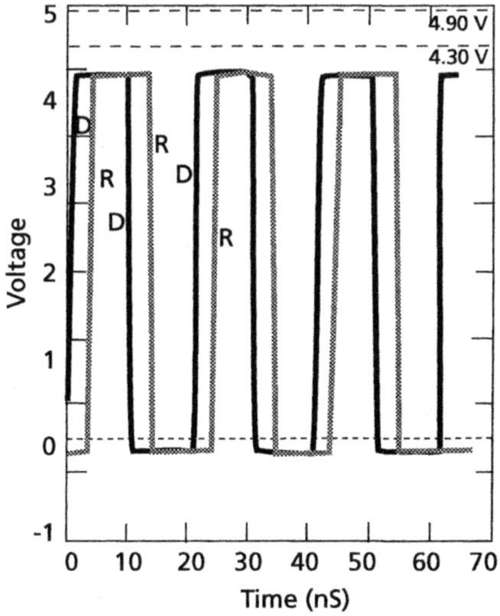

Figure 11-40. Fast default model driving 72 ohms – Vout ≈ 4.2 V

11.14 OTHER CONSIDERATIONS: TIMING AND NOISE MARGIN ISSUES

Timing information is not officially supported in IBIS. However, most simulators take the IBIS data and are able to simulate First Switch, Final Settle, Noise Margin, Buffer Delay, Propagation Delay, and Clock Skew.

Timing driven simulations references timing information from logic simulations to determine the sequence of how a set of drivers turns on and off. This information improves the simulation results for *crosstalk* and *Simultaneous Switching Noise, SSN* (ground bounce), rather than the simplistic assumption that all drivers turn on and off together. This will probably give an absolute worst case induced switching noise result. But this approach can end in an overly conservative design.

11.14.1 Vinl, Vinh, Vol and Voh

We need to have a clear understanding of the meanings of properties such as Vol, Voh, Vil and Vih. All devices, both drivers and receivers, switch at a particular threshold voltage, V-threshold. Vol-Voh, and Vil-Vih define voltage threshold windows. The explanation is that the windows come about due to the variability of *population spreads*. The Vol, Voh, Vil and Vih variables are defined as shown in Table 11-5.

Table 11-5. Switching thresholds

Threshold	Definition
Vol	The maximum driver output low voltage. All units are supposed to have a low state output voltage no greater than Vol. Vol should be defined for a given current
Voh	The minimum driver output high voltage. All units are supposed to have a high state output voltage no less than Voh. Voh should be defined for a given current.
Vil	The minimum receiver input threshold voltage. All units are supposed to require an input threshold switching voltage of at least Vil. If you rise above this value some units will switch on.
Vih	The maximum receiver input threshold voltage. All units are supposed to have switched on by or below this voltage. If you drop below this voltage some units will turn of

Note that a device at the low state has inputs below Vil and at the high state has inputs above Vih. The actual switching will occur between Vil and Vih at unique voltages.

Note that Vil and Vih are IBIS model properties while Vol and Voh are not. However, most simulators can make use of Vol and Voh in calculating noise margins and switching speeds.

11.14.2 Switching Delay Times

Four popular switching delay measurements for determining whether a physical implementation will meet its logic timing constrains are: buffer delay, propagation delay, first switch delay, and final settle delay.

Table 11-6 shows the switching delay-time definitions.

Table 11-6. Delay time definitions

Delay Time	Description
Buffer	Buffer delay is the time for a driver to turn on or off, otherwise known as rise and fall times. The IBIS Spec contains dV/dt or equivalent data that inherently gives buffer delay. See comments below about buffer delay.
Propagation	Transmission line wire delay. The time required for a wave to propagate from driver to receiver. Propagation delay is usually arrived at by taking the absolute value of first switch delay and compensating for buffer delay. "Compensating for" means subtracting from.
First Switch	First switch delay is found by measuring the delay when an input rising (falling) wave first crosses the Vil (Vih) threshold. This is when a few members of the receiver population will first be able to switch high (low) at their output. First switch measurements reported by the simulator may be compensated or uncompensated (absolute).
Final Settle	Final settle delay is found by measuring the delay to when an input rising (falling) wave last crosses the Vih (Vil) threshold. This is when the last few members of the receiver population will finally be able to switch high (low) at their output and stay there. Final settle switch measurements reported by the simulator may be compensated or uncompensated This measurement is especially sensitive to ringing that causes the input waveform to cross and re-cross the Vih (Vil) threshold

These delay times are illustrated in Figure 11-41.

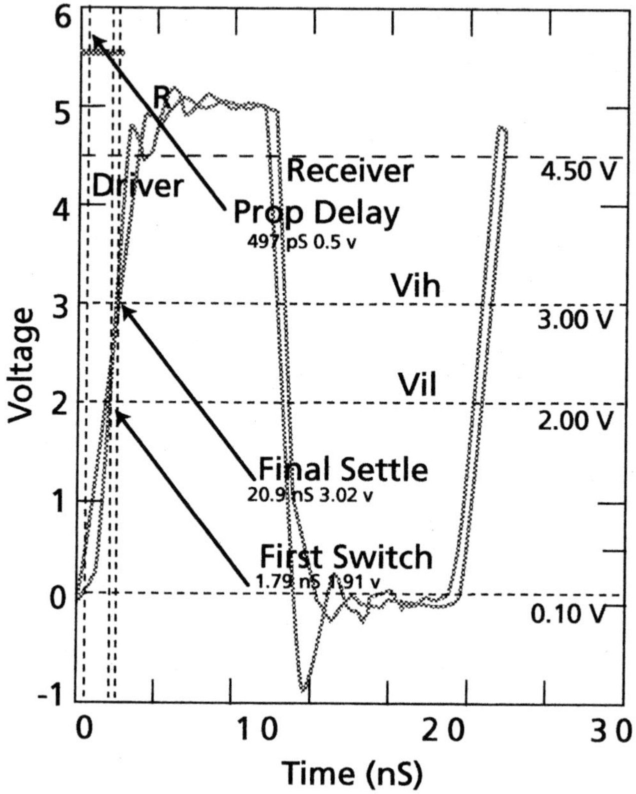

Figure 11-41. Illustration of delay time measurements

11.14.3 Buffer Delay

IBIS ignores propagation delay (= flight time) from input to output through an IC device. But the rise and fall times of a buffer (buffer delay) can affect time-of-flight measurements from driver to receiver. An event[6] occurs that tells the output buffer to turn-on and turn-off. Because of its finite slew rate, the turn-on and turn-off time of a buffer behaves nearly identically with propagation delay. Thus, the term *buffer delay*.

Propagation delay of a buffer is assumed to be entirely due to turn-on and turn-off delay in the output cell itself. Actual flight time from input pin to output pin of a device is assumed to be negligible.

The turn-on and turn-off delay of a buffer can be re-simulated using the values of the buffer delay test fixture given its element values per the IBIS

[6] In logic terms, a state-variable change.

Spec. This can be used to verify that the simulator produces the same answers as the data sheet. Likewise, buffer delay can be corrected for the effects of loading.

In simulating First Switch, Final Settle, and flight time measurements, it is wise to correct for buffer delay. If the original buffer delay was for a heavily loaded case and the current simulation is a lightly loaded case, you can get negative delay times when extracting wire propagation delays! Such answers are obviously math artifacts caused by the buffer switching faster under light loads than the data stored in the simulator.

Our simulations allow us to look at rise and fall time of a buffer. There are a few factors we need to understand about rise and fall time:

- First, the rise and fall time of a buffer varies according to the definition of how it is being measured. Is it being measured per the traditional 10%-90% rule, the IBIS 20%-80% guideline, a manufacture's "transition time" definition of 30%-70%, or what?
- Second, the actual rise and fall of a buffer varies according to the loading placed on it. IBIS provides for supplying rise and fall test fixture data, which the simulator can use to adjust buffer delay to correlate with the loading conditions we are actually simulating versus the test fixture in which the rise and fall data was taken (simulated).
- Third, rise and fall time loses its dependability if the measurement is taken on a badly ringing waveform. It can be almost impossible to verify measurements under such conditions.

11.14.4 Noise Margin

Noise margin is one of the most important considerations in Signal Integrity. It is a measure of the difference between an input switching threshold and the noise voltage on the input pin. Excessive noise may cause the input to switch again. There are two noise margins:

- Noise Margin High (logic state) corresponding to Vih
- Noise Margin Low (logic state) corresponding to Vil

Figure 11-42 shows how noise margin is measured.

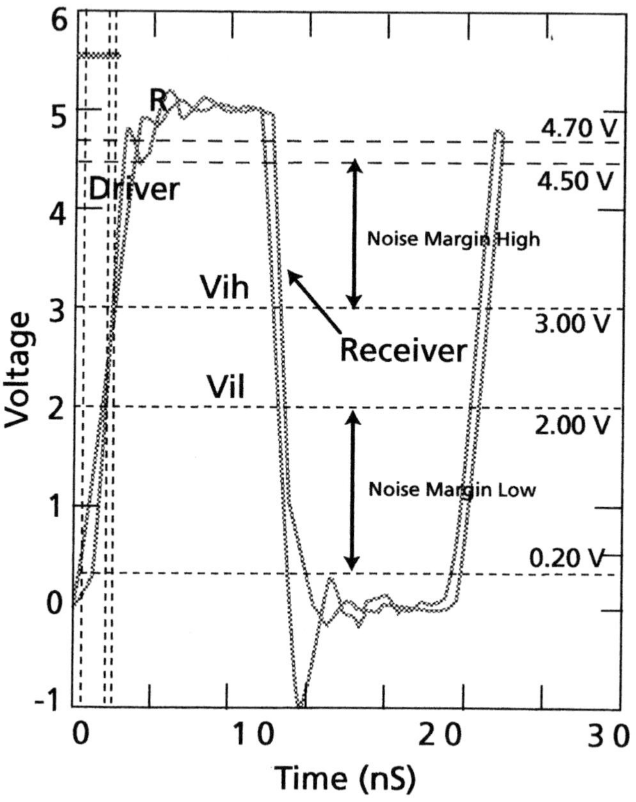

Figure 11-42. Noise figure measurements

11.15 EXPERIMENT 7: VOL FROM SIMULATION VERSUS DATA SHEET

Experiment #7 is concerned with the question: *Can we compare simulation results with a component data sheet?*

When trying to verify the accuracy of models and simulators, we need to understand modeling and simulation, versus *specsmanship* with regard to the data sheet.

Verification, as used in this book, is the process of correlating simulated results with measured results with sufficient accuracy and precision to meet the design needs.

"Specsmanship," as used in this book, is the practice of guard-banding specification limits with sufficient margin, while maintaining the perception of a well-performing part.

The next example shows how some technical issues can be caused by specsmanship, where guard-banding specification limits can blur actual part performance. Some of the kinds problems that are caused need to be solved even before starting the process of turning a design into hardware.

11.15.1 Example of I-V Curve Verification

The following circuit, in Figure 11-43, was simulated to verify a manufacture's data sheet value of Vol versus the results of using their IBIS model.

A pullup current of 100mA was established to match their data sheet conditions. The transmission line to the receiver was kept too short to enter into the result thus duplicating the discrete test fixture conditions. In arriving at the 11ohm pullup resistor value to set up the 100mA current, the simulator was used to empirically arrive at the result. This is to be expected, since a simple, linear saturation resistance does not represent the output.

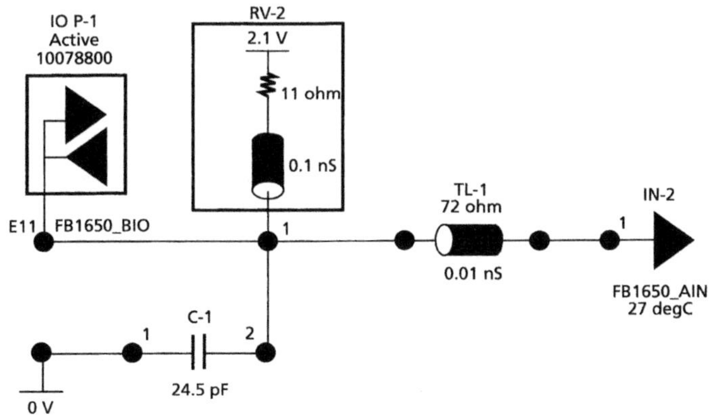

Figure 11-43. Net for simulation and verification of Vol

The pulse waveform is displayed in Figure 11-44. It shows an output low state voltage, at 100mA, (Vol) of approximately 1.0V versus the data sheet specification of 1.15V Vol.

This represents an increase of 47% in low state noise margin for the IBIS model over the data sheet. Is the IBIS model as received too optimistic, or is the data sheet too pessimistic? What result should we design to?

Calling the manufacturer revealed that the data sheet spec was guard-banded by 200 mV.

The conservatism of the data sheet derives from the suppliers' fear of being sued. Unfortunately, it tends to subvert the whole modeling process. It's obvious that good supplier-user relationships, trust and communication could avoid this situation. Beyond that, the authors advocate the application of Statistical Design and Robust™ Design[7] techniques and due diligence by all involved.

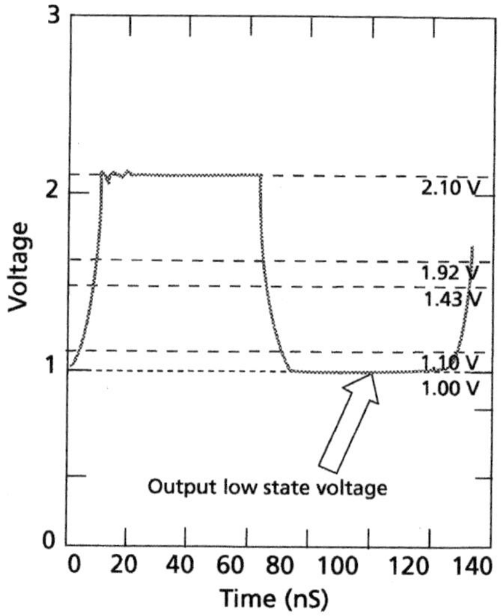

Figure 11-44. Results of simulating and verifying Vol

11.16 HOW IBIS HANDLES SIMULATOR ISSUES

11.16.1 Data Monotonicity and Convergence

Most simulators fail when non-monotonic data is fed to them. While these simulators often produce Signal Integrity results that are non-monotonic due to ringing, noise, and the like, they do not work well with IBIS models that are non-monotonic (often due to a SPICE simulation to produce the IBIS model). Some simulators filter a non-monotonic model and smooth out or just eliminate the offending data points. Some simulators can

[7] Robust Design is a registered trademark of the American Supplier Institute, Inc.

tolerate small model non-monotonicities. But in general, they all have serious convergence problems with non-monotonicities.

The IBIS Spec recognizes this and includes a serious write-up of the allowable non-monotonicity. Because non-monotonic I-V data can exist in real devices, it is not disallowed in IBIS. However, not all simulators are able to process the data. Non-monotonic V-T tables are allowed as well. Some simulators will reject this; others will process the tables. Running the IBIS Golden Parser on the IBIS file provides warning messages that call attention to non-monotonicity.

11.17 SUMMARY

Virtual experiments, in other words simulating "what-if" questions with the aid of EDA software tools, builds experience and confidence with models and simulation. Following up by correlating simulated results with laboratory measurements is an even better learning experience.

Engineers need to understand how a model performs in a simulator. The authors present seven virtual experiments in which model variables are changed. Simulation results are calculated and explained with circuit theory. Virtual modeling and simulation experiments lead to what-if *statistical* design and debug of a virtual prototype.

Virtual modeling and simulation is much more efficient and cost-effective than debugging a physical prototype. A PCB's physical debug is becoming just as difficult as a large IC's physical debug. Today, engineers using modeling and simulation in their design process regularly achieve first prototype success on 90% of new designs. Engineers using the old methods of physical debug usually take more than three or four prototype re-designs to get the product to work. But physical measurement of a prototype is still the final check against errors, false assumptions, and model limitations--which are part of any design process.

Chapter 12

FIXING ERRORS AND OMISSIONS IN IBIS MODELS

For anyone doing simulations, the abilities to validate, fix, and create models are essential skills

Abstract: The details of the IBIS Specification support most analog behavioral aspects of most digital ICs sold. Consequently, the IBIS Specification and IBIS model files are large and complex. The IBIS Specification will grow even larger and more complex as it incorporates better modeling of complex I/Os. IBIS models from suppliers often contain mistakes. Therefore, engineers must have the necessary skills to get the models ready for use in a simulation. It takes time and effort to learn how to create, validate, and fix IBIS files correctly.

12.1 IBIS MODEL VALIDATION STEPS

Quality is defined as *fitness for use*. Model quality cannot be taken for granted–not in IBIS, SPICE, or any other model format. Users who can detect and fix model problems have an advantage over those who cannot. They can proceed with simulation-aided design when others cannot. Most of this topic discusses process steps and process management for improving the quality of IBIS model files. This topic also provides guidance on fixing errors in IBIS files, plus overall management of the IBIS model validation and verification process.

12.2 PROCESS AND PRODUCT IMPROVEMENT STEPS

Figure 12-1 illustrates a four-step universal process for process and product improvement. As applied to IBIS models, the steps include: 1. Detect and acknowledge the problem. 2. Diagnose the problem's root cause. 3. Design a fix based on the root cause. 4. Verify the fix.

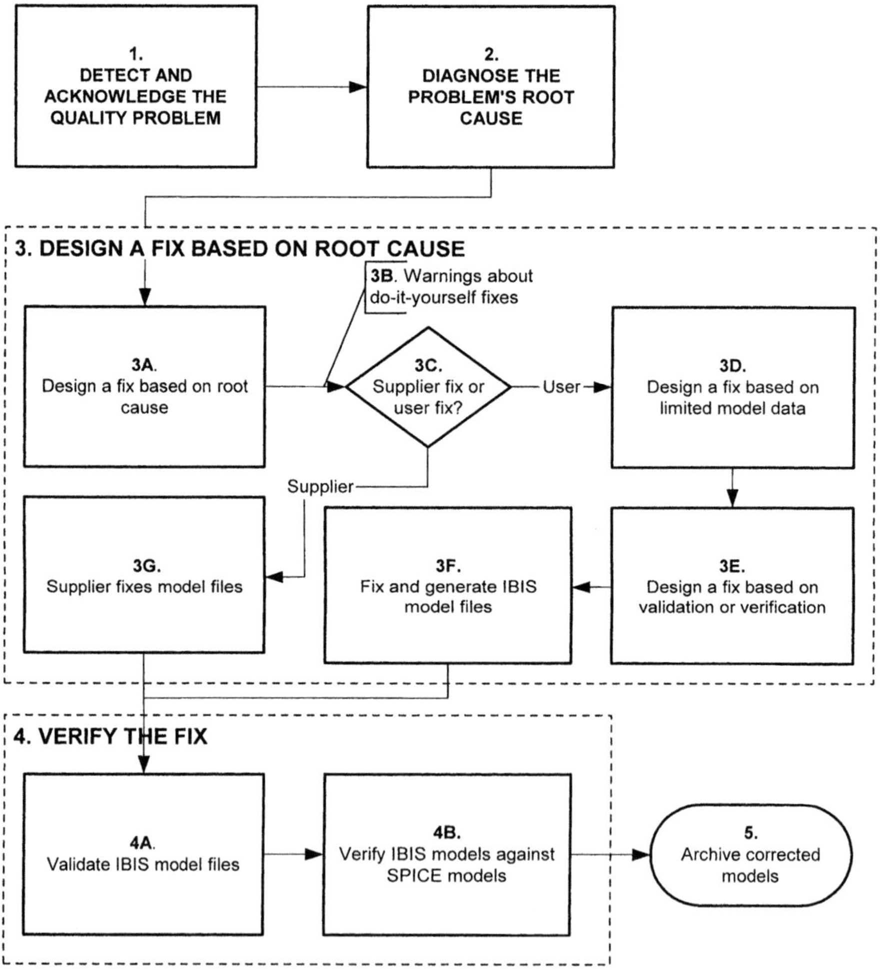

Figure 12-1. Process flow for fixing and generating IBIS models

The next topics explain the numbered boxes.

12.3 STEP 1: DETECT AND ACKNOWLEDGE THE QUALITY PROBLEM

Step 1 is the detection and acknowledgment of a quality problem with IBIS I/O buffer models. Detection is fairly automatic: we detect the problem when the model (as received) will not pass the IBIS Golden Parser or work in the simulator.

How common is the IBIS model problem? Based on the experience of one author [Leventhal] with validating approximately 800 IBIS models we concluded that approximately *70%* of these IBIS models had errors. When models first arrived from the model suppliers, these errors prevented their immediate use in simulators. Before the models would run in a simulator, it was necessary to fix these errors, which took a substantial amount of time.

Many other Signal Integrity engineers had the same experience[1] at that time. We discussed our findings with the IBIS committee. As a result, starting in 2002 the IBIS quality subcommittee compiled a good checklist and incorporated many of those checks into the IBIS Golden Parser (for automated checking). This subcommittee made a commitment to make the IBIS Spec as clear and readable as possible. They are currently working on an IBIS model Quality Specification.

When model problems are reported, a few suppliers (usually those with good overall quality) respond effectively within 2 to 3 months. But it took over two years to approach completion on the first group of reported problems. Many supplier responses were ineffective and did not fix the problem.

Why were there such poor results on model quality? A historical perspective provides clues. Problems with SPICE models have always existed. It seems that semiconductor suppliers are business people. They respond to customer needs communicated through purchasing decisions. So perhaps, modeling needs are not being made clear through that communications channel.

Model quality has recently improved, but it's still not as good as users would like. Fixing some errors use up too much of a user's time. It seems that even with effort and knowledge some model problems will remain. The user must then make a judgment call about model accuracy and take prudent risks.

---

[1] "A Critique of IBIS Models Available for Download on the Web – Part 1" Jim Bell & Dan Grogan, SiQual, Inc., paper presented at the IBIS Committee Summit Meeting 2002. http://www.eda.org/pub/ibis/summits/index-bytitle.htm Findings: At least 74% of IBIS files downloaded from the web required some fixing. At least 40% had serious errors.

12.4 STEP 2: DIAGNOSE THE PROBLEM'S ROOT CAUSE

12.4.1 Model Quality in the Semiconductor Industry

When we seek the root cause of a problem, it sometimes helps to refer to past occurrences of the problem. The record shows that model quality and availability in the semiconductor industry has never been good whatever the type of model being discussed. Before semiconductors, models for vacuum tube amplifiers were so unreliable that early work in network synthesis purposefully avoided addressing vacuum tube circuits at all.

In the earlier days of poor semiconductor process control, modeling and simulation did not work as well as it should have. For predictions made with models to be accurate, a process needed to be *in statistical process control.* In those days, users did not do a lot of modeling and simulation and suppliers did not do a lot of process control. Since computational power and PCs did not exist, there was little statistical design done. Modeling and simulation was not used effectively and its early reputation suffered.

The Japanese Quality Revolution changed American semiconductor suppliers' attitude towards quality. Sourcing a given IC device from two entirely different chip geometries and product lines was one thing that used to be done.[2] Today that is unlikely. But just as uninformed, is the common practice of die-shrink without customer notification. Suppliers often do not inform their customers after die-shrink, and may not even update the data sheet. The increased edge-rate due to smaller die sizes often causes in-circuit performance problems.

It costs money for good modeling data, and customers do not pay for it. Today, two-thirds to three-fourths of all IBIS models do not work as initially downloaded. Model quality problems have their root in poorly communicated customer expectations and traditions rooted in industry.

12.4.2 IBIS Models Are Derived From SPICE Models

Most IBIS models, perhaps over 95%, are generated by simulating the SPICE model netlist (circuit model) plus the foundry model library (device-

---

[2] In the 1950s through the 1970s, sometimes a given JEDEC device would be assembled from two entirely different chip geometries (chip process lines). Sometimes two different wafer starting resistivities and epi resistivities would be part of a single chip device process line. Furthermore, customers were never informed when incoming lots jumped around between and betwixt such disparate source material. No modeling technique will have much chance of succeeding in such circumstances

transistor model) to produce the behavioral curves for IBIS. Many problems seen in IBIS models begin with the SPICE model. For example, many mistakes involving unrealistically high clamp currents in IBIS models result from clamp-diode resistances set to zero in the SPICE model. This simplification works fine for process control, but not for circuit modeling.

12.4.3 Supplier Documentation Traditions

Tradition 1: Suppliers guardband their specifications

Suppliers normally guardband (pad)[3] the published data sheet min-max specs on their parameters. This tradition has little to do with measurement or modeling issues, and more to do with suppliers protecting themselves from lawsuits. Suppliers do not want to be legally responsible for user results.

Tradition 2: Suppliers disclaim the accuracy of their data

Suppliers disclaim the accuracy of their data. The usual phrase is: *"This information is for modeling purposes only and is not guaranteed."* Whatever that means!!! This is equivalent to the text used on a data sheet for the same legal reasons. Note that this phrase has little to do with how good or bad the model data actually is. The reluctance to reveal true process spreads is understandable. It is expensive for a supplier to generate, provide, and control statistical spreads. It is cheaper to automate testing for min-max data sheet parameters. Also, the user, if adversarial, can easily tighten the tolerance window guardband, making it very difficult to consistently ship product.

Tradition 3: Suppliers sell parts—not models

This statement is obvious on the face of it. If you asked OEMs, the overwhelming majority would say, who ever heard of paying a semiconductor supplier to produce a model of a part they're selling? Is it true that models are free? Well, no. Not if the semiconductor supplier has to hire staff and buy equipment to produce such data.

An interesting exception is the automotive industry, which requires models from their suppliers. Incoming parts are tested against those models. There are no separate data sheets. The automotive industry uses VHDL-AMS as their standard modeling language. This new paradigm must be getting accomplished with both supplier and customer still making a profit. Otherwise it would not be done. So, modeling and simulation can be done well. It just takes commitment.

---

[3] For an example, see "Chapter 11, Using IBIS Models in What-If Simulations," experiment #7.

12.4.4 Customer Reactions to Poor Models

Poor model data makes customers frustrated, angry, and adversarial towards their suppliers. But not only does the above root cause analysis apply to models; it also applies to data sheets. Plus it applies to and incoming inspection history and results.

Still, supplier and customer need to be less dysfunctional and adversarial. It's to their mutual benefit. Based on history, the problem of poor model quality is fixed only when it is important (money wise) for customers to want it fixed.[4]

12.5 STEP 3: DESIGN A FIX BASED ON ROOT CAUSE

Until the systemic problem of poor model quality gets fixed, users need to fix much of the problem on their own. Since over half of IBIS model problems are simple syntax mistakes, there is a good chance that users can fix many of them.

12.5.1 Step 3A: Design a Fix Based on Root Cause

Fix #1: Work out the solution with the supplier

Work constructively with the semiconductor suppliers for better models. "Chapter 21, Feedback to the Model Provider Improves Model Accuracy," focuses on working with suppliers to get better models. Modeling and simulation help in creating a cooperative relationship with suppliers because they involve the exchange of data rather than opinions.

Fix #2: Create and fix an IBIS file

Users who want to do modeling and simulation are often forced to innovate. This includes borrowing I/O cells of parts from similar (as closely matching as possible) technology, putting together a pin list from the data sheet and creating an experimental model—including estimates of typ-min-max. This kind of innovation may be good enough if:

- There is adequate design margin (tolerance) and/or:
- What-if models are sent to the semiconductor supplier. The supplier is asked to make the parts like the models—or to respond with what the supplier can make.

---

4 This point emphasizes why the workplace must support modeling and simulation.

12.5.2 Step 3B: Warnings on Do-It Yourself Fixes

We need to decide whether we should fix the models ourselves or ask the model supplier to fix them. We need to make these decisions both systematically, and in individual cases. Lack of access to process parameter spreads and statistical data are reasons for caution regarding making a model without supplier input. How cautious we should be, depends on how tolerant our design is to parameter variability.

12.5.3 Step 3C: Supplier or User Fixes Model?

There is no fixed formula to help in deciding whether to get a model supplier to fix and generate the IBIS model file, or whether to fix and generate it ourselves. But here are some factors to consider in the decision:

- The complexity of the problem with the model file.
- Resources available for fixing the problem. Resources include knowledge, process and procedures, time available, and sources of information.
- The urgency of the need, lead-time in the product development schedule, and criticality of the circuit.
- The track record of the model supplier.

From time-to-time, we will probably need to fix and generate IBIS models on an ongoing basis. That being the case we should strongly consider setting up processes, procedures and resources to do so. Among the options to consider, both in individual situations and as preferences, are: doing the work ourselves, relying strongly on our CAE tool provider's library, relying strongly on our semiconductor supplier's abilities, hiring a 3rd party modeling service company, or some combination of the above. If we decide to fix or generate IBIS models ourselves, we need to consider two important process steps:

- Designing the fixes based on limited model data
- Designing the fixes based on the goals of validation or verification or a combination of them.

12.5.4 Step 3D: Design a Fix Based on Limited Model Data

Much can be done with common model problems. In the standard 80%-20% rule: 80% of problems can be solved with 20% of the effort. The remaining 20% of problems require the remaining 80% of the effort. *Most IBIS model problems are syntax errors or errors of omission. Few involve model data accuracy.* Syntax errors and errors of omission are model validation issues. So fixing the errors involves devising a good model generation and checking process that ensures that the models can:

- Pass the screen of the IBIS Golden Parser
- Pass a check for certain obvious mistakes
- Run in a simulation program
- Be consistent with the data sheet[5]

Before we set out to make and fix many IBIS model files, it pays to think about the benefits of standardization: This is especially true for commodity-type devices in commodity-type circuits. Another consideration is to re-use designs where possible. Re-using commodity components and designs conserves resources.[6]

As we launch into making and fixing many IBIS model files, it pays to think about the benefits of prudent design. Early on, try to implement the design as if only ideal models are available. If possible, avoid tight, parameter-sensitive designs and avoid making critical decisions based on the simulation model's accuracy,

An even more common problem is lack of models, which is perhaps the biggest single barrier to design simulation and analysis. Model coverage is about 35% (on average) when starting a new design. Given a random choice of components with no preconditions, about 35% of them will come with good, working models available early in a design cycle. To get simulations started early in the design cycle, it is a good idea to set up a strategy of using the simulator software's default models and the schematic's pin use assignment.

12.5.5 Step 3E: Design a Fix Based on Validation or Verification

Model validation and verification are often confused with each other.

---

[5] For an example, see "Chapter 11, Using IBIS Models in What-If Simulations," experiment #7.

[6] One effort that promoted design re-use was the Defense Advanced Research Projects Agency (DARPA) Rapid prototyping of Application-Specific Signal Processors (RASSP) Program: http://www.eda.org/rassp/ The underlying philosophy was to choose a system architecture that would be stable for 10 years and to redesign 10% of a system's hardware every year to gain advantage of new technology by incremental improvements. The idea was to avoid a total redesign every few years.

Model *validation* is the process of running an IBIS model through some quality checks to verify that the model:

- Is consistent with the designated version the IBIS standard.
- Is self-consistent with the data sheet of the device, and some common sense checks of its properties.
- Will run in the simulator chosen for the purpose.

Regarding the last bullet in the above, note the following from IBIS Committee:

> "Sometimes a simulator deviates from the expected implementation of a specific IBIS feature. Since IBIS is a work-in-progress that is evolving, a given EDA supplier can get out of sync on some features, especially new ones, from time-to-time. The model may be correct, but the handshake with a specific simulator may require the user to adapt the model to the simulator."

Even before the start of the actual design, it is prudent to do model validation as soon as possible. It is a good idea to get such preparatory work done before design development schedule pressures get intense. The time between new product design projects is a good time to work on getting and testing models.

Model *verification* involves the process of checking the accuracy of the model's simulation predictions against hardware measurements. This accuracy checking is done either directly, or by the substituting a comparison of simulations with a previously hardware verified SPICE model. Verification must wait until a working hardware prototype is available. Verifying an IBIS model simulation against a SPICE model simulation is addressed in Step 4B.

12.5.6 Step 3F: Fix and Generate IBIS Model Files

We used an ISO 9000 Process at 3Com to control a procedure for fixing and generating IBIS models.[7] The documentation for the process included:

---

[7] See "Chapter 13, Using EDA Tools to Create and Validate IBIS Models from SPICE," for a procedure for generating an IBIS model from a SPICE model. See "Chapter 14, Sources of IBIS Models," for making an IBIS model by adaptation, inference, cross-referencing, and cut-and-paste. Also see "Appendix E, Key Concepts of the IBIS Model."

- A "Simulation Model Creation and Updating Process (ISO document)"
- An "IBIS model Syntax Guide (non-ISO document)"
- A "Simulation Model Verification Process (ISO document)"

When making or fixing a model, we must consider model accuracy requirements. Accuracy requirements are a tradeoff with device capability, logic speed, timing constraints and complexity, tolerance to variation, circuit topology, and termination.

To help the user determine the model elements and accuracy needed, we generated some additional documentation available on the IBIS website: http://www.eda.org/pub/ibis/training/3com-docs/

- "How to use IBIS in circuit design, test measurement"
- "Signal Integrity - board design and simulation techniques guideline"

These types of documents should be non-ISO documents available on the company Intranet.

12.5.7 Step 3G: Supplier Fixes Model Files

We can use either our semiconductor supplier to fix or generate the IBIS model file or hire a 3rd party modeling service company to do the work. Or we can use a combination. We need to screen every model file we receive—regardless of how many times it has been worked on. Half of the IBIS files sent to the model providers for correction, and thereafter returned, still came back wrong.

12.6 STEP 4: VERIFY THE FIX

In step 4, the file is checked to see if it now conforms to the IBIS Specification.

12.6.1 Step 4A: Validate IBIS Model Files

The validation step starts by checking the file with the *IBIS Golden Parser*.[8] When run on an IBIS file, the parser checks for correct syntax, format, the presence of required model elements, and a number of certain consistency conditions, for instance monotonicity of I-V curves, and realistic

---

[8] The IBIS Golden Parser works similarly to a grammar/spellchecker in a word processor. It also checks for the presence of required model elements.

maximum clamp currents. Error messages and warning messages are produced as appropriate.

If IBIS models were consistently screened through the IBIS Golden Parser before release, at least half of the potential errors could be caught at the source and corrected.

The IBIS Golden Parser is a free software program downloadable from the IBIS Committee website. The parser can be incorporated in a for-sale software product (commercial EDA tool) under a licensing arrangement with the Committee. The license fees fund further improvements to the parser. The parser can be a platform to which model suppliers and software providers can add their own enhancements.

12.6.2 Step 4B: Verify IBIS Models Against SPICE Models

IBIS quality level 2 specifies that model developers must simulate identical test loads in SPICE and in a behavioral simulator of their choosing. SPICE correlation assesses the degree to which the IBIS model simulation agrees with the SPICE model simulation.

By careful attention to detail and understanding the behavior of the I/O circuit, it is possible to achieve extremely close correlation between behavioral IBIS simulation and physical SPICE simulation. Be aware that not all behavioral and physical simulators are created equal; discrepancies may be an artifact of the simulators rather than the IBIS model extraction process. An example would be double-counting the ramp rate effects of C_comp.

The best recipe for success in SPICE correlation with IBIS is a healthy sense of curiosity. Before beginning the attempted correlation, we need to take the time to answer these questions. How does the circuit work? What circuit elements are necessary and sufficient to model the behavior of the circuit? What test loads might stress the accuracy of the behavioral model? How will the circuit be used?

Should the package model be omitted from the SPICE and behavioral simulations—at least in the first attempt at correlation? Most SPICE models do not include package effects. Will buffer I/O cell artifacts be separable from package artifacts? Some IBIS simulators calculate and display waveforms at both external package pins and internal die pads. The real issue is comparing the same buffer with or without package effects.

12.7 STEP 5: ARCHIVE CORRECTED MODELS

When we are finished with the IBIS model file and it is fit for use, the completed model may represent a considerable investment. To ensure that the file is safe and available for re-use, it should be stored in the company's component and model library CAE tool database under control of the Component Librarian.

12.8 BEYOND PARSERS AND CHECKLISTS: SIMULATIONS AND REALITY CHECKING

This topic explains the kind of reality checking a model maker needs to apply to IBIS models to contribute to the success of their design project.

The IBIS Spec has always made it clear that model I-V (current versus voltage) data is to be entered in the file under four [keywords]: [Pulldown], [Pullup], [GND Clamp], and [POWER Clamp]. All simulators assume this and will add the appropriate currents back together, when driver voltage is in the appropriate range. This is done to arrive at a rising-edge and falling-edge composite table that represents the behavior of the total I/O cell.

"Appropriate" is also determined by the structures that are actually present in the I/O cell, the voltage across the I/O and the state the device is being held in. Thus, the problem explained below will not occur if the cell is an input or if it does not contain one or both of the clamps.

As an example, consider this problem: A simulation program is supposed to add the power-clamp current to the pulldown table (and/or the ground-clamp current to the pullup table). What happens if the power-clamp current has not been subtracted from the pulldown table (and/or the ground-clamp current has not been subtracted from the pullup table) in the IBIS model file? The composite table will then double count the clamping action, resulting in a much more robust (and therefore erroneous) clamping action on high (and low) overshoots.

This error is readily seen on a graphical plot. The composite I-V table curve will have *double the slope* of the applicable clamp in the overshoot region as shown in Figure 12-2. This can occur in a real device, but it is not common.

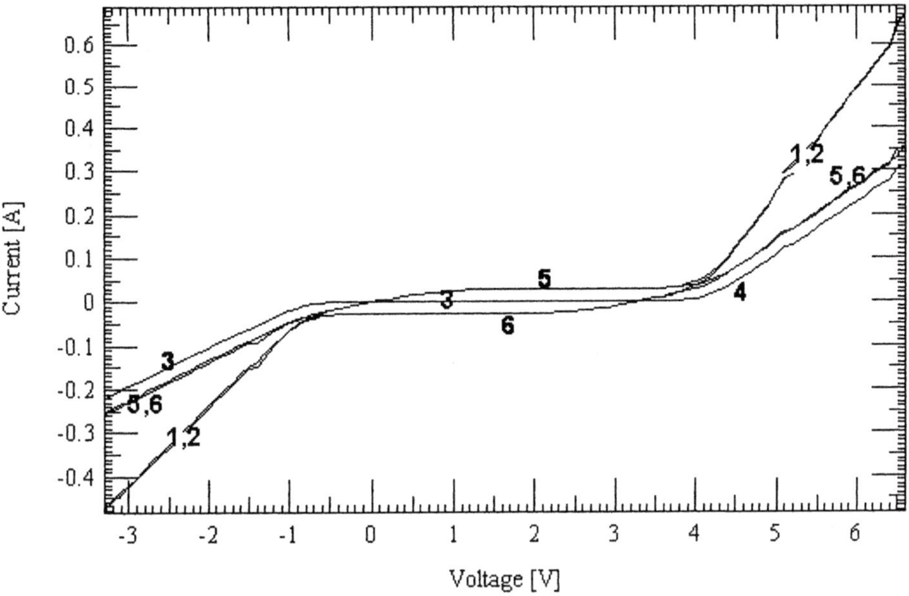

LEGEND: **1**=Fall Edge Composite Typical
2=Rise Edge Composite Typical
3=Ground Clamp Typical
4=Power Clamp Typical
5=Pull Down Typical
6=Pull Up Typical

Figure 12-2. I-V curves with incorrect double counting of the clamp currents
Courtesy of Cadence Design Systems

By contrast, look at the curves of a correctly created IBIS file as shown in Figure 12-3. While there may be offsets in the composite curve due to the magnitude of the pullup or pulldown curves, the slope of the composite curve is close to that of the clamp curves alone. This is because the clamp currents are normally significantly larger and increase much faster than that of a pullup or pulldown. Figure 12-3 shows the V-I curves modeled and added together in the simulator correctly:

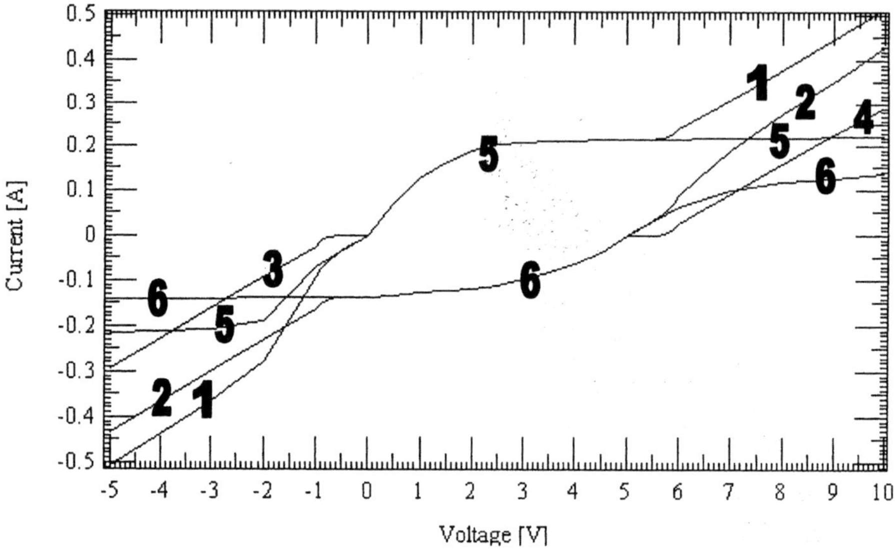

LEGEND: 1=Fall Edge Composite Typical
2=Rise Edge Composite Typical
3=Ground Clamp Typical
4=Power Clamp Typical
5=Pull Down Typical
6=Pull Up Typical

Figure 12-3. I-V curves modeled correctly.
Courtesy of Cadence Design Systems.

12.9 TOOLS PROVIDED BY THE IBIS COMMITTEE

The IBIS Committee, concerned about the quality of IBIS models being supplied by industry, gave a subcommittee the task of making recommendations and improvements. For the past few years, the IBIS Quality Committee has been working on an IBIS Quality document. The document details how each item in the IBIS Quality Checklist is interpreted and validated, and defines four levels of Quality. This document is now ready for use by IBIS model makers, model users, and PCB librarians in their model validation flow.

The subcommittee took the following approach to the quality problem:

- Recommended changes to the writing and formatting of the IBIS Spec to make it easier to read, and thus less of a contributing factor to mistakes in producing IBIS models.

- Recommended changes to the content of the IBIS Spec that will catch more mistakes in producing IBIS models.
- Incorporated more screens and error detection for IBIS requirements into the IBIS Golden Parser so that less error detection is left to humans.
- Provided a framework for indicating the quality level of an IBIS model so that users will know what work has been done for quality assurance.

12.9.1 Checklist For IBIS File Errors

Common IBIS file errors are listed in the IBIS Committee's Checklist tables. The checklist is a work-in-progress so be prepared to check up on recent changes to it at the website [57]:

http://www.sisoft.com/ibis-quality/

New and improved checks may require the IBIS 4.0 parser. Tables 12-1 and 12-2 are not part of the Committee's official recommendations, but are addition guidelines we suggest. Note:

- One asterisk (*) indicates an optional [keyword]
- Two asterisks (**) indicate an optional [keyword], but one recommended for quality models.

Table 12-1. Further interpretation of IBIS quality checks

Keyword and subparameter	**Interpretation and Quality checks**
FILE HEADER SECTION The file header provides documentation of the IBIS file. Note that user scripts check a number of the strings associated with these keywords. For this reason, it is strongly recommended that standard use be followed.	
[IBIS Ver]	The highest version for which a parser is available should be used (presently 4.1). Even if only IBIS 2.1 features are used in the model, the version should be set to at least 3.2, since ibischk4 contains added checks for IBIS 2.1 files.
[Comment Char]	If you are using the default comment character, then this keyword is not needed.
[File Name]	The *exact* name of the file. If the file is copied to a new name, the new file will <u>not</u> pass the parser without an error until this is also changed. Users should feel comfortable changing the name of the file and/or the corresponding [File Name]. File names should not contain spaces.
[File Rev]	Each value has an explicit meaning, as defined in the IBIS Specification.

Table 12-1. (cont.)

Keyword and subparameter	Interpretation and Quality checks
[Date]**	The date the file was created. It is recommended that the month and year be written out completely to aid scripted searches.
[Source]**	This tells the end user how the model was obtained. Normal sources are SPICE and bench measurements.
[Notes]**	Additional information, such as SPICE product/version or test equipment vendor/product, should be included, since these are helpful if there are later problems with models in the file. It is also helpful to provide a contact for support calls. This is also the place to document any known issues with the file, such as parser messages that can be safely ignored by the end user. Documenting known issues can reduce the number of calls for model support.
[Disclaimer]*	Some form of legal disclaimer, similar to what might appear on a data sheet.
[Copyright]*	Normally, the company that created the file. Alternatively, the company for which the model was created, if done as a "work for hire".
[Component] SECTION	
[Pin]	At least one pin connected to POWER At least one pin connected to GND All pins require a model name or POWER, GND, or NC
[Package]	R_pkg, L_pkg, C_pkg (all required)
[Model] SECTION	
[Model]	The name of the model. This is case sensitive, and must match the name used in the [Pin] section exactly. Note that a diff pair can be built by using the same model for both pins, if both pins use the same I/O circuit design. The [Diff Pin] section determines the inverting and non-inverting relationships.
[Model]	Sub-Params Model_type and C_comp required. Vinh and Vinl required for ECL receivers (I/O_ECL, Input_ECL). [Voltage Range], if [Pullup Reference], [Pulldown Reference], [POWER Clamp Reference], and [GND Clamp Reference] are not present [Pullup Reference], [Pulldown Reference], [POWER Clamp Reference], and [GND Clamp Reference] if [Voltage Range] is not present [Pulldown] required for Output, I/O, 3-state, Open_drain, I/O_open_drain, Open_sink, I/O_open_sink, Output_ECL, I/O_ECL, 3-state_ECL [Pullup] required for Output, I/O, 3-state, Open_source, I/O_open_source, Output_ECL, I/O_ECL, 3-state_ECL
Note that typ-min-max, Model_type, and C_comp are sub-parameters. Not [keyword]s themselves	

Table 12-1. (cont.)

Keyword and subparameter	**Interpretation and Quality checks**
Typ-min-max	Min corresponds to the conditions for weak/slow buffers; max corresponds to conditions for strong/fast buffers.
Model_type	The type of the model. Case sensitive. Must be one of the listed types from the IBIS Specification.
C_comp	The TOTAL output capacitance NOT including the package capacitance. MUST include the capacitances of the output transistors at the attachment to the output wiring, (for example, MOSFET pulldown Cds and pullup Cgs), the ESD structures, clamp diodes, on-die wiring to the bond pad, and the bond pad itself. Although capacitance varies with voltage and frequency, only one value is allowed; the value should be the one that gives the best match in V-T tables and reflection characteristics (either SPICE or bench tests). A number of papers describing ways to measure C_comp, and the effects of bias and frequency, are posted on the IBIS web site.
[Temperature Range]**	Chip temperature. NOT ambient temperature. typ temperature is usually higher than room temperature. Normally, CMOS has Temp (min)>Temp (max), while Bipolar has Temp (min)<Temp (max). In a high-quality model, this should be specified, even if the default values are intended.
[Voltage Range]**	Operating voltage. Required unless ALL FOUR of the other voltage reference keywords are supplied. Normally, Vcc (min) < Vcc (max). When this keyword is used alone, the other keywords default to: Pullup reference = voltage range value; Pulldown reference = 0V, Power clamp reference = voltage range value, GND clamp reference = 0V.
[Pullup Reference]	The reference voltage for the [Pullup] table. This is the offset from GND. For CMOS, value > 0 is normal.
[Pulldown Reference]	The reference voltage for the [Pulldown] table. This is the offset from GND. For CMOS, value = 0 is normal.
[POWER Clamp Reference]	The reference voltage for the [POWER Clamp] table. This is the offset from GND. For CMOS, value > 0 is normal. For dual-supply I/O, may be different from [Pullup Reference].
[GND Clamp Reference]	The reference voltage for the [GND Clamp] table. This is the offset from GND. For CMOS, value = 0 is normal.
[External Reference]	New in IBIS 4.0. Value is normally between [Pulldown Reference] and [Pullup Reference].
[Ramp]	All values must be > 0. Voltage should not exceed 80% of [Voltage Range] or [Pullup Reference] - [Pulldown Reference]. Improve consistency checking between I-V and V-T tables, including checking against combined I-V tables.

Table 12-2. Further commentary on IBIS file error checking

Error	Description
typ/min/max values not in that order	The parameters of different keywords for min typ max are in the correct order/sequence. For example, rpkg, cpkg ,lpkg, ccomp, (exception temp, which states only the conditions)
Inputs and Outputs	Require Vinh and Vinl for all receivers. Recommend Cref and/or Rref for all drivers. If Rref is used, then also require Vref. Require Vmeas for all drivers. If Cref and/or Rref not given default of zero/infinity may be used possibly generating warnings.
Hole between gnd/power clamps and IV Tables	The sweep for Gnd-clamp should be made between -Vccmax and +Vccmax. The sweep for power-clamp should be made between +Vccmin and 2*Vccmax
Monotonicity	Check combined I-V tables for non-monotonicity. Check for double counting of clamp currents.
Ramp > 100% swing	Extrapolated voltage swing for 0% to 100% must be smaller than Vcc. Improve checking between Ramp values and V-T tables when Rload = Rfixture.
Ramp and waveform (but same Rload)	The ramp-value in the ibis file and a calculated ramp-value from the rising falling waveform show the same result
V-fixture # of waveforms (that is RISING/FALLING to pullup/pulldown)	The values of each rising falling waveform belong to the fixture values, for example V-fixture 3.3v, r-fixture 50 ohm waveform starts at 2.0v and ends really at Vcc
GTL-model with different Vref, Cref, Rref, and Vmeas	The appropriate values of these 4 parameters are added referring to the min typ max conditions
Differential models only with ramp	Differential models like LVDS, PECL must contain at least one rising falling waveform, switching to vtt
Tristate-current not 0ma	The tristate current must be checked and must be smaller than 1uA, exceptions are inputs/outputs with internal pullup-pulldown behavior
Excessive currents	All excessive current above 1A should be removed from the I-V-Tables
Too few points in waveform	Time-step in rising falling waveforms. The time step in V-T-waveform should be about 1/10 of ramp-time (the smaller of both rising and falling)
C_comp same value for in and io	The values for C_comp must be checked for plausibility. Usually each input/output/IO has different values
Numbers of internal stages	The output stage was stimulated with a number of internal stages/inverters for the simulation of the rising falling waveforms. The number was …..
Stimuli description	The stimulus used for the V-T-waveform of the output had a rise-time and a fall-time of ___.ps. The period was ___ ns.
Lead in time	The V-T-waveforms are referenced to a common input-stimulus-time. See ibis 4 specification at the keywords [rising falling waveform] There are no incorrect dead-times' in the waveforms (for example 15ns and no start of a transition

12.9.2 IBIS Quality Checklist

The IBIS Quality Checklist was developed by a subcommittee of the IBIS Committee to help model makers and model users evaluate IBIS files. The IBIS Quality Checklist is a work-in-progress and contains about 100 items as of this writing.

The first section of the checklist addresses the IBIS file header. It includes a check for [File Name], which is also checked by the parser. There is a visual check for [File Rev], where the values described in the IBIS Specification cannot be checked for correctness by a parser. The documentation that accompanies the checklist describes how visual checks are performed, based on the IBIS Specification and common use by EDA tools.

The next section of the Quality checklist contains checks for components, such as verifying that all pins on the physical component are included in the [Pin] section. The third and largest section of the checklist addresses I/O models. For example, when table data points are taken at evenly spaced intervals, the model can have too few data points in transition regions. The IBIS Quality checklist recommends at least 10 points in the transition region. Quality levels set by the IBIS Subcommittee on Model Quality are listed below in Table 12-2 in summary.

A summary of the quality of the IBIS models contained in an IBIS file should be included in comments at the end of the Header information as prescribed in the IBIS Quality Checklist. These comments can appear after the Notes section in the Header. The IBIS Quality Summary comments must begin with the string - '|IQ' (that is, the comment character followed by "IQ" for IBIS Quality. Each buffer model in an IBIS file can be separately characterized by its quality. The summary should contain a list of each buffer in the file, followed by the quality level for that buffer, followed by any comments the model creator deems necessary.

An overall quality level should be indicated for each component in the file. This will be the minimum of the quality levels of the buffers used by the component. If the IBIS file contains more than one component, the overall quality of the IBIS file itself should be indicated. This will be the minimum of the quality levels of components contained in the file. If the file contains only one component, the component and file quality are synonymous, so only the component quality level needs to be indicated.

12.9.3 IBIS Quality Levels

The Checklist defines *four levels of quality* for an IBIS model. The levels are listed below and are shown in Table 12-3.

- Level 0 checks can be automated. The IBIS 4.0 parser now does most checks.
- Level 1 checks require visual inspection of the file.
- Level 2 checks compare model data against SPICE and/or hardware test data, defining a numeric Figure of Merit (FOM) for comparison.
- Level 3 checks includes both SPICE and hardware comparisons.

Table 12-3. Recognized quality levels per the IBIS Committee

Quality Level	**Includes**	**Description**	**Notes**
0		Can be checked by software IBISCHK 0 Errors and Warnings: Warnings must be documented Vil/Vih on receivers Standard Load on drivers Overshoots defined	
1	Level 0+	Checks for correctness and completeness All model spec parameters defined and validated All pins defined All models run All I-V/V-T tables inspected Ramp Data validated	
2a	Level 1+	Spice Correlation Cross Reference IBIS Accuracy FOM	See row 2b
2b	Level 1+	Lab Correlation Cross Reference IBIS Accuracy FOM	Full Verification
3		Level 2a + Level 2b	

Some IBIS file problems are created not by the model maker, but by later changes. For example, one company moved the [Notes] section to the top of the IBIS file for web posting; this file then violated the IBIS requirement that the first keyword be [IBIS Ver]. Another common error is changing the file name without updating the [IBIS File] Keyword inside the file itself.

When correlating waveform data, use the FOM and "I/O Buffer Accuracy Handbook" from the IBIS committee. An alternative approach is outlined in the FSV method.

12.9.4 IBIS Quality Specification

The subcommittee is now formulating an IBIS Quality Specification. The latest versions of the checklist and specification can be found at: http://www.sisoft.com/ibis-quality/

12.9.5 IBIS Accuracy Handbook

The IBIS Accuracy Handbook (http://www.eda.org/pub/ibis/accuracy/) [55] goes into detail on how to compare two waveforms for accuracy. One waveform being lab data, or a SPICE simulation previously verified against lab data. The readme.txt summarized in Table 12-4 explains what is available:

Table 12-4. Summary of readme.txt file

File	Description
handbook.pdf	I/O Buffer Accuracy Handbook. Covers SPICE models and includes useful checklists.[9]
ack.tar.Z	C-source code to facilitate correlation of two waveforms. Computes three figures of merit that quantify correlation. Stored in Unix compressed tar format. To unpack, type: uncompress ack.tar.Z tar -xvf ack.tar
ack.zip	Ack program stored in MS Windows zip format using WinZip program.
checklist.txt	A suggested checklist for reviewing IBIS Model data by Greg Edlund.
designconibis.zip	"A Tour of the IBIS Accuracy Specification" (MS PowerPoint) Robert Haller and Peter LaFlamme. Presented at DesignCon99, February 2, 1999
report.pdf	I/O Buffer Accuracy Report using methods outlined in the Handbook for IDT's ALVCH16831 buffer.
testbrd.tar.Z	Complete Cadence Concept and Allegro files for a small test board that demonstrates lab techniques useful in the verification of model accuracy. A companion to the Handbook.
trailer.txt	Example of a correlation data table that can be included at the end of an IBIS data file using comments.

12.9.6 Golden Waveforms

An IBIS file may contain any number of [Test Data] sections representing different driver and load combinations. Golden Waveforms are a set of SPICE waveforms simulated using known real world and ideal (standard) test loads. They are useful in verifying the accuracy of behavioral simulation results against the SPICE model from which the IBIS model parameters originated.

Inclusions of IBIS Golden Waveforms in an IBIS file are a suggestion at this point. EDA tools may not yet incorporate this feature.

[9] The old "Accuracy Specification" material is found under the archive directory.

12.9.7 Figures of Merit

The Figure of Merit (FOM) computation method is used to derive a numerical measure of the agreement between two waveforms that is based on the area between them. Thus the FOM is based on the Curve Overlay Metric. The Curve Overlay Metric applies to cases in which the measured and simulated data should theoretically lie directly on top of each other.

FOM is currently a work-in-progress and is undergoing an editorial tune-up by the IBIS Committee. FOM is covered in greater detail in "Chapter 17, Verifying Model Accuracy by Using Laboratory Measurements."

12.10 IBIS COMMON ERRORS CHECKLIST AND CORRECTION PROCEDURES

Experience with IBIS file errors led to the creation of a common errors checklist. This list appears below in Table 12-5. Automated checking for many of these errors will be built into the IBIS Golden Parser, available free for downloading from the IBIS Committee website.

Table 12-5. IBIS common errors and corrections

Error	Correction
Numerous [End] statements: Improper use of SPICE2IBIS tools?	Text edit the IBIS file and remove the extra statements
Syntax errors of all sorts. **For example:** Tab characters following keywords, lines longer than 80 ASCII characters, non-ASCII characters, (/) \ % @ # * etc. Data item names too long. Missing keywords and data items. Literal (actual) keywords inserted in the text of uncommented-out notes. File names that are different than the [File Name] keyword. Only spaces and underbars "_" are allowed in names. Syntax errors in supplied IBIS models are so common as to hardly bear mentioning.	To diagnose and fix problems, use the latest version of the IBIS Standard, an IBIS Model Syntax Guide, and the IBIS Golden Parser downloaded from the IBIS Committee website. Also, use the parser provided by the Signal Integrity software. It is usually composed of the Golden Parser plus enhancements added by the simulator company. Otherwise, have the IBIS model supplier send a corrected file.
Missing units in [Package]. H instead of nH, F instead of pF.	Verify the correct units with the IBIS model supplier and text edit the IBIS model file to add them if the work is not excessive. Otherwise, have the IBIS model supplier send a corrected file.

Table 12-5. (cont.)

Error	**Correction**
No minus signs on current out of a device ([Pullup], [GND Clamp]	Verify the correct current convention with the IBIS model supplier and ask them to send a corrected model file. Because of the amount of work required, text edit the IBIS model file to correct the problem only if the supplier is not forthcoming in a timely fashion.
[Pullup] and [POWER Clamp] not Vcc relative. **Comment:** The IBIS convention is Vtable = Vcc - Voutput. So, for a Vcc of 5 volts when: Voutput = -5, then Vtable = 10 Voutput = 0, then Vtable = 5 Voutput = 10, then Vtable = -5	Verify the correct voltage convention with the IBIS model supplier and ask them to send a corrected model file. Because of the amount of work required, text edit the IBIS model file to correct the problem by adding or subtracting a constant amount (usually the Vcc value) only if the supplier is not forthcoming in a timely fashion.
Range of I-V tables do not correspond to -Vcc to +2Vcc	Verify the correct voltage range with the IBIS model supplier and ask them to send a corrected model file. Because of the amount of work required, text edit the IBIS model file to correct the problem only if the supplier is not forthcoming in a timely fashion. Data points can be added or subtracted. Linear extensions of tables are reasonable when adding points
No clamps in [Model]—possible, but suspicious. **Comment:** One possible explanation for missing clamp tables is that their currents were included in the pullup/pulldown tables of a measured device model. Particularly in a non tri-state-able device where they cannot be separated out of the total output current. Another possible explanation is a 5-volt TTL compatible input on a low voltage device where they are not present. But, most high-speed digital devices incorporate ESD protection.	After verifying problem, require generation of a corrected model by the model supplier. Otherwise, not correctable
Non-unique [Model] statements, that is, two cells called "Output."	Consult with device supplier regarding which cells are the correct ones. Contact model supplier and ask them to send a corrected model file. Text edit the IBIS model file to correct the problem only if the supplier is not forthcoming in a timely fashion. To correct the file, delete all the incorrect cells

Table 12-5. (cont.)

Error	Correction
Unrealistic currents in clamp tables, that is, Giga-Amperes and Nano-Picoamperes.	Consult with device supplier regarding realistic maximum and minimum currents. Contact model supplier and ask them to send a corrected model file. Because of the amount of work required, text edit the IBIS model file to correct the problem only if the supplier is not forthcoming in a timely fashion. To correct the file, delete all the unrealistically high data entries. Also, clamp tables should eventually clamp. Instead of ridiculously small values at the end of the table, just set the current to zero. Most simulators want to see at least two consecutive points of 0 mA at the end of all clamp tables.
[Ramp] dt does not represent 20% to 80% into 50 ohms.	Get the corrected slew rates from the model supplier and text edit the IBIS model file to correct the problem. If the supplier is not forthcoming in a timely fashion, do the following: The slew rate, dV/dt_original, will have to be re-sized to the correct definition of 20% to 80% of the output switching range. First, establish the correct output switching range, call this V. Most often this will be Vcc = 0 volts. But, not always. For instance, the output swing of a BTL driver is 2.1 to 1.1 volts or 1 volt. Then, calculate 60% (80% - 20%) of V = dV_new. Next, find dt_new given that dV/dt_original = dV/dt_new. Finally, enter the corrected slew rate(s) in the IBIS file
Slew Rates ([Ramp] - dV/dt_r and dV/dt_f) are ridiculous, for example 1.5E-10nS.	Get the corrected slew rates from the model supplier and text edit the IBIS model file to correct the problem as in the previous problem.
Data table typ-min-max in wrong order.	Contact model supplier and ask them to send a corrected model file. Because of the amount of work required, text edit the IBIS model file to correct the problem only if the supplier is not forthcoming in a timely fashion. To correct the file, all the data will have to be re-entered in the correct order

Table 12-5. (cont.)

Error	**Correction**
Big discontinuities (large spikes) and non-monotonicity in I-V or V-T tables when viewed visually. **Comment:** Most simulators nowadays can handle small discontinuities in the input data fed them. But, large discontinuities usually prevent convergence. Large discontinuities are also not realistic and are usually caused by measurement artifacts and/or simulation of unrealistic models The IBIS golden parser, ibischk4 (and probably many other versions) and checks for non-monotonicity in the IBIS file and gives warnings. Any non-monotonicity data can be removed when the parser gives a warning.	Contact the model supplier and ask them to send a corrected model file. Because of the amount of work required, text edit the IBIS model file to correct the problem only if the supplier is not forthcoming in a timely fashion. To correct the file, delete all the unrealistic data entries and re-enter smoothed data for the offending points.
Double counting of clamp currents, that is, [GND Clamp] magnitudes in both [GND Clamp] and [Pulldown]. **Comment:** The I-V tables of an output, as measured, will include the action of the pullup/pulldown tables and the clamps, if present. In many devices it is possible to disable the pullup/pulldown and then measure the action of the clamps alone. Sometimes model suppliers forget to subtract the clamp currents from the total output behavior and present the behavior of the pullup/pulldown itself.	Contact the model supplier and ask them to send a corrected model file. Because of the amount of work required, text-edit the IBIS model file to correct the problem only if the supplier is not forthcoming in a timely fashion. To correct the file, subtract the clamp currents from the total output currents and enter the remainder as the pullup/pulldown table(s).
Common Differential Pin-Pair Mistakes: Making the non-inverting pin inverting, or vice versa, on drivers or receivers. Entering Launch Delay data for a receiver (Input) cell. Entering Vil or Vih data for a driver (Output) cell.	

12.10.1 Commentary on Errors

About half of all IBIS files examined contain obvious mistakes:

- Incomplete pin lists (very common)
- No IBIS version specified
- Input cells with no switching thresholds or only one threshold

- File names that do not match the Keyword [File Name] inside the file
- Data not in the typ-min-max sequence
- Files with input cells only - no outputs or I/Os (Like an alligator with a head at each end – a very mean beast.)
- Missing model cells, duplicate Model_name, etc.

In defense of model suppliers having seen so many simple mistakes—IBIS is complex; IBIS files are long, detailed and not particularly interesting reading, and computer data input is unforgiving. IBIS is not designed to be human-readable. That is why IBIS files must be run through the IBIS Golden Parser before release!

A software-based parser is much better than a human being at catching syntax mistakes and the like. IBIS model suppliers should habitual use a good parser and do at least an eyeball check in a graphical user interface. The parser will make syntax mistakes obvious. The GUI will make odd curves, missing pin lists, and other mistakes obvious.

Excellent parsers are available from EDA tool vendors. One can only hope that they will provide them free of charge to semiconductor suppliers. No models, or bad models = no software tool sales. Plus parsers can be designed to do reality checks on the data itself, not just syntax and inclusion-of-required-elements tests.

People developing products at the cutting edge of technology often design with developmental devices. A word on getting models from semiconductor suppliers–developmental models may never be obtained in time to do any good in meeting a development schedule.

Lastly, for 2/3rds of all IBIS models returned to their supplier for correction, do not expect a reply in a reasonable time frame. Over half were never fixed and returned. So learn to fix what you can on your own.

12.11 3COM'S ISO 9000 PROCESS FOR IBIS MODELS

To sum up thus far: we have discussed a process flow, a file parser, a checklist, simulations and figures-of-merit, and some reality checking on an IBIS file. In this section, we will explain an ISO 9000[10] process developed at 3Com by one author [Leventhal] for checking and improving the quality of

---

[10] ISO 9000 is a certifying method if OEMs want to sell electronic equipment in the European market. These standards are developed and administered by the International Organization for Standardization [59].

IBIS models as first received from the semiconductor supplier. Regulators inspect and certify companies as compliant with these ISO standards.

Why is this material included? Because:

- Companies selling electronic equipment in Europe need ISO 9000 certification.
- The certification process leverages a company's internal skills. A company has to think through and describe the processes they use as part of certification.
- ISO 9000 certification provides a framework for process and product improvement.
- Some of the forms and correction procedures we developed at 3Com may be useful to readers and their companies. Through the generosity of 3Com this work was made available through the IBIS Committee website at: http://www.eda.org/pub/ibis/training/3com-docs/

Individual organizations must decide for themselves whether they want to support this framework for acquiring robust IBIS models. We developed this process at 3Com to be certified as complying with ISO 9000. The ISO 9000 family of standards is primarily concerned with quality management—what the organization does to fulfill:

- The customer's quality requirements and applicable regulatory requirements,
- while also aiming to enhance customer satisfaction,
- and achieve continual improvement of its performance in pursuit of these objectives.

Other companies, notably Siemens, National Semiconductor and Intel also developed similar processes.

12.11.1 3Com Model Request Form

Model acquisition and verification requires a significant investment of time and resources, plus significant lead-time to get and fix model files. One intent of the ISO 9000 model-validation process was to conserve resources by defining the model elements and level of verification needed. The user determines which model elements he or she needs in the IBIS file, and specifies it on a Simulation Model Request Form.

This form is reproduced in Figures 12-4 and 12-5. Also specified on the request form is the level of verification that will be needed. The elements requested determine the model checklist and verification needed. We developed this form because every IBIS model is different, every circuit application is different, and simulators can differ in their capabilities to use the information in an IBIS file. The designer or user of the IBIS model usually knows best what they will need in the IBIS file. This form facilitates communicating those needs to others.

Simulation Model Request Form

Requester Name____

Component 3Com Part # (if known) ______________________________

Supplier's Part # (required) __________________________________

Phone # of Supplier ________________________

Submittal Date _______________ Start Date ________________

Type of Model Required: ___IBIS ___SPICE ___S_Parameter __Other (2_port, etc.)

For non_IBIS models attach your requirements:

IBIS Model:

Refer to 3Com IBIS Model Standard (E0173) and IBIS Model Standard 3.2

3Com Minimum Standard IBIS Model (required)[1] ____________________

(requestor's comments)

IBIS Model "Add_Ons" You Can Request (Optional):

V_T Curves[2] ____________ Signal_name(s)[3]____________________

Buffer switching test fixture parameters[4]_ Individual pin R_L_C parasitics[5]______

Differential pin properties: vdiff [6]______ tdelay[7]______

Large package Pin Mapping[8]__________

1. The "3Com IBIS Model Standard (E0173)" defines a minimum acceptable model as containing V-I curves, slew rates, pin (list) connections, package parasitics, voltage and temperature ranges, model name and type, input and output threshold voltages, die capacitance, all necessary "boiler plate" such as IBIS version, file revision, etc., and any "if present in device" elements.
2. Voltage Vs time rise and fall waveform data for parts difficult to model with simple slew rates such as those with soft turnon/turnoff characteristics.
3. Optional signal names by pin number.
4. Switching test fixture data when different than 50 ohms. Buffer delay can be adjusted (recalculated) for different loads when test fixture is specified.
5. Individual pin R-L-C parasitics override default package parasitics when supplied.
6. Heads-up that differential pin-pairs will be modeled. You can request data on voltage threshold (output or input) and launch delay offsets.
7. Ibid.
8. Used when ground and power bussing for individual/groups of pins is employed.

Figure 12-4. Simulation model request form, page 1

Large package voltage bussing[9] Pullup Reference ____ Pulldown Reference ____

Power Clamp Reference ____ GND Clamp Reference ____

Package Model[10] __________________ (R_L_C matrices, mutually coupled pins, etc.)

Model Selector[11]_____ (required for programmable devices)

Driver Schedule[12]______ (for multi_stage drivers, soft turn_on, etc.)

Model Spec[13]____ (for hysteresis)

Add Submodel & Submodel Spec[14]_____

(special purpose functionality, i.e., dynamic clamp, bus hold, etc.)

Series Pin Mapping ________ & Series Switch Groups[15]_______

(Diode) Transit Time parameters[16]____

Board (abstracted) Description (model)[17]___

Model Verification Options[18]

Level 4 ____ Level 3 ____ Level 2 ____ Level 1 ____

Refer to 3Com IBIS Model Standard (Xxxxx)

9. Can define reference voltages different than assumed high rail of Vcc and low rail of zero.
10. Used for defining RLC matrices describing pin parasitics. You can use this data to for mutual coupling between pins and high frequency effects. Overrides individual pin parasitics.
11. Used to pick a Model from a list of Models for pins with programmable buffers.
12. Describes the relative model switching sequence for referenced models to produce a multi-stage driver.
13. Used for defining refinements in threshold voltages that describe input switching hysteresis effects.
14. Adds special purpose functionality such as dynamic clamping or bus hold to an existing model.
15. Used to associate two pins joined by a series model and allowable state combinations of series switches.
16. Diode transit time parameters for power and ground clamps used for estimating diode capacitance.
17. Used to describe a "board level component" for an additional level of abstraction for ease of use in board - board simulations.
18. The "3Com IBIS Model Standard (E0173)" defines the four approval levels as:
    Level 1: By direct measurement: High level of approval.
    Level 2: By correlation with a correlated SPICE model.
    Level 3: By correlation with the device data sheet.
    Level 4: No parametric verification - will run in simulator: Low level of approval IBIS model file checked for correct syntax with IBIS Golden Parser.

Figure 12-5. Simulation model request form, page 2

For example, is dV/dt good enough, or will V-T waveform data be needed? Is conforming to the data sheet good enough, or will lab measurement verification be needed? This is done either at the time a new component is requested or when a design is to be reused and re-simulated or when troubleshooting a production problem.

12.11.2 Internal IBIS Standard

In the discussion below, model element is equivalent to keyword. To guide the user, 3Com developed an internal IBIS standard, which draws heavily on the current IBIS standard. It also:

1. Organized and separated model elements for better understanding and readability
2. Separated model elements into a *minimum standard group required of all models and an optional (not required unless specified by the user) group.*

   The IBIS Spec itself, being a compromise between semiconductor suppliers, EDA tool users, and OEM users does not always well represent what OEM users truly may need on a given design. The internal spec should offer guidance on what is truly needed. This will aid in the formulation and negotiation of a purchase specification when such an instrument is needed.

   For example, min-max data, and V-T curves need to be requested when appropriate, which is almost always. Also, for example, it makes no sense to simulate a GTLP driver that has only [Ramp] data, and not its V-T data in the IBIS model file, and that is known by an experience engineer at the start of using it in a design.

3. Defined several levels of verification
4. Explained methods for creating, verifying, and detecting and correcting common IBIS model mistakes

12.11.3 Make or Buy the IBIS Model?

If a model is unavailable, it will need to be made or bought. In 3Com's ISO 9000 procedures, we used a document, "Simulation Model Procurement Process," which spelled out responsibilities, documents to use, process steps, and approvals.

The decision to spend money to buy an IBIS model implies that decision makers know the all the costs–purchase costs, and the cost of the lost opportunity to do modeling and simulation. A model may be bought by:

- Hiring a 3rd party to make it
- Hiring a company staff to make it
- Paying the semiconductor supplier to produce it
- Licensing an expanded model library from an EDA tool provider.

The costs and risks of not getting the model must also be known. What are the chances that the prototype will not work unless simulated and what are the lost opportunities that would result?

12.11.4 Verification and IBIS Quality Levels

The results of a quality level determination can be documented in the IBIS file itself once an OEM company has invested resources in establishing and improving the quality level of an IBIS model file. This can be done as part of the [Notes] keyword and/or as "|" comments in the file. The quality level can be updated as the model progresses from-to:

- Ready for simulation
- Verified against the data sheet
- Verified against a SPICE simulation
- Verified against prototype lab measurement data

12.12 IBIS MODEL ACCEPTANCE AND LEGITIMACY

12.12.1 Acceptance

IBIS has achieved a lot of acceptance and legitimacy because of its ability to do Signal Integrity simulation in a balanced and accurate fashion. Still, arguments remain over whether SPICE is better. Advocates for different modeling methods make arguments that border on religious faith. But in our view, which model is better depends on the goals of the simulations.

IBIS does not succeed because it is firmly mandated by a Regulatory body like ISO 9000, the FCC, or emissions requirements. The IBIS Spec itself is full of escape clauses, oughts, and caveats. This is because such voluntary industry standards are developed by consensus. This consensus is among parties with sometimes competing interests: Semiconductor suppliers, simulation program suppliers, and end-user product designers.

Individual IBIS files are usually full of the usual manufacturer's disclaimers about the model not having any resemblance to the device it is modeling. The past history of poor quality of IBIS models, in general, has hurt their legitimacy just as similar results originally hurt SPICE.

People ask why the IBIS Committee cannot enforce discipline. That is not how EIA Standards committees work. The 1970s was last time that a standards committee attempted to enforce discipline, when JEDEC tried it on 1Nxxxx, 2Nxxxx, and 3Nxxxx data sheet, and the effort did not work.

IBIS only provides guidance on a format for the exchange of data to be used in modeling Signal Integrity. *Its discipline comes from its usefulness and successes.*

12.12.2 Legitimacy

What hurts the legitimacy of IBIS, and SPICE, and modeling and simulation in general is poor quality models. The following results are presented to reinforce the points about quality made in the beginning of this chapter.[11]

The results reported are these: in a six-month period in 2000, about 225 IBIS files were subjected to the following screening process:

1. Convert (parse) to a Cadence Design Systems .dml file to run in simulation
2. Correct mistakes or return to source for correction.
3. Iterate.
4. Apply an eyeball reality check in the Cadence model viewer GUI:
   - Check for a complete pin list
   - Observe whether I-V and V-T tables seem realistic—Look for ridiculous clamp currents, for example, $1xE^{23}$ amps
5. Simulate a test net and observe whether reflections seem realistic.
6. Iterate.

---

[11] The experiences reported here mirrored similar experiences in 1977 when dealing with quality problems found by burn-in (similar to a "Highly Accelerated (HASS) Stress Screening" test) on ICs. In that instance the quality referred to was the occurrence of latent manufacturing defects in transistors and ICs. The response curve in the mid 1970s of semiconductor suppliers to latent defect problems revealed by burn-in exactly matched the response curve for fixing IBIS model problems some 30 years later, in every respect.

The IBIS files were mostly gathered from supplier web sites or e-mailed to us. They were almost always generated by simulating a SPICE deck to obtain tables, filling in the pin connections (sometimes), and adding the other required information. Here are the results:

49 failed but minor fixes cured the problem
47 additional failed and were returned to their supplier for correction

This gives an initial quality of 57%!!!!

- Of the 47 models returned to suppliers, the suppliers fixed and returned only 15. Of these 15, roughly half failed *again*. Few of these second failures ever got fixed.
- An additional 100 important-to-critical models, that were unavailable, were also requested directly from suppliers. In three months, we received about 32 of them. After an additional 10 months at least 50 remained undelivered.

Essentially, when problems arc found, a few good suppliers will get back with solutions quickly, several more will do so in a reasonable time, and then it begins to tail off to never, or almost never, in an exponential decay fashion. Likewise, the best suppliers have very much fewer problems than those with poor turn-around times on fixing problems.

After the original report, about 800 model files were checked in a two-year period, with similar results. These results led directly to lobbying the IBIS Committee to get involved, which led to the IBIS QC checklist and additions to the IBIS Golden Parser. Just think about the negative impact this model quality issue has on high-speed board development and the cost of doing modeling and simulation.

Rather than wait many months to see whether a supplier will fix an incorrect IBIS model or provide a new one, the engineer can visit the supplier and assess the quality of the supplier's operation. The engineer should evaluate whether the company has a competent modeling staff and a long-term commitment to support the staff.

12.12.3 A Last Word on Quality

In quality, companies generally get what they are willing to pay for and give attention to. It is common practice for purchasing agents to drive down the costs of purchased items. The benefits of lower costs are immediate and obvious. In the case of modeling and simulation, benefits are usually significantly delayed in time and sometimes require effort to measure. Modeling and simulation are intended to improve the quality of company products and lower their costs. What is a good way to assess these quality and cost benefits?

Consider adopting an approach like regular employee performance reviews and applying it to policies, procedures, suppliers, and design processes. If ICs from the lowest-cost producer (who probably does not supply models) work as well as those for a higher-cost producer, then everyone is happy. If a low-cost simulator works as well as a higher-cost simulator, then that simulator should be used. If an expensive simulator license is not being used to advantage, management must address the problem and fix it. Otherwise they have made a poor investment.

When done on a regular basis, performance reviews of policies, procedures, suppliers, and design processes ensure that desired results are achieved. If the review shows that they do not achieve desired results, management needs to fix them.

12.13 SUMMARY

The quality of circuit models provided by most (not all)[12] semiconductor suppliers is never entirely trustworthy—no matter what type of model is provided. IBIS is no exception. Fortunately, the great majority of IBIS model mistakes are in syntax, format, and completeness.

To catch mistakes, engineers should always run their IBIS model file through the IBIS Golden Parser and model quality checklist. Engineers should also examine both the IBIS file and their characteristic curves, plotted with EDA tools.

Many semiconductor suppliers, especially those with poor track records, may not be capable of quickly and effectively fixing mistakes. Consequently, engineers planning to do much simulation should possess the necessary skills in order to fix most of their own models.

---

[12] Noteworthy exceptions are five large semiconductor suppliers of very complex state-of-the-art ICs. They rarely make mistakes on an IBIS model. When they do, they respond quickly and effectively. It is the authors' opinion that the best state-of-the-art companies pay attention to the quality of what they do.

Chapter 13

USING EDA TOOLS TO CREATE AND VALIDATE IBIS MODELS FROM SPICE

Commercial tools, such as Model Integrity® from Cadence Design Systems, are available to assist in creating and validating IBIS models

Abstract: Model validation prepares a model for release. Validation checks both model syntax and model data. The final step is using the model in a simulation to check for unexpected interactions between the model and the simulator. The validation methodology described in this chapter can be applied to any model.

13.1 INTRODUCTION

There are several ways to create and validate IBIS models. The easiest way is to identify a pre-existing IBIS file from a similar part (assuming it is close enough), and adapt it as needed. But, if IBIS models are not readily available, they can be created from SPICE simulations. IBIS models can be validated using commercial EDA tools, such as Cadence's Model Integrity® tool and Mentor Graphics' Visual IBIS Editor®.

This chapter explains how to create and validate IBIS models generated from SPICE using Model Integrity. We also discuss some "free" tools available on the IBIS web site. For additional information on a specific tool, please contact the respective company. The opinions expressed here are those of the authors.

In this chapter, we use an I/O buffer example to show how to:

- Prepare for the SPICE to IBIS conversion. This includes validating the SPICE models for BJT and CMOS devices in the netlist for the I/O buffer that will be converted.
- Convert SPICE netlists to IBIS models.
- Validate IBIS models converted from SPICE by using the IBIS Quality Checklist.
- Identify and fix common model errors.
- Model passive interconnections in IBIS.

One example will be used throughout this chapter to illustrate these steps. This example includes a SPICE netlist for a simple CMOS I/O buffer with NMOS and PMOS model parameters. The same steps would be used for BJT and BiCMOS buffers. These steps can also be used for a wide variety of I/O circuit topologies. After discussing our example in detail, there is a discussion of how to handle advanced I/O buffer designs, such as differential buffers.

13.2 I/O BUFFER EXAMPLE

To illustrate several steps in the IBIS modeling flow, we use the I/O buffer in Figure 13-1 as an example.[1] This I/O buffer is later used as part of a complete IBIS file to illustrate IBIS model file validation.

The steps illustrated using this I/O buffer example include:

- SPICE simulation setup.
- SPICE-to-IBIS buffer model extraction.
- IBIS buffer validation.
- Assigning the I/O buffer model to component pins in an IBIS file.

13.2.1 Circuit Design and Netlist

The I/O circuit selected for the examples in this chapter is a CMOS push-pull stage, shown in Figure 13-1. The I/O pad serves as both the input to the receiver stage and the output from the driver stage. This buffer includes a pre-drive stage that combines the Enable and Data inputs to generate the buffer *pad* (output) signal. The pre-drive stage uses PMOS and NMOS FETs to introduce a timing offset between the gate signals of the two output transistors. The timing offset minimizes the crowbar current that

---

1 This I/O buffer circuit was originally created by Arpad Muranyi of Intel, and can be found on the IBIS web site under "Training."

flows through the output stage between Vcc and GND when the buffer toggles.

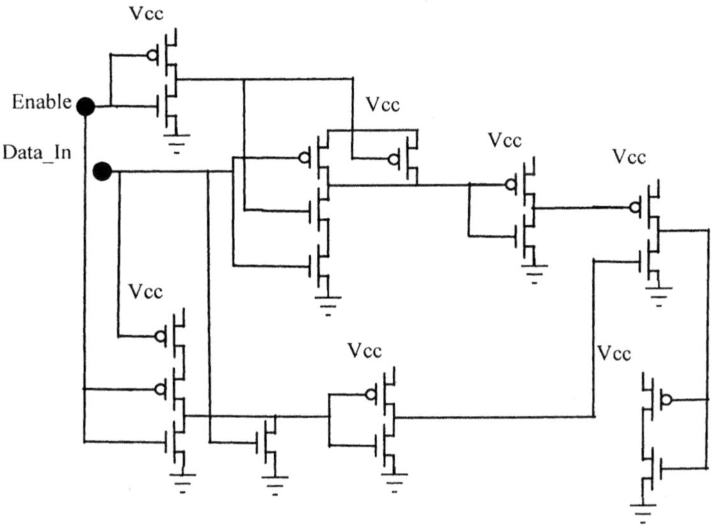

Figure 13-1. I/O buffer circuit for the examples

Clamp currents arise from a number of physical effects:

- First, of course, are any actual clamp diodes.
- Second is any ESD structure attached to the I/O buffer's output pin.
- Third are all of the parasitic diodes between the source and drain and the substrate or well in the MOSFET structure, as shown in Figure 13-2.

The I/O buffer example, shown in Figure 13-1, does not contain any diodes or ESD circuit. Only parasitic diodes are clamping the output. The parasitic diodes provide only weak clamping.

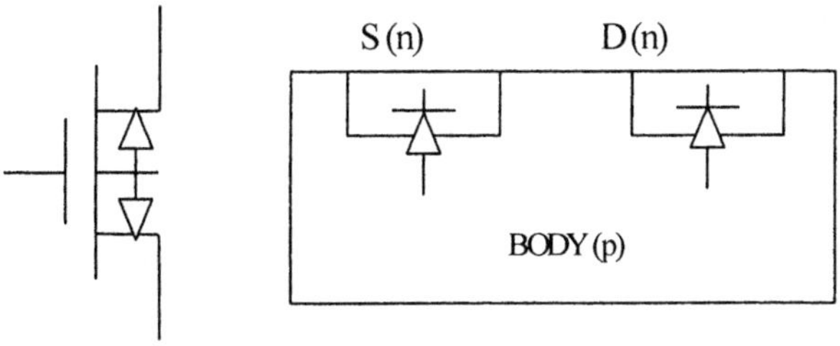

Figure 13-2. NMOS parasitic diodes.

The netlist used to generate IBIS model data must also include all of the capacitance loading on the output pad of the buffer. Total pad capacitances of 2-5 pF are not uncommon. Some of this capacitance comes from the transistor and diode capacitances, which are normally included in the SPICE models for these devices. Because the output drive transistors are relatively large, their capacitance is usually large. Additional capacitance is found in the buffer's ESD structures, interconnection wiring, and the I/O pad structure itself.

The simple buffer used for the examples in this chapter can operate as an Output buffer when the Enable is active, and as an Input when the Enable is off. This buffer does not include any ESD protection. A small amount of clamping is provided by the parasitic diodes. Other buffer designs use separate circuits for driver and receiver functions.

To generate a netlist for SPICE-to-IBIS conversion in tools such as Cadence's Model Integrity, all buffer currents must pass through the nodes explicitly listed on the I/O buffer's subcircuit. There must not be any global nodes in the buffer's netlist. Global nodes can cause internal nodes of the buffer subcircuit to be unintentionally shorted to nodes in the main circuit. Global nodes also allow currents to bypass corresponding subcircuit pins, making it impossible to monitor these currents in a SPICE simulation.

There are two common ways of using global nodes in subcircuit netlists. The first is to use a SPICE-defined global node. For example, node "0" is always a global node; "GND" is also a global node for many versions of SPICE. The other way is to explicitly define a global node, such as Vcc, using the .GLOBAL statement.

Since automated extraction from a schematic or layout generally names the ground node with a global name, it is often necessary to rename the ground node to something like "GND1" after extraction. Similarly, explicit .GLOBAL statements must be commented out. These global nodes, such as GND1 and Vcc, must then be added to the subcircuit's node list, as shown in the following example.

Extracted netlist:

```
.subckt example 1 2 3 4
.global vcc
M1 1 0 vcc NMOS
* etc.
```

*After removing global nodes by, using * to comment out:*

```
.subckt example 1 2 3 4 vcc gnd1
*.global vcc
M1 1 gnd1 vcc NMOS
* etc.
```

13.2.2 Model Parameters

The example's transistor SPICE-model parameters represent an older 3.3V, 5-μm CMOS technology. The SPICE model parameters have not been verified against hardware measurements, since this CMOS process is no longer in production. Consequently, our IBIS model example can be validated and verified against the SPICE model but it cannot be verified against test data because parts are no longer available.

It is important to use appropriate transistor parameters for the model. In this example, we chose SPICE LEVEL=3 NMOS and PMOS, to allow geometric parameters, such as width and length, to be used. This model is simpler than the BSIM3 model, making it more suitable for examples.

Using the wrong model parameters for a model produces invalid results. For example, using SPICE MOSFET LEVEL=3 parameters in a BSIM3 model would generate different simulation results than using them in a LEVEL=3 model. Similar problems would occur using BSIM3 parameters in a LEVEL=3 model. Furthermore, model parameters, such as BSIM3, are generally not portable between simulators, since implementation of the device equations may vary. Some simulators have even added parameters to the original BSIM3 model. Because of the way SPICE is implemented, a simulation would still run with incorrect parameters. To do this, SPICE would use default parameter values for any unassigned parameters and ignore any extra parameters not used by the specified model.

In a real model, simulation results are sanity checked against the appropriate data, such as device specifications, functional specifications, chip prototypes, or a data sheet. In some cases, the SPICE model serves as the data sheet for the I/O buffer, so there is no separate data sheet.

13.3 SPICE-TO-IBIS CONVERSION METHODOLOGY

Generating an IBIS file from SPICE models consists of three major steps:

- Simulating each I/O buffer in SPICE.
- Extracting simulation results to generate an IBIS buffer model.
- Combining IBIS models into a complete IBIS file.

In IBIS, three population corners are used to represent the statistical variation. These corners are referred to as typ, min, and max. Min corresponds to transistors with low current, and max to transistors with High current. It is possible for an I/O to have its fastest edge rate at some other statistical corner, depending on the circuit design.

Each step needs information that is not contained in the buffer's SPICE netlist or transistor parameters. For example, when setting up the SPICE simulations, we need the Vcc and temperature for min, typ and max corners. When extracting the SPICE simulation results to create an IBIS buffer, we must add data such as Vinh and Vinl. Creating an IBIS file also requires adding the pin list and header information.

The documentation provided with Cadence's Model Integrity goes into the specific requirements of that tool in detail. The files used for examples in this chapter are:

- template.sp (the SPICE template for the main circuit)
- template.lis (the SPICE output listing)
- template.buf (the IBIS buffer model generated from template.lis).

Relevant sections from each file are included in examples in this chapter, although the files are too long to include in their entirety here. For those who have access to Cadence SPECCTRAQuest® (Allegro SI), the files are under <install_dir>\share\pcb\modelintegrity\Spice_Templates. Alternatively, similar templates for use with IBISCenter, a SPICE-to-IBIS tool, can be found on the IBIS web site under "Training."

13.3.1 Step 1: Simulating Each I/O Buffer in SPICE

The first step in any SPICE-to-IBIS conversion is setting up and running the SPICE simulations for each I/O buffer. There are many ways to do this, such as a graphical user interface, a script, or simple text editing.

Any version of SPICE that produces an output listing (*.lis file) can be used for SPICE-to-IBIS conversion. Generally, the SPICE flavor chosen is based on the SPICE model library. For example, if the SPICE transistor model library provided by the foundry is targeted for HSPICE, then HSPICE must be used to obtain correct simulation results. Major foundries support the major simulators (HSPICE, Spectre, and Eldo). They may also offer a choice of BSIM or EKV MOSFET models. Captive (in-house) foundries can support these, or they can support proprietary SPICE models.

Users must set values for TEMP (temperature), Vcc (Power supplies), and device models (libraries) for each of the three IBIS corners of typ, min, and max. *TEMP is the die temperature, which is higher than ambient.* A common typ value for TEMP is 50 degrees Celsius.

In Cadence's Model Integrity, the parameters are all set together, once for the V-T simulations, and once for the I-V simulations. This is described in detail in the Model Integrity documentation. For any SPICE-to-IBIS conversion, the same settings must be used for the two sets of simulations (I-

V and V-T). Any additional user-defined parameters must also be set in the same manner. The only exception is .TEMP, which must be set separately for each corner for most versions of SPICE.

Figure 13-3 below shows how SPICE parameters are set in the SPICE template provided with Model Integrity. The user changes the values as necessary when generating a main circuit netlist for the I/O buffer. These parameters illustrate the settings for a 3.3V buffer with 5V clamp supply. This type of buffer is often referred to as an "over-voltage safe" buffer.

```
.TEMP 50      $ Temperature of typical case
*-------------------------------------------------------------*
.PARAM PUref_typ = 3.300V $ Pullup reference voltage, typ.
.PARAM PUref_min = 3.135V $ Pullup reference voltage, min.
.PARAM PUref_max = 3.465V $ Pullup reference voltage, max.
.PARAM PCLref_typ=5.00V $ Power clamp reference voltage, typ.
.PARAM PCLref_min=4.750V $ Power clamp reference voltage, min.
.PARAM PCLref_max=5.250V $ Power clamp reference voltage, max.
*-------------------------------------------------------------*
.PARAM PDref_typ = 0.000V $ Pulldown reference voltage, typ.
.PARAM PDref_min = 0.000V $ Pulldown reference voltage, min.
.PARAM PDref_max = 0.000V $ Pulldown reference voltage, max.
.PARAM GCLref_typ = 0.000V $ GND clamp reference voltage, typ.
.PARAM GCLref_min = 0.000V $ GND clamp reference voltage, min.
.PARAM GCLref_max = 0.000V $ GND clamp reference voltage, max.
**************************************************************
```

Figure 13-3. Example 1: SPICE parameter settings for a 5V-safe 3.3V buffer.

I-V and V-T simulations are run at each of the three corners. These simulations generate the following simulation outputs:

- Ipad vs. Vout (data=0, enable=on)
- Ipad vs. Vout (data=1, enable=on)
- Ipad vs. Vout (data=0 or 1, enable=off)
- Vpad vs. Time (data rising, R_fixture to Vcc)
- Vpad vs. Time (data falling, R_fixture to Vcc)
- Vpad vs. Time (data rising, R_fixture to GND)
- Vpad vs. Time (data falling, R_fixture to GND)

The SPICE output for data extraction is generated using SPICE .PRINT statements. Node names and their ordering must be consistent between the .PRINT order and the extraction tool. If print nodes are not in the correct order, then the wrong data could be extracted to one of the tables in the buffer model. Different SPICE-to-IBIS conversion tools use different .PRINT ordering, so SPICE output files are not interchangeable.

In SPICE, nodes between subcircuit calls and subcircuits are matched by order, not by name. This can present a challenge to the engineer, since the node order of a subcircuit is not always predictable or well documented. Figure 13-4 shows SPICE node matching between a subcircuit and a subcircuit call. The node "pdref" in the main circuit is automatically connected to node "1" inside component X1.

```
* subckt numbered nodes, with functional order of
* nodes =  in out puref PCLref pdref GCLref /en
.SUBCKT BUFFER 1 2 3  4  5  6  7
* . . .
.ENDS
X1 pdref pulldown puref PCLref pdref GCLref Von BUFFER
```

Figure 13-4. Example 2: Node matching in SPICE

The main circuit contains the subcircuit call(s). It also contains the bias and transient voltage sources, with zero-volt voltage sources frequently used as ammeters. The netlist could contain one main circuit for each IBIS table column (24 main circuits), or there could be one circuit for each I-V table (4 circuits) and one for each V-T table (4 circuits). Cadence's Model Integrity tool uses the latter approach, and provides a template that contains all of the circuits in one file. In either case, SPICE simulates each main circuit in sequence.

As I/O buffers get faster, the transition time for the output signal becomes shorter and the maximum toggle rate increases. Thus the default V-T simulation time settings might need to be changed. The print step must be small enough to have at least 10 points in the transition region of the fastest V-T table. Also, some Signal Integrity tools do not allow buffers to toggle sooner than the end of the V-T table. In those tools, models created using default V-T end times might not toggle properly.

The settings for generating V-T tables must meet two requirements.

- First, the slowest transition needs to settle before the end of the simulation time.
- Second, and perhaps more important, there should be at least 10 points in each transition, including both min and max corners.

Using the defaults in s2ibis2, which simulates exactly 100 points in 20 nsec, the time step in the V-t table becomes 0.2ns, or 200ps. Today, high-speed I/O buffers can switch faster than 50psec. For these buffers, using s2ibis2 defaults would create a model with only one or two points in the transition region, which will not provide sufficient accuracy in the V-T table. Also, sudden changes in slope associated with too few points in the V-T table can lead to slower simulation time, as the simulator often must work harder to converge.

Once the simulation netlist and main circuit settings have been set up, the next step is to run the actual SPICE simulations using the appropriate SPICE simulation tool (such as SPICE 3f5, HSPICE, or PSpice, as required using a transistor model library). If one is creating many buffer models, one can create a script to do the netlist and main circuit editing, and then submit the SPICE simulations. When doing only one or two, it might be easier to submit the SPICE jobs manually. Whether using a tool such as s2ibis2, a script, or a manual process, one needs to know the path to the SPICE executable (often aliased to a command name).

After the SPICE runs have finished, it is good practice to do a quick scan for error and warning messages in the SPICE output file. This detects common problems, such as convergence failures or transistor model parameters not being used by SPICE. Scripting could include these checks, while also screening out do not care warning messages.

Viewing the original SPICE curve for voltage vs. time and current vs. voltage is also helpful in troubleshooting. The original SPICE I-V curves contain the total current for the I/O pin under each driving condition. Thus the SPICE Pullup curve includes the I/O buffer's pullup current as well as both the power and ground clamp currents. Similarly, the SPICE Pulldown curve includes the I/O buffer's pulldown current as well as both power and ground clamp currents.[2] When the I/O buffer is not enabled, the SPICE current includes just the two clamp currents.

Separating these total pad currents into the respective I-V tables is done by subtracting the clamp currents from the total current, which generates the "Subtracted Pullup" and "Subtracted Pulldown" data. These two subtracted current tables are then formatted into IBIS tables. The total clamp current is separated into two tables, one for power clamp current and one for ground clamp current. This is done by assuming the ground clamp current goes through 0A at 0V and the power clamp current goes through 0A at Vcc. This assumption also allows the detection of pullup and pulldown load currents (such as non-linear resistors implemented with MOSFETs).

A good check on the IBIS model extraction process is to view the original SPICE output I-V and V-T tables graphically, using a tool such as

---

[2] Accounting for clamp currents is very important and often confusing – BOTH clamp currents are ALWAYS present. Incorrect algorithms appear in a number of SPICE-to-IBIS tools, such as s2ibis2 from North Carolina State University.

Model Integrity. Note that the power clamp and ground clamp data look identical at this stage, since both clamps are always active in the circuit. Likewise, the power and ground clamp currents also show up in the pullup and pulldown circuit characteristics. When seen on the same scale, it is hard to tell the pullup and pulldown curves from the clamp curves in a SPICE file. Zooming in is usually needed to see the non-zero pullup and pulldown currents between 0V and Vcc.

During extraction, Model Integrity allows viewing of the subtracted I-V curves. In these curves, the power and ground clamp data is subtracted from the pullup and pulldown data. The power and ground clamp data is separated into separate power and ground clamp curves. Viewing these curves serves as a check on the correct identification and subtraction of table data.

Figure 13-5 shows the SPICE ground-clamp data for the I/O buffer example. At this point, the power clamp data has not been subtracted yet. In this example, the largest ground clamp current occurs at the minimum corner conditions for our MOSFET I/O buffer example, due to device physics (FETs have higher current at lower temperature, while diodes have higher currents at higher temperature).

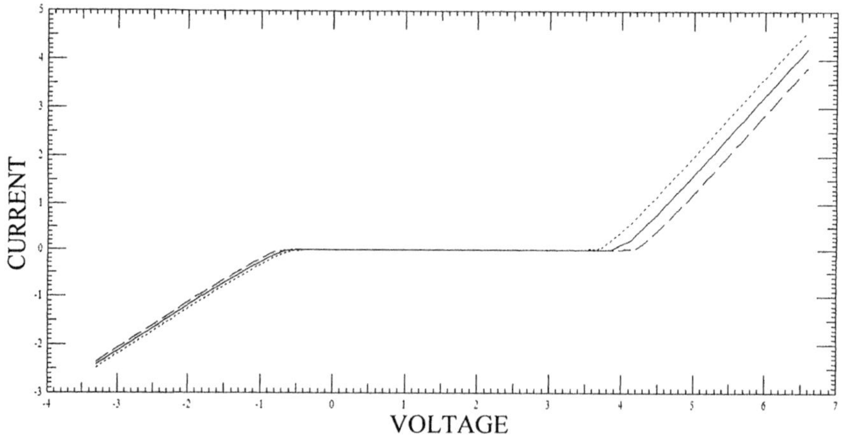

Figure 13-5. SPICE data for the buffer's GND clamp curve

Figure 13-6 graphically shows one of the four V-T tables for the I/O buffer. A quick visual check for this I/O buffer shows that there are more than 10 points in each transition, and that all transitions have settled before the end of the simulation time at 15 nsec. As expected, the MAX curve has the highest voltage swing and fastest transition edge rate. The curves also exhibit an initial peak due to forward voltage coupling caused by the capacitance from the FET gates to the output pad (C_{ds}).

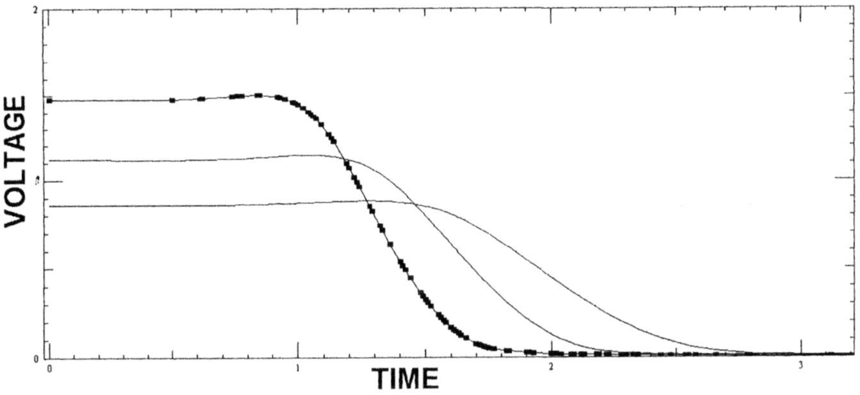

Figure 13-6. The falling set of the buffer's V-T curves
Data points are marked for the MAX model.

13.3.2 Step 2: Extracting Simulation Results to Generate an IBIS Buffer Model

Once the SPICE runs have all been completed, data extraction can begin. At this point, there should be 12 I-V tables (4 tables times 3 corners) and 12 V-T tables (4 tables times 3 corners). Depending on how SPICE simulations were run, this could consist of 24 separate files (one for each table corner), two files (one for I-V and one for V-T results), or a single file containing all results. But whether run individually or with a single output file, all results must be combined (concatenated) into a single file before extracting an IBIS model in Cadence's Model Integrity tool.

13.3.2.1 Obtaining C_comp

C_comp is one of the most difficult parameters to obtain for an IBIS model. C_comp is the effective loading capacitance for signal reflections at the I/O pad for a driver or receiver. This is greater than the fixed wiring capacitance at the pad, since it also includes junction capacitances. Also, C_comp is usually less than the pin capacitance of a packaged part.

Physically, there is capacitance to each DC voltage. This capacitance also varies with voltage, as well as with buffer state (driving high, driving low, or receiving). However, there is only one parameter in the IBIS 3.2 model for this capacitance. To account for capacitor current caused by DC power voltage changes, IBIS 4.0 introduced four capacitance terms (C_comp_pullup, C_comp_pulldown, C_comp_power_clamp, and C_comp_gnd_clamp). Because all of the C_comp parameters are fixed values, the value of C_comp is sometimes tweaked to get better agreement between the IBIS model and SPICE and bench measurements.

Because C_comp is the dominant effect in receiver reflections, it is commonly characterized by measuring the input current due to a changing voltage. Since I = C*(dV/dt), and C = I/(dV/dt), C_comp is easily found using a voltage ramp. dV is the typical signal swing and dt is the typical transition time. Similarly, values for C_comp can be determined for the min and max corner conditions. The current I is not measured instantaneously, but is averaged over a rising and falling ramp. This averages out the voltage dependence of the capacitance just as a signal would see it averaged.

Sometimes an averaged capacitance is not sufficient (for example, if reflections from a driver or the capacitance varies with driver state, and this is significant in a design). In this case, a table-based IBIS model could be replaced with a SPICE, Verilog-AMS, or VHDL-AMS model. Support for these model languages was added in the IBIS 4.1 specification.

13.3.2.2 IBIS Model Creation in Cadence's Model Integrity

IBIS data extraction from the SPICE output file begins with identifying the available data. For example, if only one corner is found for results, that corner is assumed to be the TYP corner. If there is more than one corner, then each must be assigned to be typ, min, or max. This assignment is normally based on the value of the power supply voltage. At the same time, a check is run to make sure the same conditions are used for each I-V and V-T corner pair.

The next step is to extract all of the parameters from the SPICE file. This includes voltage and temperature for each corner, as well as the load for the V-T curves (typically 50 ohms).

Some of the parameters required for an IBIS buffer model are not used in the SPICE simulations. These include Vinh and Vinl, and the timing conditions (Vmeas, Vref, Rref, and Cref). C_comp is also obtained separately. In the Model Integrity tool, these parameters are entered on a GUI (graphical user interface), as shown in Figure 13-7.

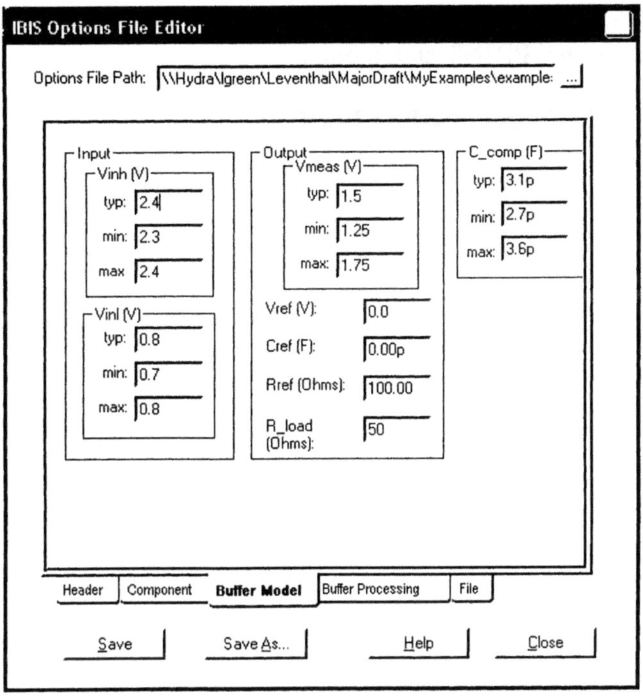

Figure 13-7. Setting buffer model parameters

The next step is to set the extraction options. Model Integrity provides several extraction options, shown in Figure 13-8. Some of these options control the number of points (rows) extracted for each table. IBIS 3.2 allows up to 100 rows, but up to 1000 are allowed beginning with IBIS 4.0. Another important parameter is the tolerance setting. A loose tolerance allows fewer than the specified number of points to be extracted, while a tighter tolerance requires more points to achieve the required accuracy.

Model Integrity can also detect linear or non-linear resistance, controlled by the "R Detection Tolerance" value. Non-linear resistance can, for example, be designed using a FET biased in the ohmic region. The current from a resistor internal to the I/O buffer is assigned to one of the clamp tables. Model Integrity can also use just the leading portion of the V-T table. This is particularly helpful when all of the V-T curves settle much sooner than the end of the SPICE time. All V-T tables start at Time = 0.0, since there is only one toggle event in a SPICE V-T simulation.

Once the three typ-min-max corners have been identified and the options set, the actual table data can be extracted from the SPICE output file (*.lis). In Model Integrity, the beginning of a table is identified by the label TIME, since all data was generated from transient simulations. The remaining column headers are ignored, since Model Integrity assumes data was in the correct order in the .PRINT statements.

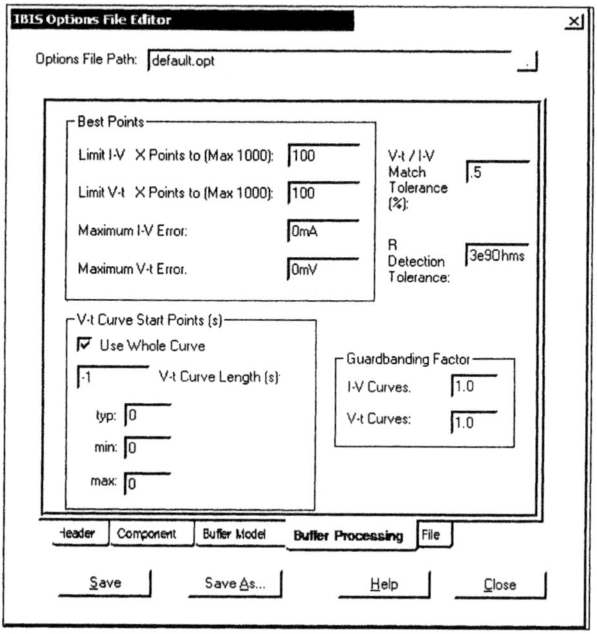

Figure 13-8. Options for buffer model extraction

Each data column is extracted to one IBIS table column, and then the columns are aligned for the three IBIS corners for each table. Since clamp currents are included in every I-V table in the SPICE output file, the clamp currents must be subtracted from the pullup and pulldown SPICE tables to form the subtracted_pullup and subtracted_pulldown tables. The subtracted tables are then used for the IBIS [Pullup] and [Pulldown] tables.

A comparison of the SPICE curves before and after subtraction is shown in Figures 13-9 and 13-10. Numeric resolution makes the difference between two large numbers close to zero, so the subtracted curve returns to zero at the left edge (-Vcc). This computation artifact is seen in many SPICE-to-IBIS tools. The total (summed) current is unchanged, since the remainder of the current will be assigned to a clamp table.

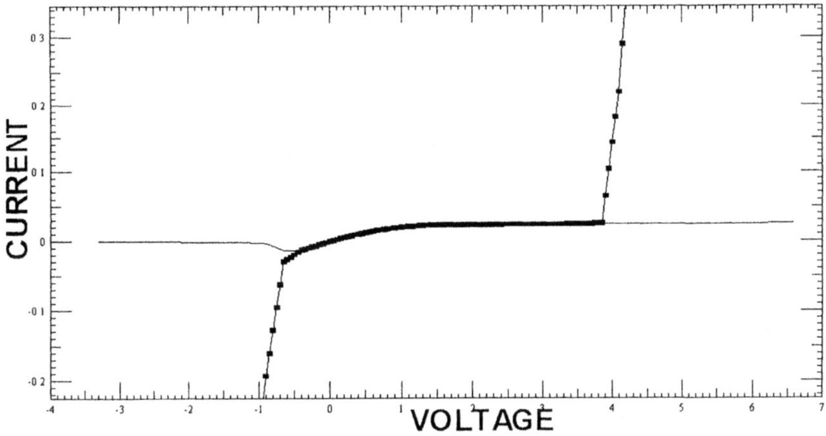

Figure 13-9. Pulldown table before (dots) and after subtracting clamp currents

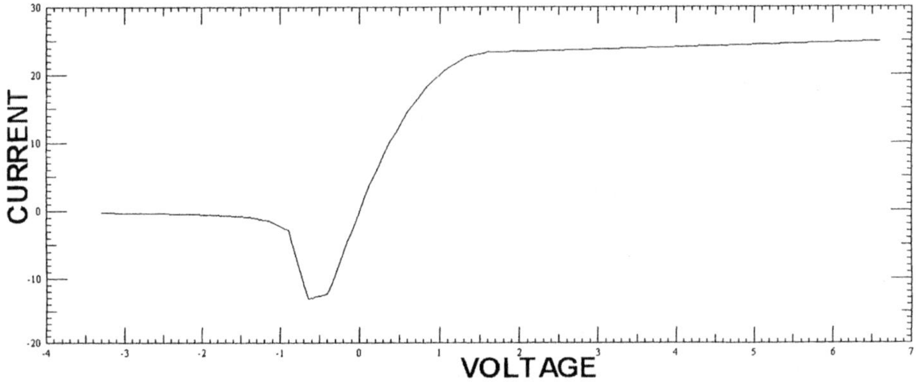

Figure 13-10. Subtracted_pulldown curve
Note the inversion point near V=-0.7V

At this point, each table still has as many points as the original SPICE output, so the number of points needs to be reduced. Model Integrity does this using a "Best Points" algorithm. The algorithm chooses the next point for each table from the available points, selecting the point furthest from the linear fit based on the points chosen so far. Figure 13-11 shows an early stage in the fitting process. Once a value has been selected, then data is included on all three corners so that no "NA" values appear in a table. The best points algorithm is used for both I-V and V-T tables, and has proven to give excellent fitting results.

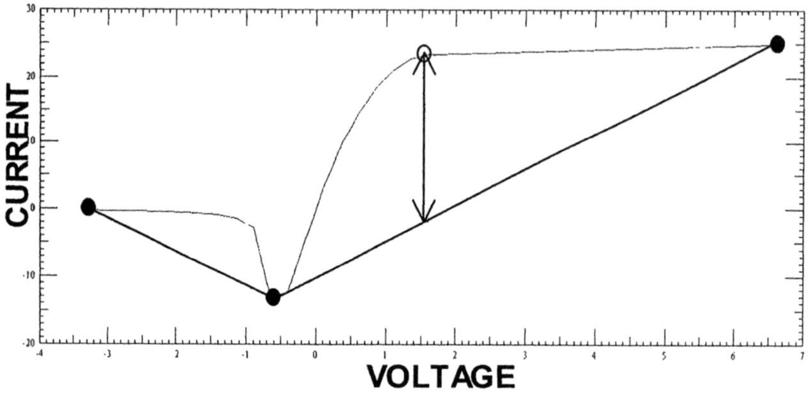

Figure 13-11. Selecting the next point using the "Best Points" algorithm

Once the desired data points have been selected, the table is ready to be added to the model, and the next table can be processed. This continues until all IBIS tables have been completed for the model.

All the information required to create the IBIS model is now ready. The parameters and tables are then used to create a syntactically correct IBIS model. At this point, one can validate the buffer model, process additional models, or go on to create a complete IBIS file.

13.3.3 Step 3: Combining IBIS Models into a Complete IBIS File

An IBIS file describes interfaces at the component level. A complete IBIS file contains one header section, the package model and pin list for one or more components, and one model for each unique I/O model. Figure 13-12 shows an example of a header section, while Figure 13-13 shows an example of a component section.

```
[IBIS Ver]       3.2
[File Name]      example.ibs
[File Rev]       0.0
[Date]           9-8-2004
[Source]         Green Streak Programs.
[Notes] Extracted from io33v.lis using Cadence Model
Integrity. This model is for demo use only.
[Disclaimer] This model is for DEMO use only,
and does not represent any actual part from any manufacturer.
[Copyright]      August 2004.
```

Figure 13-12. An IBIS header section

```
[Component]    CompName
[Manufacturer] Nobody
[Package]
|      typ       min        max
R_pkg 120mOhm  100mOhm    150mOhm
L_pkg 3.0nH    2.5nH      4.0nH
C_pkg 5.0pF    4.0pF      5.5pF
|
[Pin] signal_name model_name R_pin L_pin C_pin
A1  data1     io33v
B2  data2     io33v
C1  input     in33v
12  output    out33v
98  Vcc      POWER
99  Gnd      GND
```

Figure 13-13. An IBIS component section

As a practical example, consider an FPGA family, such as the XiLinx Virtex II. Even a reduced file can be 80 or more pages of data. The FPGA vendor provides an IBIS file containing a single component with a dummy pin list that includes one pin for every I/O buffer available with that FPGA family. Users create a new pin list for each FPGA design, using I/O buffers appropriate to each design. The user should remove I/O buffer models that are not used in a design, since removing unused models significantly decreases the size of the IBIS file.

The IBIS header describes the IBIS file contents. Some of the keywords, such as the [Disclaimer], are similar to wording used on data sheets. Other keywords, such as the [File Rev], provide information to the IBIS file user that is not documented elsewhere. While none of the header information impacts simulation, it does provide useful information to the user. The example in Figure 13-12 shows a model with a [File Rev] of 0.0, indicating that the IBIS models in the file are preliminary and have not yet been verified against hardware (this is normal for demo models).

The IBIS component section describes the interface between each integrated circuit pad and the corresponding package pin connecting to the printed circuit board. The package model default is a lumped RLC circuit. The package parameters are required in the IBIS file, even if a more detailed IBIS package model file is also used. For a real device, every pin on the component is included in the pin list, including all power and ground pins, as well as any unconnected pins. IBIS reserves models names of “POWER”, “GND”, and “NC” (No Connect) for these three types of pins. All other pins have model names assigned to them.

The IBIS parser and simulation tools check to make sure the referenced models are in the same IBIS file as the component. Within an IBIS file, there can be only one model with a given name. Thus, if two components each call model "io33v", then they must use the same io33v I/O buffer. (If there are two buffers with the same name in the file, the parser will issue an error.) If one needs two different buffers with the same name, they must be in different IBIS files. Most tools can then uniquely identify each model.

The header and pin list are normally generated by the model maker, although the user might change the file name and [File Name] to follow their favorite naming convention. Model Integrity includes a sample pin list file and header text. The header text is edited through Model Integrity's IBIS Options editor, as shown in Figure 13-14, while the pin list can be edited in Model Integrity or any text editor.

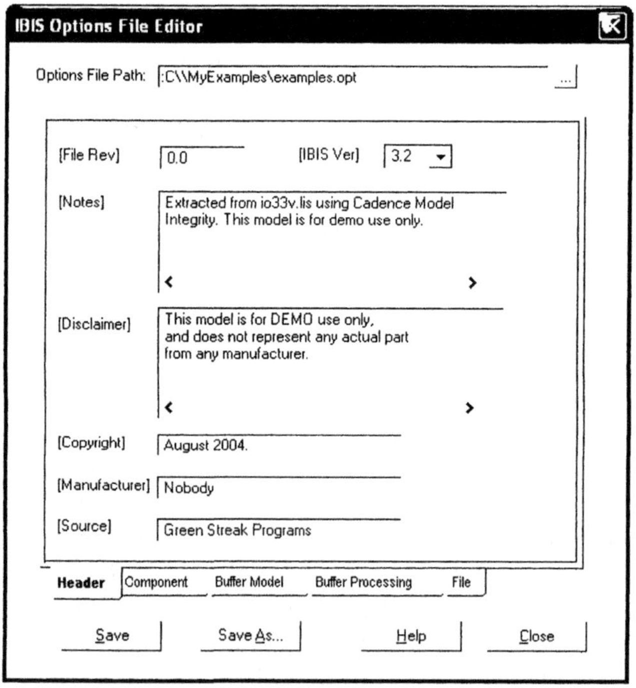

Figure 13-14. IBIS header editing in Model Integrity

Once the component information, including the pin list, and the header information are complete, an IBIS file (IBIS model) can be generated. First, the header information must be formatted according to the IBIS specification given for [IBIS Ver] in the header (this also defines which version the IBIS parser will check against). This is followed by the component's package information and pin list. Next, all models referenced in the pin list are concatenated into the file. Finally, the [End] statement is added to the end of the IBIS file.

To concatenate the various buffer models from their individual files, or from other IBIS files, one must identify the buffer files. In the simplest case, buffer files have been created in the same directory as a pin-list file. In other cases, there might be more than one buffer with the same name, uniquely identified by a directory path rather than by a name. In that case, the model maker must uniquely identify the path as well as the buffer model file. If the buffer model is already contained in another IBIS file, then both the model name and IBIS file must be uniquely identified.

In Model Integrity, one can point to a specific directory path to search for the buffer models, assuming the buffer model name matches the buffer model's file name. For example, in the pin list in Figure 13-13, io33v is called by more than one pin, so Model Integrity would search for a file named io33v.buf in the specified directory. The user can also manually set the file name and path–particularly useful when buffer model files were previously saved in more than one directory.

Figure 13-15 shows the path settings for the pin list of Figure 13-13. Note that models are not needed for POWER, GND, or NC (No Connection), since these are predefined. All other models must be defined to create a complete IBIS file. If a [Model] statement is not found for a model used in the pin list, then the parser will report an error. The model maker can then use cut-and-paste to add that model to the IBIS file.

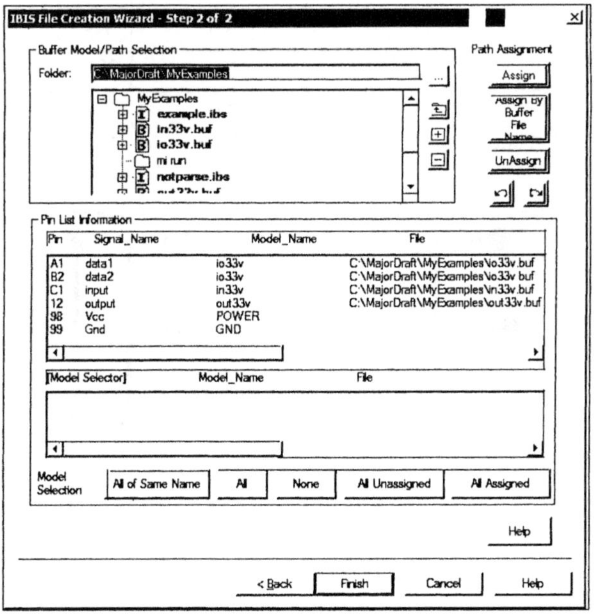

Figure 13-15. Identifying a buffer model file for each pin

13.4 MODELING PASSIVE INTERCONNECTIONS IN IBIS

IBIS has a number of formats for modeling passive interconnections. The EBD format was developed to model modules and plug-in boards, while the PKG format was targeted more at package modeling. Both EBD and PKG were part of the IBIS 3.2 specification, while ICM was developed later.

The ICM specification was developed separately from the IBIS 3.2 and 4.0 specifications. Originally targeted at connector modeling, it was broadened to allow generic interconnects to be modeled, including packages, modules, and connectors. ICM accepts models in a number of formats, including nodal and matrix formats.

Table 13-1 compares some of the key capabilities of these IBIS formats. In particular, the ICM specification includes all of the features of the EBD and PKG formats, although there are some syntax differences. The ICM specification adds support for S-parameters, which allows modeling of any frequency-dependent interconnections. S-parameter models can be generated by field solvers, from simulations, and from bench measurements. Figure 13-16 shows examples of EBD and PKG topologies.

Table 13-1. Comparison of IBIS interconnection model formats.

Type of Model	EBD	PKG	ICM
Lumped (R, L, C)	Yes	Yes	Yes
Coupled traces	No	Yes	Yes
Matrices	No	Yes	Yes
S-parameters	No	No	Yes
Branching (forks) and dangling connects	Yes	No	Yes

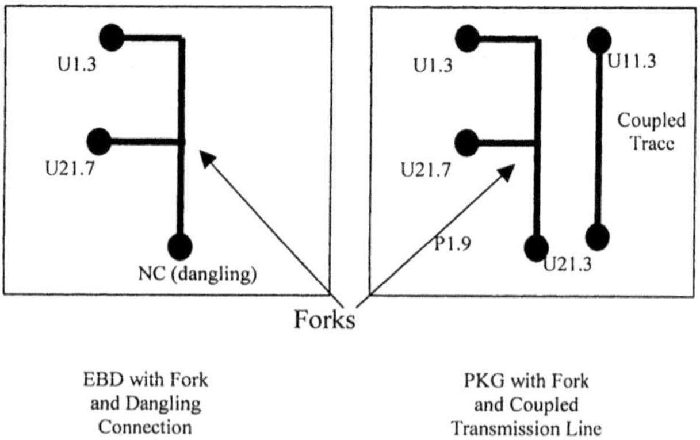

Figure 13-16. EBD and PKG topology example

EBD and PKG models can be incorporated directly into an IBIS file, or they can be separate *.ebd and *.pkg files. They are linked to IBIS component information. For an EBD model, each pin on the interconnection is associated with a component pin (the interconnection usually interconnects several components). PKG models, on the other hand, belong to a single component, replacing that component's default package model.

ICM files are presently independent of any other IBIS file. Connections are made at the EDA tool level. The IBIS committee is working on ways to connect ICM models in a similar way to EBD and PKG models.

Since EBD and PKG interconnection models can be included in an IBIS file, they are parsed with the standard IBIS parser (ibischk4). On the other hand, ICM models are parsed with a separate parser (icmchk). At this time, there is no IBIS Quality checklist for interconnection models, although passivity and causality checks should be made before an interconnection model is released.

13.5 IBIS MODEL VALIDATION

An IBIS model must be validated before it is released for design use. This accomplishes many things. For the model user, it prevents wasting time simulating with bad models, and gives a higher level of confidence in simulation results. For the model maker, it acts as a preventative measure, reducing the effort required to support model users and to calm customers upset over inadequate models.

The IBIS Quality Checklist, covered in detail in "Chapter 12, Fixing Errors and Omissions in IBIS Models" and "Appendix I, IBIS Quality Checklist" provides an excellent validation checklist. In the next topic, we discuss validating buffer models, with examples of both good and bad models.

Many of the same validation steps are also used for other models, such as interconnection models and SPICE models. For example, a SPICE model's syntax is checked by running an operating point simulation (.OP) in SPICE. Simulation in SPICE is used to check things such as continuity of current vs. width (an issue in early BSIM models).

13.5.1 Validating an IBIS Buffer Model

There are six major steps in validating an IBIS buffer model. These checks are performed before a model is released by a model maker, and are often also performed by model users and model librarians. These six steps can uncover potential problems early, and avoid wasted design effort caused by bad model data.

1. Parse to check syntax (ibischk4, icmchk).
2. Examine parameters.
3. View tables graphically.
4. Perform other data checks (including IBIS Quality Checklist).
5. Simulate each buffer and check results.
6. Formally release for design use.

After a buffer model has been validated for use, a model user might add two more steps.

- The first step is to verify the models against hardware, including the suitability of the IBIS models for the design environment.
- The second, and final, step is to close the loop, updating the modeling process, from creation and validation through verification.

The first six steps are discussed in more detail in the following sections. "Chapter 16, Methodology for Verifying Models," discusses verification of a model against hardware in more detail. Closing the loop is a methodology that varies greatly from one company to the next, and is not covered in detail in this book.

Note that some valid IBIS models will fail one or more of the standard checks, due to advanced buffer features. For example, LVDS buffers are designed to drive a 100-Ohm differential load, while the IBIS V-T syntax allows a single resistor to a fixed voltage. Therefore, these buffers might have "DC Mismatch" errors reported by the parser. Where there are known acceptable parser errors or warnings, this information should be added to the [Notes] section. This assures the model user that the problems have been checked out, and in turn reduces the support effort for the model maker.

There are many tools for IBIS model validation, more than there are for model creation. Common features include a GUI (graphical user interface), linking of parser messages to lines in the file, and graphical viewing of IBIS I-V and V-T tables. Many tools allow viewing of summed I-V tables. Some show only [Pullup] + [Power Clamp] or [Pulldown] + [GND Clamp], while others show the correct summing: I_up = [Pullup] + [Power Clamp] + [GND Clamp] and I_down = [Pulldown] + [Power Clamp] + [GND Clamp].

Step 1: Parsing an IBIS Model

The first step in validating an IBIS file is to run the latest IBIS parser, which is ibischk4 version 4.0.2. Ibischk3 ended at version 3.2.9 and no new bugs are being fixed. The latest version is always available on the IBIS web site.

The latest version of the IBIS parser should be used even when the IBIS file (models) uses only older IBIS features. For example, if an IBIS file has an [IBIS Ver] of 2.1 or 3.2, it should still be checked with ibisck4. The latest parser checks the file against the IBIS specification corresponding to the [IBIS Ver]. The latest parser is always compatible with all IBIS versions. When bugs are found in a parser, only the latest parser version is fixed.

One should evaluate all errors and warnings reported by the parser. Frequently, a somewhat arbitrary decision has been made as to when something is called an error vs. a warning. For example, DC mismatches between 2% and 10% are warnings, while mismatches over 10% are errors. Figure 13-17 shows a typical error message for this mismatch. Why 2% and 10% one might ask–only to discover this is an arbitrary cutoff.

```
ERROR - Model XX1: Thc [Falling Waveform]
    with [R_fixture]=50 Ohms and [V_fixture]=1.8V
    has TYP column DC endpoints of 0.78V and 1.80v, but
    an equivalent load applied to the model's I-V tables yields
    different voltages ( 0.82V and 1.54V),
    a difference of 5.94% and 36.89%, respectively.
```

Figure 13-17. Parser DC mismatch message.

Figure 13-18 shows how Model Integrity marks lines that correspond to warnings and errors in a parser report. The complete parser report is presented in the lower window, so the user can troubleshoot parser messages that do not have a corresponding line number (such as DC mismatches).

Step 2: Examining Parameters

Viewing the parameters will check for two things: typos in entering information, and missing values. For example, if Vinh and Vinl are omitted from a model, then TTL values are used by default.

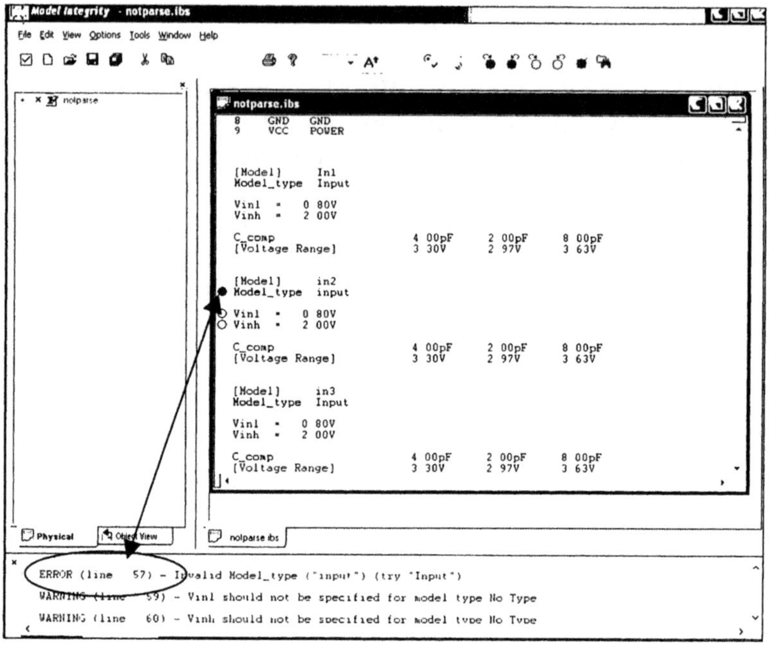

Figure 13-18. Matching a parser message to a line in the IBIS file

[Temperature] is another important parameter to check. This is the operating temperature of the IC (not ambient), and it should be significantly higher than Min and Max ambient temperatures. In particular, the TYP temperature should always be higher than 25C, with 50C being a common value. If the temperature does not cover the die's operating temperature range, then the corresponding Min and Max IBIS corners will not represent I/O behavior over the buffer's specified operating range. If [Temperature] is not set correctly, then it may be necessary to edit the main SPICE netlist, rerun SPICE, and extract the model again.

Step 3: Viewing Tables Graphically

It is often easier to examine a data table graphically than to view four columns of numbers. For example, the parser will report non-monotonic tables (slope changing sign), but it does not report whether the issue is numeric noise or an actual problem. For I/O buffers with internal feedback, I/V tables sometimes have large dips in the middle of the active operating region, and the model is also likely to have DC mismatch problems. Sometimes it is necessary to break feedback when generating I-V tables, or other advanced IBIS modeling techniques (such as submodels) may be required.

Figure 13-19 shows an I/O model with problems in the [Pulldown] data table. When viewing the data table, it is not immediately clear which corner or voltage is associated with the parser error message. The parser report will report at least one slope inversion, but it could report an inversion caused by numeric limitations during table extraction rather than the problem point. A graphic view of the table data makes the problem area stand out.

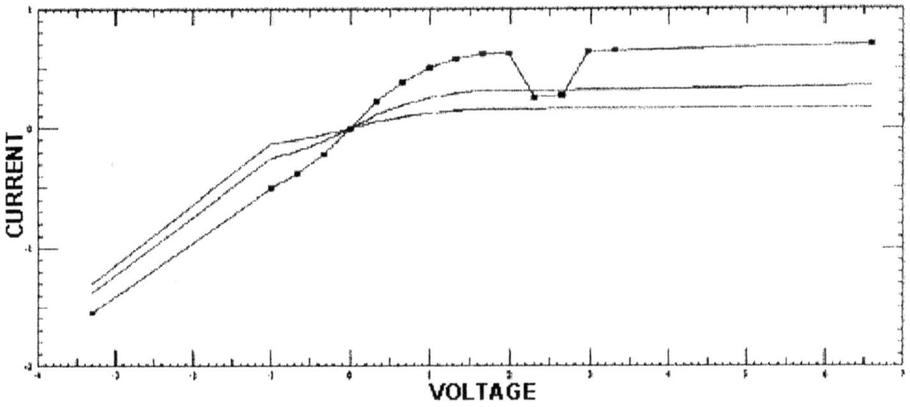

Figure 13-19. Non-monotonic problem in [Pulldown] data

Step 4: Other Data Checks

There are many checks that are possible for IBIS data quality. Many of these are included in the IBIS Quality Checklist, which is covered in "Chapter 12, Fixing Errors and Omissions in IBIS Models."

Since an I/O buffer is designed for a specific interface (such as PCI or LVDS), the model should be checked against the appropriate interface specification. Similarly, the V-T curves and [Ramp] values should be checked against each other, as well as against design specifications.

While [Ramp] values are not used during simulation, they are used in various pre-simulation algorithms. Thus it is important to have values that are at least close to the values obtained for the 20%-80% transition times based on the V-T tables.

The IBIS parser checks against the IBIS specification corresponding to [IBIS Ver], but it is not capable of making all of the checks in the IBIS Quality Checklist. Many tools add their own checks, beyond what the parser reports. In Model Integrity, for example, additional data checks are made during translation to DML (Device Model Language, Cadence's proprietary format).

Step 5: Simulate Each Buffer

Each I/O buffer model should be simulated. When appropriate, both Input and Output operation should be verified by simulation. Often, one can design a generic test circuit, and pass parameters into it. Figure 13-20 shows how Model Integrity sets up a typical validation simulation. Note that Model Integrity passes in the values of this test circuit's components from the IBIS model.

While it is not necessary to simulate in every available simulator, it is important to simulate in at least one. Simulation will verify that the model is parsed and used correctly by at least one simulator. Different simulators use models slightly differently, and some Signal Integrity tools support model features beyond what is supported in IBIS. If a model user reports a problem with a model, it is sometimes necessary to debug simulator interactions as well as the model data itself.

Simulating and checking results validates a number of model features. First, of course, it verifies that the model will function as expected in at least one simulation tool. Simulation using the test parameters (Cref, Rref, and Vref) verifies that the I/O buffer can switch past the Vmeas voltage during bench testing. Simulation using the V-T table load checks the waveform and edge rate against the IBIS tables.

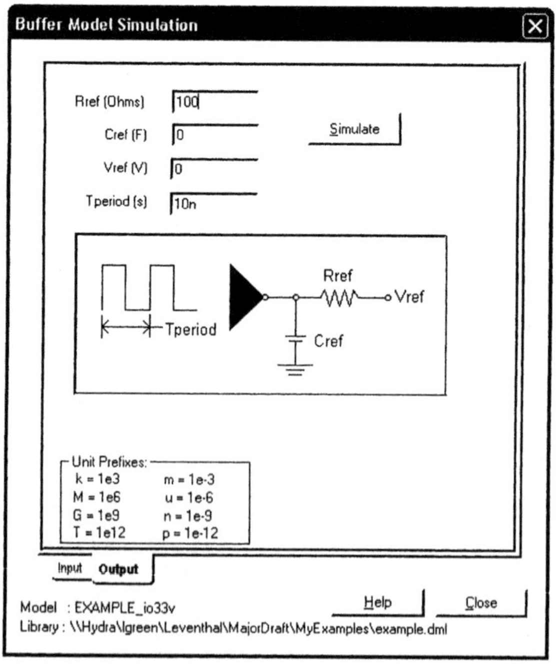

Figure 13-20. Simulation setup in Model Integrity®

Simulation comparisons are normally done on unpackaged devices, to make it easier to separate package effects from I/O model effects. For example, a complex package can introduce impedance mismatches and signal ringing, making it more difficult to extract the buffer's edge rate. If more than one buffer is simulated, crosstalk in the package can further complicate the waveform.

Sometimes, it is not sufficient to simulate a buffer by itself. For example, a differential pair requires simulating two buffers. In a general purpose pre-layout tool, such as Cadence's Signal Explorer, the user can create a complex interconnection that includes more than one buffer (such as a differential I/O pair driver and receiver), as well as complex terminations and other loads.

A separate simulation is used to validate the package model. For example, if the default package model is used with R_pkg set to 0 Ohms, a simulation will show whether this value is accepted. Some simulators require non-zero values for the package parameters. Excessively large package model values can lead to a severe slowing of the edge rate at the PCB connection, indicating that the selected package is unsuitable for that component. For packages using EBD, PKG, or ICM models, simulation can also be used to validate the package model.

Step 6: Formally Releasing an IBIS File

A model should not be released simply because it was automatically generated, because there may have been undetected problems in data creation and extraction. Releasing an IBIS file should assure model users that all of the models in the file have been validated. A formal release could be as simple as posting the IBIS file in a shared directory, or as complicated as adding it to a shared model library (a complex database).

A paper trail provides an opportunity to track and improve your IBIS validation and methodology. This trail could be maintained as a file, rather than on actual paper, and could even be added to the IBIS file itself under an IBIS documentation keyword, such as [Notes].

Posting the file to the web, or making it available on request, sometimes under a Non-Disclosure Agreement (NDA), normally follows release of a model by a semiconductor vendor. Usually an internal release includes an email announcement to all interested design groups. Either way, a release is a statement that the model has passed its quality checks.

13.5.2 Verifying the Validation Methodology

Validating the methodology for IBIS model creation and validation is important. When things are going well, both designers and management can trust the process. When things go wrong, it is easier to debug the problem, whether it is with the original SPICE model, the model creation process, or the data extraction process.

"Chapter 16, Methodology for Verifying Models" goes into model verification against hardware. When both validation and verification are part of the overall design methodology, users and managers have solid foundations for their trust in the overall IBIS modeling and simulation flow.

13.6 SUMMARY

Model creation and validation can, to some extent, be automated using EDA tools. However, visual examination of data, including graphical viewing of data tables, is an important part of validation. Since a model might perform differently in different simulators, simulation is used to validate a model for use with a particular simulator.

Users must pay careful attention when generating a model. Each model creation tool has tool-specific requirements for its input and output files. For example, s2ibis3 uses a separate input and output file pair for each IBIS table column, while Model Integrity uses a single input file and a single output file. Output files generated by s2ibis3 cannot be used by Model Integrity, and output files generated by Model Integrity cannot be used by s2ibis3.

During model creation, some model parameters are entered manually. It is important to understand the parameters, so appropriate values are used. For example, the TEMP (temperature) used in IBIS model generation is the die temperature, and thus higher than ambient. During visual data checking for validation, this can be verified as being around 50 degrees, rather than 25 degrees.

Model validation has a well-defined methodology. The model file is parsed to check syntax (either by the target simulator or by a stand-alone parser), the model parameters are visually examined, table data is viewed graphically, and other checks for data consistency and correctness are made. The last step is validation is to run the model in the target simulator to make sure the model produces the expected simulation results. After a model is verified, it can be formally released to design groups. Following this methodology allows each design group to work with a "known-good" model.

Model validation can be applied to any type of model. For example, SPICE models and their parameter sets can be verified. Since SPICE models are generally not portable, there will be a separate parameter set for each supported SPICE simulator.

Sometimes, a model will not be adequate. For example, early BSIM models had a problem with discontinuities in MOSFET current vs. channel width plots. In this case, changing model parameters cannot fix the problem. Instead, the underlying model equations must be changed by the simulation tool developer. An example of this is the EKV MOSFET model, which improved the modeling of moderate inversion operation over the BSIM models.

The validation methodology presented in this chapter can be applied to both structural (transistor-level) models and behavioral models. Behavioral models can be expressed in many ways, including SPICE controlled sources, IBIS tables, VHDL-AMS, and Verilog-AMS. In all cases, the validation consists of making sure the model data is correct, and that it produces the expected simulation results.

PART 5: MANAGING MODELS

Chapter 14

SOURCES OF IBIS MODELS

IBIS models are available from a variety of sources, but it takes perseverance to find and prepare models that are fit-for-use

Abstract: Half the total modeling and simulation expenses are the labor costs involved in obtaining, fixing, creating, and archiving models. Engineers and others need to understand the sources and strategies for obtaining IBIS models in sufficient quality and quantity to meet their needs. They also need to know how to create and validate IBIS models when none are readily available.

14.1 MODEL NEEDS CHANGE AS A PRODUCT IS DEVELOPED

In "Chapter 13, Using EDA Tools to Create and Validate IBIS Models from SPICE," the authors explained how to convert a SPICE model to an IBIS model. In general, that process is neither easy nor quick. This chapter explains a process for getting IBIS models from other sources. In general, this process is easier and sometimes quicker.

The need to get models is ongoing throughout the product development cycle. At the start of the development cycle, we are likely to have only a fraction of the models that we eventually will need. As the design project develops, models need to be more accurate. This is because simulations need to be more accurate as design margins get tighter. Also, before releasing a PCB to prototyping and production, we should simulate *all*[1] switching nets.

---

[1] Modeling and simulation are a defense against unintended errors occurring in the product design.

14.2 LIST OF IBIS MODEL SOURCES

Table 14-1 identifies most sources for obtaining models, starting with the most convenient and quickest ways. The sources at the bottom of this table take more time and effort. From the top of the list to its end, the acquisition cycle time changes from a few minutes to perhaps months. It is very common to find errors in models at every stage of working with them. Asking the model supplier to fix those errors can add additional months to the model acquisition cycle. These sources are discussed throughout the chapter.

Table 14-1. Sources of IBIS models with benefits and issues

Source	Benefits	Issues
EDA tool library (default or generic models)	Models are almost always validated. This library includes default models and part-specific models.	Models are not often verified. For example, a 1.2V deep submicron generic model in the former case, or a verified SN74HCT245A in the latter case.
Company library	Models are already validated or verified.	
EDA tool library (models of specific components)	Models are almost always validated.	Models are not often verified.
Downloaded from the semiconductor supplier's website	Quick, convenient, and most likely to represent true process spreads.	Models are rarely validated or verified. The models usually require fixing before they will run in an EDA simulator.
Requested directly from the supplier's representative and emailed	Most likely to represent true process spreads and not as labor intensive as making the model ourselves.	Models are rarely validated or verified. They usually require fixing before they will run in an EDA simulator.
Library purchased from a third-party modeling service	Quick and convenient if the third-party modeling service knows what they're doing. Not labor intensive.	Uncertainty about representing true process spreads. These models may, or may not, be validated or verified. Models should be treated as not validated and not verified until we build trust in the supplier. Can be costly.

Table14-1. (cont.)

Source	Benefits	Issues
Adapted and inferred from an existing model of similar parts	Allows the design engineer to make the IBIS model when part-specific model is unavailable.	Creating an adapted model takes a great deal of work, and has the potential for many errors. Older technology parts (for example, a SN74LVT16646DGG) often do not have an IBIS model or one that works. If they are old technology components, it is unlikely that the supplier will now model or fix the errors in the model. Validation and verification should be part of our process.
Purchased from a third- party modeling service	A custom model that is made-to-order when needed. Not labor intensive.	Uncertain about representing true process spreads. These models may or may not be validated or verified. Should be treated as not validated and not verified until we build trust in the supplier. Can be costly.
SPICE-to-IBIS conversion (if the semiconductor supplier sends us the SPICE model)	Allows OEM to make the IBIS model when nothing is available.	Reluctance of supplier to share proprictary SPICE model. Validation and verification should be part of the process. Creating models from SPICE takes a great deal of work, and has the potential for many errors.
Developed by the semiconductor supplier (in parallel with the development of the IC itself)	The supplier who knows the process spread is doing the modeling. The user should get them to share preliminary models for early simulations, if the part is being built on the same technology process. It is best to ask the supplier for I/O characterization model cells from similar, already manufactured parts. If the I/O cell characterization has been done, it is not that hard to complete the rest of the IBIS file data.	It is quite common for component and model development to not be completed early in a project. Particularly when the supplier is developing a custom IC. Often, the board must be laid out, even prototyped, before the part or its model even exists. Validation and verification should be part of this process.
A request to the SI-LIST email reflector for help in pointing to a possible source for the model	When we are really stuck, fellow engineers can sometimes point us in the right direction.	Too many cries for help should be avoided as they abuse the system of mutual aid. The list moderator may shut off a participant's access if there are too many requests.

14.3 USING DEFAULT MODELS TO GET STARTED

When we begin a project, we often need to quickly run some simulations. Sometimes we may not be able to find the models for doing the simulations, but we still need to get quick answers. To proceed, we need to set up the simulation program to automatically default to a generic model. The default model allows the simulation of a virtual circuit to get a few preliminary results, including an answer as to whether a net will or will not switch. We can quickly discover whether some essential information is missing from the netlist, BOM, or layout symbol. There may be many initial database problems to correct in parallel with efforts to obtain simulation models. These issues will include undefined pin uses, undefined parts, connectivity problems, and many startup database problems.

Once we can run a simulation on a net, we need to start replacing the generic default models with models that are more accurate.

14.4 USING THE COMPANY'S MODEL LIBRARY

A company's model library is a valuable investment. If models from previous projects are archived and retrievable, they can be leveraged for current projects. The best time to improve and add to the model library is when the company is between design cycles. For details of model acquisition, storage, and retrieval see "Chapter 15: Working with the Model Library."

14.5 USING THE EDA TOOL PROVIDER'S MODEL LIBRARY

EDA tool providers often include a library of IBIS files with their software. The libraries from some providers are fairly extensive, and may include parts like 1.2V CMOS generic default model, or more specific component part numbers like SN74LVDS04DT. We should find out whether the provider has *expanded* model libraries available. Two providers with extensive libraries are Cadence Design Systems and Mentor Graphics. Other EDA-tool-provider libraries may be less extensive, leaving users to fend for themselves.

IBIS model libraries supplied by EDA software providers should have had due diligence exercised on them, should be nearly error-free, and should run in the tool provider's simulator. The EDA-tool-provider libraries will not take care of all model coverage problems, but they should help substantially.

A key question to ask the EDA provider is how the models are generated and validated. An integrated EDA software tool from a first-class provider should be able to:

- Integrate well with a schematic-capture (creation) tool at a very early design stage.
- Provide a large library of models of readily available generic and common parts.
- Allow the creation of nets and topologies on the fly for what-if simulations, given only a scratch pad sketch of the logic net.
- Add and define components.
- Create simulation models given only a data sheet or similar input.
- Integrate well with a board layout tool.

Many Signal Integrity programs allow use of the schematic-assigned pin use (IN, OUT, BI, POWER, GND, and UNSPEC) from the netlist to point to an appropriate default generic model available in the model libraries. Search paths to previously validated and verified models in the component library should also enable their use. Easy re-setting of defaults to generic model types, from among a selection of them, should also be a feature of the EDA tool.

The strategy of using default generic models has some drawbacks. We must think about the assumptions that the program has been fed and the results of the simulations. Otherwise initial simulations may be well off the mark on ringing and switching times. A quick look at whether a net is hooked up properly, and can switch at all, may be all that is needed very early in development. But that quickly changes and we should not automatically assume it indicates whether or not there are Signal Integrity problems.

14.6 SEARCHING THE WEB FOR THE SUPPLIER'S MODEL

14.6.1 Ways to Search the Web

Many semiconductor manufacturers post their IBIS models directly on their websites. From there they can be downloaded quickly and conveniently. Other semiconductor manufacturers post information about how to obtain models on their websites. These directions range from a fairly easy pinging of a listed model, which is then quickly e-mailed, to a much more difficult process. In these cases, the model requestor is no longer anonymous and the supplier can follow up with a sales inquiry. In some cases the directions are a bit onerous. This includes registering with the supplier and signing an NDA. There are three strategies for using the web:

- Go to the IBIS Committee Model website
- Use a search engine to look for a specific IBIS model
- Use a search engine to look for a specific semiconductor manufacturer's website, or go directly there if the address is known.

Each strategy has its advantages. Multiple methods should be used when there is a problem locating a model. Models available directly from the makers of the device offer the advantage of being the most likely to have the latest, most accurate models and parameter spreads. However, the track record of most semiconductor supplier provided models is that they are full of syntax errors, have not been validated, and will not initially run in an EDA tool simulator.

14.6.2 Using the IBIS Committee Website

Information on suppliers who do provide simulation models, third-party test, measurement, and modeling services can be found on the web. The most stable web address is at the IBIS Committee website as shown in Figure 14-1: http://www.eigroup.org/ibis/ibis.htm

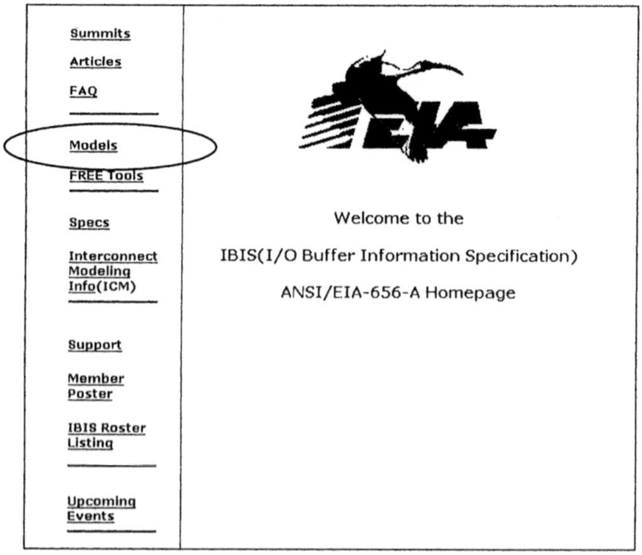

Figure 14-1. IBIS Committee home page

To display a list of links and contact information of over 65 semiconductor suppliers, click Models (shown in the left navigation bar).

The Model page, shown in Figure 14-2, lists the web links for about 65 semiconductor manufacturers. Some suppliers provide IBIS files on their web sites; others must be contacted directly.

Caution: Since the web is a dynamic environment, some links may be out-of-date. However, they are updated periodically.

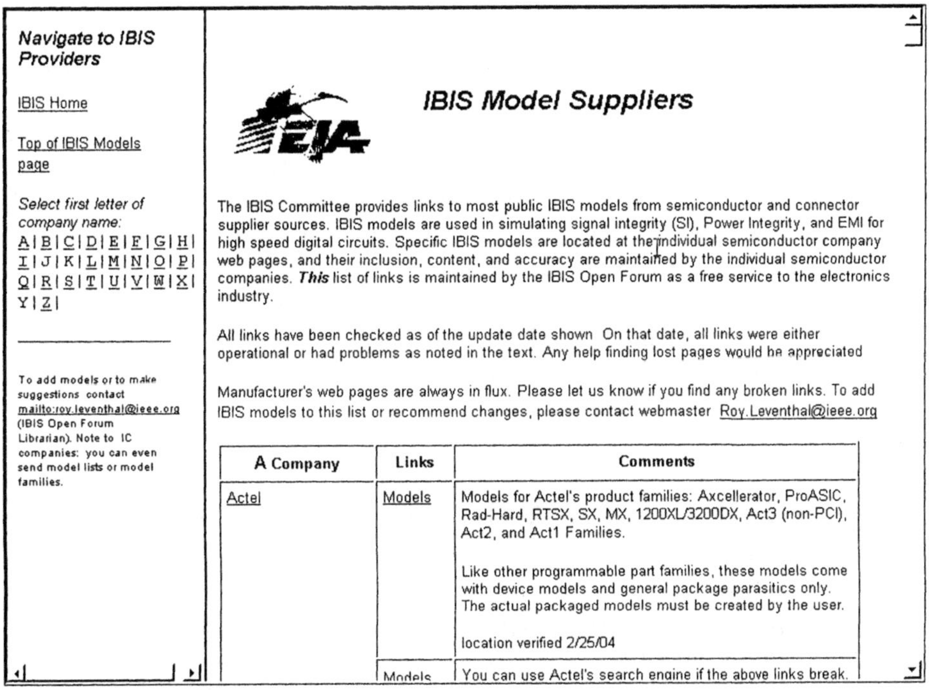

Navigate to IBIS Providers

IBIS Home

Top of IBIS Models page

Select first letter of company name:
A | B | C | D | E | F | G | H | I | J | K | L | M | N | O | P | Q | R | S | T | U | V | W | X | Y | Z |

To add models or to make suggestions contact mailto:roy.leventhal@ieee.org (IBIS Open Forum Librarian). Note to IC companies: you can even send model lists or model families.

IBIS Model Suppliers

The IBIS Committee provides links to most public IBIS models from semiconductor and connector supplier sources. IBIS models are used in simulating signal integrity (SI), Power Integrity, and EMI for high speed digital circuits. Specific IBIS models are located at the individual semiconductor company web pages, and their inclusion, content, and accuracy are maintained by the individual semiconductor companies. ***This*** list of links is maintained by the IBIS Open Forum as a free service to the electronics industry.

All links have been checked as of the update date shown On that date, all links were either operational or had problems as noted in the text. Any help finding lost pages would be appreciated

Manufacturer's web pages are always in flux. Please let us know if you find any broken links. To add IBIS models to this list or recommend changes, please contact webmaster Roy.Leventhal@ieee.org

A Company	Links	Comments
Actel	Models	Models for Actel's product families: Axcellerator, ProASIC, Rad-Hard, RTSX, SX, MX, 1200XL/3200DX, Act3 (non-PCI), Act2, and Act1 Families. Like other programmable part families, these models come with device models and general package parasitics only. The actual packaged models must be created by the user. location verified 2/25/04
	Models	You can use Actel's search engine if the above links break.

Figure 14-2. IBIS Committee model suppliers page

14.6.3 Using Other Websites with Lists of Supplier Links

Other lists of links can be found at the TeraSpeed, Zuken, Sintec, and Mentor websites, and at the EDA Group:

http://www.eda.org/fmf/wwwpages/IBIS_models.html

14.6.4 Using Search Engines on the Web

Website addresses continually change. Companies may go in and out of business, and frequently rework their websites. Currently, the IBIS

Committee website is stable and is updated periodically. But it too will probably change in a couple of years. Google and Yahoo are currently the dominant search engines. But things seldom stand still in technology, and there are several competitors including Clusty® at http://vivisimo.com/.

As of this writing, searching the web with Google for the phrase "IBIS Model" produced 81,900 hits, including links to semiconductor manufacturers and to third-party modeling services. We can do a search with a web search engine for a particular IBIS model, for example: 74HCT00 ibis.

14.6.5 Searching the Web for the Semiconductor Supplier

Even though many semiconductor companies do not maintain a web page for models, some have many IBIS models, while others have only a few. A few semiconductor suppliers barely know what an IBIS model is. For all these cases, we need to contact the supplier or their sales representative directly, and request the models we need.

Many links to semiconductor suppliers maintained on the IBIS Committee's page of model links are, in fact, to suppliers that have to be contacted directly. The advantage of finding a supplier's web page is that it often contains contact information.

14.7 REQUESTING MODELS DIRECTLY FROM THE SUPPLIER

The supplier, or their sales representative, can be contacted directly by telephone with a model request — especially if the Purchasing Department asks the supplier to supply models as part of potential or actual business.

A growing number (perhaps now a majority) of semiconductor suppliers require direct contact from potential users who want an IBIS model of one of their parts. Previously, the majority of suppliers would simply post their model files on the web. But now they'll use this contact information as a potential business lead.

Some suppliers require users to sign a Non-Disclosure Agreement (NDA) before getting the file. It is not immediately obvious why an NDA is needed, since IBIS cannot be easily reverse-engineered and does not contain any obvious intellectual property information.

Before official production release, Intel, for example, requires NDA control. The models are valuable for new designs, but the models may change, so Intel wants to control who has these early models, and when they got them, so they can update them when necessary.

Sometimes suppliers want to hide, or try to hide, basic technology details from competitors. But such attempts only slow down the leakage of such information–they do not stop it. Unfortunately, the NDAs and direct contact requirements slow down the communication with potential customers too.

For customers, the NDAs and direct contact requirements sometimes hide the supplier's lack of staff and expertise to properly support their released models, and to correct any mistakes that have been released with them. As a rule of thumb, suppliers with a good track record on released-model quality *do* have the resources to support them, and suppliers with poor released-model quality *do not* have the resources to fix problems. Normally, when available, the requested model file will be sent via email.

When a model cannot be found, it is a good idea to contact the supplier through their sales representative or support group. Posting a model file to the web often lags behind generating the file. The sales rep will advocate and follow up for the user. They want the business. Lastly, the supplier's technical support group can help us identify the best similar part and process to use, by inference, in the interim.

Semiconductor manufacturers are in the best position to understand their process parameters and population distributions. Additionally, they often have the measurement technology and experience necessary to take model measurements. Historically, On, National, and Texas Instruments have provided SPICE models.[2] Some models can be downloaded from their web sites. Many suppliers are now providing SPICE models for complex IO.

Semiconductor process engineers are not always disciplined about informing their applications engineers about changes to their process and re-characterizing devices if warranted. In economic downturns, companies downsize their model support staffs. When things improve, they are too busy to do much modeling.

Conclusion: Just as with part quality, good, up-to-date model availability is not driven by the state of the economy, previous experience, complaints from the customer, or anything else. Customers setting clear expectations, and making purchasing decisions based on those expectations, drive good, up-to-date model availability. For example, the automotive industry insists on VHDL-AMS models and gets them.

---

[2] SPICE models, which have been available since the 1950s, are not particularly better or more available than any other type of model.

14.8 PURCHASING A COMMERCIAL THIRD-PARTY MODEL LIBRARY

There are third-party companies that make IBIS models for users from parts (measurement) or simulation of SPICE models provided by semiconductor manufacturers. Some third parties are good, and some are not. With the downsized economy, resources and staff in modeling groups in the EDA and semiconductor industries (and OEMs) have been cut. Before assuming that the models will give good simulation results, establish the third-party company's model quality capability.

Some third-party sources are:

- Cadence Design Systems: http://www.cadence.com/
- Mentor Graphics, (Hyperlynx/ICX):
  http://www.mentor.com/highspeed/resource/ibis_modeling.html
- SiSoft: http://www.sisoft.com/services.asp#6
- TeraSpeed: http://www.teraspeed.com/modeling.html
- Zuken: http://www.zuken.com/

For ordinary, commodity-type parts, doing business with a third party provider may not be cost effective. A few dozen models may cost many thousands of dollars from third parties. A third-party model costs more than getting a model from the semiconductor suppliers, or having a model bundled in the model library of simulator providers.

On the other hand, third-party libraries can be a bargain when the hidden costs of creating a the model are considered–Particularly, when including the cost of the diversion of critical resources from the job of designing and simulating a product.

For high-pin-count full-custom components with several types of technologies, fast switching speeds, and other challenges, it is quite common for model production to lag part production by several months. More importantly, critical decisions and board place-and-route will lead both by several months. Getting an accurate IBIS model made quickly by a third-party may be expensive, but cost-effective.

It is hard to convince management to spend any money on models, third-party or otherwise. Industry believes that models should be available free. However, models are not made for free. Model providers need to be paid for the staff and other resources that go into making the models.

Nevertheless, modeling and simulation are very cost effective. When getting a product to market, just one mistake not caught by modeling and simulation can cost months, and many millions of dollars. Engineers must sell management on the idea of paying for models, if necessary. Again, workplace support is necessary for modeling and simulation.

14.9 USING MODELS ADAPTED FROM OTHER MODELS

14.9.1 Deciding Whether to Adapt an Existing IBIS Model

Where will a model come from if none is available, as is often the case? Should we request that the semiconductor supplier create the model when they do not have one readily available? Or should we try to generate the model ourselves?

Semiconductor manufacturers may lack the expertise and EDA tools required to generate IBIS models. Perhaps they have not seen sufficient reasons to produce them. They may take a considerable amount of time to respond even if we convince them of the need for providing IBIS models. If they are first starting to consider producing IBIS models, it may be years before they can respond.

Once models are made available, suppliers need to spend money to provide modeling support for customers. And they may be drawn into interactions with customers that may discourage them. Customers with real modeling expertise may argue about some modeling issues. Therefore, some suppliers will simply be unable to provide quality IBIS models upon request for a long time.

In the end, we still need an IBIS model when faced with a design that needs to be simulated. The ability to create one is an essential skill. The job of creating one may be time-consuming and involve much effort. There are possible ways to save some of that time and effort. Those possibilities include adaptation, inference, and cross-referencing to an existing IBIS model. Adaptation, inference, and cross-referencing of an existing IBIS model file all require some manipulation of the existing file.

14.9.2 WARNINGS!

When changing an existing IBIS model file, we should always have a backup copy of the original file. We also need to watch out for legal prohibitions against file modifications as written in files, NDAs, and other agreements. Some prohibitions claim that the model provider owns the

OEM's design if the OEM modifies the model.[3] Additional items users need to watch out for in making a model are:

- The components, or model information, may not be a truly representative sample of the component population.

  For this reason, we do not recommend that customers characterize and model parts themselves. But sometimes there is no other choice than for a user to do their own measurements or make their own models.

- It takes real skill to configure an in-house modeling system, not just a knowledge of IBIS.

  Measuring models in the laboratory requires test and measurement expertise. Instrument makers, such as Tektronix and Agilent, can provide some assistance. We should ask them to recommend appropriate computer-aided instruments for taking modeling data. National Instruments and others can recommend software for data acquisition, virtual instrumentation, and curve fitting algorithms.

Additional skills needed for making IBIS models include converting SPICE models to IBIS (s2ibis) models. Suppliers sometimes provide a SPICE model for their components. There are times when they will provide a SPICE model more readily than an IBIS model despite the possibility that the SPICE model can be reverse-engineered.[4]

Both measured and simulated models require managing the activity to assure a statistically significant sample of devices has been characterized. Support personnel for quality assurance, quality control, and manufacturing can help. They can explain the statistical significance of a small sample of parts early in the design process, and the level of confidence attached to the sample. Personnel in the Quality department can also help by explaining some of the tools used to design for insensitivity to variation. For example, how to use Design of Experiments (DOE) and RobustDesign® methods. The infrastructure devoted to measurements, quality control, screening parts, statistical significance, and supplier qualification can be easily adapted to provide data input and design information for modeling and simulation.

---

[3] Such clauses rarely standup in court, but no one wants the aggravation of a lawsuit. See "Chapter 21, Feedback to the Model Provider Improves Accuracy" for more details about this issue.

[4] Supplying SPICE models saves suppliers the effort of converting the SPICE models to IBIS. A user may want a SPICE model for simulating complex I/O. That issue is discussed in detail in "Chapter 20, The Challenge to IBIS." When using SPICE models, there is a steep price to pay in terms of increased simulation run time.

14.9.3 Adaptation, Inference, and Cross-Referencing

If we cannot find or make a model using the straightforward approaches discussed earlier, we should consider using adaptation, inference, and cross-referencing. These approaches allow us to get a workable model from an existing similar component that does have an IBIS model. Adaptation, inference, and cross-referencing are essentially what we are doing when we let our simulator default to an ideal or generic model.

Modeling and simulation using adaptation, inference and cross-referencing are good for lower-accuracy simulations. These approaches can be very useful, especially when applied early in the product development cycle. Getting an approximate answer is valuable for initial simulations as long as the adapted and generic model represents the fundamental characteristics of technology, speed, and impedance. For instance, an analog voltage source generator (pulse or clock) with the correct swing, speed, and impedance, plus capacitive loading can drive a transmission-line topology in making such estimates.

Adaptation, inference, and cross-referencing are similar and somewhat overlapping ideas. They are separated here for the sake of discussion.

Adaptation

Adapting an IBIS model file means to use one that currently exists as a substitute for one that does not currently exist. An example is to use a Fairchild Semiconductor 74ABT126 for a Fairchild Semiconductor 74ABT125 part by changing the pin list.

Inference

Inference means that one device, or model, is similar to another. Thus a 74LVT240 device from Philips Semiconductor would be inferred to be similar to a 74LVT240 device from Fairchild Semiconductor. Likewise, inferring a 74LS00 is similar to a 54LS00 is a fairly good guess.

It makes sense to use, for example, a 3.3 Volt CMOS generic default model for a first approximation of performance if that is the type of part in the circuit. Better yet, use a 3.3 Volt CMOS generic default model from the intended supplier of the new part being developed. Use the established process for the new configuration being developed.

But it would not make sense to use a 3.3V CMOS default model for a low-voltage GTLP controlled-turn-on and turn-off Open-Drain-type part. A CMOS Totem-Pole part would be totally wrong as a default technology model. If we only need to verify that the receiver will switch, a generic default model of the right technology will do the job initially.

Cross-Referencing

IBIS buffer libraries from ASIC and FPGA suppliers are a good example of useful *cross-referencing*. Here the basic [Model] sections are provided and the user only has to configure a [Pin] list and the proper header keywords and other standard templates. An example of this type of adaptation is included at the end of this chapter,

Therefore, if a supplier says that they do not have an IBIS model for a part, it's worth asking whether they have one for a part with the same IO cell characteristics. The required pin list, C_comp, and other required items can be defined and an IBIS file can be created.

In the case of SPICE models for parts using the same underlying semiconductor technology process, we should assume their SPICE models are similar as a first approximation.

Measuring a sample of parts with the aid of a program like IC-CAP® from Agilent on a vector impedance meter, and making some estimates of variability, we can develop a S-Parameter[5] model.

Warning: When we do device test measurements, we sometimes need to correlate the measurements with the component supplier. Correlation can be exacting and tedious work.

14.10 REVIEW

14.10.1 Using the Data Sheet When Building a Model

When building an IBIS model, the data sheet of the component is almost always needed. The data sheet provides information for:

- Pin lists that tell the simulator what I/O models are connected to which pins.
- Switching threshold voltages.
- Sometimes I-V curve and switching speed data is provided.

[5] See "Chapter 4, "Measuring Model Parameters in the Laboratory" for a description of S-Parameter models.

14.10.2 Using Software When Building a Model

Some of the free and commercial tools available for IBIS-model extraction are s2ibis, ModelCenter, Model Integrity, ApSim, INSA, and ibisinf. Hyperlynx and Sintec have useful IBIS-model viewers. A more complete survey of EDA modeling and simulation software providers is included in "Chapter 22, Future Trends in Modeling."

14.11 PURCHASING CUSTOM MODELS FROM A THIRD-PARTY

Contractors, consultants, EDA companies, and third-party modeling service companies can be hired to build models, especially for custom ICs. These custom models tend to be the most expensive. If the quality of the models from these sources is satisfactory, they can save a lot of time and effort. This allows us to focus on the product design itself, especially when we are first learning about IBIS models.

14.12 CONVERTING SPICE MODELS TO IBIS MODELS

A supplier is often willing to provide a SPICE model, especially if a signed NDA has been negotiated. From that SPICE model, we can create an IBIS model. "Chapter 13, Using EDA Tools to Create and Validate IBIS Models From SPICE" covers the conversion process in detail. Suppliers commonly use SPICE models to control their semiconductor processes.

14.13 USING A SUPPLIER'S PRELIMINARY MODELS

14.13.1 Motivation and Discussion for This Example

If the supplier is developing a custom IC, it is quite common for the part and model development to be late. Often the board must be laid out and prototyped before the part and its model even exist. Examples include large FPGAs under concurrent development, especially custom ICs that will use new semiconductor technology processes.

Most semiconductor suppliers today do not produce their own silicon. This is referred to as (wafer) fabless manufacturing. They design the chip from an architectural (IP) point of view, but then they go to an outside semiconductor fab house to make it. The fab house owns the process and the process characterization data.

Get the semiconductor supplier and fab house to share preliminary models for early simulations. If a part is being built on an existing technology process, it is best to ask the supplier for I/O characterization model cells from similar, already manufactured parts. If the I/O cell characterization has been done, it is not that hard to complete the rest of the IBIS file data.

14.13.2 Procedure for Cloning a Usable IBIS Model

This short procedure provides a work-around set of steps for cloning a useable IBIS model for ASICs and FPGAs. It is not a full procedure for creating a model from SPICE to IBIS. Instead, it is a procedure to follow when creating a model by adaptation, inference, or cross-referencing. It basically consists of cutting and pasting from an existing IBIS file, and checking the results with the IBIS Golden Parser.

The cutting, pasting, and cross-referencing starts out from a more generalized IBIS file of a part downloaded from the semiconductor supplier web site.[6]

The generalized files are incomplete and unusable because they need pin lists and package models that are device-specific. Those specifics are unknown until the user selects a specific logic, technology, and package for the device. Rarely do more than a fraction of the I/O cells and package data available in the file get used in a single specific device. By selectively cutting and pasting from the general file, a user can put together a usable IBIS model early enough to be of value in their design process.

Rules for writing IBIS files

- Use only a text editor. Do *NOT* use a word processor.
- ASCII alphanumeric characters only.
- | (vertical stroke: the character that identifies comments).
- Do *NOT* use special characters, such as / & @ %, except inside comments, _ (underscore) is allowed.

---

[6] The generalized FPGA file can contain over 100 I/O cell models, a dozen different sets of package parasitic models, and account for devices of over 1000 pins.

The file resulting from cutting and pasting follows next. However, the paste-up is shorter than a real-world example would be. It eliminates many pages of repetitious data. Only one I/O cell has been included, many of the I-V and V-T table data points are deleted, and it has only three pins. See "Appendix C: Sample IBIS Model" for the extra detail.

14.13.3 Example IBIS Model

This topic presents a FPGA IBIS paste up file with the following characteristics:

- Just a few pins called out
- One I/O cell model included to keep the example short
- V-T waveform data taken under several different loading conditions
- Much of the I-V and V-T data deleted as unnecessary detail

Figure 14-3 shows the required information. [File Name] must match the name on the file.

```
|*****************************************************************
|                          ABC Manufacturing, Inc.
|                            IBIS Models for FPGA#1 device
|*****************************************************************
[IBIS ver]        2.1
[File name]       abcdefg.ibs
[File Rev]        3333.3
[Date]          2/22/1905
[Source]        Hspice simulations (18-07: TT)
[Notes]         ABC IBIS model for abcdefg I/O.

        abcdefg LVDS drivers require the use of external
        resistors to provide the correct voltage levels.
        Please read ACME application note XXXXX "Abcdefg LVDS
        Drivers& Receivers:  Interface Guidelines" for details
       COMMENTS: Models include process corners for MIN and MAX.
[Disclaimer]     The data in this model is derived from SPICE
   simulations using modeling information extracted from the target
   process. While a great deal of care has been taken to provide
   information that is accurate, this model is considered preliminary
   as it has not been verified by actual silicon measurement. Treat the
   data in this model as preliminary until actual silicon verification
   is performed.
[Copyright]       Copyright 2002, ACME Inc., All rights reserved
```

Figure 14-3. Header information in IBIS

The second set of information includes essential information ([Component], [Package], and [Pin]) shown in Figure 14-4. [Component] is usually the commercial name that the semiconductor supplier gives to the marketed or data sheet device. Example: SN74ABT16244DN. The [Pin] list requires manual entry because pin number, signal_name, and model_name have to be matched from different sources. The data sheet aids with pin number and corresponding signal_name information. But data sheets are not oriented to matching up signal_name with an IBIS-meaningful model_name as found in an IBIS file. Making that match may be the single largest task when constructing the file in this manner, particularly for an FPGA with its multiplicity of choices.

```
|*************************************************************************
|                      Component abcdefg
|*************************************************************************
[Component]        ABCDEFG
[Manufacturer]     ACME Inc.
[Package]
| variable         typ                min                    max
R_pkg              0.258              0.1870                 0.3280
L_pkg              5.1nH              2.4000nH               7.8000nH
C_pkg              1.1pF              0.8000pF               1.4000pF
|**************************************************************************
   *
|
[Pin]   signal_name        model_name         R_pin     L_pin     C_pin
1        XYZ+                 XYZ
2       PGND                GROUND
3       VCCO                POWER
|Etc.
```

Figure 14-4. Component, package, and pin information in IBIS

The list of pins, shown in Figure 14-4, points each pin to the associated model (shown in Figure 14-5). In this example, there is just one model_name, "XYZ." Following the pin list, there might be many [Model] sections. The simulator would next scan down the file until it found the matching XYZ name. The simulator would start reading the model information from that point. Model_type is one of the most important subparameters in IBIS, and should be matched exactly for the selected pin and buffer. For example, Open_drain versus Output. A shortened version of Model XYZ follows:

```
|**********************************************************************
  *
|                              Model XYZ
|**********************************************************************
  *
[Model]           XYZ
Model_type        I/O
Polarity          Non-Inverting
Enable            Active-Low
Vinh  = 1.4V
Vinl  = 1.0V
Vmeas =   1.2000V
Cref =  10.0000pF
Rref =   1.0000M
Vref =  0.000V
C_comp            5.0000pF            4.0000pF            6.0000pF
[Temperature Range]     25.0000             100.0000            0.000
[Voltage Range]        2.5000V             2.3750V             2.6250V
[Pulldown]
| voltage      I(typ)              I(min)              I(max)
  -2.5000      -88.8600mA          -60.2580mA          -0.1064A
  -2.3000      -88.8600mA          -60.2580mA          -0.1064A
  -2.1000      -88.8600mA          -60.2580mA          -0.1064A
| The data between these points is left out. It's extra detail that
   isn't needed for
| explanation.
   4.6000        0.2315A             0.1791A             0.2755A
   4.8000        0.2563A             0.1978A             0.3053A
   5.0000        0.2820A             0.2171A             0.3362A
[Pullup]
| voltage      I(typ)              I(min)              I(max)
  -2.5000        0.2955A             0.2311A             0.3407A
  -2.3000        0.2697A             0.2106A             0.3114A
  -2.1000        0.2437A             0.1899A             0.2816A
| The data between these points is left out. It's extra detail that
   isn't needed for
| explanation.
   4.6000       -1.3640A            -1.4610A            -1.2360A
   4.8000       -1.5990A            -1.6740A            -1.4770A
   5.0000       -1.8340A            -1.8880A            -1.7200A
[GND_clamp]
| voltage      I(typ)              I(min)              I(max)
  -2.5000       -1.8030A            -1.7360A            -1.8280A
  -2.4000       -1.6860A            -1.6300A            -1.7060A
  -2.3000       -1.5680A            -1.5230A            -1.5850A
```

Figure 14-5. Model information in IBIS (1 of 3)

```
| The data between these points is left out. It's extra detail that
  isn't needed for explanation.
  2.3000          1.0190nA             16.8200nA              1.4260nA
  2.4000          1.0770nA             19.3500nA              1.5130nA
  2.5000          1.1930nA             51.4000nA              1.6590nA
[POWER_clamp]
| voltage      I(typ)                I(min)                I(max)
  -2.5000         7.3920nA             67.8700nA             67.5200nA
  -2.4000         6.9270nA             63.8500nA             64.0300nA
  -2.3000         6.4030nA             60.4200nA             60.7700nA
| The data between these points is left out. It's extra detail that
  isn't needed for
| explanation.
  -0.2000         1.3000nA             22.0000nA             19.0000nA
  -0.1000         1.2000nA             20.0000nA             17.5800nA
   0.0000         1.1640nA             19.1500nA             15.1900nA
|
[Ramp]
| variable        typ                  min                   max
dV/dt_r           1.4070/0.3280n       1.3026/0.4908n
  1.4946/0.2610n
dV/dt_f           1.4472/0.3288n       1.3535/0.4905n
  1.5291/0.2665n
R_load = 0.1940k
[Rising Waveform]
R_fixture = 0.1940k
V_fixture = 0.000
| time            V(typ)               V(min)                V(max)
  0.000S          24.7500uV            13.6400uV             77.8600uV
1.000e-10S       -51.8250uV           -34.5700uV             27.1050uV
   0.2000nS        0.2840mV            10.4050uV              0.8903mV
| The data between these points is left out. It's extra detail that
  isn't needed for
| explanation.
   4.8000nS        2.3450V              2.1710V               2.4910V
   4.9000nS        2.3450V              2.1710V               2.4910V
   5.0000nS        2.3450V              2.1710V               2.4910V
[Rising Waveform]
R_fixture = 0.1940k
V_fixture = 1.2000
V_fixture_min = 1.1000
V_fixture_max = 1.3000
| time            V(typ)               V(min)                V(max)
  0.000S          41.3000mV            53.1100mV             37.3300mV
1.000e-10S        41.2200mV            53.0700mV             37.2850mV
   0.2000nS       41.5600mV            53.1100mV             38.1400mV
```

Figure 14-5. (2 of 3)

```
| The data between these points is left out. It's extra detail that
   isn't needed for explanation.
    4.8000nS        2.4210V              2.2680V              2.5580V
    4.9000nS        2.4210V              2.2680V              2.5580V
    5.0000nS        2.4210V              2.2680V              2.5580V
[Falling Waveform]
R_fixture = 0.1940k
V_fixture = 0.000
| time            V(typ)              V(min)              V(max)
  0.000S          2.3450V              2.1710V              2.4910V
1.000e-10S        2.3450V              2.1710V              2.4910V
    0.2000nS      2.3450V              2.1710V              2.4910V
| The data between these points is left out. It's extra detail that
   isn't needed for
| explanation.
    4.8000nS      0.2219mV             0.4934mV             0.2078mV
    4.9000nS      0.2170mV             0.4525mV             0.2031mV
    5.0000nS      0.2067mV             0.4555mV             0.2367mV
[Falling Waveform]
R_fixture = 0.1940k
V_fixture = 1.2000
V_fixture_min = 1.1000
V_fixture_max = 1.3000
| time            V(typ)              V(min)              V(max)
  0.000S          2.4210V              2.2680V              2.5580V
1.000e-10S        2.4210V              2.2680V              2.5580V
    0.2000nS      2.4210V              2.2680V              2.5580V
| The data between these points is left out. It's extra detail that
   isn't needed for
| explanation.
    4.8000nS     41.5100mV            53.7067mV            37.4833mV
    4.9000nS     41.5100mV            53.6700mV            37.4450mV
    5.0000nS     41.4800mV            53.6300mV            37.5200mV
| End [Model] XYZ
[End]
```

Figure 14-5. (3 of 3)

14.13.4 Validating the IBIS Model

After constructing an IBIS file, it is essential to check for syntax mistakes. Most IBIS files are too long and too detailed to check manually. The IBIS Committee provides free, downloadable software to help check IBIS files. It is called The IBIS Golden Parser and can be found at: http://www.eigroup.org/ibis/tools.htm

After reviewing the instructions about platform, the user can download and use version 4. It is backward compatible with previous versions of IBIS. The easiest way to use it is to:

1. Place the ibisfile.ibs and the parser ibischk4.exe in the same subdirectory.
2. At a command line prompt, type:
   ```
   >ibschk4 ibisfile.ibs > ibisfile_parser_results.txt
   ```
3. Check the results for errors and warnings.

   Errors will prevent the file from working in most simulators. Warnings might keep us from getting good results. When the parser is run on the IBIS file, it will check model file syntax for correctness. It will also check for missing information, data-point monotonicity, and inappropriate numbers like diode clamp currents of 1×10^{23} or 1E23 amps. The parser catches many careless mistakes, which account for most known model problems.
4. Perform a visual check, in a visual editor (GUI). For example, missing pins in the pin list are common. Odd-looking pullup curves and other anomalies can be spotted.
5. Correct the mistakes or return the file to the source for correction. What the user can and will do depend on experience level, desperation, and distractions.
6. Iterate the above. The user should not expect to catch every error on the first attempt to parse a model.
7. Associate the model with a Company Part Number. This is discussed in "Chapter 15, Working With the Model Library." Cloning the original IBIS file and giving it a corporate part number will allow storing it in the company's component library along with schematic symbol files, and CAD layout footprints.
8. Send the model file to the Component Database Librarian for archiving and release. The user should agree to methods for passing released model files to each other. Suggested procedures for doing this are discussed in "Chapter 15, Working with the Model Library."
9. Verify simulations as a board moves through the design process. Capture the results of the verification work that was done by entering comments into the IBIS file itself. For example, note whether:
   - The lab results look like the simulation
   - Any discrepancies were not resolved
   - A verification study was done, data was captured, and accuracy was calculated

14.13.5 Running ibischk4.exe (Golden Parser) on the File

Key points to know about running the IBIS Golden Parser on a model file are:

- The parser checks *everything* down to the last detail. IBIS syntax is very rigid. For example, GND is correct for a model name, not GROUND.
- Since an IBIS file can be very complex, never release it to a corporate database or a customer without validating it by running the parser on the file.

Using the example file (shown in Figures 14-3 through 14-5), the IBIS Golden Parser found many mistakes on the first try. This was a result of using Microsoft Word instead of a text-only editor. On the second try, after the user made corrections in a text editor, with the results shown in Figure 14-6:

```
IBISCHK4 V4.0.1
Checking abcdefg.ibs for IBIS 2.1 Compatibility...
ERROR - Component 'virtual-E': Model 'GROUND' for Pin '3' not
defined.
Errors  : 1
File Failed
```

Figure 14-6. Golden Parser error message #2: Failed

The IBIS file was then edited with the ASCII text editor. On pin 3 the word GROUND was changed to GND. The Golden Parser was then run on abcdefg.ibs again, with the results shown in Figure 14-7:

```
IBISCHK4 V4.0.1
Checking abcdefg.ibs for IBIS 2.1 Compatibility...
Errors  : 0
File Passed
```

Figure 14-7. Golden Parser error message #3: Passed

14.14 ASKING SI-LIST AND IBIS E-MAIL REFLECTORS FOR HELP

14.14.1 Protocol

The SI-LIST and the IBIS Users email lists are a source of last resort. They can be queried for the availability of models. Most model files have language forbidding their further distribution by the original recipient. Therefore, it is *not* appropriate to ask members to send any IBIS file. Participants in the email lists do not have much time to act as clearing houses of IBIS-model availability information.

14.14.2 WARNINGS!

The Si-List and IBIS E-mail reflectors are like moderated chat rooms. It is easy for their moderators (gatekeepers) to revoke privileges, which they will do if a participant:

- Is overtly self-promoting, such as in marketing messages.
- Is a source of offensive messages, SPAM, and viruses.
- Asks for help so often that it appears the participant is expecting others to do their work for them.

Most model files contain language constraining further distribution by recipients. The best use of email chat rooms for IBIS models is to inquire about availability and sources.

14.14.3 Signal Integrity List Email Reflector (SI-LIST)

The Signal Integrity List email reflector is moderated email expanders list that functions one step removed from a chat room. Currently it has over 3800 participants worldwide discussing all aspects of high-speed modeling and simulation. Questions about IBIS models in general, or an IBIS model in particular, can be sent to the list. The responses are often quite helpful. But the list should not be overused. The responders are taking their own time to help.

To subscribe to Signal Integrity List, send an email message with your name and email address to:

- si-list-request@freelists.org
- Put SUBSCRIBE in the SUBJECT

- List archives are viewable at: http://www.freelists.org/archives/si-list
- Old list archives are viewable at: http://www.qsl.net/wb6tpu
- For gateway to archives and subscriber controls: http://www.freelists.org/webpage/si-list

To email an inquiry or response to the list: Si-list@freelists.org

14.14.4 The IBIS List Email Reflector

The IBIS email reflector has 400 plus participants[7] worldwide. It is similar to SI-LIST. The IBIS reflector has a policy of accepting **NO** email attachments because of the continuing problem of viruses and Spam. Contact information follows:

For help or to subscribe/unsubscribe, email: majordomo@eda.org

With the appropriate command messages in the body:

```
help
subscribe ibis <optional e-mail address, if different>
subscribe ibis-users <optional e-mail address, if different>
unsubscribe ibis <optional e-mail address, if different>
unsubscribe ibis-users <optional e-mail address, if different>
```

Or email a request to ibis-request@eda.org.

IBIS reflector archives exist under:

http://www.eda.org/pub/ibis/email_archive/ Recent

http://www.eda.org/pub/ibis/users_archive/ Recent

http://www.eda.org/pub/ibis/email/ E-mail since 1993

Participants on the IBIS email reflector need to be even more reticent about asking for specific IBIS models.

14.15 MODELING TOOLS ON THE IBIS WEBSITE

Getting comfortable with the IBIS model and methodology can be slow. The IBIS Committee has posted additional information on their website. Following is some of this information from the "Free Tools" page that is relevant to creating an IBIS file:

---

[7] Companies joining the IBIS Committee open forum can vote on specification modifications and improvements.

1. *IBIS Golden Parser*

   The IBIS Golden Parser executables for various OS platforms are available for download. Click ibischk4 for the IBIS ver4.0 parser and select the correct OS type.

2. *ICM Golden Parser*

   The ICM parser executables for various OS platforms are available for download. Click icmchk1 for the ICM ver1.0 parser and select the correct OS type.

3. *IBIS Cookbook*

   The Cookbook describes all the steps necessary to generate an IBIS model. This is a downloadable Microsoft Word file. IBIS Specification versions 1.1, 2.1, and 4.1 are available at:

   http://www.eigroup.org/ibis/tools.htm

Many other free tools, some provided by commercial EDA suppliers are available at this website.

14.16 SUMMARY

Engineers and others must learn the strategies for obtaining good IBIS models and component coverage. These strategies include starting with approximate models and progressing to more accurate models. As the design cycle progresses, approximate models must be replaced with more accurate ones and the simulations must be re-run.

Sources of IBIS models include: semiconductor suppliers, third party modeling services, and conversion from a SPICE model. Another source is a similar part with an IBIS model that can be adapted for use with the current part.

Chapter 15

WORKING WITH THE MODEL LIBRARY

A well-managed library helps conserve and leverage engineering resources

Abstract: A well-managed library of component choices *and* their models improves the chances of successful modeling and simulation activities. Standardization of commodity components reduces the proliferation of unnecessary parts and data and is a philosophy that should be employed in managing the library. The company's component library serves as the central database for all component-related information. If model files are part of this component library, the component library can also provide for model storage, retrieval, and use.

15.1 THE BEST WAY TO MANAGE MODELS

Engineers who use the company's simulation software packages need to be able to access and retrieve models. The best way to make the models accessible is to create a model library. The best place to put that library, which is a database of model files, is within the component database.

Over time, the model library usually increases in size to hundreds, or even thousands, of active device model files. Models for resistors, capacitors, connectors, and other passive circuit elements will number many additional thousands of data items. To manage the huge number of data items involved, we need to develop an efficient filing and retrieval system.

In many ways, a model library is similar to a typical book library. For example, to assist users in finding models, we can classify models in a system similar to the Dewey-decimal system. We may want to set up a check-in/check-out system, control model circulation, and limit the resources invested–while providing a well-rounded selection of models.

15.1.1 Benefits of Model Libraries

If a library of models (and their associated parts) is well managed, it can provide several benefits. These benefits include:

- Easy identification of the best parts available for a particular job.
- Good match between the demands (for example, power, speed, and fan-out) on a part and its capabilities.
- Good characterization of the model properties and their population distribution.
- Easy retrieval and reuse of parts and their simulation models.
- Previously verified accuracy of simulation models.
- Prime and alternate supplier availability.
- Consolidated purchases and good pricing.
- Avoiding part number proliferation.
- Data ready, and parts available.

Approximately half the resources consumed in modeling and simulation activities are related to acquiring and fixing models. As a result, corrected models may represent a substantial investment. It is important to avoid unnecessary proliferation and duplication of components and their models by standardizing components and models. Standardization helps to:

- Enable a sparse but wide selection of models
- Conserve the resources expended for modeling and simulation.

One way to standardize is to use commodity-type parts. Commodity implies several sources of the same–or nearly the same–thing. For details on this way of standardizing, see Section 15.2.

15.1.2 Establishing a Model Library

The best people to lead and drive the modeling simulation effort are the people in the Engineering department. To build the model library, the Engineering department should leverage the company's pre-existing component library.

The company's component library was probably created earlier by the CAD and PCB layout departments. If simulation is being implemented for the first time, the many thousands of resistors, capacitors, and other components already in the library may need to have their numeric electrical property values (ohms, farads, and henries) entered into the database. Since these values may be of no interest to the average PCB mechanical designer, the required fields may not have been reserved for these values or have had data entered. In fact, the people involved in maintaining the component library may not realize that resistors, capacitors, and capacitance even have electrical properties. Just entering these values can be a substantial effort and may discourage those involved.

Before embarking on modeling and simulation, engineers must ensure that passive elements like resistors and capacitors have their electrical data included in the library–not just their package outlines and electrical symbols. Getting started in modeling and simulation may require adding these thousands of data elements to the pre-existing component library. After the models are added to the library, the library can be inundated with hundreds of model files in various stages of usability and with potential errors. Before long, correcting and verifying the models can be an on-going task.

The company's internal part numbers, which are already listed in the Bill-Of-Materials (BOM), are used to identify data in the component library. The part numbers are commonly used to archive and retrieve the models. Most component libraries already include subdirectories with schematic symbols, board layout outlines, and pin locations (also known as footprints). It is common to add model files to that subdirectory.

15.1.3 Who Controls the Model Library?

Component librarians should be responsible for storing and retrieving component data, and for making the board layout footprints and schematic symbols accessible to EDA tool users. They also manage the component library's check-in/check-out procedure for modifying and maintaining component data. By incorporating models into the component library, those same component librarians can also handle the model files with procedures similar to those already in place.

Product Design Engineer. When a company is small, the Product Design Engineer does most of the work himself. This includes:

- Designing the logic functionality, timing, and product.
- Obtaining and validating most of the model files.
- Documenting quality and fixes to models.
- Running most of the simulations.

Signal Integrity Engineer. When a company becomes larger, a specialist, such as a Signal Integrity Engineer, can be hired to assist the Product Design Engineer with the modeling and simulation tasks.

15.1.4 Suggested Process for Acquiring Models

The Signal Integrity Engineer is responsible for obtaining, testing, correcting, documenting quality and fixes, and parsing model files. These files are sent to the Component Librarian along with appropriate instructions. A check-in/check-out procedure is used to access or revise model files previously stored by the Component Librarian.

Figure 15-1 shows a suggested process for acquiring model files and preparing them for use. One objective is to make the models available to the widest possible set of EDA tool users. A second objective is to avoid setting up a supporting Signal Integrity Engineer to become a bottleneck or a policeman in the team effort. A third objective is to provide EDA tool users with some protection against using unproven, non-functioning models, or those models still in development.

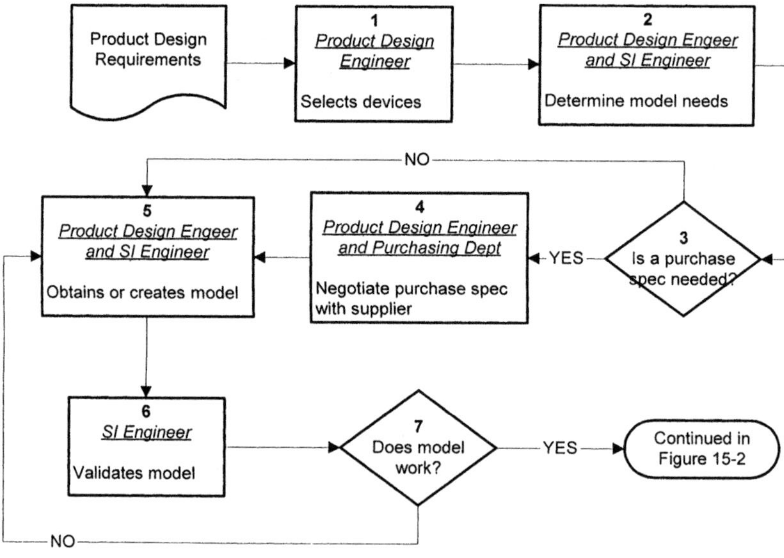

Figure 15-1. Model acquisition flowchart

When models are acquired and placed in the library, models should undergo a process of validation. Validation and verification are not the same. Validation is a very specific type of verification. In the case of IBIS models, validation is the act of confirming that the IBIS file conforms to the IBIS Spec. Verification, as used by the authors, is the act of confirming that the simulation using the model correlates with laboratory measurements. Verification is done late in the design cycle when prototype hardware is available for measurement. If engineers need to wait until the models are verified before using them, the design process would be impossible implement.

Process model verification against a previously verified SPICE model simulation, or against hardware measurements, can (and must) happen in parallel with the design development process. The verification process is shown in Figure 15-2.

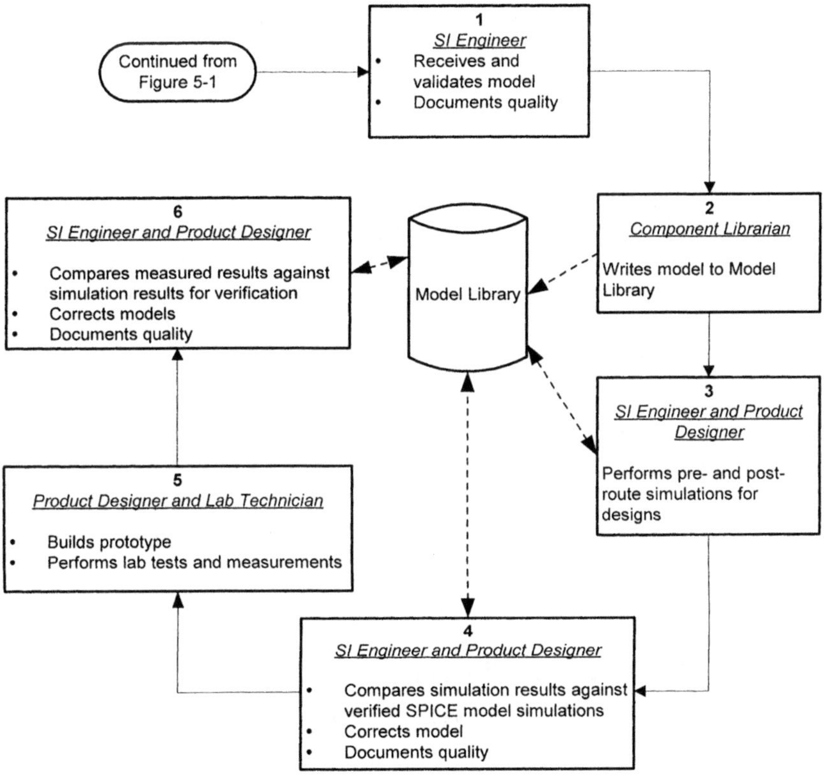

Figure 15-2. Model verification flowchart

15.1.5 Controlling Models in the Library

Some control over the placement, maintenance, and modification of model files in libraries accessible to all tool users is needed–given the variable quality of their accuracy and readiness for use.

We recommend storing validated and verified model files in the component library, while making new, unvalidated model files inaccessible to the general user. When users request permission to use the restricted models for simulations, they can be informed of model issues. Here is a suggested process for library management:

- A library of components, organized by internal company part number, should be created for storing schematic symbols, board footprints, and model files for use in the EDA tool. The Component Librarians should be the only people given file write privileges to this database. Untested and experimental files should be placed in an "unapproved" location with more limited user access. Files that are tested and released should be accessible by all EDA tool users.
- The IBIS file, as received, is stored under the semiconductor supplier's part number with its .ibs file extension. The same IBIS file, with corrections, plus quality-level test results and comments, is stored under the company part number with a .ibs extension. IBIS files from more than one semiconductor supplier for the same generic part number (for example 74ACT00) can be combined in one file under the company's internal part number.[1] The different models are identifiable by their unique [Component] names (for example SN74ACT00DT) as IBIS provides for, and as EDA tools can handle.
- After parsing in an EDA tool, the IBIS file is often converted to the EDA tool vendor's native language and given a new file extension in place of .ibs. For example, files parsed by the Cadence EDA tool get a .dml extension. The company part number with this new extension, and the library search paths, leads the EDA layout and simulation tool to the model files.

15.2 COMPONENT STANDARDIZATION AND LIBRARY MANAGEMENT

Commodity-type parts are easy to standardize and classify. Examples include "glue logic" and surface mount resistors, capacitors, and inductors. From the many commercially available commodity components, there are only a few combinations of package size, wattage, etc., that should be needed for most designs. There will always be a few exceptions to the standard choices, but most engineers will go along with a reasonably administered standardization policy.[2]

---

1 The internal part numbered IBIS file is a *derived file.* In addition to containing model files from different suppliers (for the same generic industry number), it can also contain IBIS quality level results, lab verification data, and more.

2 Engineers are more open to standardization as long as they do not confuse *distinctiveness* for *creativity* or *economy*. See also ASPOVEPA in the glossary and in the index.

Custom, specialized, and state-of-the-art components are harder to standardize and classify.

15.2.1 Classification Trees Aid Component Standardization

The first strategy in component and model library management is to develop a classification scheme. We can classify components in several ways, such as by:

- Specialized semiconductor technology and construction
- Number of individual transistors (gates)
- Package type
- Application end-use
- Electrical property values

A classification tree for components can help component specialists in choosing standardized parts.

Classification by end-use application and electrical properties is closer to how an engineer views a component and how supplier selection guides are organized. This classification is an alternative to the company's part numbering system. If users understand the library's classification scheme, using a search engine to find parts is relatively easy. A component classification tree can provide the following benefits:

- An aid in providing a sparse, but wide selection of parts by offering only a few choices per end-use category and property range. Thus we can avoid proliferation and gain the leverage of volume purchasing and production.
- An aid in identifying parts that have alternate suppliers. Comparing end-use and component electrical properties helps in choosing alternate suppliers.
- An aid in searching the library. Using electrical properties and end-uses to classify parts aids in finding parts that satisfy those parameters.
- An aid in finding information in the supplier-provided selection guides. Both the classification and selection guide are based on end-use and electrical properties.

15.2.2 Sample Component Classification Tree

The classification tree shown is based on application or circuit use with a secondary consideration of semiconductor technology. Figures 15-3 through

15-12 are some examples of how classification trees and selection guides are created.

All Electronic Components —
- Electromechanical Devices
- Non Mechanical Transducers (sound, chemical, etc.)
- Electromechanical Devices
- Electro-Optic Devices
- Electronic & Electromagnetic Devices

All Electronic & Electromagnetic Devices —
- Semiconductor Devices (*Continued in Figure 15-4*)
- Passive Devices
- Electronic Modules
- Electron Tubes

Figure 15-3. All electronic components

All Digital ICs —
- Basic Logic
- Timing
- Bus Interface
- Micros
- Computational
- Memory
- Special

Continued in Figures 15-5 and 15-6

All Interface ICs —
- Drivers
- Switches
- Converters
- Data Capture

Continued in Figure 15-7

All Linear ICs —
- Linear Amplifiers
- Feedback Amplifiers
- Response Shapers
- Comparators
- Generators
- Controllers
- Computational
- Converters
- Specials

Continued in Figures 15-8 and 15-9

All Transistors —
- Darlington BJTs
- General Purpose BJTs
- BJT Modules
- RF & High Frequency BJTs
- JFETs
- MOSFETs
- FET Modules

Continued in Figures 15-10 and 15-11

- Other Discrete Semiconductors *Continued in Figure 15-12*

Figure 15-4. All semiconductors (cont. from Fig 15-3)

- All Basic Logic
  - Gate
  - Inverter
  - Flip-Flop
  - Latch
  - Schmitt Trigger
  - Decoder
  - Multiplexer
  - DeMultiplex
  - Mux/Demux
- All Timing
  - Time Delay
  - Counter
  - Clock
  - One-Shot
  - TTL Oscillator
  - Frequency Divider
  - Multivibrator
- All Bus Interface
  - Buffer
  - Bus Driver
  - Receiver
  - Transceiver
  - Memory Driver
  - Clock Driver
  - UART
  - Modems
  - Parity Gen & Checker
- All Micros
  - Microprocessor
  - Microcomputer
  - Microcontroller
  - Micro Support
- All Computational
  - Adder
  - Multiplier
  - Divider
  - Multiply/Divide
  - Magnitude Comparator
  - Word Comparator
  - Arithmetic Logic Unit
  - Priority Encoder
  - Look-Ahead Carry Gen

Figure 15-5. All digital ICs (1 of 2)
(cont. from Fig 15-4)

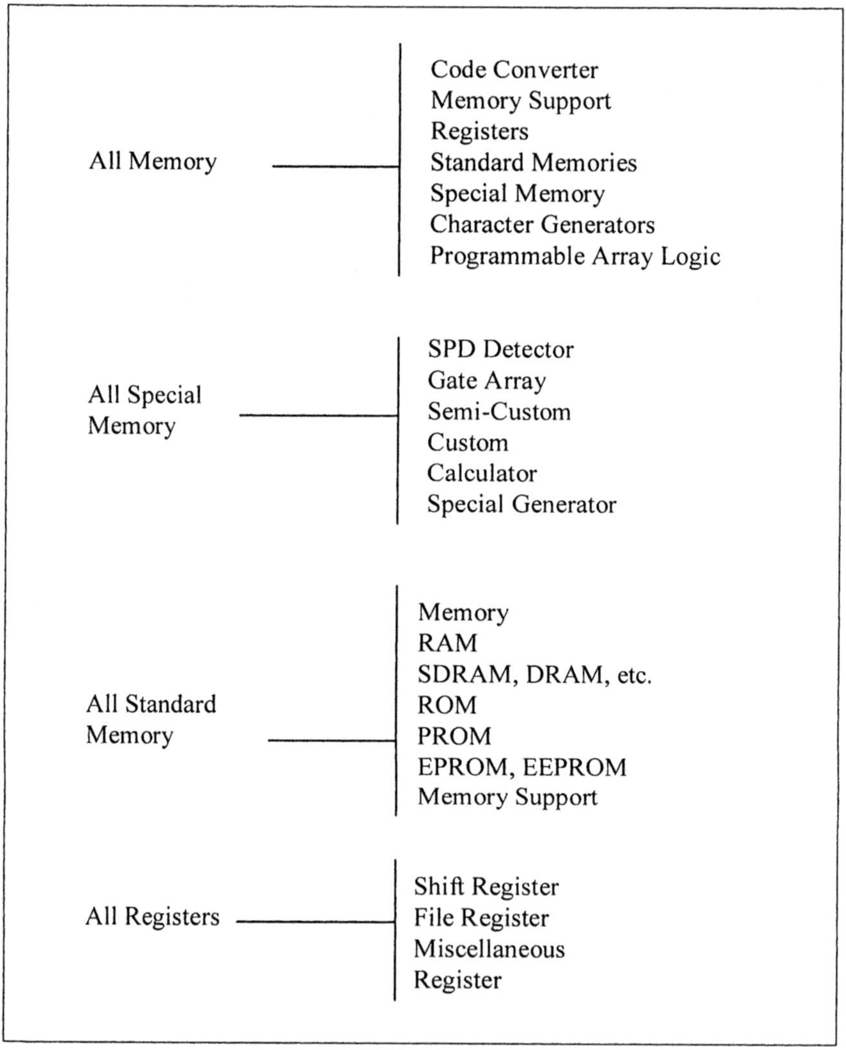

Figure 15-6. All digital ICs (2 of 2)
(cont. from Fig 15-4)

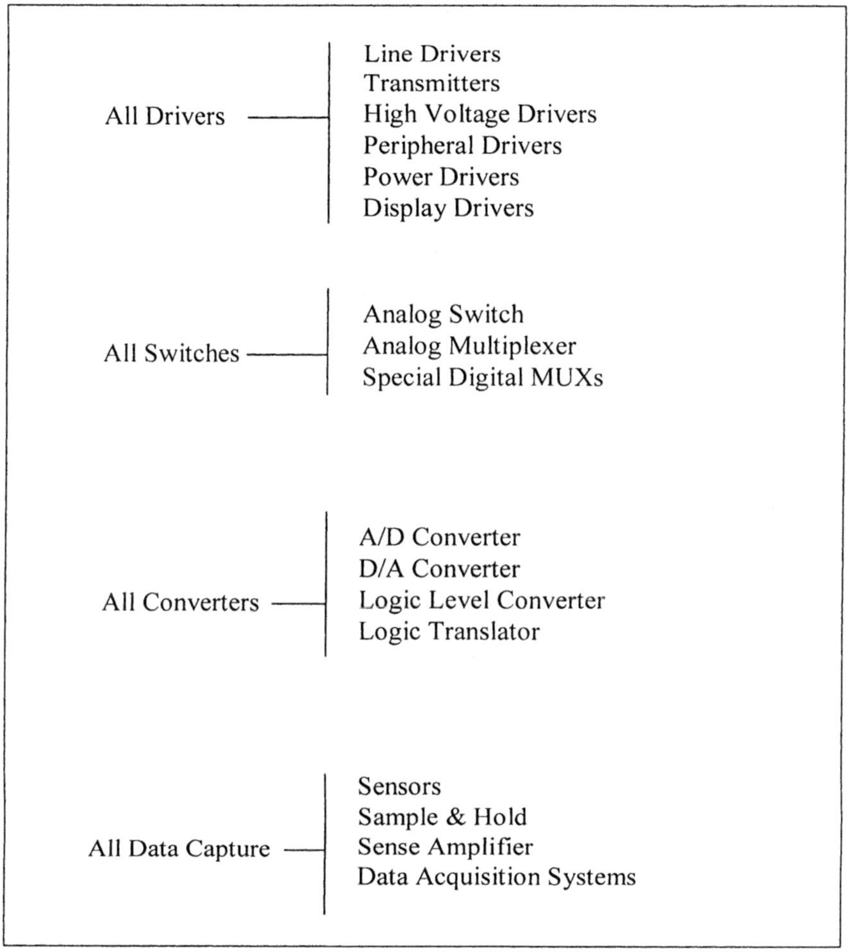

Figure 15-7. All interface ICs
(cont. from Fig 15-4)

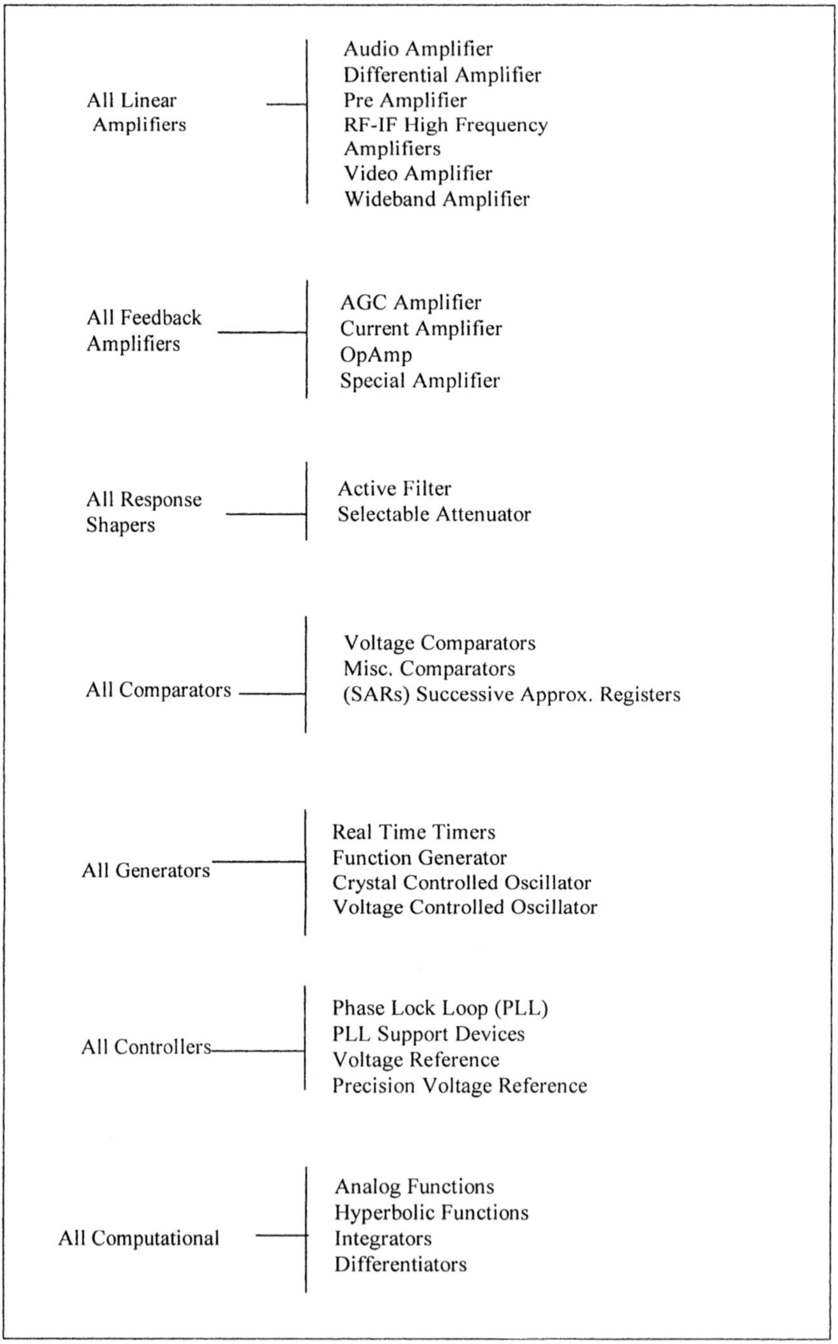

Figure 15-8. All linear ICs (1 of 2)
(cont. from Fig 15-4)

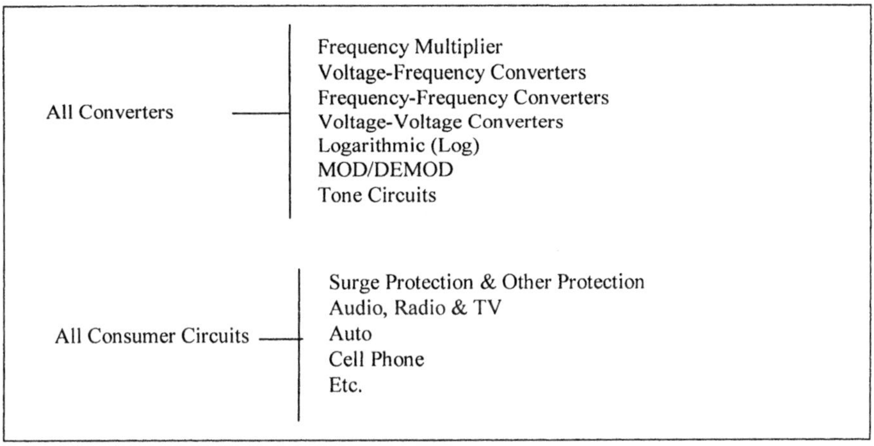

Figure 15-9. All linear ICs (2 of 2)
(cont. from Fig 15-4)

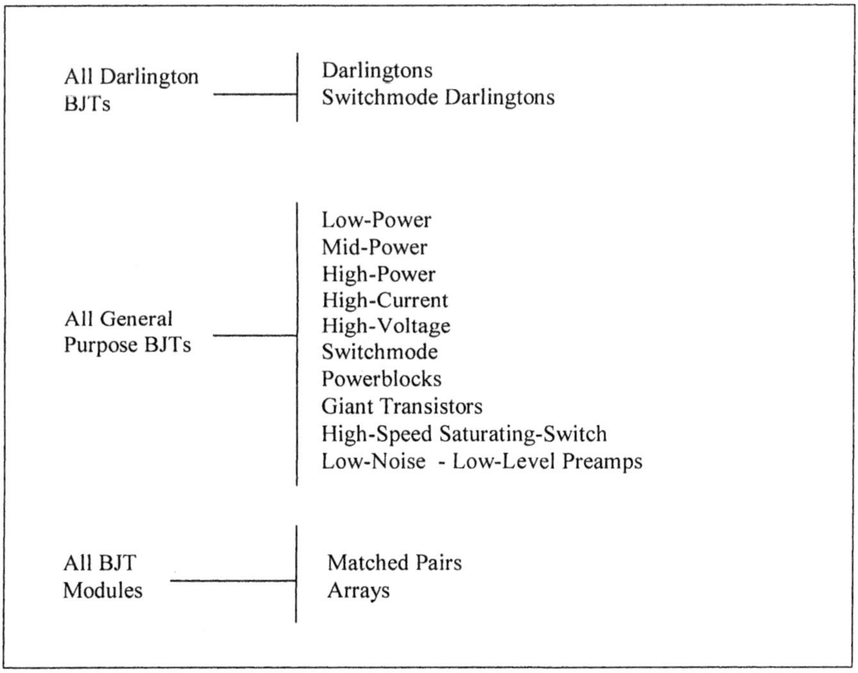

Figure 15-10. All transistors (1 of 2)
(cont. from Fig 15-4)

All RF & High Frequency BJTs	RF Small Signal Types RF Linear Power Receive-Transmit Types Video Monitor Drivers
All JFETs	General Purpose Switches RF Amplifiers OpAmp Preamp Inputs
All FET Modules	Dual JFETs Quad MOSFETs Quarter Bridges Half Bridges
All RF Small Signal BJTs	Low-Noise Preamps IF-AGC Amplifiers Local Oscillators Mixers Frequency Multipliers
All RF Linear Power Receive-Transmit BJTs	CB/SSB CATV/Amateur Radio Mobile Communications UHF-FM Radio Avionics Power Amplifiers
All MOSFETs	Low- to Mid-Power Power MOSFETs RF-Power MOSFETs GaAs FETs Dual-Gate MOSFETs

Figure 15-11. All transistors (2 of 2)
(cont. from Fig 15-4)

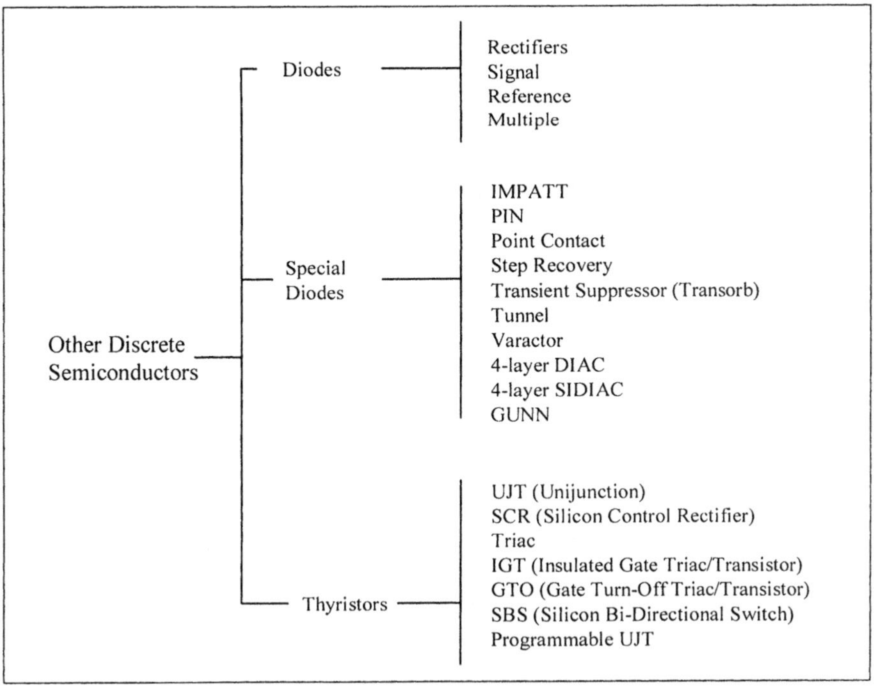

Figure 15-12. Other discrete semiconductors
(cont. from Fig 15-4)

15.2.3 Standard Parameters Aid Component Selection and Standardization

How can we compare and contrast devices on the basis of their properties? We can best compare and contrast components by selecting the same standardized test conditions and parameters. Standardization allows straightforward device-to-device and model-to-model comparisons. Once parts are classified into the same category, a set of standard test parameters, models, and model parameters can be devised and applied to the part categories. An example set of standard parameters is shown in Table 15-1.

What if the categories are by end-use, but we need to compare a BJT and a CMOS transistor? We can compare similar measures like switching speed. But the comparisons will be more abstracted. For an example involving the same kind of device in the same kind of application, standardized test parameters do help.

As an example, consider a general purpose, small-signal discrete BJT. For a discrete BJT transistor, a smart choice of a parameter to measure is one that can be used in many ways. For example, we can choose a parameter to do all of the following:

- Describe a basic behavior, especially a range of behavior
- Relate back to the physics and design of the semiconductor chip
- Relate one part to another, such as a bit bigger or smaller chip
- Control the manufacturing process of a semiconductor technology
- Used directly in modeling, design, and simulation

A good example parameter is the beta of a BJT. Beta can be directly used in each of the modes just described.

Table 15-1. A parameter set for general purpose BJTs

Parameter	**Description**
Type	For example, general purpose
Subtype	For example, low power
Pc (max)	Power dissipation, case @ 25°C
Pa (max)	Power dissipation, ambient @ 25°C
Max Voltage	Max voltage that is safe for all configurations
Max Current	Max continuous DC collector current
BVceo (min)	Max allowed collector-emitter breakdown voltage, base open
BVcbo (min)	Min allowed collector-base breakdown voltage, emitter open
BVcbo (max)[3]	Max allowed collector-base breakdown voltage, emitter open
BVces (min)	Max allowed collector-emitter breakdown voltage, base shorted to emitter
BVebo (min)	Max allowed emitter-base breakdown voltage, collector open. For characterization and control only--do not test on shippable product.
hFE low (min)	DC small signal beta, low-current region
hFE mid (min)[4]	DC small signal beta, mid-current region
hFE mid (max)	DC small signal beta, mid-current region
hFE hi (min)	DC small signal beta, high-current region
Vbe(sat) (min)	Base-emitter saturation voltage
Vbe(sat) (max)	Base-emitter saturation voltage
Vce(sat) (max)	Collector-emitter saturation voltage
ft (min)	Frequency of small signal beta zero crossing, or Gain-Bandwidth
Ccb (max)	Collector-base capacitance - guarded
Ceb (max)	Emitter- base capacitance - guarded
Icbo (max)	Collector-base leakage current,[5] emitter open
Ices (max)	Collector-emitter leakage current, base shorted to emitter
Iebo (max)	Emitter-base leakage current, collector open
t(on)[6] (max)	Switching on-time = t(d) + t(r)

---

3 Specifying a window on BVcbo helps to constrain the structure of the collector region.

4 Specifying windows on a number of DC parameters helps to constrain the structure of the transistor.

5 Leakage currents are an indicator of the quality of a semiconductor's processing.

6 Saturated switching times are specified with test fixtures that set up forced collector, base turn-on and base turn-off currents.

Table 15-1. (cont.)

Parameter	Description
t(off) (max)	Switching off-time = t(s) + t(f)
t(s) (max)	Switching storage time
t(r) (max)	Switching rise-time
t(f) (max)	Switching fall-time
Vbe (on) (max)	Base-emitter on-voltage
Package	For example, T092, SOT23, T03, etc.
Rth(j-a) (max)	Junction-ambient thermal resistance degrees C/watt
Rth(j-c) (max)	Junction-case thermal resistance degrees C/watt
Tj(op) (max)	Maximum junction operating temperature
T(stg) (max)	Maximum storage temperature
NF, narbnd7 (max)	Noise figure, narrow-band
NF, widbnd (max)	Noise figure, wide-band
Mask set #	For examples, see the topside geometries in "Chapter 7, Using Data Sheets to Compare and Contrast Components."
Technology	For example, planar double-diffused, planar triple diffused, etc.

15.2.4 Test Conditions for Component Standardization

After we create a classification tree, the next step in standardization is to set test conditions, including bias, frequency, and test fixture to a fixed set of values within each of the selection guide cells. For example, the low-current-general-purpose, BJT cell, both npn and pnp. That way direct comparisons can be made for the same and similar devices from different suppliers, and in different packages.

Standardized test parameters, and process limits on them, can more readily be made to "swim upstream" in the form of a proposed purchase specification.[8] Particularly if the semiconductor supplier has been engaged in the setting of those parameters. There is also more of a chance to include simulation model parameters and limits in the specification.

Standardization makes it easier to get agreement between a given circuit application and how the parts targeted to the application get characterized. For example, if the intended circuit application calls for 10 mA of drive, it makes little sense to characterize a part delivering 100 mA of drive. For a population of such parts, DC parameters are more convenient and cheaper to

7 Noise figure measurements are taken in test fixtures.

8 Specifications that "swim upstream" is a concept verbalized by D.C. Sessions of Philips Semiconductor. It means that a user can ask a supplier to manufacture a semiconductor that meets the proposed specification and model. The user and supplier can negotiate the possibilities of meeting the specification. The user's proposal is often the result of doing "what-if" simulations.

measure than SPICE, IBIS, or other models. For a process that is in Statistical Process Control, DC parameters can be a cost-effective surrogate set of quality assurance measurements on a production basis, and a supplier can use them to diagnose process drifts.

15.3 STORING AND RETRIEVING MODEL FILES

15.3.1 Naming Model Files for Archiving

Model files come from suppliers already identified by the supplier's part number. For example:

- 2N3904.spc (2N3904 SPICE model)
- hct245sc.ibs (IBIS model for the Fairchild 74HCT245SC) and (RF Micro Devices RF2301)

Most EDA tools are able to find such model files with the assistance of a file browse window in their graphical user interface (GUI) tool. Most EDA GUIs also provide some kind of a model assignment tool or window, where the model file can be associated with the BOM part number for the purpose of simulating the part with the model.

However, at some point, it is usually an efficiency-enhancing step to rename the model file with the company part number and an appropriate *file extension* such as *.ibs* or *.spc*. When the model file bears the same part number as the BOM and the correct file extension, many EDA tools automatically search model libraries and the load model files. In the simulation tool, the model will be associated with the part number and the object in the CAD database. This is especially handy with large boards undergoing many iterations of simulate→change the layout→resimulate. Otherwise the constant manual reassignment of models to modified board databases can waste a lot of engineering time.

How is the file renaming done?

- The best time to do file renaming is after receiving the file and after some validation is done. File renaming should certainly be done before the model is stored in the library as a validated or verified model.
- For IBIS, we must make the IBIS keyword "[File Name] hct245sc.ibs" the same as the renamed file "[File Name] 1004320.ibs, for example."

- It is always best to rename the file on a *copy* of the original file and to add a comment into the renamed file referencing the old, original supplier file name.

In the case of the Cadence EDA tools, two steps get combined. The model validation step associated with running the IBIS Golden Parser on the file gets combined with the step of converting the IBIS file to the Cadence native language, which is DML. This conversion applies the parser plus some additional model checks programmed in by Cadence, and strips out items like comments, notes, and disclaimers, and produces a *.dml* extension file. For example, 1004320.dml. The part number 1004320 in the board database then gets matched to the 1004320.dml file when the board gets loaded into memory for simulation.

15.3.2 Model Retrieval

This topic discusses attaching models to symbols, netlists, BOMs and board layouts. Two ways of attaching models to components for use in an EDA tool simulator are:

- Attaching the model to a schematic symbol that has user input-variables associated with the symbol.

  The schematic is basically a netlist and a Bill Of Materials (BOM). It begins to describe the components, their properties, and their interconnections as a structured (SPICE) netlist for simulation. At low analog frequencies or slow digital edge rates, simulation from the schematic without board layout information can be sufficiently accurate. At higher edge rates and frequencies the physical layout enters into the simulation.

- Attaching signal models to the board layout BOM part number.

  At high frequencies and fast edge rates, layout and interconnections significantly affect simulation results. Therefore, simulation is usually run from a board layout tool with EM field solvers extracting models of the board interconnections.

When simulation is run from a schematic, there is no confusion about the connections of a circuit and what the designer sees. But when simulation is run from a board layout tool, there is a perception problem. The net connections can be seen in a schematic layout GUI, but cannot be seen easily from the board layout GUI. Therefore, most simulator companies provide a topology extraction tool that is annotated with etch length information, characteristic impedance, etc. User-controlled variables are then fed to the simulator either from the layout tool, the topology tool, or both.

Most design and debugging are done with the aid of some type of schematic GUI. A netlist is hard to visualize for design and debug purposes.

15.3.3 Search Paths to Model Libraries

Simulators use many kinds of model files, such as an underlying lumped-element SPICE model, an IBIS behavioral model, or a high-frequency 2-Port Scattering-Parameter model. Whatever the model type, the simulator needs to know where to look for the needed model and how to recognize that it is the correct model file. File extensions, such as: *.ibs, .pkg, .spc, and .dml,* differentiate between the types of models.

The common approach is to put the model files in subdirectories and then to provide the EDA tool with the necessary search paths. The search paths can be set up in a configuration file. The library implementation often needs provisions for differentiating between verified and unverified models and other related issues.

15.3.4 Model Acquisition

For simulation tasks to move product design and development forward, access to model files from the model library is crucial. Also crucial is who maintains the model files and when they are maintained. Most users should have read-access to the files, while only a few need read- and write-access.

Providing control of model files is similar to providing control of EDA board-layout footprint files. Board layouts will not be correct unless correct part dimensions and pin locations are in the component directory. Likewise, simulations will not be correct without correct models. At today's multi-gigabit switching speeds, correct information on board layout, models, and a component's physical dimensions are all required for correct simulations. Plus, qualification applies as much to models as to the physical components themselves–both must work!

Today, it is simply impractical to build enough prototypes to accomplish what can be done with simulation. With qualified models, we will do simulations covering the probable range of performance to verify that a circuit will operate properly within an expected range of variables.

As the design simulations progress, engineers have many opportunities to verify IBIS files against previously verified SPICE model simulations (given adequate resources). Also, engineers have many opportunities for direct verification of the IBIS model simulations against measurements.

Before products are released to manufacturing, taking laboratory measurements on prototypes is an almost universal testing activity The automatic forwarding of laboratory test measurements to the person responsible for simulations should be part of hardware design processes and procedures. Also, this type of verification data is very valuable collateral when partnering with semiconductor suppliers.

15.4 ASSIGNING MODELS TO COMPONENTS IN EDA SIMULATORS

When we are done adding the model to the model library, we now need to work with the EDA tool to assign that model to a component. How do we assign SPICE models in SPICE netlists to components? In a SPICE netlist, the user directly modifies the netlist entries.

To start, the user enters the netlist into the EDA tool. The netlist provides the EDA tool with certain information, as listed below:

1. The SPICE netlist tells the EDA tool the component is connected in the circuit.
2. The component's SPICE model call-up is done with a sub-program, or submodel program call.
3. Certain model property values may be set in the netlist, entered from the schematic, or entered from the simulation tool GUI.

For IBIS models, the EDA user interface usually provides mechanisms for accomplishing this task and the tasks associated with changing model parameters for a simulation run. In all simulation work, a *signal model* (such as SPICE, IBIS or S-Parameter) must be *assigned* to a component, just as a value for a resistor must be assigned to the resistor before the simulation can be run. In many respects, IBIS model assignment for simulation is similar to SPICE model assignment.

The EDA CAD database provides components pin connection information and signal trace routing information to the simulator. The assignment of the model to a component plus the library path information allows the EDA tool to load the data for the model. Certain model property values may be set in the simulation program, and/or entered from the GUI. An example is: Choosing the typ, or min, or max process corner—or all three—to run in the simulation.

IBIS model assignment for simulation is different from SPICE in that the assignment work is done from an EDA tool GUI. Many tools are available in the marketplace; although they all handle things a bit differently, they mostly handle things nearly the same. The EDA tool GUIs vary in their user-friendliness, and the tool providers are always improving them.[9] The Cadence EDA tool GUI from their v13.0 release is shown in Figure 15-13. It is fairly typical.

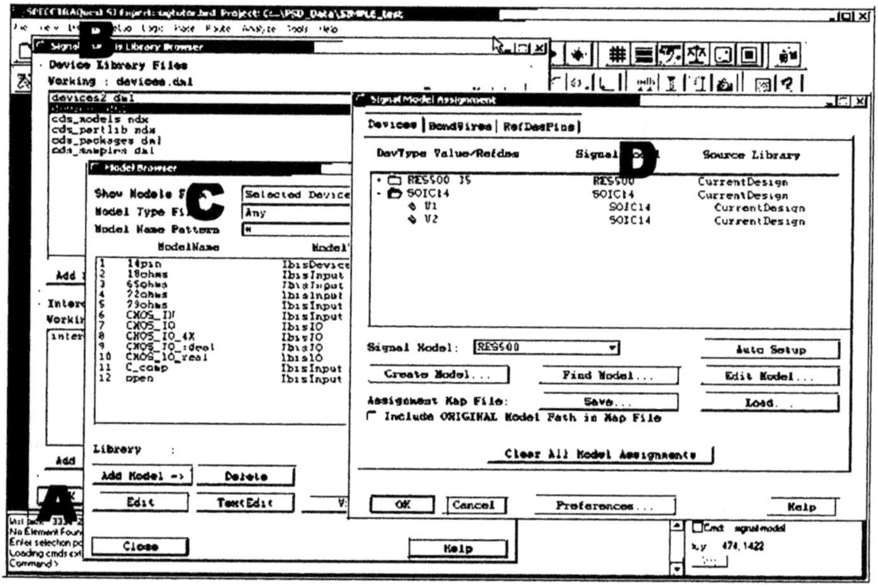

Figure 15-13. Model assignment in the Cadence v13.0 EDA tool GUI

The windows in Figure 15-13 are a tiled-over view of the board layout. They could just as well be a tiled-over view of a net topology. Shown are windows for:

- **A**: Barely visible in the background is the GUI window for the board layout tool. The board cross-section can be controlled from this GUI, simulations can be run from it, and nets can be extracted into a schematic view tool and GUI.
- **B**: The library browser window shows the model libraries available, including:
  - *Local.* Models available in the design subdirectory

9 EDA tool providers are always ready to explain to users why their EDA tool is best. Much can be learned from listening to their sales speeches.

  - *Default.* Generic, and supplier specific, particular models packaged with the EDA provider's program
  - *Company.* Both released model libraries and pre-released libraries used by the Signal Integrity Engineer for newly received and experimental model files.
- **C**: The model browser window, which lists the individual component models and some instances of individual I/O cell models. Model files can be added, edited (maintained), and deleted from this window. Individual pin-signal model data can be edited and maintained.
- **D**: The Signal Model Assignment window, where signal models are attached to physical parts. This can be done by component type. Or the list of individual parts (each one denoted by a board reference designator — U1, U2, etc.,) can be expanded, and different instances of the same component type can get different models. The list of pins of a particular part can be displayed and manipulated.

Modern EDA tools have great flexibility for browsing and manipulating (add, delete, convert, and edit) model file libraries and individual models plus, the particular assignments of models to physical part pins in the PCB design.

Before, during, and after this signal model assignment activity, the Signal Integrity Engineer should run audit reports in the EDA tool GUI if the tool provides the capability. These can be audits of the BOM, netlist, model assignments, components, and net scheduling to check what the program actually sees as the setups for simulation.

In Cadence, a Map File is referred to in the signal model assignment GUI. The Map File facilitates the re-application of model assignments to later version board databases. This facilitates the simulate→redo layout→resimulate iterative loop during virtual board debug. In later versions of the Cadence EDA tool, this memorization of model assignments was made easier by incorporating model assignments initially done by the Signal Integrity Engineer in the primary board database under the control of the CAD group.

Most efficient of all for re-iteration of simulations is when the CAD Department PCB Designer incorporates the signal model in the board design database itself. This is illustrated in window D, where the GUI shows that the signal model is obtained from the "Current Design" library. The reason that the CAD Department PCB Designer does this job is that he or she has to be the gatekeeper on the prime board design database. That leaves the Signal Integrity Engineer free to simulate design changes until solutions are found to problems without worry about wrecking the baseline database. Plus solutions are *not* solutions until the layout and route is modified and *those* changes re-simulated.

15.5 FLEXIBILITY IN MODEL CHOICES AT RUN TIME

It is a good idea to provide flexibility at simulation run time to allow for easy switching to another supplier of a part number. Models for the same part number from different suppliers can differ significantly. In IBIS, different models and package types from the same supplier can be included in a single file, and often are by the model supplier. For example, one often sees IBIS files 3.3V and 5.0V glue logic model versions in a couple of different packages. During model assignment and simulation these different components can be selected for a particular simulation run.

Each unique model and package combination gets a unique [Component] keyword. Each unique [Component] can be selected at simulation run time. Use the company's part number on the model file when combining models from different suppliers. Give each unique supplier/model/package combination a unique [Component] name. For additional flexibility we can use a – (dash) number or suffix to differentiate different model versions.

When simulating from schematic (topology) prior to layout and routing, we need to accommodate simulating the same part in different package configurations. When simulating the same part in different package configurations from a layout tool, we are making some inappropriate assumptions. The different package sizes and pin locations will, in reality, change the circuit behavior. If simulations indicate that a change in package type is appropriate, be sure the PCB layout department changes the layout, and then resimulate for realistic results.

15.6 SUMMARY

A library of cleaned-up model files represents a considerable engineering investment. One way to conserve and leverage engineering resources is to make sure the library is well managed. Another way is to standardize components that are commodity items. A component classification tree helps users get started with standardization, which makes parts easier to find. Making the library readily accessible to EDA tool users is very important.

PART 6: MODEL ACCURACY AND VERIFICATION

Chapter 16

METHODOLOGY FOR VERIFYING MODELS

The final step in modeling is to verify models in simulation against hardware data

Abstract: Verification compares model simulation against hardware test data. Comparisons must take into account all test circuit effects, such as probes and power distribution systems, as well as tolerances and corners. Verification can be done against a single model (stand-alone) or against a model within a design. This verification methodology can be used for SPICE, IBIS, and other model types.

16.1 OVERVIEW OF MODEL VERIFICATION

Model verification is the act of ensuring that the model file is both complete and correct for *its intended use.* Verification compares the model against the hardware being modeled. Thus verification cannot be performed until the functional unit (such as a device or circuit) represented by the model has been implemented by building a prototype. This makes model verification the final step in modeling.

Ideally, models are verified against the hardware data before the models are released. However, in practice, models are usually released earlier—after validation. Validation is the act of ensuring that the model file is complete and correct *per a standard.* For example, an IBIS model is often released once it passes the IBIS parser. This process is described in "Chapter 13, Using EDA Tools to Create and Validate IBIS Models from SPICE."

It can take several weeks to have a physical prototype built and tested. For example, verification of an I/O buffer requires one to get silicon fabricated, a test board built, and test data collected. The cost of the PCB and the packaged chip can easily exceed several thousands of dollars, not to mention the cost of testing. To cover the effects of process variation takes dozens of parts from different die lots, and even more parts are needed if a statistical distribution is required. Only after the test data are available can one compare the model and its simulation data against the measured test data. Figure 16-1 shows an example.

Verification is strongly recommended for all model types. Model parameters are often obtained by optimizing many parameters together. In the BSIM3 model in SPICE, for example, there are over 60 important parameters. The newer BSIM4 has over 200 parameters. Some of the BSIM parameters are correlated (that is, changing one parameter changes another during optimization); nine independent parameters can account for over 90% of the data fit. When parameters are correlated, optimization can result in one parameter having a non-physical value (not physically possible for the given technology), reaching an "optimal" solution by having another parameter with its own non-physical value.[1]

Another common model problem is extrapolating model characteristics beyond the normal operating range. For example, SPICE models created based on data between 0V and Vcc might or might not be valid outside of this range. Model verification needs to done over the complete operating range, including the appropriate operating regions (DC, transient, and AC) for all operating conditions, including all bias and temperature combinations.

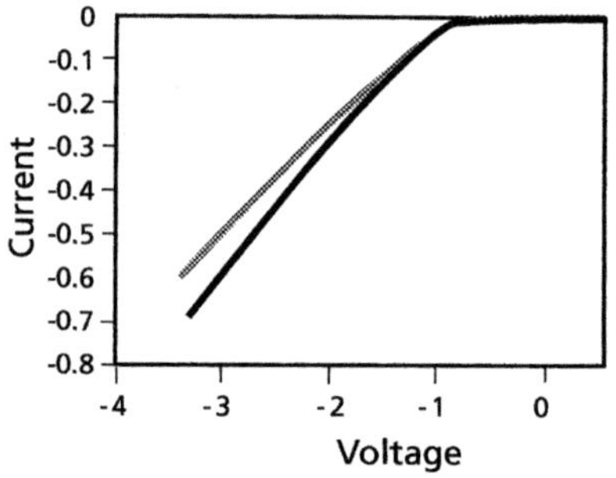

SPICE (line) and IBIS (dots) data overlaid for 3 corners. There is a good match at each corner.

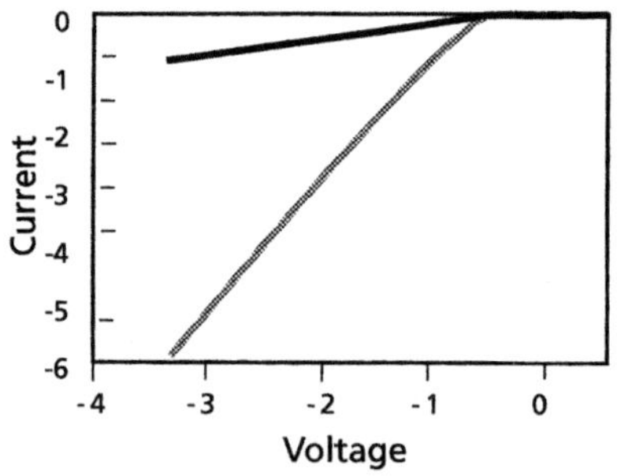

Mismatch between SPICE and IBIS (upper dots) and test data (lower dots) for the max corner.

Figure 16-1. SPICE and IBIS models compared to measured data for a clamp diode
Data courtesy of Cypress Semiconductor

Signal integrity and EMC are particularly sensitive to operation outside of the range [0V, Vcc]. Voltage in the models needs to be accurate over the

[1] Non-physical should be understood as not physically realizable. One example is a silicon device (Eg=1.1eV) having a model parameter of Eg=2.1eV.

full range of [–Vcc, +2Vcc] to cover potential ringing and overshoot. (Alternatively, a model could cover –Imax to +2Imax, where Imax is the maximum transient current from the signal source.) Accurate modeling over the full range allows accurate simulation of reflections from a transmission line load, including short circuit and open circuit loads.

Verification can also be done against another verified model for the same part. For example, a verified SPICE circuit model might be available. The test data serves as the primary model, and each model verified against test data serves as a secondary model for verifying additional models. The advantage is that the part's new model could be compared under more conditions against the secondary model, including over the anticipated range of statistical production variations. The risk is that the secondary model might not have been compared to test data under some of the conditions used to verify the new model, as shown in Figure 16-1. If one is verifying against a secondary model, it is important to know over what range the other model was verified.

A model can (and often must) be used before verification data becomes available. At the same time, IBIS models historically had problems in data extraction during model creation.[2] Other model types are not immune to similar problems. It is a wise person who incorporates some form of model validation and verification into their overall design flow.

16.2 MODEL VERIFICATION METHODOLOGY

Verifying a model proceeds in three steps:

- Determining what model characteristics to verify
- Determining the test setup required to verify the model
- Determining whether the model simulation and test data compare within the allowed tolerances

A number of questions must be answered before verifying a model. For example:

- What are the intended operating conditions for the part in the system, and which system parameters are important to model accurately?

[2] Many models created with s2ibis2 had data problems due to algorithm choices. For example, only one of the two clamp currents was subtracted from the pullup (and pulldown) current, resulting in a double counting of clamp currents. These and other problems have been fixed in s2ibis3, which is available on the IBIS web site.

- What degree of accuracy is required to verify system functionality – is a 10% tolerance acceptable, or are eight significant digits required?
- What are the process variations (particularly in integrated circuits), and what statistical properties can be verified using a limited number of parts?

A second set of questions concerns the test setup. Items here include such things as the selection of the test circuit, the spread in impedance for a trace on the test board, the tolerance and bandwidth of the test setup, and the effects of test equipment loading on the measured results. One must also decide how the edge time is to be defined. Some data sheets use 10-90%, while others use 20-80%.[3] For frequency, the usual convention is to use the 3dB bandwidth.

High-speed models are particularly hard to verify. One of us (Green) once encountered a project where a part on a prototype board having an edge faster than 1 nsec was measured with a 500-MHz oscilloscope. The rise time for the scope was long enough so the part's edge time appeared slower than the data sheet claimed. For today's high-speed interfaces, the edge rates are comparable to the very best oscilloscopes. This means that the actual edge rate sometimes needs to be calculated from the measured edge rate and the scope's edge rate using:

$$T_{actual} = \sqrt{(T_{obs}^{\ 2} - T_{scope}^{\ 2} - T_{probe}^{\ 2})} \tag{16-1}$$

where:

T_{actual}	=	actual $T_{10\text{-}90}$
T_{obs}	=	measured $T_{10\text{-}90}$
T_{scope}	=	$T_{10\text{-}90}$ of scope (test equipment)
T_{probe}	=	$T_{10\text{-}90}$ of test probe
$T_{10\text{-}90}$	=	$0.338/f_{3dB}$ for a Gaussian pulse [60]

Verification is also sensitive to the type of test heads being used. A passive test head can often be modeled as an RLC circuit. An active test head requires a more complex model, particularly at higher frequencies, where the frequency content of a fast edge is still significant. Models for test heads are used in the simulations during model verification, thus matching the circuit being tested in the simulation circuit. Test head manufacturers often post the data sheet, including the probe bandwidth, on the web. For an example from Tektronix, please see

---

[3] IBIS uses the 20% to 80% edge times under the [Ramp] keyword.

http://www.tek.com/site/ps/0,,60-12025-SPECS_EN,00.html.

16.2.1 Test Boards and Circuits

There are two types of test boards. The first type is designed explicitly to test the part. For example, a board is often designed for manufacturing testing of chips (integrated circuits) when they go into production. The second type of board is a design prototype that includes the part. The manufacturer of the part typically develops the first board, while a product design group develops the second.

There are a number of advantages to using a simple board to test a part. The test board usually has a simple load, such as a trace (transmission line) and a test head, perhaps with termination at the end of the trace to reduce reflection. This makes it easier to measure the behavior of the actual part for comparison to the model, particularly if the test trace has a constant width (constant impedance).

Figure 16-2 shows an actual test board. This test board incorporates a realistic circuit, including bypass capacitors. Special pads with SMA connectors were included in the layout for scope probe attachment.

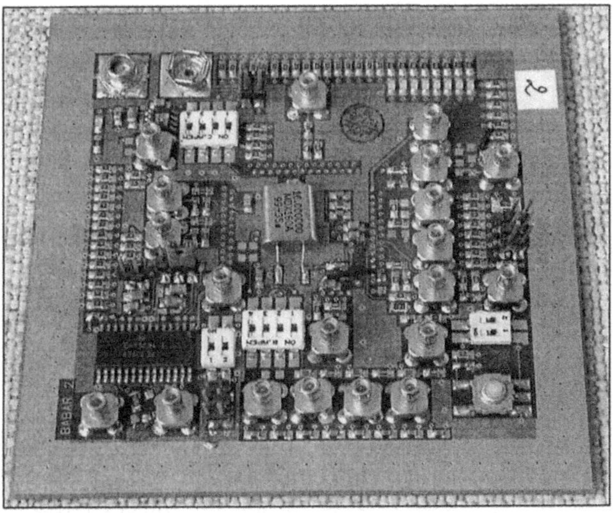

Figure 16-2. A realistic circuit on a test board
Photo courtesy of Etienne Sicard, INSA-Toulouse, France

There are also advantages to verifying a model using a prototype board. The prototype board represents an actual design, including traces that branch and change layers. Impedance changes cause signal reflections, which travel back to the driver as well as forward to receivers and tri-stated I/Os. The prototype board allows the model to be verified for both the first transition (before reflections arrive) and for reflected signals. A product prototype board also allows one to verify that the actual circuit will work, making it somewhat easier to justify under the project budget. On the other hand, using a product prototype rather than a dedicated test circuit board further complicates the already complex task of model verification.

A sample schematic and test board file, testbrd.tar.Z, can be found on the IBIS web site, under http://www.eda.org/pub/ibis/accuracy/. The files in this directory are described in the file readme.txt. Although the IBIS Quality Checklist has superseded the checklist found on this web page, the Handbook is particularly interesting for its description of the many test conditions (such as open circuit and impedance loading) that can be used for I/O buffer verification.

Many silicon manufacturers provide development boards or design kits. These are midway between production test boards and product boards, containing a simple yet realistic test circuit layout. The list may also include simulation examples for signal integrity simulation.

Excellent examples from XiLinx can be found at http://www.xilinx.com/ and from Altera at http://www.altera.com/products/devkits/kit-index.html.

Collecting I-V (DC current vs. voltage) data is straightforward. Measuring fast transients, such as V-T (voltage vs. time) data, requires careful attention to power and ground returns. Measurement techniques for high-speed transients are well documented in books and application notes.[4] Similarly, specialized equipment such as TDRs (time domain reflectometers) and VNAs (vector network analyzers) are available for characterizing high-bandwidth interconnects. For example, a TDR can be used to obtain capacitance, based on $I = C^*(dV/dt)$, as long as the reflected current is not affected by other reflections. Capacitance and inductance can also be measured using frequency domain measurements, such as resonance frequencies and resonance bandwidth.

16.2.2 Comparing Data Sets

Verification requires comparing the test data, from any type of test board, against a simulation that includes the model being verified. The simulation includes the loading of the traces and the test equipment. Ideally, the part on the board has the same process corner as the model used in the simulation. Since the process variation is statistical in nature, it might take many parts from different production runs to fully verify a statistical model. On the other hand, an IBIS model with only three corners can be verified against the

---

4 For information, see [1, 2, 26, 39, 60, 76, 101, 106, 112, 114, 119, 124, 133, 134, 136].

statistical extremes (such as three-sigma or six-sigma), if the parts manufacturer can provide parts sorted by process corner. During testing, the part also needs to be at the same temperature and voltage conditions as the simulation.

Since simulation tools, such as SPICE, specify the die temperature rather than the ambient temperature, care must be taken during testing to collect data at the correct die temperature. Otherwise, if the die and its package and mounting can be modeled in a thermal simulator, and the power dissipation in the part is known, adjustments can be made to the electrical simulation in SPICE-like models. But remember that IBIS models cannot be manipulated in that way, since IBIS models are specified at typ-min-max die temperatures and the measurement data needs to match with those conditions in the model file. Adjustments to the test environment (ambient temperature and/or power level) can be made to achieve that match if die-package-mounting thermal resistance is known and other thermal conditions are specified.

Similarly, the simulation voltage at the part needs to match the voltage at the part during testing. To do this, one can test a part before it is packaged (bare die), or include the package parasitics in the simulation.

It is important to have enough data points for comparison. For example, the capacitance for varactor tuning diodes is not well described by the SPICE diode model, as shown in Figure 16-3. Note that the simulation and test data sets can be fit at any two points, or the slope and value can be matched at one point, but not both. This is why it is not sufficient to check just one or two points when verifying a nonlinear model.

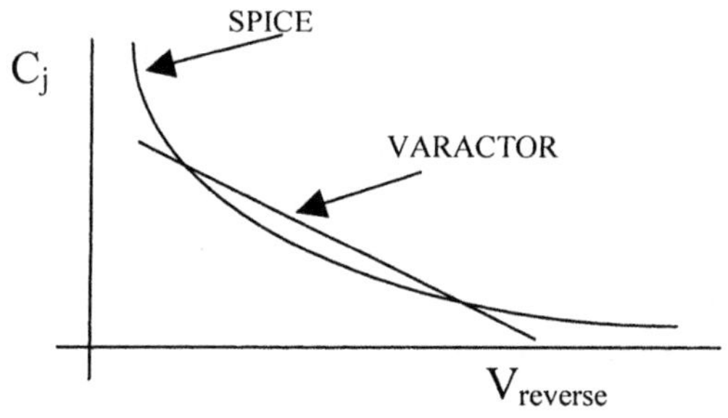

Figure 16-3. SPICE model and varactor diode capacitance (C-V) curves

The bottom line is that the test set and simulation must match when the data is collected to have any hope of the two data sets matching. Every variable that matters (voltage, temperature, statistical corner, and others) must be controlled. Everything in the circuit must be modeled correctly, from a trace that is modeled as a lossy transmission line to the test head loading on the circuit.

16.2.3 Accuracy

Both accuracy and precision are important when comparing measured and simulated data values. Accuracy measures the separation of the true value from the average of the sample values, while precision measures the separation between the sampled values. Figure 16-4 illustrates graphically the difference between accuracy and precision.

(a) accurate but not precise (b) precise but not accurate (c) accurate and precise

Figure 16-4. Accuracy and precision

In the days of slide rules and visually reading oscilloscope displays, 10% precision was acceptable. With the advent of digital meters, scopes, and digital computers, one can obtain 5 digits using 17-bit representations, and even more using 32- or 64-bit numbers. Problems in putting together a test setup, such a large wire loop in a test probe or instability in a digital-to-analog converter's voltage reference, can limit the accuracy of test data. "Chapter 17, Verifying IBIS Model Accuracy by Using Laboratory Measurements," discusses further how the test setup affects the accuracy of the test results. Test equipment manufacturers provide articles and application notes on optimizing test setups [1, 2, 124].

Simulation accuracy, on the other hand, is limited by the convergence options rather than by the number of bits used for calculations. For example, when SPICE options are set to RELTOL=0.001 and VNTOL =1u, then a voltage of 100pV could have an error of 1uV, 10 times larger than the actual voltage; larger voltages would be accurate at best to 4 significant digits (0.1%). Setting SPICE to print out 8 digits does not improve accuracy, so the last 4 digits would represent numeric noise. As a general rule of thumb, most simulators will agree within 1% to 5% (given the same convergence tolerances).

16.2.4 Evaluating Verification Data

Given the limitations in measurement and in simulation, how does one compare test and simulation waveforms? Since the comparison is intended to verify the model for use in a design environment, the model is not expected to be accurate to 10 decimal places. However, the model must be accurate enough to make a go/no-go decision whether to build a physical design prototype.

Figure 16-5 shows three waveforms. While it is clear that the first two waveforms are identical except for a time shift, it is not as obvious how well the first and third waveforms are correlated. The third waveform shows that ringing, such as that caused by transmission line effects, has caused a mismatch.

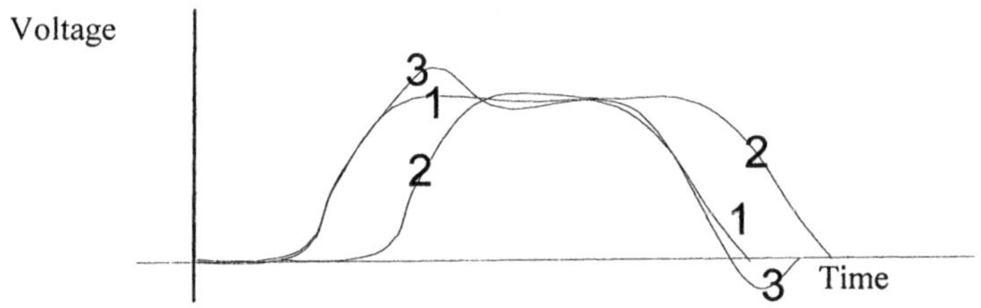

1=Unshifted waveform

2=Shifted waveform

3=Ringing waveform

Figure 16-5. Three waveforms for comparison

For a good match when performing waveform comparisons, one must have identical circuits for test data and for simulations, including all probes and terminations. Any mismatch in the test circuit model and the physical test circuit will lead to mismatches when simulation and test data are compared. While some physical objects can be neglected (such as via impedance for 1kbit/sec data on a printed circuit board), these same physical objects can be significant under other conditions (such as the same vias for a 10 Gbit/sec system). The test circuit model could include many effects, such as package parasitics, via models, and the power distribution system. It requires engineering experience and judgment to determine when physical objects must be accurately modeled, and which physical objects are not of concern for the part and test system being modeled.

Waveform comparison usually has two phases. First, the waveform is shifted, and then a visual comparison or Figure of Merit (FOM) calculation is made. For example, in Figure 16-5, the shifted and unshifted waveforms will have a poor FOM unless the two waveforms are aligned in time. To correlate waveforms numerically, a least-squares calculation can be used, or a sum of absolute errors. References [5, 9, 16, 31, 70] present a number of methods. Interpolation is often required to obtain data values for comparison, since X-values often do not align between test and simulation data, or there may be more points in one data set than the other.

The IBIS Accuracy Handbook uses the absolute error for its "curve overlay metric" FOM (in percent):

$$FOM(\%) = 100\% \times \left[1 - \frac{\sum_{i=1}^{N} \left| Xi(golden) - Xi(DUT) \right|}{\Delta x \times N} \right] \tag{16-2}$$

where:

golden = the simulation data
DUT = the device-under-test measured data
Δx = the step size along the x-axis (usually voltage or time)
N = the total number of points being compared

Any FOM or visual comparison depends on matching the simulation and physical conditions, including the statistical variations of the parts. For a small sample size (fewer than 25 parts), the statistical variation can make it difficult to see whether there is a good waveform comparison. In SPICE, where models can be set to match a specific process statistical state, one can more easily compare waveforms. In IBIS, where only typ, min, and max are defined (including voltage and temperature as well as process state), an FOM has significance only for parts from those three process states, and not for process states in between.

At minimum, non-statistical models require identifying which process corner particular parts are from, and accepting the associated increase in simulation-to-measurement data correlation errors. Non-statistical models lack the flexibility to allow a complete model-to-part verification. Verifying any non-statistical model, such as an IBIS model, is similar to a unit-by-unit verification, where the model parameters (such as IBIS corner tables) are specific to particular parts within a population spread. A truly complete IBIS model verification would include measuring individual part model parameters, simulating with those parameters, building prototypes with the measured parts, measuring performance of the prototyped parts, correlating results with the simulations, and keeping track of data on a unit-by-unit basis. Verification of the same model for use in another simulator would require repeating the simulation step, and comparing to the already measured test data.

Statistical coverage varies from model to model. In some cases, a model might have a distribution function, while in other cases it might have min and max models. In some cases, only a typical model is provided by the manufacturer, and the user is left to estimate minimum and maximum performance.[5] If statistics (or min and max) modeling are provided, one would expect the part's test data and model's simulation to both lie within the statistical envelope.

In some models, the min and max represent not absolute best and worst case performance, but rather performance corners taken from a wider distribution. For example, a Gaussian distribution extends to infinity on both sides, although physical parts at the outer ends of the distribution would be rejected during production testing. The min and max are often taken at three-sigma or six-sigma points of the broader distribution.

Figure 16-6 shows how measured data for a random part (one that does not correspond to any of the IBIS corners) lies within the boundaries defined by the min and max table data in the IBIS model. If this IBIS model were generated using three-sigma limits for the I-V tables, then one would expect some physical parts to lie outside these corners.

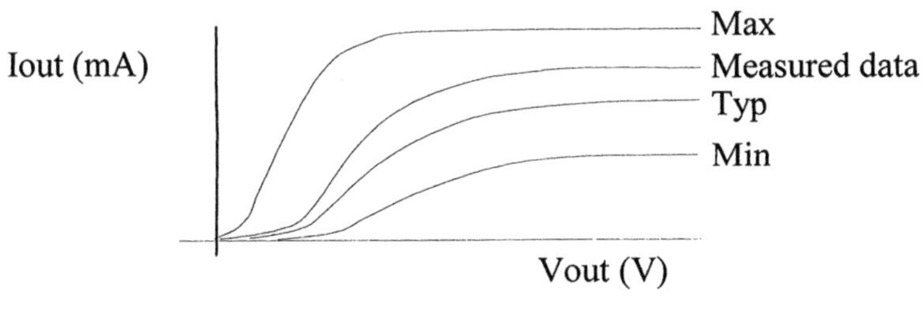

Figure 16-6. Curve overlay for one test part and model's IBIS corners

16.3 VERIFYING SPICE MODELS

SPICE (Simulation Program with Integrated Circuit Emphasis) was originally developed at the University of California at Berkeley in the 1960's. It was quickly adopted by industry, and today there are many commercial tools based on the original SPICE algorithms. Furthermore,

5 The estimation factor depends on both circuit design and process. The "rule of thumb" factor between typ and the other corners can be as high as 2.5, or as small as 1.3.

many simulation tools support the Berkeley SPICE syntax and models (SPICE 3f5 being the lowest common denominator). Some tools also include proprietary models and algorithm features, which are not portable to other tools. SPICE 3f5 is also available as open source from a number of web sites.[6]

SPICE must be viewed as both a simulator and as a set of models. In SPICE, model equations are coded, and are part of the simulator executable. The user generally does not have access to the source models (SPICE 3f5 source code is available on the web). The good side of this is that the simulator and models are verified together. The flip side is that it is almost impossible to modify the model equations or to add a new model type without the source code.

Let's take a look at some of the BSIM parameters for SPICE. Some of the parameters, such as TOX (gate oxide thickness), have physical meaning. Other parameters, such as K2 (the drain/source depletion charge-sharing coefficient), are purely curve-fitting parameters.

There are some effects that are not modeled in the BSIM equations. The source/drain current parameter, JS, uses only the diode's area (Amps/m^2). There is no parameter for the perimeter's diode current. Similarly, one of the junction capacitance models requires two parameters to physically fit the data, but the BSIM model provides only one parameter.

There are many limitations in extracting SPICE model parameters. As mentioned previously, some of the parameters are correlated. That is, they are not independent. For example, TOX interacts with VFB (flat band voltage) and gate capacitance parameters (CGDO and CGSO). Interactions make it difficult to arrive at a best-fit set of parameters, since having one parameter at a non-realizable physical value can be balanced by another parameter at a balancing non-realizable physical value. Another limitation is that some parameters are dependent on the test circuit for the FET. For these parameters, a simple inverter might produce one value, while a 4-input NAND will produce a different value, due to different values of Vgb for the stacked N-FET devices.

Another limitation in model parameter extraction is simply making a parameter fit without enough data. Foundries usually do not collect much data outside of the operating range (0V to +Vcc). This makes it difficult to accurately fit parameters such as the source/drain diode current JS. Also, extracted model parameters often do not give much statistical information, and model parameters are often generated only for three corners for each device type. The BSIM model parameters are unable to indicate the

---

6 Google uncovered sites such as http://www.et1.tu-harburg.de/private/kb/download.html (several operating systems) and http://www.rocklinux.net/packages/spice.html (Linux).

interactions between NMOS and PMOS parameters, although some of the parameters are physically correlated (such as TOX).

There are a number reasons why SPICE continues to be such a widely used simulator, in spite of its inherent limitations. The first, of course, is that foundries provide SPICE model parameters to their customers. Next, the averaging of SPICE parameters between on and off states is acceptable for large-swing transitions. Circuit designers use "16-corner analysis"[7] to cover all combinations of the best and worst process and environmental parameters (best/worst NMOS, best/worst PMOS, hot/cold temperature, min/max Vcc). While corner analysis usually results in functional designs, it can also overly constrain designs. Those simulation tools that support statistical modeling reduce over-constraints by allowing engineers to design to a statistical spread, such as three-sigma or six-sigma.

Remember that SPICE was developed for digital integrated circuits. The goal was to accurately predict timing and power consumption of a set of transistors in logic gates and I/O buffers. In the early days, modeling the low-bias, weak-inversion transition region accurately was not required. Analog designers who tried to use SPICE quickly discovered the model limitations. For example, SPICE MOSFET equations usually give good fits to data in strong inversion and in depletion but not in weak inversion, and BJT equations do not correctly model base charge during fast transients. Today, with low-voltage and low-power integrated circuits, MOSFETs are often operating in the weak-inversion region when they are fully on. This led to development of the EKV model [35, 40, 131], which has been implemented in a number of SPICE products. Foundries also support EKV models, generating EKV parameters for a number of simulator tools.

16.3.1 Example 1: Diode Model Verification

A SPICE diode model would appear to be a fairly simple case to verify against measured data. Yet, even this simple case presents complexity. The SPICE diode model represents a p-n junction device. As discussed above, there are some underlying assumptions in the model equations. Understanding those assumptions helps us to understand what accuracy might be obtainable with a model, and where simulation data and test data are likely to diverge.

Forward bias (DC) operation is determined by the parameters IS, N, and RS. On a semilog plot of I versus Vd, the y-axis intercept of the straight line is IS and the slope is (1/N*Vd). RS is then determined by curve fitting at high currents.

---

[7] The total number of simulation corners is 2^N, where N is the number of independent variables.

Figure 16-7 shows a typical semilog plot for a diode. Physically, the slope is changing, from N=1 at low currents to N=2 at high currents. A typical parameter value for N is 1.2 to 1.4 in the forward bias range (about 0.5 to 0.8V for silicon diodes). This means that the model will overestimate current at low voltages and underestimate current at high voltages. If the reverse breakdown voltage is left at its default value, then the model will also underestimate leakage current (reverse bias current).

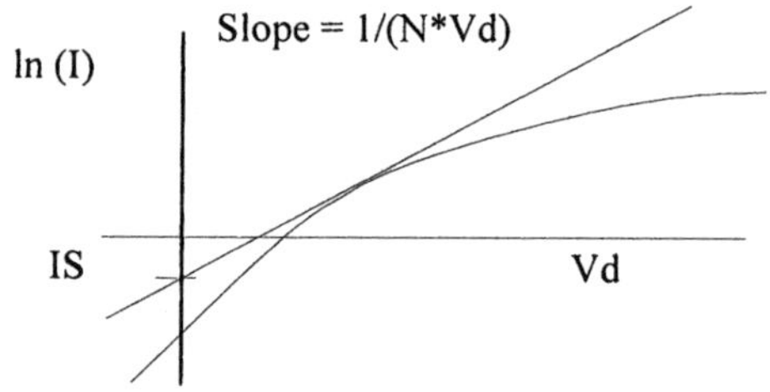

Figure 16-7. I versus V for a small signal diode

The diode transient response is modeled using both depletion layer capacitance (CJO) and charge storage time (TT, related to carrier recombination). The model equations incorporate an underlying assumption that the junction is diffused; with a dopant grading profile that is step, linear, or exponential, making the gradient parameter, M, between 0 and 1. This assumption was valid for early technology. However, this can be violated for advanced technologies that use implantation, which peaks the dopant concentration below the surface. This mismatch is possible for both discrete and integrated circuit diodes.

For diffused diodes, the transient response is well predicted by the combination of CJO and TT [79]. For hyper-abrupt diodes, the prediction is less accurate. For all types of diodes, the frequency domain model could lose accuracy, due to an early bug in hand-coded derivatives preserved in many SPICE implementations. (Since fixing the bug would make the model equations incompatible with existing parameter sets, this bug has sometimes been retained at the request of customers!) Some simulators offer both the UC Berkeley model bug and the corrected diode AC model.

Given these limitations, what diode characteristics should be used in verification? If the diode is to be used for large signal (on/off) switching, then the DC I-V data sets should be compared at the two ends of the switching transient, while transient V-T and I-t data sets (waveforms) should be compared over the entire length of the rising and falling transitions. On the other hand, if the diode is to be used in the frequency domain, then a frequency-domain simulation should be used instead of a time-domain simulation to generate the simulation data set for comparison.

Clamp diode characteristics need to be verified in both forward and reverse operation. The reverse direction represents normal circuit operation, where diode reverse current contributes mainly to power consumption and output capacitance. The forward direction represents the clamping action, where the circuit is outside its normal operating range. As mentioned previously and shown in Figure 16-1, forward bias parameters might not be well characterized by some fabs.

Verification should always cover the full range of conditions to which the diode might be exposed. This includes both minimum and maximum die temperatures, as well as maximum forward and reverse bias conditions.

16.3.2 Example 2: MOSFET Verification

There are dozens of MOSFET models available. Furthermore, the parameter values (and even the number of parameters) depend on the target simulator. For example, HSPICE [10] supports over 50 models just for MOSFETs. Remember that verification verifies the model, the parameter set, and the simulator together.

The SPICE BSIM models[8] were developed for transistor-level modeling of digital circuits, with lesser emphasis on analog characteristics. Consequently, the models do well for circuits having large signal swings, such as inverters and comparators. The BSIM models perform best in the MOS ohmic and strong inversion regions of operation.

The BSIM model performs less well in the weak inversion region. The model transitions between the ohmic and strong inversion regions by applying an exponential turn-off of one and an exponential turn-on of the other during a transition between the two regions of operation. There is no attempt to model the physical characteristics. During verification, this is seen as a good DC fit at low and high currents, with a less accurate fit between the model and test data between these two regions, near where Vgs=Vds. This mismatch becomes larger at higher Vgs.

FET capacitances, which determine frequency characteristics, are of concern to both analog and digital designers. The limitations in the BSIM MOSFET model can lead to mismatches when validating a circuit. The capacitance associated with the charge sharing between the source and drain is modeled by a single parameter, while it physically depends on both Cgs and Cds. This charge sharing parameter can be different for different

---

[8] BSIM is actually a family of models, currently BSIM1, BSIM2, BSIM3, and BSIM4. There are also levels within each model, such as BSIM3v3 and BSIM3v4 [32, 38, 63, 117].

transistors in the same circuit, such as those in the voltage divider shown in Figure 16-8.

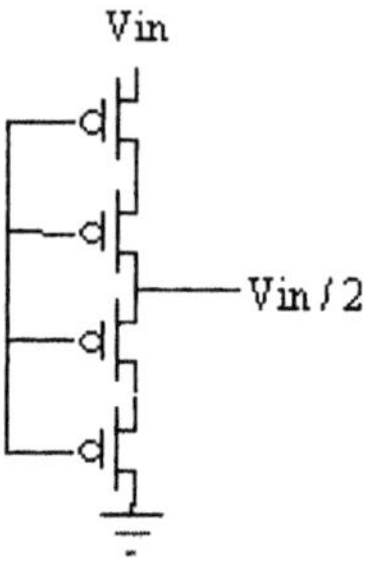

Figure 16-8. MOSFETs with different charge-sharing, but the same charge-sharing parameter.

Validation mismatches appear in many forms. In Figure 16-9, Cgs is frequency dependent. When only the high-frequency behavior is modeled, both transient and frequency validation mismatches can occur for low-frequency circuits. Similarly, other transistor effects that are incompletely modeled will lead to mismatches in circuits that depend on those particular transistor effects.

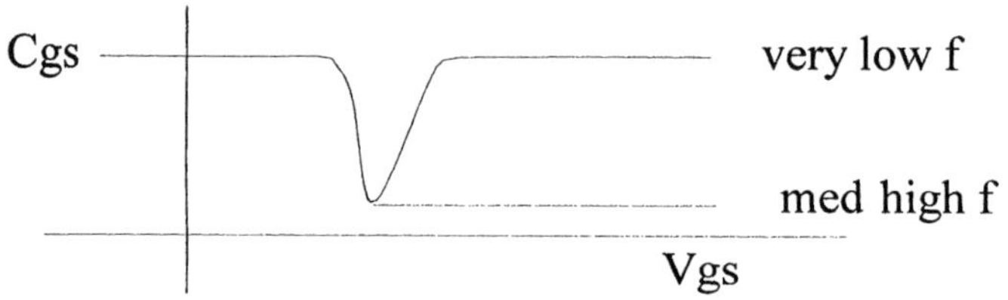

Figure 16-9. Capacitance versus frequency

In modern integrated circuits, built with 90nm or shorter gate lengths, model accuracy is becoming a major concern. Differences of 50% have been reported between simulation and measured circuit performance.[9] In part, this is due to approximations made in BSIM, because these circuits operate in weak inversion. This is further exacerbated by inadequate modeling of the interconnects, such as interconnect inductance, voltage-dependent wiring capacitance, and other high-speed on-chip effects.

In modern integrated circuits, both bipolar and MOS devices are sensitive to perimeter effects, such as junction leakage and high electric field effects. These are generally difficult to model. Many model makers have not put a lot of effort into this area, since circuit operation is normally less sensitive to

---

[9] Discussions at meetings of the Fabless Semiconductor Association (http://www.fsa.org).

these effects. However, as device sizes continue to shrink, it will become more important to include these effects for accurate modeling. Failure to model these will lead to increasing mismatches during device and circuit verification.

Recently, efforts have been stepped up to improve device models. While BSIM4 continues to use a curve-fitting approach, other models have gone back to fundamental physics. Models such as the EKV MOSFET model incorporate such things as conservation laws and Maxwell's equations for all operating regions. These improved models result in improved verification for both transistor devices and circuits incorporating them.

16.3.3 Example 3: Macromodel Verification

Subcircuits and macromodels represent physical parts. Both subcircuits and macromodels are built from simulation components. Generally, a subcircuit represents the part's structure (such as a collection of transistors for an op amp), while a macromodel includes non-physical components, such as controlled current source to represent a transfer function. A subcircuit or macromodel for a functional block, such as an I/O buffer, is also called a model of the part.

The netlist for a subcircuit is usually generated in one of three ways:

- Manual editing of a text file
- Extraction from a schematic
- Extraction from a layout

If parasitics are included when extracting the netlist from the layout, the simulation results are more likely to match the measurement test data. In some cases, such as power devices, neither the SPICE model nor the macromodel approach can provide a sufficiently accurate model [71, 79]. In this case, verification fails, and an equation-based or table-based behavioral model should be considered.

Global nodes might exist in the subcircuit netlist (both ground and nodes declared using ".global"). Using global nodes in model extraction or simulation can prevent simulation results from matching test data in the presence of ground/power noise. For circuits with significant impedance in the power and ground connections, the bounce can be large enough to disrupt system operation. For example, one of us (Green) once saw a board with enough power supply noise to reset the system memory on each clock cycle! Verifying power distribution models for use in model verification is discussed in more detail in Section 16.4.

Verification of a subcircuit requires knowing the process corner for the transistor parameters. It is rare that an actual part corresponds exactly to one of the process corners in the device model library. Sometimes, it is possible to obtain the device model parameters for the physical part. At other times, one must be satisfied with an envelope test, showing that the measured performance data for the subcircuit lies within the best case and worst case (such as 17-corners) simulated data boundaries.

In some cases, the measured data will not lie inside the simulated data boundaries, due to interactions between stages of the circuit. When this happens, it is helpful to run statistical (Monte Carlo) simulations to obtain an improved envelope. Obtaining parts from several production runs can be used to create statistical device models for simulation use, and for comparing simulation data and measured test data envelopes.

Verification for functionality includes all operating regions that can be measured. Each input and output pin is checked. If a pin is bi-directional, then it is checked for both input and output responses. Digital functionality is validated by checking data output bits. Analog functionality could include:

- DC and AC input impedance (Iin/Vin)
- DC and AC transfer functions (Vin or Iin versus Vout or Iout)
- Transient response (transfer functions) under loading.

As a general rule, test data should be collected in all of the domains for which operation is specified. Fast Fourier transforms can be used to convert between transient and frequency responses. Particular care should be taken when the subcircuit or test circuit exhibits resonance behavior, making sure there are enough data points near the resonance peaks.

Subcircuits can contain a wide variety of devices, with different characteristics. Consider a BiCMOS circuit, containing both BJT and MOS devices. The BJT current will increase with increasing temperature, while the CMOS current will decrease. As a result, the circuit's I versus T is non-linear, and might even be non-monotonic, peaking at a temperature other than Tmin or Tmax.

As mentioned earlier, non-linear behavior should be verified over a range for each variable (such as Vcc and temperature), rather than at just the endpoints. The range and number of data points will depend on the subcircuit, but should include enough data to verify curve shapes and trends as well as endpoints.

External conditions are also variables that need to be considered when verifying a subcircuit. For example, the switching time of an I/O buffer is strongly dependent on the load type (resistor or capacitor), while the endpoints depend only on the load resistance. Obviously it is impossible to simulate or collect measurement test data for all possible load conditions. It is an engineering judgment call as to how many points are needed. Good engineers follow statistical guidelines and use DOE[10] [41, 95, 107, 122] techniques to assure adequate coverage of a design's performance envelope.

Simulation data can be collected under conditions where measured test data cannot be obtained because the part would physically fail. For example, simulations can cover voltages out to (and beyond) where circuits latch up and devices break down. Since circuits cease to function at these voltages, it is not necessary to verify a model in this region.

Many circuits can, and will, operate outside of their normal range without ceasing to function. This happens, for example, when a reflection arrives on a transmission line, taking the output voltage above Vcc or below ground. This case will be discussed in more detail in the next section.

16.4 VERIFYING PDS MODELS

The power distribution system (PDS) plays an important role in model verification. Most models are created based on ideal node voltages generated by ideal power supplies. For example, during model creation, the ground voltage is always exactly 0.000V everywhere in the circuit. Similarly, a node named Vcc has a constant voltage everywhere.

A physical implementation always has non-deal connections to the power supply. The resistance of the power supply, power planes, vias, routing traces or wires, and the part's package — all of these generate voltage drops, often referred to as IR drops. When the current is changing (di/dt), the inductance of these interconnects produces additional voltage changes. In high-speed designs, the inductive drops are much larger then the IR drops. The result is a decrease in voltage difference between the local "power" and "ground" at a part. These voltage changes can be large enough to lead to design failure.

Terminology becomes particularly important when discussing power distribution. For example, a physical design can have multiple "grounds," such as digital, analog, chassis, and earth grounds. Depending on the system, the difference between a pair of ground points can be anywhere from millivolts for a precision analog circuit to several kilovolts for a power station.

The first rule of verification is that there is only one physical ground (reference) point in any system. Often this point is taken as the lower voltage (return) pin of the power supply, but it can be any point in the system. *All*

[10] Design of Experiments: a technique for designing tests to obtain maximal information.

other voltages must be measured relative to this single reference point. For high- speed parts, where the distance across the PDS is greater than 10% of the wavelength, only "local" voltage differences have physical meaning.

The second rule is that performance of any part (such as an I/O buffer) depends only on the local voltage difference between Vcc_local and Gnd_local. The output current of an I/O buffer (driver) depends on the local difference at the on-chip Vcc and GND metal for that I/O buffer circuit. Similarly, the input threshold voltage for a receiver depends on the Vcc and GND voltages on-chip at that receiver. Anything that causes a local voltage difference to change will affect the operation of the part.

Figure 16-10 shows how interactions occur in a PDS: the current from both I1 and I2 will affect the local voltage (Vcc-GND) seen at I1. Any time there is a shared current path, even if it is a thick plane, the power distribution currents affect all of the parts on the board, including integrated circuits. This is why precision analog designers often use "single-point grounding" to prevent inadvertent coupling through power and ground interconnects.

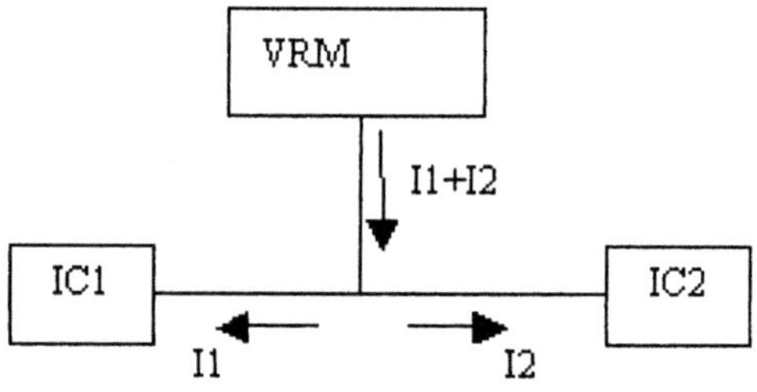

Figure 16-10. Power distribution currents leading to voltage drops

A power distribution system must be analyzed for all connections from the power reference to the ground reference. This power distribution model is later used in simulation, improving the quality of the verification for the part being modeled. A case of particular interest to printed circuit board designers is shown in Figure 16-10, where the wires represent all current paths for the power distribution. A similar set of wires would represent the ground distribution, but are not shown in the figure. Power is supplied by a voltage regulator module (VRM) mounted on the board. The board also contains integrated circuits and passive devices. In this design, the power distribution system has two reference points: the GND reference at the negative pin of the VRM, and the Vcc reference at the positive pin of the same VRM.

Normally, the passive interconnect effects are modeled separately from the IC currents, allowing one to do what if tradeoff studies on I/O buffer selection and signal scheduling. The power distribution across the printed circuit board to an I/O buffer on IC1 has several major components. These include:

- The power supply itself, including its internal resistance and inductance, and its package parasitics.
- Connection from the power supply to the board (including such things as via fields, connectors and solder bumps).
- The interconnection from the board to the power pins of IC1 (planes, traces, and vias).
- Passive components, such as decoupling capacitors, including their package parasitics.
- The chip package pins on each IC for power and ground.
- The on-chip routing on IC1 (metal traces on the silicon).

Some of these voltage drops are more significant than others. The di/dt effects dominate over the IR effects even at relatively low frequencies. For example, one of us (Green) once measured a 2V drop on a local 5V supply and a 2V rise in local ground on a 10 MHz board lacking bypass capacitors, as a result of di/dt effects. As a general rule, the longest and narrowest parts of the interconnect contribute the most to the power distribution drops. In most board designs, the on-chip metal is the smallest contributor, the planes and vias are intermediate, and IC packaging and bypass capacitors are generally significant enough to require careful modeling during verification.

The effect of package pins can be quickly estimated. Consider the case of a 1.5nH pin supplying an I/O buffer that switches 10mA in 50psec. The voltage drop across the pin is L*di/dt = 1.5e-9*10e-3/50e-12 = 0.30V. During a rising output, local Vcc will droop by 0.30V, and during a falling output local GND will rise by 0.30V. Low-inductance [106, 136, 137, 138], bypass capacitors (including power/ground plane pairs) are used in many designs to reduce the fluctuations in local Vcc and GND voltages.

It is harder to estimate the effect of via fields. Clearly, if the power plane has overlapping via cutouts, then the current must go around the entire cutout region, greatly increasing the plane's impedance. Dense via fields (similar to Swiss cheese) also have high plane impedance. The higher the percentage of a region that is blocked with vias, the higher the plane impedance. This implies, for example, that the power distribution under a large flip-chip will have a high impedance. To reduce the impedance, a layout should use small vias with large separations. Where it is necessary to have dense via fields, the power distribution impedance can be improved by having a parallel path for the current, such as a region of metal flooding tied to the same voltage. Another alternative is to use blind or buried vias to reduce the via density in the power planes. Likewise, in the production of the board, how well the supplier plates out the ground and power supply planes in manufacturing them is going to have a big effect on actual performance.

Contributions from core current and I/O buffers are both significant to voltage fluctuations on the planes. To save simulation time, an entire IC (other than the I/O pin(s) being verified) is often modeled as a single current source, and often approximated using a triangular or trapezoidal waveform.

If the rail voltage (power or gnd) changes significantly, then the I/O buffer will deliver less current, slowing the output transition. If there is large noise in the PDS, the signal could even fail to switch completely at the receiver. Careful modeling of the bypass capacitors, including their package and internal series resistance and inductance, is needed when modeling the rail bounce and the resulting droop in I/O currents.

I/O buffers are not the only devices sensitive to rail bounce. Memory devices have been known to automatically clear themselves in the presence of large ground bounce. Power substations can see over 1kV of ground voltage from one side of the station to the other. Satellites and airplanes can have chassis voltages above 1kV relative to the earth ground below them.

A related effect is SSN or simultaneous switching noise. (This is also known as SSO, or simultaneous switching outputs). In this case, an I/O buffer drives normal current when it switches and all other I/O buffers are not switching. However, when other I/Os are switching in the same direction, the rail voltage changes by N*dV, where N is the total number of switching I/Os and dV is the droop caused by one switching alone.

SSN can cause several effects: for example, increased coupling between I/Os, and slower delay. When there are large package inductances, large numbers of signals switching simultaneously, or high-current buffers, then it is important to model the power distribution system during the model verification process. Things that must be modeled:

- Power distribution from the power supply to the chips involved in the verification simulation;
- Power distribution within the package;
- Crosstalk from power/ground traces on the package and PCB to the signal being simulated;
- Number of simultaneously switching buffers;
- Buffers that switch near the time of the signal of interest;
- At very high speeds, power distribution on the chip should be modeled.

There are specialized tools for extracting the parasitics associated with a PDS. The distribution can be extracted as a set of transmission lines or as lumped elements. When using lumped elements, each lump must represent no more than 1/3 of the distance a signal travels during Tedge. For example, a 100-psec edge in FR-4 (160psec/in) requires a lumped model for each 0.21-inch (1/3*100psec/160psec/in).

Verification of the PDS model is done by comparing simulation and test data for various Vcc and GND pins on the board. Once the PDS model has been verified, it can be integrated into the validation flow. The PDS model can then be used to represent the PDS in an SI simulation.

16.4.1 PDS Interconnect Extraction and Testing

During PDS extraction and testing, careful attention must be paid to global nodes such as Vcc and ground. As we have seen, the voltage varies across the power and ground planes, so there is only one reference point in a system. In particular, there is only one point for "ground." All other ground voltages will differ from this reference point. The reference point is often taken at a physical location that is easily reached, such as the pin of a connector or a voltage regulator.

During PDS model extraction from a layout, care must be taken with use of the global node "0". Some simulators treat GND as another name for the global ground node. There are two ways to address the problem, depending on how one wishes to treat the concept of ground. At least one ground node is required, since it is the reference node for the simulation circuit.

The first, and most common, method is to treat all ground points as equivalent, and place all PDS effects into the power distribution. The ground plane is then treated as an ideal plane. In this case, inductance loops are calculated for the ground and power plane pair, and the equivalent inductance is placed into the power supply network. At the package level, the individual power and ground parasitics could be separated to calculate the internal voltages for a part.

The second PDS modeling method is to treat the ground plane parasitics separately from the power plane parasitics. This is used when the two planes are not continuous, such as when one plane has a cutout area or a split. In this case, each ground connection point (such as a pin or via) has a unique name. For example, the two sides of a connector pin could be named "GND_Pin1_A" and "GND_Pin1_B". Similarly, Vcc uses a unique name for each connection point. The extracted PDS model must not have any statements declaring node names as global (for example, a .GLOBAL statement in HSPICE), to avoid accidentally shorting nodes inside a subcircuit to other nodes of the same name.

With either method, the plane pair needs to be subdivided. There are two fundamental approaches to model the interconnect's points. The points may be taken at physical points across the physical design, or only at component pins and vias. In some cases, it is preferred best to use a combination of the two approaches. The coarser the grid, the faster the extraction of the equivalent circuit and the simulation speed for the model. The finer the grid, the better the accuracy. As a general rule, the length of a grid cell should be smaller than 1/6 of the edge distance (Tedge*velocity) for lumped plane models. Larger cell sizes would need to be modeled as transmission line equivalents. In SPICE, for example, RLC circuits simulate much more efficiently than transmission line circuits, so a moderate grid cell size could actually reduce simulation time.

The connection between adjacent pairs of points can then be modeled as discrete (RLC) equivalents, or as distributed equivalents (transmission lines); with an RLC model simulating faster than a transmission line model for the same number of points. Once a model has been extracted, it can be translated to other formats, such as Touchstone S-Parameters, to meet tool requirements.

While automated extraction tools default to naming all ground pins the same, and all power pins connected to the same voltage supply the same, this can often be disabled. This forces the tool to give each node (point) a unique name. In other cases, it is necessary to manually edit pin names for those chips or devices that are being used in your simulation.

16.4.2 Using the Verified PDS Model

A verified PDS model is often required for verifying a model against test data. Test data is obtained using a physical design, which usually includes power distribution as well as the part being tested. When the PDS model is not included in the simulation of the part's model, there will be a lower-quality correlation between the part's simulation and hardware test data.

A test board might contain more than one part. As a result, the current in the other parts can interact with the PDS, changing the local Vcc and ground voltages at the part under test. If the current-versus-time characteristics are known for this second part, the current could be modeled as a time-dependent current source for faster simulation.

Some models, such as IBIS, do not include information on sensitivity to PDS voltage noise. In this case, the part's response to SSN (simultaneous switching noise) or other PDS noise cannot be verified. However, it might be possible to show that PDS voltages stay within tolerances, which in turn limits PDS noise on that part of the test board.

When checking simultaneous switching noise (SSN), it is important to model the interaction between switching devices and the PDS. To do this, each model must be connected to local power and ground supply nodes. These nodes allow the simulator to calculate the current flowing from the

PDS into and out of the part. For example, to see SSN effects, IBIS models would need to be in IBIS 4.1 using either SPICE or an AMS format, with local power and ground pins defined on the subcircuit. Since node "0" (ground) is a global node, the model must not contain node 0; a unique node name would be required (such as node "gnd0"). This unique node would be connected to the PDS node at the part in the netlist, while the simulator's node "0" would be located at the voltage regulator. The simulator is then able to calculate the time-dependent value of Vcc_local-GND_local for each part in the design.

16.5 VERIFYING IBIS MODELS

There are thousands of IBIS models available. The majority of the models are for I/O buffers. Models for terminators, packages, and modules (such as daughter cards) are also available. IBIS models for interconnections can use SPICE, matrix, and S-Parameter formats, among others. While IBIS models are portable, in the sense that signal integrity simulators support IBIS, the models might not perform the same way in every simulator. For example, some simulators automatically add the IBIS package model to the simulation netlist, while other simulators expect that to be done by the user.

Since the vast majority of IBIS models are developed from simulations, no account is taken of power supply effects. The only power supply effects in an IBIS model are contained in the three corners (typ. min, and max). Because these corners also contain temperature effects, there is no way to separate voltage and temperature effects.

When validating an IBIS model lacking data on power supply sensitivity, it is desirable to minimize the power supply fluctuations at the part. This can be done by minimizing the PDS impedance over the signal bandwidth, using bypass capacitors, decoupling capacitors and coupled power and ground plane pairs.[11]

When performing hardware testing for validation, it is important to minimize the effects of the power distribution system (PDS). Since SSN (simultaneous switching noise) can cause strong voltage variations in the PDS, an IBIS model should be based on test data made with only one I/O buffer switching. It is possible that this might not be feasible for some designs.

---

[11] Cadence Power Integrity.

16.5.1 IBIS Verification Tests

The IBIS verification test suite includes DC tests for I-V currents and transient tests for V-T waveforms, made at each of the three IBIS corners. Reflection tests can be used to determine the value of C_comp; measurement results can also be used to fine-tune the value of C_comp. Since IBIS does not model the currents in the power and ground pins, these currents can be measured but not verified. The currents can be verified for other models, such as SPICE or ICEM.[12]

The DC I-V test measures the total I/O current. When verifying the I-V tables in the IBIS model, one checks the I/O currents as follows:

I_up = [Pullup] + [POWER Clamp] + [GND Clamp]
I_dn = [Pulldown] + [Power Clamp] + [GND Clamp]
I_recvr = [Power Clamp] + [GND Clamp]

For transient tests, it is important that the same effects be included in the test system and in the model. If the measurement is done on a packaged part, then the IBIS simulation must also include package effects. The simulation must also include the effects of the measurement probe, including its loading on the test circuit and its bandwidth limitations. Bandwidth limitations of test equipment are particularly important for I/O buffers. For example, a 500-MHz test system is unable to capture the details of a sub-nanosecond transition waveform.

For V-T testing, the test circuit's load should match the load specified in the V-T table being verified, including the impedance of any transmission line connecting the I/O buffer to the load and probe. The model is verified at each corner. IBIS models have data only for three corners (typ/min/max). The IBIS specification does not specify how model tables are used with independent changes in voltage, temperature, and process, so comparisons are not possible at all possible statistical corners. Waveforms can be compared using the Figure of Merit (FOM) equation (after shifting the time axis to maximize the FOM).

16.5.2 Comparison Issues

There are a number of causes for comparison issues. As mentioned above, issues often arise in the setup for high-speed testing. Sometimes there are interactions (crosstalk) between the traces to the I/O being tested and

[12] The latest ICEM 62014-3 specification draft can be found at: http://www.lesia.insa-toulouse.fr/~emccompo/download/report/icemcdv.PDF The parent organization page can be found at http://www.ic-emc.org/.

neighboring traces on the test board, but the interactions were not included in the simulation. Another common problem is having the ambient at the model's temperature, rather than the die; the die temperature can be tens of degrees above ambient. The test probe and equipment can pick up noise from the environment, as well as load the circuit being tested. The issues related to test setups are covered in more detail in "Chapter 17, Verifying IBIS Model Accuracy by Using Laboratory Measurements."

The parts provided for testing are from a statistical spread, and thus the transistors and diodes might have different model parameters from the SPICE process corner models. If enough parts are tested, the statistical spread can be determined. More important, from a verification point of view, is that simulations with min and max IBIS models should bound the measured DC currents, the edge rates (dV/dt), and the V-T waveform shapes.

In addition to the loads specified for the V-T tables, the model can be verified under various reflection conditions. This is used to verify such things as ringing being sufficiently controlled by the clamp diodes.

The I-V tables in IBIS cover the range –Vcc to +2Vcc to cover all possible reflection voltages. In a transmission line system, a strong driver switching low-to-high could launch a voltage Vs (nearly Vcc) into a transmission line. The high-impedance receiver at the other end of the line reflects the voltage back towards the driver (reflection coefficient $\approx +1$). If there are no clamps on the receiver, then the voltage at the receiver doubles, becoming +2Vs (nearly 2Vcc).[13]

Similarly, a strong driver switching high-to-low launches a voltage Vs (nearly -Vcc); the high-impedance receiver reflects the wave; and the voltages at the receiver and driver could approach -Vcc. If the edge rate is fast enough, the transient voltage (at the driver or receiver) could potentially reach 2Vcc or –Vcc before the clamps switch on far enough to limit the voltage. Performing transient open-circuit/short-circuit load testing allows verification of the IBIS driver model for worst-case reflection conditions.

In some test setups, for example product prototype boards, the test board contains traces with long travel times, compared to the edge transition delay. This can lead to unexpected delay measurement results if the ends of the traces not properly terminated, as shown in Figure 16-11. For example, if the end is unterminated, reflections can take several round trips to damp out, leading to a measured delay of several nanoseconds for a sub-nanosecond buffer.

---

[13] In the PCI interface standard, the driver output is near Vcc/2, and this reflection-driven voltage doubling is used to generate the necessary switching voltage at the receiver.

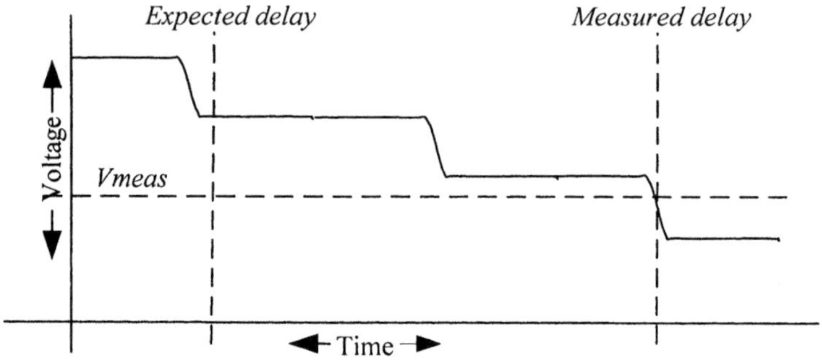

Figure 16-11. Measuring transition delay with reflections

Verification correlates measured data against the model and the simulator together. (The same test data can be used for verifying the SPICE models and SPICE simulations.) If a simulator does not use some of the data available in an IBIS model, such as leaving out the package model, it causes a discrepancy between measured and simulated waveforms. In some documented cases, the IBIS Specification is not clear about how models are to be used in simulation; this includes differential models. If the differential halves of the buffer are completely independent, there is usually good matching with test data. However, if the two halves of the buffer are tightly coupled (example: LVDS), then the matching might not be as good, depending on the way a given simulator handles the coupling.

The 17-corner analysis that can be done with SPICE or test data reduces to three corners in an IBIS model. The IBIS max corner is often chosen based on the conditions for the strongest DC currents (strong PMOS, strong NMOS). This usually corresponds to the same corner conditions for the fastest output transitions (edge rates). In some circuit designs, however, the rising and falling edges might be fastest at different corners, since the PMOS is turning off as the NMOS turns on and vice-versa. For example, a max rising edge might correspond to a process having strong PMOS and weak NMOS, while a max falling edge might correspond to a process having strong PMOS and weak NMOS. Depending on how the IBIS model corners were chosen, the worst-case corner during testing might or might not match the worst case IBIS corner.

To get the best correlation between measurement test data and model simulation data, one must test the test board as well as the I/O part. One can section a printed circuit board to check the manufacturing stackup. It is possible (although admittedly difficult) to obtain S-Parameter data for connectors, vias, lossy traces, and other interconnects using RF or microwave techniques. The interconnect data can then be included in the next simulation, improving the match between measured test and simulation data sets.

A number of studies have been presented at IBIS Summits[14] on the IBIS web site as well as the web sites of the EDA toolmakers. As a general rule, the correlation between measured data and IBIS models is good enough to allow the model to be used for simulation during design. The IBIS comparison to test data might not be as good as to SPICE, due to process and manufacturing variations, as well as the limitations of high-speed test equipment. On the other hand, an IBIS model can correlate better to test data than SPICE does, particularly if the model was generated from test data

16.5.3 Verifying a Design

An I/O buffer is used as part of a design. The I/O buffer interacts with other elements of the design, such as clock and memory signals. In addition to verifying waveforms, it is often desirable to verify timing features such as signal delay or clock jitter.

Unfortunately for the system designer, timing is divided into different domains: part of the timing is due to the internal logic of an integrated circuit (often several clock cycles), the delay through the I/O buffer, the delay along the trace to the receiver, and the internal delay of the I/O receiver. The clock, which starts a timing cycle, has jitter. The logic delay depends on the logic operations and the use of latches to time data flow. The delay though the I/O buffer depends partly on how much it is loaded by the trace. The delay through the trace depends on what other nearby traces are transitioning; this can change delay by as little as 30%, or as much as a factor of 5 or 10. Finally, the receiver's internal delay depends on its loading due to traces and gates on the integrated circuit. The driver and receiver delay can also be degraded by PDS and SSN noise.

All of these considerations regarding timing variations are critical for state-of-the-art parts, such as microprocessors that clock faster than 1 GHz. A few tens of picoseconds difference in timing delays can be very important in the timing budget.

All these variations make calculating the timing budget complex. Simulations can be used to observe the data eye pattern and the BER (bit error rate). The eye pattern can be verified using a fast waveform measurement. The BER can be verified using a BER tester, or by saving the received data to memory and later comparing it to the original data.

Often, data-dependent bit errors occur on just one line of a bus. This is usually associated with crosstalk or delay mismatches. Verification of the

---

[14] http://www.eda.org/pub/ibis/summits/, particularly in recent January and June Summits.

design is used to determine the underlying cause. For example, if the I/O buffer's internal delay and trace delay were not summed correctly, part of the delay could have been double-counted, leading to an error in predicting setup and hold margins; and the longest or shortest bus delay bit would show errors. Crosstalk with other bits on the bus (or any other signals switching at this time) could advance or delay the signal until it was outside of the setup and hold window.

In other cases, the boards might work intermittently, or the yield of functional boards might be low. In this case, verifying the measured test data against simulation can be used to track down problems. One common problem is a statistical spread of I/O buffers, other parts, or trace impedances that is outside of the models' spreads.

Using verification methods to troubleshoot boards benefits both the model maker and the end user. The model maker can correct the model if the model's corner characteristics do not match the physical parts. The end user picks up additional techniques for addressing signal integrity with statistical spreads, and gains confidence that a verified model leads to a verifiable design.

16.6 VERIFYING OTHER MODEL TYPES

All of the techniques described in this chapter can be applied to other model types. This includes interconnect models, as well as other electrical models, mechanical models, sensor models, optical models, and models that combine more than one of these domains. Many of these models are written in mixed-signal behavioral languages, such as Verilog-AMS [37, 69, 129], VHDL-AMS [8, 130] and SABER-MAST [81, 105].

A common requirement is validating an interconnection model. Interconnect models can be expressed in numerous formats, such as RLC circuits, transmission lines, frequency-dependent matrices, or S-Parameters. As described previously, the simulation circuit must include all of the test setup effects; such as the input edge rate and test probe loading. In validating a passive circuit, one can check the transient characteristics, such as a step response, by using an input edge rate that is fast compared to the time constants of the interconnect. A TDR (time domain reflectometer) can be used to gather both reflected and through transmission characteristics. One can also validate a passive circuit by performing a frequency sweep, covering a range from near DC to the highest frequency of interest. Note that the highest frequency of interest will usually be determined by the fastest driver edge rate, rather than the clock frequency. Fourier transforms can be used to convert between time domain and frequency domain characteristics, as well as to deconvolve the effects of the input edge rate and test setup bandwidth, so it is not necessary to collect test data in both domains.

Next, let us consider an example of a mixed-signal, mixed-domain model written in VHDL-AMS. Traditionally, an electrical engineer would simulate and verify the various electrical sections separately: such as a digital control, an analog section, and a power supply. Meanwhile, a mechanical engineer would be simulating the motor for torque and speed, and an aeronautical engineer would be simulating the airframe and control surfaces. The interfaces would be manually checked, but not necessarily simulated. If one can afford over-design, this is not a bad approach – just ask the pilot who did the barrel roll in a Boeing jet over Lake Washington!

On the other hand, this approach can also lead to dismal failure. For example, one of the Mars landers missed Mars completely, simply because one engineer worked in meters while the other worked in miles. When models use only numeric values, as with SPICE and IBIS, it is possible for mistakes like this to occur. A model in VHDL-AMS or Verilog-AMS can include units as well as values, making it easier to catch errors in units and in conversions between electrical and other domains.

Another common challenge in model verification is comparing model data involving feedback from one subsystem to another. This challenge is greater if not all of the subsystems are electrical in nature. For example, modeling of human beings, and verifying the model for a statistical range of persons, is difficult. Recently, models that include human interactions have been created, verified, and used in simulation to design [8], a successful project.

The VHDL-AMS model allows each designer to write a model for their subsystem, and its interactions at its inputs and outputs. One of the advantages of simulation and verification of VHDL-AMS models is that interactions between subsystems, even in different domains, can be checked.

For all model types, whether SPICE, IBIS, VHDL-AMS, SystemC, or even models written in a programming language such as C++, the verification process is the same. First, one determinates the conditions to be used for both simulation and for hardware testing. This includes environmental variables, such as temperature, and system inputs, such as voltages or logic states. Next, one determines which system responses are to be used for verification; this could include a temperature, a voltage, or any other measurable output.

Next, one selects the hardware test bench. This could include a stand-alone part, a single printed circuit board, or something as complex as an automobile (sensors, CPUs, mechanical brakes, gas pedal, wiring harness, etc.). The simpler the test bench, the easier to verify a single model. As design proceeds, a simple model might be replaced with a more complete model, allowing additional verification. A more complete test bench makes it possible to verify interactions between subsystems.

16.7 SUMMARY

Verification compares model simulation results against hardware test data, making model verification the final step in modeling. In traditional engineering design, the prototype provided test data that verified both the component models and the overall design. The verification methodology described in this chapter can be used for SPICE, IBIS, and other model types. Verification can be done against a single model (stand-alone) or against a model within a design.

Verification verifies a model for its intended use. This means collecting test data from a test board with a typical circuit, rather than testing the part in isolation. The test board must be designed so that unintended artifacts, such as reflections, do not "hide" the data needed to verify the model. The test system must have low probe parasitics and a sufficient bandwidth to minimize unintentional impact on the test circuit and test data.

During simulation, the simulation circuit model must take into account all test circuit effects, such as probes and the power distribution system, in order to get a good match between the model and the measured performance. Of course, if a model (or a physical component) is used outside its intended operating region, the model verification can fail completely.

Model verification can be done at many stages, from component design to final engineering system implementation. Verification can be done for transistors, components, modules, boards, and even entire assemblies. Although each of these models might be implemented in a different modeling language, from SPICE to IBIS to VHDL-AMS, the methodology is the same - design the hardware test, run the test, and compare the results to an equivalent simulation.

The earlier verification is done, the faster design can proceed. Less redesign is needed if a problem is found. Once a model has been verified, it is not necessary for each design group to independently verify that model.

Chapter 17

VERIFYING MODEL ACCURACY BY USING LABORATORY MEASUREMENTS

Laboratory measurements are used for checking models and simulations against the hardware prototype

Abstract: Potential sources of errors in both measurement and simulation are many, and should be resolved in order to achieve verification. In the verification process, engineers ensure that their model and simulation results accurately compare to independent laboratory measurements.

Engineers traditionally verified models in the laboratory with the following process: obtain physical units, measure their model parameters, simulate their behavior with their measured parameters, build circuits with the measured units, measure the circuits in the laboratory, and correlate the measurements with the simulation. This process is expensive, tedious, and time-consuming work. A more efficient alternative is to model and simulate a population of devices, assemble a representative sample of the devices into multiple test circuits, measure the population of test circuits, and then compare it to the population of simulation results.

17.1 INTRODUCTION

Correlation means to compare and demonstrate that two items agree. When comparing simulation to laboratory measurements, the measurements are judged to be *true*, but that is not always the case. When we compare two items, we provide supporting data on the measurement accuracy and say whether the measurements agree or disagree within a known percent. What is considered an acceptable percentage of accuracy depends on the design's needs and objectives. When we want to achieve tight accuracy, we need to understand the possible sources of error.

In this chapter, we consider:

- Several sources of common errors. They include errors caused by incorrect measurements, process variations, and incorrect definitions.
- Two ways of verifying models:
  - By correlating simulations and measurements on a device–by-device basis.
  - By correlating the statistical spread of the measurements with the statistical spread predicted by modeling and simulation.
- An example of the development history of semiconductor modeling. Knowing how modeling actually developed leads to a more practical and forward-looking attitude about model accuracy.

17.2 INSTRUMENTATION LOADING AS A SOURCE OF ERRORS

This topic discusses a what-if simulation concerning oscilloscope loading and measurements. The what-if simulation illustrates some of the effects that instrumentation has on physical circuits [9, 39, 76, 112, 114]. Signals with considerable high-frequency content are especially sensitive to instrument loading. These loading effects are easily observed in a test laboratory. The loading effects can be modeled and simulated in the virtual, computer-aided-engineering design world. Understanding the computed results provides an excellent learning experience.

This simulation shows that the act of measuring an electronic signal actually changed the electronic signal.[1] The change is significant and measurable unless care is taken.

17.2.1 Simulation 1 (Baseline)

The topology in Figure 17-1 is simulated to illustrate the effect an oscilloscope probe has on measurements of switching waveforms in the circuit. The simulations are run to show rise-time portions of the waveform. Clock frequency is 50 MHz and the driver model used is a CMOS_IO_ideal model that contains V-T tables with sharp corners. The fast-switching selection of the model population is used.

The circuit in Figure 17-1 consists of a driver and a 72-ohm line with a 0.5ns delay and a 72-ohm ideal load. The driver's rise time is the same

---

[1] Heisenberg's Uncertainty Principle says that measurements change the quantities being measured.

magnitude of the delay (about 0.5ns). The driver is nearly ideal with a very low output impedance and the Vcc at 10% high (5.5 volts). This experiment illustrated an interesting case where any mismatch reflections became intermingled with the rise time.

The simulation results without the oscilloscope probe attached at the receiver are shown below in Figure 17-2.

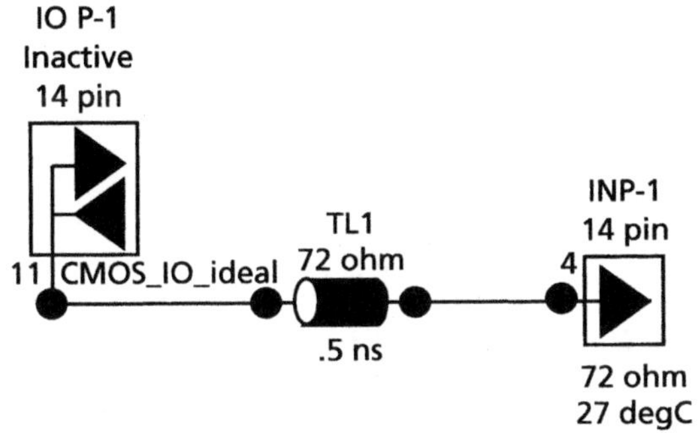

Figure 17-1. A simple net topology

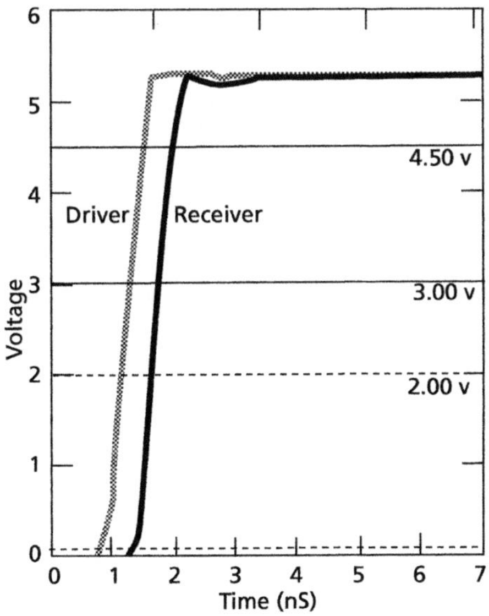

Figure 17-2. Reflections for simple topology of Figure 17-1

17.2.2 Simulation 2 (Heavily Loaded Probe)

A virtual oscilloscope probe is next modeled to show how a passive input probe of 2pF input capacitance (7pF total loading capacitance) and a 10:1 divider ratio might look on the oscilloscope screen. The oscilloscope screen is represented as the IN-1 receiver. This model is created by adding the extra receiver and its input circuit. Probe lead inductance for a 6-inch ground lead is modeled, but not the EMI noise pickup effects that could also result. The circuit topology for this setup is shown below in Figure 17-3.

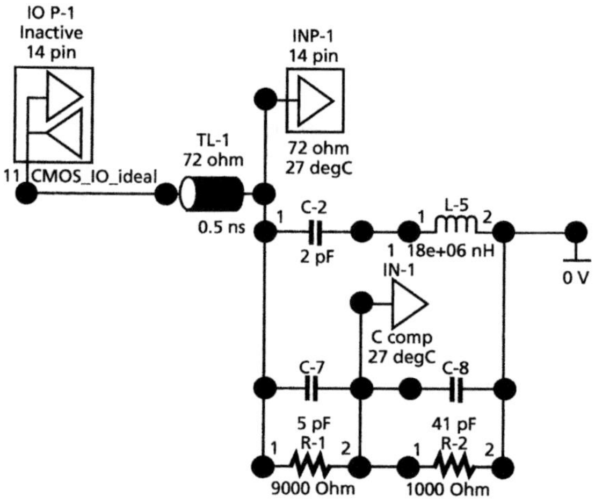

Figure 17-3. The simple net for Figure 17-1 with a modeled oscilloscope probe

Consider first the loading that the insertion of different probe capacitances presents to the circuit. This loading is shown by the capacitive reactance |Xc| listed in Table 17-1.

Table 17-1. Loading introduced by probe capacitance

Frequency (MHz)	**Loading: |Xc| (ohms)**		
	2.5 pF	**5 pF**	**10 pF**
625	101.92	50.96	25.48
1250	50.96	25.48	12.73

In this case, a real receiver would present very high real input impedance in parallel with its input loading capacitance, which would then be in parallel with the loading of the probe. If the real receiver had clamps (switched low-impedance shunts) and was driven into the non-linear region by line reflections, we would have a highly non-linear behavior captured by the I-V tables of the IBIS model.

Next, we calculated Reflection Coefficient, ρ (rho), given by:

$$\rho = (ZL - Zo)/(ZL + Zo) \tag{17-1}$$

Example: 2.5pF @ 625 MHz

$$ZL = 1/[1/72+j/101.92] = 58.8\angle{-35.18}° = 48.09\text{-}j33.83 \tag{17-2}$$

$$\rho = [48.09\text{-}j33.83 - 72]/[48.09\text{-}j33.83 + 72]$$
$$= [-23.9\text{-}\ 33.83]/[120.1+j33.3] \tag{17-3}$$

$$|\rho| = \sqrt{1715.68/15568.48} = .33 \tag{17-4}$$

These results are shown below in Tables 17-2 and 17-3:

Table 17-2. Composite load impedance

Frequency (MHz)	**Load Impedance, ZL (ohms)**		
	2.5 pF	**5 pF**	**10 pF**
625	58.8∠-35.18° 48.09-j33.83	41.7∠-54.67° 24.14-j33.97	24∠-70.48° 8.04-j22.66
1250	41.7∠-54.66° 24.14-j33.97	24.0∠-70.48° 8.04-j22.66	1.25∠-80° 0.217-j1.23

On our 72-ohm line we get:

Table 17-3. Resulting reflection coefficients

Frequency (MHz)	**ρ (magnitude)**		
	2.5 pF	**5 pF**	**10 pF**
625	0.33	0.577	0.817
1250	0.577	0.817	0.988

How does the simulator present this result? Remember that we have a 10:1 probe so the trace at the bottom of Figure 17-4 is what appears on the oscilloscope screen. The traces show the driver, receiver, and probe voltages to the same voltage scale. The oscilloscope trace is attenuated 10:1, and is rolled off due to scope bandwidth. The simulation also shows some additional ringing at the actual receiver. Looking at the ringing now present at the receiver, notice the additional mis-matching of the line introduced by the probe. However, the rolloff due to the oscilloscope bandwidth shows a very different picture.

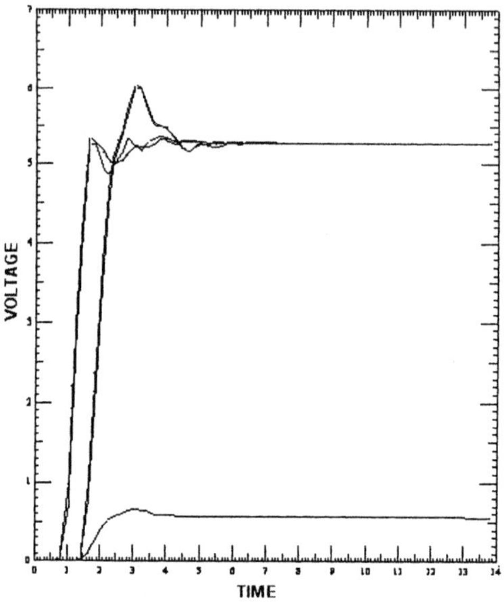

Figure 17-4. Reflections with passive oscilloscope probe

17.2.3 Simulation 3 (High Frequency Probe Load)

The second virtual oscilloscope looks more like how an active input probe of 1pF total loading capacitance appears. A 2:1 divider ratio gives a better look at what the viewed waveform looks like. Probe lead inductance for a ½-inch ground lead with no EMI noise pickup is modeled in Figure 17-5.

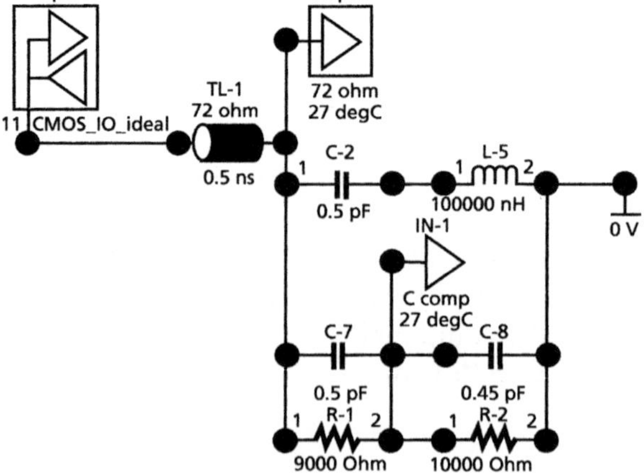

Figure 17-5. Reflections with active high-frequency oscilloscope probe

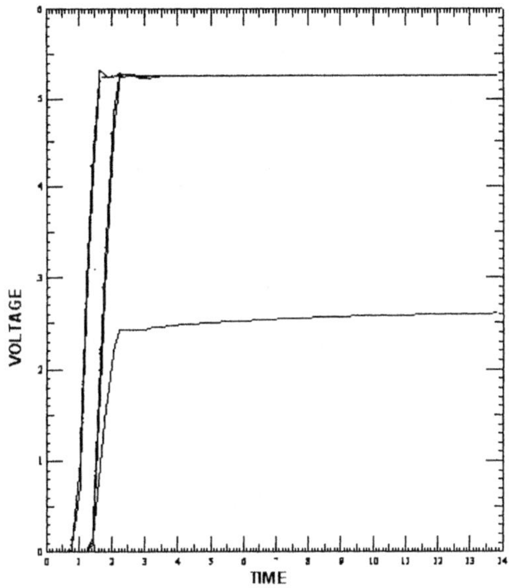

Figure 17-6. Reflections on a simple net with high-frequency oscilloscope probe

As indicated by a reflection free and flat oscilloscope waveform, that probe may not be perfectly compensated. But compensation is close to perfect, as shown by the simulation results in Figure 17-6.

17.3 OTHER TEST SETUP ERRORS

The effects of attaching an oscilloscope test probe to a circuit may be familiar to readers as a source of introducing possible measurement errors. Additional errors may be caused by inaccuracies in the test equipment due to offsets, non-linearities, ambient temperature, and more. Modern test equipment from reputable instrument manufacturers is usually well designed to give accurate results. However, it pays to read the instrument's specifications on expected accuracy and to ask when it was last calibrated!

Sources of less familiar errors may cause additional problems in the test setup. The frequency content of today's high-speed signals is a particular problem. A newer, but just as important problem is the high switching surge currents demanded by today's large ICs. Consider:

$$v = L\frac{di}{dt} \tag{17-5}$$

It does not matter whether dt is becoming very small or di is becoming very large. If L is a leakage coupling inductance, a significant interfering noise voltage can be induced in a measurement probe.

The signal and power-bounce edge-rate frequencies are well into the range where coupling between cable signal wires and power supply wires can be quite a problem. Good signal and test cable layout for high-speed testing is a science in itself. Even coaxial cables have leakage through their braided outer ground sheath to nearby cables. Bends caused by routing increase the leakage. Also, in a test setup, there may not be room for a lot of coaxial cables. Alternatives to coaxial cables are shielded and twisted pairs, ribbon cables with ground shield wires, and much more. Straight ribbon cables with no shielding provisions are usually a pretty poor choice.

Whether a shielded cable is grounded at both ends, one end, and which end also makes a difference. So too does the frequency of the signals being measured and connected. A cable-grounding scheme that is good at very high frequencies is usually not the correct choice at lower frequencies.

Connectors are often a problem. They can couple signals internally, or from a PCB net routed near them. Those unwanted signals can be coupled onto the ground sheath of a cable. Once the signals are there, the relatively long cable lengths can form a very efficient radiating (or pick up) antenna. Signals on ground sheaths are common-mode. That is, the outside world usually does not see a nearby return current that differentially balances the tendency to radiate (or pick up) energy.

The signal and power-bounce edge-rate frequencies are well into the range where resonances can be supported—on a power plane, inside an enclosure or cabinet, or with a test setup tabletop. Hence, how a test board gets placed, enclosed, and oriented in a test setup can affect its operation and measured test results.

Ask EMI engineers how many times they have struggled to get repeatable measurements with a given test setup. If someone walks across a room or waves a hand around a test setup and sees measurements change, then coupling and pickup effects are present. On a dry day, a Styrofoam cup (with or without liquid) picks up a static electric charge that affects measurements and may even cause ESD damage.

17.3.1 Conclusion

Modeling and simulation of a test setup is much easier to control than an actual test setup. Consequently, modeling and simulation is more repeatable and "reliable" than laboratory measurements. If real world laboratory effects are not included, simulated results will be "wrong." That is why we need the check of test measurements plus resolution of discrepancies to ensure that simulations and measurements make the same assumptions and represent each other well.

It would be desirable if most EDA simulation software suppliers provided a library of virtual test instruments to use with extracted topology simulations. This would be an excellent aid in debugging correlation test setups. Additional applications notes about test and measurement can be found at:

Agilent
http://www.measurement.tm.agilent.com/index.cgi?CONTENT_ID=3606 &User:LANGUAGE=en

Alex's Electronic Resource Library
http://htdconnect.com/~alexx/lib/test.htm

RF Café
http://www.rfcafe.com/references/app_note_links.htm

The first objective of this study of oscilloscope probe loading is not to model probes accurately, but to justify using higher frequency probes where needed. Modeling the probe is a fallback solution if higher frequency probes are unobtainable.

17.4 SIGNAL NOISE AS A SOURCE OF ERRORS

Signal noise can be a source of measurement error in both simulations and measurements. For example, consider correlating rise-time. A portion of the ringing due to power- and ground-bounce and non-zero supply-point impedances will appear on the output of the driver. So will crosstalk and noise from nearby drivers. These effects can be small, or they can significantly change the measurements [114].

How well measurements correlate with simulation depends on whether the above effects are present in the physical prototype, and how well the effects have been modeled and simulated.

For example, if rise time is defined from a steady state low (0%) and high (100%), what happens if the steady state is never achieved or interpreted differently by different observers?

17.5 MEASUREMENT DEFINITIONS AND TERMS AS A SOURCE OF ERRORS

17.5.1 Rise Time Definitions

Table 17-4 is an example of discrepancies involving rise time measurement and the effects of different *definitions* of rise time. All measurements at a particular process corner are of the same waveform. Only the definitions have changed, leading to different results.

Table 17-4. Rise- and fall-time definitions applied to a sample device

Definitions		**Process Corner**		
		Slow (ns)	**Typical (ns)**	**Fast (ns)**
Traditional 10%-90% of low-high state	Rise	5.99	3.84	3.0
	Fall	5.84	3.14	1.23
IBIS, 20%-80% of low-high state	Rise	3.72	1.99	1.6
	Fall	3.58	1.43	0.886
"New" t_t, 1.3V-1.8V 30%-70% of low-high state	Rise	1.99	1.23	1.16
	Fall	1.69	0.913	0.627

Table 17-4 shows the results of measuring the rise- and fall-times of a single, well-behaved waveform using the different definitions of *historical, IBIS and "New."*[2] The measured rise and fall times depend on which definition is used. The rest of this section presents the different definitions for rise time and other switching speed terms. The differences arise because the definitions *are* different. Understanding measurement definitions is important to avoid error.

17.5.2 [Ramp] Definitions

We start the IBIS definition of *[Ramp].* [Ramp] is expressed as a ratio of the voltage swing versus the time to transition this swing. Full on is zero volts, and full off is the power supply (Vcc) voltage for the conditions when the ramp is measured. In the case of 5-volt CMOS under nominal conditions, the power supply voltage is 5 volts. So in this case, 0% corresponds to 0.0 volts on a rising transition, and 100% corresponds to 5.0 volts when the output is terminated by R_{load}. Per the IBIS definition:

---

[2] As seen on an IC data sheet: t_t, transition time, was documented instead of rise and fall time.

- The ramp is measured from 20% to 80% of the full swing, *absolute*. If a reduced voltage swing (less than 20% to 80% of the full swing) is used, the measured slew rate does not meet the IBIS definition. [Ramp] while expressed as a slew rate (dV/dt) is correct only for a 20% to 80% swing. It is *not* to be reduced.
- If R_{load} is not specified, it defaults to 50-ohms.
- R_{load} is connected to 0.0 volts for rising edge and Vcc for falling edge.
- ECL and PECL technologies—R_{load} is connected to 2.0 volts less than the highest rail for rising and falling edges.
- Other terminations apply for Open_drain and Open_source.

17.5.3 [Ramp] Data Versus Slew Rate Data

Slew Rate is dV/dt. There are two slew rates, dV/dt_r for slew rate rising edge and dV/dt_f for slew rate falling edge. The [Ramp] data is presented as min-typ-max data on slew rates. However, the first source of confusion arises when IBIS files present slew rate data that is measured over ranges different than the 20%-80% range of IBIS. The next source of confusion arises when we realize that slew rate is widely used in many different contexts than IBIS. Slew rate can even refer to the instantaneous rate of change of the voltage waveform.

In IBIS file format, slew rate data (and all data) arranges the data columns typ-min-max left-to-right, not min-typ-max like most simulators and data books follow.

17.5.4 Slew Rate and Rise (Fall) Time

Some simulators use *fast, typical and slow* instead of min-typ-max. Minimum rise-time corresponds to maximum rise slew-rate. Minimum rise-time also corresponds to fast switching (and minimum fall-time, and so forth).

The typ-min-max columns are based on buffer conditions defining typ-min-max (voltage, temperature, and process) and they usually correspond to typ-slow-fast transition times, typ-weak-strong drivers, and typ-low-high rails. Temperature effects are another matter as typ-low-high is the usual case for BJTs while typ-high-low is the usual case for CMOS.

17.5.5 Alternate Definitions of Rise- and Fall-Time

Historical (Data Sheet) Definition

Historically, the definition of rise-time (and similarly slew-rate) was the time to go from 10% to 90% of the rail-to-rail (usually 0 to Vcc) swing at the output of a device–similarly for fall-time. This is still the convention followed by most of the electronics industry on *data sheets*. This definition is appropriate when transmission lines are not involved. There may be some ground and power bounce to account for, but in general, swings at the driver are not from – Vcc to + 2Vcc as in high-speed transmission-line switching.

IBIS Definition

IBIS definition of rise-time (and similarly slew-rate) is the time to go from 20% to 80% of the rail-to-rail (usually 0 to Vcc) swing at the output of a device: Similarly for fall-time. This was adopted because of non-linearities, clamping, and -Vcc to +2Vcc swings on the I-V curves, and long tails on CMS switching waveforms.

"New" Definition

"New" Definition is an alternate definition of rise-time (and similarly slew-rate). It is the time to go from 30% to 70% of the rail-to-rail (usually 0 to Vcc) swing at the output of a device. Similarly for fall-time. It is a wild-card definition seen on one data sheet that used it for a controlled drive (soft turn-on and soft turn-off) device Suppliers are free to define data on their data sheets any way they wish. The problems arise when trying to correlate measurements taken with divergent definitions.

17.6 TWO WAYS TO CORRELATE MODELS WITH MEASUREMENTS

There are two basic verification methods:

- Correlation on a unit-by-unit basis—classical
- Correlation on a statistical basis—a newer idea

The following process is for correlation on a unit-by-unit basis:

1. Measure the model parameters of a device.
2. Use those measurements in the simulation of the behaviors of the devices in a circuit.
3. Use the devices in the construction of one or more breadboards of the circuit.
4. Measure the same type of behavior of the devices in the breadboard.
5. Compare the results of step 2 and step 4.

For an excellent discussion of these steps as applied to high-frequency RF test socket measurements, see [101].

17.7 INVOLVING PRODUCTION IN VERIFICATION

Verifying simulation results to laboratory measurements is very desirable. Sometimes, however, verification is done on too small a sample to give good statistical confidence. We can ask the Production Test and Measurement department to measure a reasonably large sample of results gathered over a production range of component and board populations.

Six-sigma quality programs attempt to eliminate production test and measurement by assuring that the quality is so good that it does not (and should not)[3] need to be tested. Six-sigma quality requires failure rates of less than 10 PPM. Gathering sufficient statistics to *verify* six-sigma quality requires production runs of twenty million units—not very likely with today's production runs.

The measurements and data gathering in a production test area need to be simple and clear. For an example of setting up a successful production environment, see "Appendix H-2, Test and Measurement Realities." Appendix H-2 also discusses setting up test standards devices and fixture units between a supplier's outgoing inspection and a customer's incoming inspection.

17.8 AN EMI/EMC EXAMPLE

In a recent IEEE-EMC Chicago Society presentation, calibration results on a dipole antenna at 160 MHz were reported.[4] Mike Howard of Liberty Labs discussed the calibration of a dipole antenna used on a satellite mission to the planet Mars, called Mars Express. The dipole was placed on this

---

3 Extra testing and handling of fragile six-sigma components creates more problem devices than it finds.

4 http://www.ewh.ieee.org/soc/emcs/chicago/sectfiles/pmeet.htm

satellite and operates from 1.5 to 5.5Mhz. The dipole is pulsed and is used to detect water beneath the permafrost of the surface of Mars. The Mars Express was launched atop a Russian Soyuz rocket in June 2002.[5] Significant test setup measurement differences of 6dB and more (due to antenna orientation and other issues) were concerns.

Interestingly, the speaker said that correlation between modeling and simulating the antenna agreed within 3/10ths of a dB! This means that V2 and V1 agree within a ratio of 1.0351. By contrast, differences of +/- 6 dB are common in test measurement setups, as any experienced EMI/EMC engineer knows.

17.9 CORRELATING UNIT-BY-UNIT MODEL MEASUREMENTS

Unless a component's model parameters are measured and used in the simulation, there is no way of knowing which part of the population distribution the component came from. Plus the act of measuring the parameters on today's sensitive devices and then assembling the devices in a prototype can destroy a device or cause its model parameters to drift. One strong motivation for six-sigma quality is to avoid all acceptance testing of ICs.

Before measurement, there is little chance of knowing all the important measurement issues, such as coupling through power-supply cables and instrument loading. But if correlation accuracies of +/- a few percentage points are the objective, such effects would have to be accounted for in the simulation model. An even better strategy is to reduce stray measurement effects.

Random variation in the manufacture of the PC board itself affects results significantly. For example, the characteristic impedance (Zo) of a run on a layer can vary. And there are variations in termination resistors, connector parasitics, and more. Often we discover discrepancies when a completed virtual PCB design gets processed to a set of drawings and sent to a PCB fabricator to build the board. The board fabricator may tell us that there is no way they can produce the targeted characteristic impedance (Zo) with the specified dielectric, line width, and line spacing. For example, FR4 material has glass fibers in it. With close line spacing, the dielectric close to the

---

[5] The speaker mentioned that he planned to submit changes to an antenna-measurement test-setup standard based on this project.

etched runs becomes non-uniform, and this effect must be accounted for in the simulation.[6]

Errors in understanding how to model a device and how to simulate it in a circuit *can* exist. When finding the source of discrepancies, we may discover that errors can exist in the model, the device data, the simulator, or the measurements. If progress is to be made, we must resolve discrepancies, and improve our modeling, simulation, or measurements. When resolving discrepancies, we should take the view of eliminating special test setup anomalies, if possible, rather than modeling them.

The normal process of good design and measurement tries to eliminate the effects of ground bounce, crosstalk, and poor termination—not mimic them. For an example, see Figure 11-9. The objective in a design is not to mimic such a waveform, but rather to prevent it.

We also need to be clear about our correlation objectives. Will +/- 10% do, or must we have +/- 2%? Is the objective to make a design decision, or to find the limits of computation accuracy on a new EDA tool?

The judgment standards to apply when verifying that a product design came out as intended should be similar to those that are applied in Regulatory testing. There, the 80-80 rule is applied. To qualify a product, measurement data must demonstrate that 80% of the units have an 80% confidence level of being within emission limits.

More certainty is required when defining the capabilities of modeling, simulation, or an EDA tool. It is then wise to do a correlation project *off line.* That is, not connected to new product development. Further, such studies should be done on very simplified, well-controlled PCBs specifically designed to study particular questions. If the modeling and simulation proof-of-tools study is done well, simulating complex design issues can be done with confidence.

17.10 STATISTICAL ENVELOPE CORRELATION

What can we do to avoid the difficulties, delay, and expense of one-to-one verification and correlation studies? We need to make the following adjustments to our thinking:

- Settle for correlation within a reasonable engineering approximation.
- Make use of correlation on a statistical and probabilistic basis.

---

6 See "Appendix H, More Reality Tales about PCB Fabrication," for additional practical insights about measuring an accurate value of Zo.

Statistical envelope correlation should try to simulate and measure a statistically meaningful spread of model parameters, and board manufacturing variations. Individual devices and their model parameters are not measured and tracked. Instead, histograms (or population spread and envelope of expected and measured responses) are generated and compared.

Statistical correlation [9, 16, 70] requires using a statistically significant sample of parts. Accounting for random variation due to the PWB and instrumentation should also be considered. To eliminate chance results a sample size of 25 units (DUTs) or more must be used. Even more sophisticated statistical approaches, such as DOE [41, 95, 101, 122], can be used to gather more meaningful data with fewer experiments.

Statistical correlation usually assumes that the parameters studied are independent. This is usually not the case for either DUT or prototype board. Parameters that are dependently related can be handled statistically by *mixture experiments* in both the simulation and measurement space. But statistical verification is still a good technique if parameters are not too strongly interrelated and variation is random. In fact, if both model simulation and measurements give a Gaussian distribution of results that closely overlay each other, and then the correlation would be very good. Especially if preformed in a prototype circuit that implements the intended use of the DUTs.

Success is achieved when the predicted performance distribution *envelope* shows good correlation with the measured performance distribution *envelope*.

17.11 SIGNAL INTEGRITY AND CORRELATION

The phrase "design objectives" for Signal Integrity and Logic Functionality usually means standard measures, listed in Figure 17-7.

- Rise Time
- Fall Time
- First Switch Delay
- Final Settle Delay
- Propagation Delay
- Noise Margin
- Timing Margin
- Overshoot
- Undershoot
- Crosstalk

Figure 17-7. Standard measures of Signal Integrity performance

These standard measures are commonly used to judge whether a net has met its design objectives. They are usually produced during simulation, design and development of a PCB. At the end of a simulation run, they are presented in tabular format. Also, they are often measured in the laboratory on hardware prototype boards undergoing design qualification and verification.

Bringing together the simulated and measured results for the standard measures is a straightforward instance of correlation: Particularly if the correlation is done on a statistical envelope and sampling basis. For doing the comparison, we strongly recommend using a set of Monte-Carlo simulations and a statistically significant sample of measured units [61, 94]. Correlations may be described with the following terms, depending on how it is measured. In other words, a correlation is:

- *Coarse-grained* if based on standard measures.
- *Finer-grained* if based on waveforms.
- *Even finer-grained* if based on feature selective validation (FSV).

Standard Signal Integrity measures as a means of correlation are usually a course-grained correlation. Their big advantage is that a large set of circuit networks can be correlated at one time.

17.12 WAVEFORM CORRELATION

The IBIS Committee has done some excellent work on documenting waveform-to-waveform correlation for establishing Signal Integrity correlations. This can be found at: http://www.eda.org/pub/ibis/accuracy/ [55]. Their readme.txt file reviews the items available for download from the website. The contents of the readme.txt file are also detailed in "Chapter 12, Fixing Errors and Omissions in IBIS Files."

17.12.1 Figure-Of-Merit

Of particular interest in the I/O Buffer Accuracy Handbook are the Curve Overlay Metric (Section 4.3) and its associated Figure-Of-Merit (FOM), shown in equation (17-6).

$$FOM(\%) = 100\% \times \left[1 - \frac{\sum_{i=1}^{N} \left| Xi(golden) - Xi(DUT) \right|}{\Delta x \times N} \right] \qquad (17\text{-}6)$$

where:
golden = the simulation data
DUT = the device-under-test measured data
Δx = the step size along the x-axis (usually voltage or time)
N = the total number of points being compared

A small C program or script computes the Figure of Merit defined in the above equation. The first numerical task that the algorithm must carry out is to map each set of data points to a common x-y grid by interpolation. The second task is to slide one curve against the other along the x-axis. In the case of the IV curve, this is a trivial step because the two curves are already aligned, but the x-axis origin is an arbitrary point in the case of voltage-time waveforms. Once this is accomplished, the algorithm can then perform the third and final step: comparing the data points and calculating the figure of merit.

The user must determine acceptable values for the three figures of merit used in the handbook for a given application. The model developer can attempt these goals for FOM values:

- FOM1 greater than or equal to 98%
- FOM2 less than or equal to 10%
- FOM3 goal is dependent on VDD
- Include plot of SPICE and behavioral waveform overlay

The IBIS committee also supplies "ACK," a small C program, for computing FOM, and a User Guide. The ACK software utility for computing FOM can be found at: http://www.eda.org/pub/ibis/accuracy/. This utility and other items can be downloaded free from this website.

One advantage of using FOM (plus the Curve Overlay Metric covered next) method is that verifying a model with a less-than-perfect circuit can be done. This method will work when we are not able to achieve a perfect transmission-line match with zero reflection.[7]

But there are limitations to pushing the less-than-perfect circuit idea. For instance, when correlation is less-than-perfect, how do we separate out the error portion due to the circuit that surrounds the DUT (including bias circuitry and test fixturing) from the error portion due to the device under test itself?

The example in this topic uses waveform-to-waveform correlation. However the ideas can be extended to correlating two sets of measurements,

---

[7] It seems more credible when we say that the model was verified under actual conditions. But any good Signal Integrity engineer will try to have their design produce clean waveforms.

or a set of measurements and a set of simulations. The FOM ideas parallel the concept of an error function.

17.12.2 Curve Overlay Metric

The Curve Overlay Metric, illustrated in Figure 17-8, measures how well the two curves or waveforms match each other by summing the absolute value of the x-axis (or y-axis) differences between the two data points, weighing the sum against the range of data points along that axis and dividing by the number of data points.

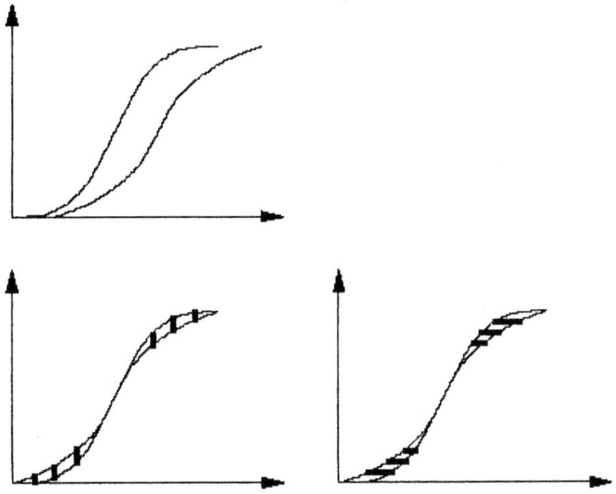

Figure 17-8. Illustration of Curve Overlay FOM Calculations [55]

From the I/O Buffer Accuracy Handbook [55]:

> The curve envelope metric applies to cases in which the measured data are, in theory, bounded by two curves (or waveforms) that represent process-voltage-temperature extremes. In general this metric is useful when the processing conditions of the sample component are unknown. The Curve Overlay Metric returns a yes or no value depending on whether or not every one of the data points falls within the envelope boundaries defined by the min and max curves. The plot below demonstrates a laboratory pull-down curve (solid line) that is slightly stronger than the typical curve (middle dashed line) and less well within the (outer dashed lines) [envelope]. . .

. . The Curve Overlay Metric applies to cases in which the measured and simulated data should theoretically lie directly on top of each other. For example, a structural simulation of 50 Ω load and a behavioral simulation of the same load should theoretically yield identical results. Another example is the measurement of a known-typical sample component and a structural simulation of the same network under identical process-voltage-temperature conditions. The Curve Overlay Metric measures how well the two curves or waveforms match each other by summing the absolute value of the x-axis (or y-axis) differences between the two data points, weighing the sum against the range of data points along that axis, and dividing by the number of data points.

Figure 17-9 is an illustration of the curve envelop metric.

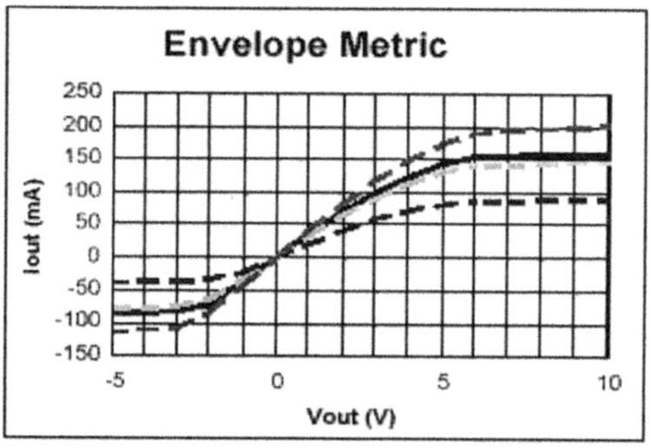

The Curve Envelope Metric presents a difficulty in the case of unterminated transmission line loads. Because these waveforms overshoot normal logic levels and ring back, the min and max waveforms intersect each other and do not define an envelope. Therefore, the Curve Envelope Metric may not be applied to the Open Transmission Line load or the Transmission Line and Receiver load.

Figure 17-9. Envelope Metric [55]

17.13 COMPUTATIONAL ELECTROMAGNETICS AND THE FEATURE SELECTIVE VALIDATION METHOD

Verification of electromagnetics simulations against laboratory measurements is a correlation task just as complex as correlating Signal Integrity waveforms. However, the time domain waveform correlation FOM method is not as appropriate for data in the frequency domain, where most Computational Electromagnetics (CEM) data is produced. For CEM data correlations, the Feature Selective Validation Method (FSV) is often more appropriate [5, 7, 18, 30, 31, 82].[8]

FSV methods of correlating CEM simulations with laboratory measurements are currently being developed. To make FSV available to any user, a dedicated standalone software interface was developed. The software can be downloaded at http://ing.univaq.it/uaqemc/.

The simulations being verified are data on resonances, emissions, Signal Integrity, and other electromagnetic computations. These simulations usually involve semiconductor models both as driver signal sources and as receiver loads on a network.

The data being correlated usually involves the amplitudes and frequencies of a signal's frequency domain spectrum or the resonances on a network. Thus, two pieces of spectrum information are usually being compared: the location of the frequencies, and their amplitudes. The FSV method develops three measures of correlation:

- Amplitude Difference Measure (ADM). ADM is calculated both on a point-by-point and single-value basis.
- Feature Difference Measure (FDM). FDM is calculated both on a point-by-point and single-value basis.
- Global Difference Measure (GDM). The Root-Mean-Square of ADM and FDM both on a point-by-point and single-value basis.

To validate the numerical approach, both quantitative (numerical) and qualitative (human-language) measures were developed and compared to each other. The quantitative measure is too complex to describe here, but is basically a magnitude and statistical calculation [31]. The qualitative assessments of specific simulation-measurement data set comparisons used 50 experienced EMC engineers, who visually examined plots of the data.

The human-language measure was developed from a six-point binary rating scale of: 1=excellent, 2=very good, 3=good, 4=fair, 5=poor, and 6=very poor. Figure 17-10 shows the decision tree set up for the qualitative method in the FSV process.

---

[8] The cited references use the term "validation," whereas the authors use "verification" to mean the same thing.

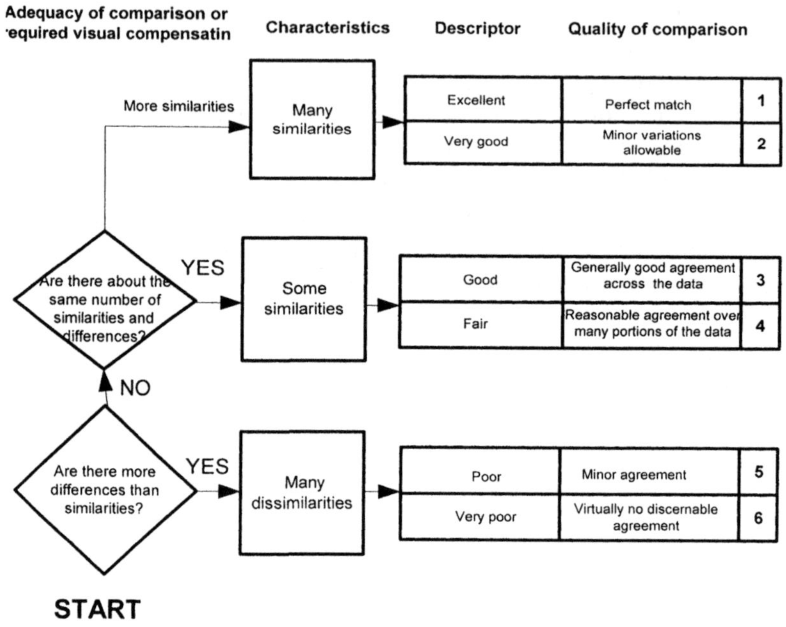

Figure 17-10. Decision tree for qualitative visual assessment of the correlation of data sets Used with permission [7].

Table 17-5 compares quantitative and qualitative measures developed over a moderately large set of correlations.

Table 17-5. Quantitative versus qualitative measures of correlation using FSV Used with permission [31]

FSV Quantitative Value	FSV Qualitative Value
Less than 0.1	1=Excellent
Between 0.1 and 0.2	2=Very Good
Between 0.2 and 0.4	3=Good
Between 0.4 and 0.8	4=Fair
Between 0.8 and 1.6	5=Poor
Greater than 1.6	6=Very Poor

The task of correlating data sets can be partially mechanized with this table of comparisons by relating objective and subjective measures. The authors of the FSV studies caution that additional work remains to be done to validate the method.

Subjective judgments are still important. The difficulty of the task and the physics involved should be part of the judgments. Group experience (as opposed to "group think") is an important subject to understand better. The experienced observations are recorded in a histogram of responses. An important question for the future is how and why experienced observers see things differently. Alternatively, the quantitative FSV values are a good measure against which individual observers can assess their own thinking.

An example of qualitative correlation results from running pairs of data sets through the FSV process are shown in Figures 17-11 through 17-13. Figures 17-11 and 17-12 illustrate the FSV method, but not all its subtleties.

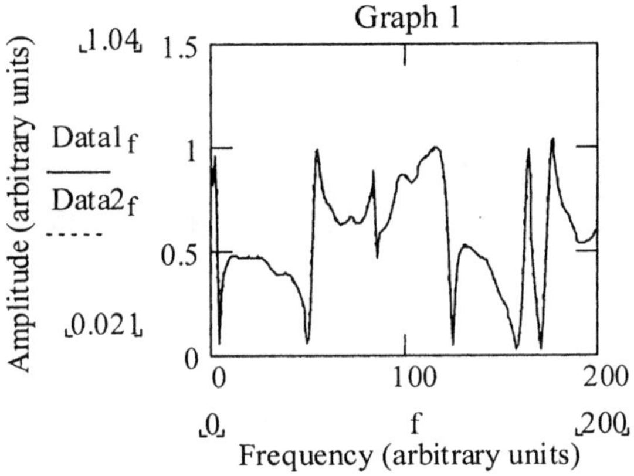

Figure 17-11. Graph 1 shows the data sets are nearly identical at this scale
Used with permission [7]

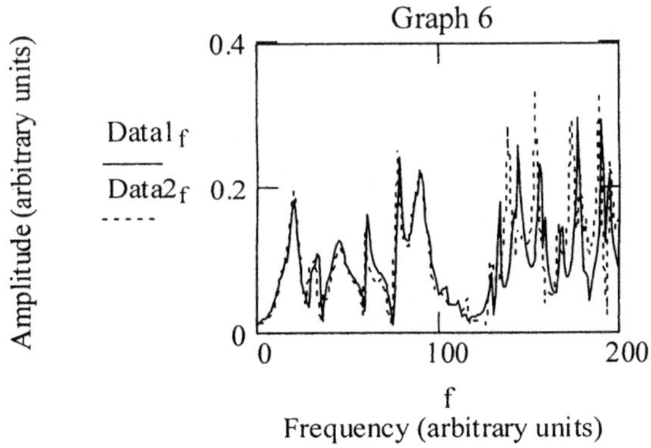

Figure 17-12. Graph 6 shows the data sets have started to diverge
Used with permission [7]

When a user wants to compare two data sets of any length, the FSV tool allows the user to apply the FSV techniques to the data. The user can obtain results and information in either graphs or ASCII files. Figure 17-13 illustrates the GDM comparison of the histogram results shown in Figures 17-11 and 17-12.

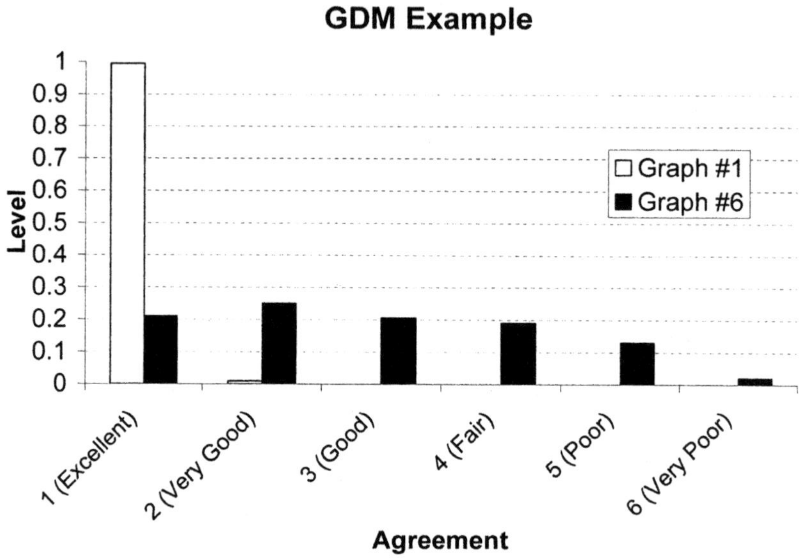

Figure 17-13. GDM comparison of histogram results from graphs 1 and 6. Used with permission [7]

17.14 IBIS GOLDEN WAVEFORMS

When provided in an IBIS model, the V-T waveform tables can be reproduced to verify the accuracy of the simulator. That is, the waveform tables can also serve as "Golden Waveforms" to verify the simulator results. Since the load conditions that produced the tables should be provided, the simulator should be able to reproduce those waveforms using the specified loads. The *simulator* is verified if the IBIS waveforms and the simulator waveforms agree.

A good verification study would have both the simulated results and measured results agreeing within a few percentage points. To accomplish this, we need an accurate model and measurements made within a few percentage points of accuracy.[9] Accounting for all primary effects, and at least the secondary effects at the switching speeds of interest, is also required.

The exercise of using the Golden Waveforms to verify a simulator *does not* verify the IBIS model. For that, a first step might be to check the IBIS model for consistency with the device data sheet, followed by correlation

---

[9] Calculations can never be more accurate than the data upon which they are based.

with a SPICE model previously verified by measurement data, or directly with laboratory measurements.

17.15 HOW UNEXPECTED ERRORS LED TO AN ADVANCE IN MODELING

17.15.1 Introduction

This topic recounts some early history in the development of the BJT transistor and the models used to represent it. The discovery of the BJT was a serendipitous event, where the transistor inventors were looking for a different phenomenon in device physics. By being aware and curious, they turned an experiment's failure into one of the crucial events of the twentieth century. In 1947, they invented the BJT as a point-contact transistor [110, 111].

Soon thereafter, a new technology, the double-diffused silicon planar transistor, was developed. Note that the base region of a point-contact transistor is very wide. The new technology is a narrow-based device. The behavior of the new device and how to model it was quite different from the first models used. The development of the new models was an outstanding example of diagnosing the cause and improving the results of how models represent components.

Unexpected errors are vexing when we are trying to meet a project schedule. But discovering and resolving a systematic modeling error can be exciting when free of such schedule pressures. Diagnosing and fixing a real modeling error can lead to an advancement of technology as this story relates. Knowing how limitations in modeling were addressed teaches us something about the attitudes we should maintain about modeling, simulation, and design.

17.15.2 Initial Theory

The first semiconductor device of real consequence was the BJT (Bipolar Junction Transistor). Up until then, only p-n diode junctions were made and theories about majority and minority carriers, recombination, energy band-gaps, and other phenomena were still being developed.

Originally, semiconductor theory had been formulated regarding surface-inversion field effects in semiconductor materials, basically the MOSFET. A MOSFET transistor was what the experimenters were originally searching for. However, the desired effect was masked by too high a level of impurities, lattice defects, and other factors that were not yet accounted for in the theories of the day.

While experimenting, the discoverers made a device with a neutral region separating two opposing p-n junctions, so narrow as to cross over into a new area of semiconductor behavior. What happened was that not all the minority carriers injected (emitted) from one forward-biased p-n junction recombined (as in a diode) before reaching the output contact. The minority carriers were reaching the second junction. That junction appears to be a forward biased n-p junction on the side where the injected minority carriers were arriving. These carriers were then swept up and collected in a region where they would again appear to be majority carriers with low recombination rates. In time, the region between the p-n ⇔ n-p regions would simply come to be called the base. With very little loss in current carriers, they were able to input current on a low-voltage forward-biased circuit side and see that current flow through a high-voltage reversed-biased output circuit.

Thus, a little power could control a lot of power. In other words, they had discovered the concept of gain (or amplification) in a semiconductor. This concept of gain is one that they were already familiar with in the case of vacuum tubes. In a vacuum tube, a small voltage on a grid element (with low leakage current) controls the current flowing in a high-voltage, high-power output circuit. The BJT's discoverers recognized they were achieving the same kind of thing. What separated the discoverers from the also-rans was the reaction "I wonder why that happened?" instead of anger–plus a lot of careful work.

The physical model in Figure 17-14 helped the discoverers understand what they were observing. Figure 17-14 shows an emitter-base collector device with current flowing from emitter to collector. Space-charge depletion layers are the shaded regions between the p and n regions. In this p-n-p arrangement (emitter-base-collector), the current carriers are holes injected from the emitter. A p-carrier (electron hole) is the absence, or deficit, of an electron. The hole sees a small potential hill to jump at the first forward-biased junction.

Therefore, a number of the carriers get injected (jump) into the base where they will tend to recombine and drop out of the conduction energy-band.

However, if the base is narrow and not too heavily n-doped, most of them will diffuse to the base-collector junction, where they fall down a potential hill and conduct out into the collector circuit.

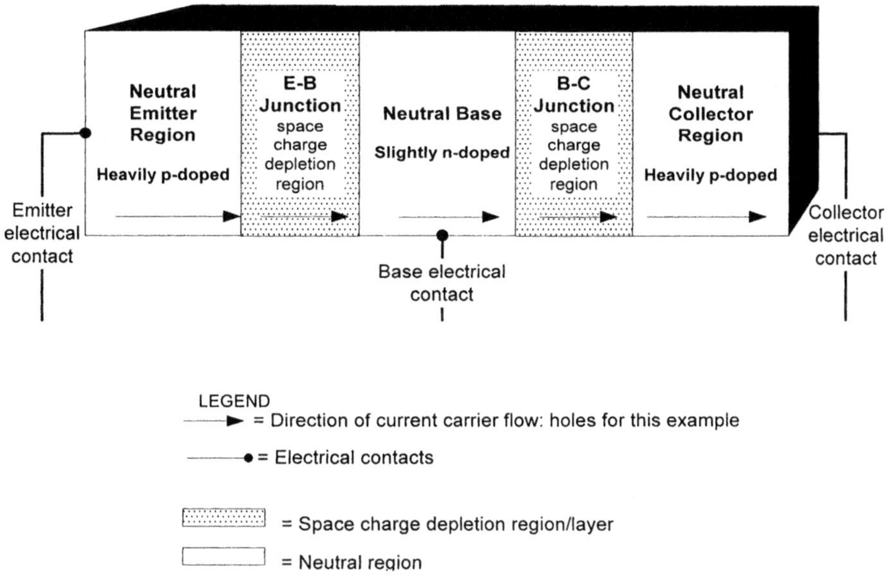

Figure 17-14. A one-dimensional structural model of a pnp transistor

To a first approximation, carrier concentration gradient drives the diffusion process. That is, a dense cloud of holes (missing electrons), near the injecting junction, drifts and diffuses (driven by random thermal motion) to the other junction, where the hole concentration is low.

First attempts to model the electrical circuit behavior of the bipolar junction transistor consisted of an input diode with its built-in junction voltage and internal resistance operating as a current sensing resistor, and a dependent output current or voltage generator controlled by the input current and the Beta of the transistor.[10] The schematic of this model is shown in Figure 17-15. The same model is used today when it is safe and necessary to highly simplify models to speed up simulation.

17.15.3 New Observations

One of the next, early advances to modeling came when semiconductor process technology was greatly improved by the development of the double-diffused silicon planar process (DDSPP) [53]. The original model of Figure 17-15 did not display much understanding about what was causing the BJT effect. Prior to DDSPP most of the BJT properties were easily and simply measured empirically. Developers soon understood that it is the input junction-voltage that controls current injection and that the base current

[10] This model mimicked models for vacuum tube devices.

simply represents a recombination loss. In a very real sense, BJTs do not amplify current. A perfect BJT would need no base current. Thus, we had the first transistor model.

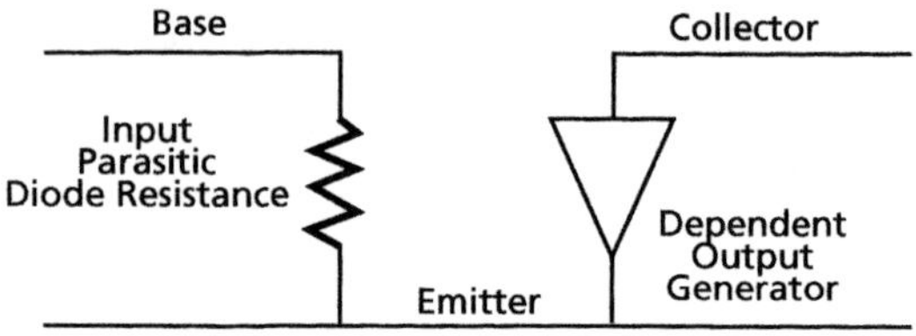

Figure 17-15. The first transistor model

The collector-current characteristic curves (Ic versus Vce for stepped Ib drive) expected from the simple 2-element model are shown in Figure 17-16.

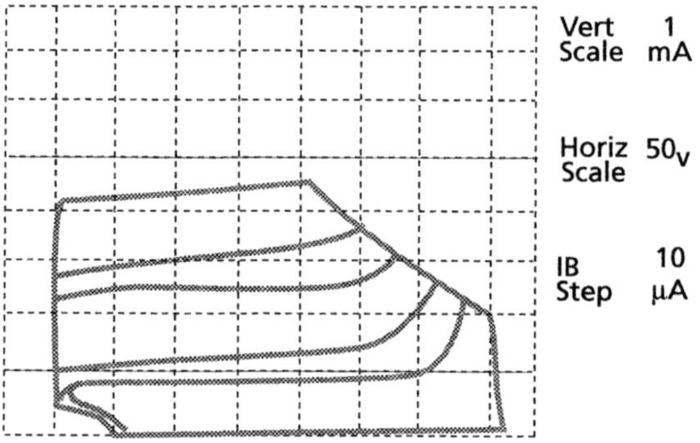

Figure 17-16. Collector characteristic curves predicted by the first transistor model. Photo from a Tektronix 576 curve tracer. [74]

Figure 17-17 shows the collector-current characteristic curves (Ic versus Vce for stepped Ib drive) seen when narrow-base BJTs were developed.

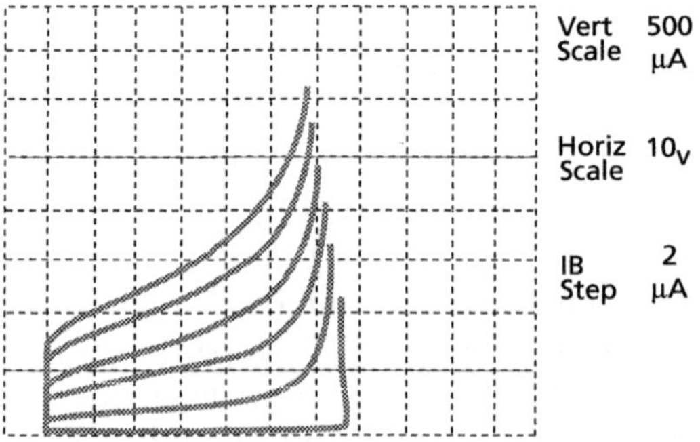

Figure 17-17. Collector characteristic curves of a narrow base silicon planar device. Photo from a Tektronix 576 curve tracer. [74]

When users began measuring and characterizing the narrower base BJTs, the simple model represented in Figure 17-15 could not explain what was being observed. It was relatively easy to ponder what was causing saturation voltage effects at very low Vce and breakdown current effects at very high Vce. It was harder to explain increasing gain with increasing Vce, and an apparent collector output admittance and impedance effect (slope of the "flat" portion).

The first step in correcting the first simple circuit model was to get it to work properly. This was done by adding some elements that could describe the new behavior seen at the input and output ports of a BJT. From measurement and observation, model developers realized that they would have to add some output conductance and a dependent feedback generator on the input. Thus was born the 4-element 2-Port matrix model in its various guises, shown in Figure 17-18 in y-parameter form.

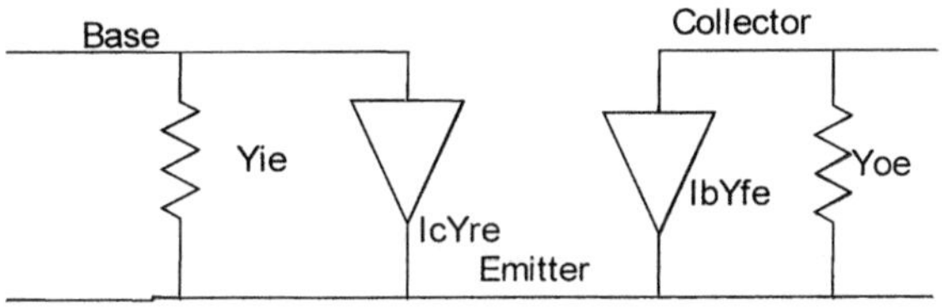

Figure 17-18. The first 4-element 2-Port model postulated to explain the behavior seen in narrow base devices

This new model allowed actual devices with the new behavior to be measured, characterized, and used in circuit analysis. In this example, we show the admittance-parameters: yie, yre, yfe and yoe. A 4-element black-box model can also be represented by the other formulations of 2-Port transistor models—including the following matrix parameter sets: impedance, hybrid, conductance, scattering, and others. Each has some form of dependent forward output- and reverse-feedback input-generator, and some form of output- and input-impedance.

17.15.4 From Observation to Improved Model

After proposing the black-box model, investigators were intrigued. What was the physical explanation for the new behavior seen in narrow-base devices?

Device physicists began thinking more about junction space-charge depletion regions widening out under the influence of increased reverse-bias electric fields. Dr. John Early (Early effect) was the first contributor to successfully explain the feedback effect of collector voltage in the base circuit.

The apparent output conductance (yoe) and feedback voltage (IcYre) shown by the curves in Figure 17-17 were due to base-narrowing effects as collector-base reverse bias increased. This increase caused the base-collector space-charge-depletion region to widen. This fit in with the growing realization of how a BJT worked on a physical level.

The new models were too complex to analyze by hand calculations of transistor input-output formulas. So the new device-physics models led to the development of a computer program, SPICE, which could simulate their behavior. The new knowledge spurred further device physics research and understanding that eventually led back to the surface-inversion MOSFET originally sought by the inventors of the transistor, plus a host of other devices such as the JFET, SCR, and GTO.

Dr. John Moll, of Ebers-Moll modeling fame, told this story to one author [Leventhal]. He was intimately associated with Dr. Early, Dr. Grove, Dr. Ebers and others. His point was to not let theory get one too detached from observation and to not let surprises make you give up the effort.

This process of observation → theory → observation → theory… continues today as we develop ever-smaller MOSFET geometries and new types of semiconductor devices such as GaAs and SiGe transistors. The development of the BSIM3 and BSIM4 models for deep sub-micron MOSFETs is just such a recent case in point.

Assumptions made for larger geometries no longer hold. For instance, for larger geometries, capacitance under a gate dominates, while for deep sub-micron geometries, sidewall capacitance dominates. Therefore, the SPICE models, and/or simplifications made to its underlying device physics equations, differ in important ways. Two-Port models are no longer popular because of SPICE. But 2-Port models live on as S-Parameter models.

17.15.5 Conclusions About Model Verification

John Moll's story about developing new models was instigated by a failed attempt at verification. Perhaps 99.99999% plus of model verification studies do not lead to developing a whole new technology, a new model, or advancing technology. Most such cases are simply inaccurate models, incomplete models that do not characterize the model parameters available, or inaccurate verifications. These obstacles in the path of modeling and simulation cause some to abandon it.

We should be careful about pushing the limits of a model (theory) into a new area of process or technology, where it just might break down.

Hopefully, the theorists and model developers are educated enough to be on the lookout for such cases as they push technology development.

We should also take a more constructive attitude about modeling and simulation. These cases arise when the model is not sufficiently detailed, does not account for observed measured effects, or was simply not accurately measured. There is always a root cause addressable by more careful measurements or more careful modeling.

> The proper response to verification problems is to diagnose the cause and improve the results.

17.16 RECOMMENDED VERIFICATION STRATEGY

The debate over which verification study approach, simplified or realistic prototype, to use is summarized in Table 17-6.

Table 17-6. Pros and cons of verification approaches

Verification approach	**Pros**	**Cons**
Simple Laboratory PCB	• Easier to separate different effects and understand what is going on. • Designed to make measurements easy. • Easier to get accurate, close correlation. IC models can be verified before a design even starts.	• Does not directly confirm that a particular design works. • Extra expense of hardware that is not shippable product.
Complex Prototype PCB	• Provides direct confirmation that a design works. • Removes doubts about surprises hidden in complexity.	• Prototype boards are generally much more expensive to build, may be damaged in making measurements, and are not designed for measurements. • Interactions are often too complex to separate out and understand. • High accuracy correlations are unlikely. • Must wait until a design is finished and hardware is built.

For many years there has been an ongoing debate about whether to verify a model in a simple, well-controlled laboratory PCB or in a complex product prototype PCB. The authors recommend the simple approach. The general principle is to include only as much detail as relevant to the verification task at hand. In other words, the more realism to the actual prototype that is included, the less accuracy the user should expect.

17.16.1 Successful Verification: Keep Things Simple

Ensure Success by Separating the PCB and Semiconductor Behaviors

Simple systems are easier to understand and learn from. If the objective is to verify a semiconductor model, too much board complexity obscures what is going on with the IC. If the objective is to verify a model extracted by a board EM simulator and model extractor, too much IC complexity obscures what is going on with the board simulator.

Consider the PDS system effects discussed in "Chapter 16, Methodology for Verifying Models." The system is very complex to model and operate. Again, to make a design work, the designer applies bypassing and other techniques to prevent any significant power plane voltage bounce and noise. If the designer is very successful, the designer can treat the voltage supply nodes as 0 Volts and Vcc Volts and simplify the model, simulation, design, and verification.

Ensure Success by Reducing Complexity

Too much complexity can prevent success in verifying the accuracy of a simulation tool or a model. Verifying a simulation tool or a model requires highly accurate correlation and a simple circuit. Verifying that a complex design will work should be done to a looser accuracy standard. The less exacting the accuracy requirements, the more detail can be included to emulate the actual design.

Consider the two waveforms in Figure 17-19. Which of these two simulated waveforms should be chosen for a model verification study requiring, perhaps, +/-2% accuracy?

The waveform on the left will not even work for a normal switching network. A conservative, prudent approach needs to be adopted regarding verification of model behavior in PCBs. Otherwise the task of achieving verification accuracies of less than a few percentage points becomes hopelessly complex and headed for failure. Poorly behaved boards are not good in products or in verification studies.

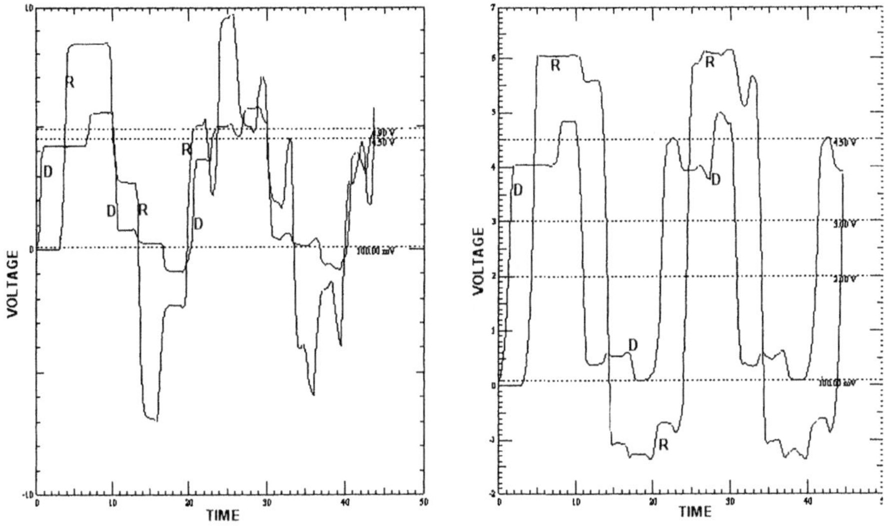

Ideal open circuit receiver with C_comp = Cin = 2.5pF, fast driver

Default CMOS input receiver with C_comp = Cin = 6 pf and clamping, slow driver

Figure 17-19. Waveform complexities for a verification study

Ensure Relevance by Having Verification Emulate Design

To avoid problems, prudent designers create clean designs and prudent modeling engineers create clean verification studies.

For example: Consider just one aspect of a sensitive RF amplifier–its power delivery system (PDS). A designer would adopt an approach that respected well established rules-of-thumb and would plan for good image-plane placement, low ground return impedance, bypassing-filtering, ground vias, shielding and more.

This is done even before considering the design of the amplifier circuit itself. Why? Sensitive RF amplifiers are easily disrupted by voltage bounce and feedback through the PDS system. Extremely small geometry, very high-speed serial busses are just as sensitive to PDS problems. Therefore, a prudent designer avoids such problems from the beginning. Otherwise there is a good chance the design will not work.

17.17 SUMMARY

In "Chapter 4, Measuring Model Properties in the Laboratory," the authors discuss verification of individual model parameters. In this chapter, the authors discuss verification of the complete model in a circuit.

Lack of correlation between models and laboratory measurements come from many sources, including errors and assumptions in the models, and errors in the laboratory measurements. If progress is to be made, engineers must resolve and diagnose these discrepancies when they occur. Comparing results from simulation and measurement involves both objective quantities and subjective (human) judgments. Tight correlation between modeling and measurement is quite expensive and should not be undertaken routinely. For routine design decisions, the proper defense against slightly inaccurate models is to use desensitized, rugged, variation-tolerant designs.

More efficient than unit-by-unit correlation is statistical envelope correlation of model simulations. Statistical correlation provides information on a population spread of components. Statistical envelope correlation of model simulations involves measurements of sample spreads of devices in multiple unit prototype runs.

To keep pace with evolving semiconductor technology, engineers should expect models to evolve. Sometimes laboratory measurements provide the first indication that something new needs to be considered. The proper response to verification problems is to diagnose the cause of the unexpected and improve one's understanding.

Chapter 18

BALANCING ACCURACY AGAINST PRACTICALITY WHEN CORRELATING SIMULATION RESULTS

Since all models contain simplifications and approximations of what they represent, engineers must exercise good judgment for deciding when a specific model is fit-for-use

Abstract: Today, circuit speeds are increasing, component sizes are decreasing, and I/O buffers are becoming more complex. Consequently, accurate modeling is becoming more difficult, expensive, and complicated. Good engineering judgment is needed when deciding whether a model is accurate enough. Judgment involves understanding accuracy and precision, the methods of correlation used, and design safety margins. Correlation has two main objectives: the proof of a new EDA tool or theory, and the release of a new product to manufacturing. Each requires different levels of accuracy.

18.1 ESTABLISHING ABSOLUTE ACCURACY IS DIFFICULT

Establishing *absolute* accuracy, perhaps within +/-2%, can be a very difficult task—especially on a regular basis. A generation ago, an experienced engineer would not have attempted to achieve such accuracy on a regular basis, with the existing limited computational power and test instrument accuracy. The rule of thumb then was that agreement within an engineering approximation of +/-10% was acceptable. Therefore, design decisions were made and designs were planned to succeed with those kinds of tolerances in mind.

The authors recommend that decision on design tolerances still be set to around the same level of +/-10%. Why? Are not computers much more powerful, modes more detailed, and test instruments much better? Yes, but:

- With today's edge rates, deep sub-micron technology, and circuit complexity, simulating everything on a regular basis on all circuits remains as problematical as a generation ago–despite the enormous increase in computational power.
- More significant for correlation, however, is to account for all parameters in a particular unit's correlation. Models have become much more complex to account for effects never seen before. More complex models mean more parameters to characterize and use. We do not even characterize the SPICE model parameters that we already have. A BJT SPICE model has had about 40 model parameters available for about the last 30 years. However, we rarely see more than a dozen model parameters, given real (as opposed to default) values. We rarely use the full analysis power of the models we already have.
- We now see a component's behavior interacting with the circuit, power busses, instrumentation, cables, and enclosure in ways not always anticipated.

We often assume that the difficulty in achieving +/-2%correlations lies with modeling and simulation. Instead it could equally be a problem with measurement. It is often very difficult for two test measurements to agree within +/-50% on a regular basis.

Still we ask, "*Do not state-of-the-art circuits require correlations between simulation and measurement tighter than +/- 10%?*" Answer: Yes, they do, but we must invest the required correlation effort in the following places when:

- Designing state-of-the-art circuits for the first time.
- Establishing the credibility of new modeling and simulation EDA tools or methods.

However, ordinary product designs should not require such tight correlation effort. We need to exercise practicality and good judgment to build affordable products.

What is the purpose of our correlation? Is the purpose trust-building that requires accurate correlation? Or is the purpose product design that should be acceptable with less accuracy? If the latter, as long as the product operates well within its safety margins and design assumptions, we have achieved our design objectives.

18.2 IS A MODEL ACCURATE ENOUGH TO BE USABLE?

Is a model accurate enough to be usable, and what does that mean? Earlier, we compared two simulations, and compared a simulation with a measurement. Additionally we looked at two methods of establishing correlation: unit-by-unit and statistically. In this chapter, we discuss the issue of whether our modeling and simulation is good enough to produce a good design (usable) versus how close we can come to real life (accurate). We want to know how well we can trust modeling and simulation. Also, we want to understand the terms affecting our perceptions of "good enough."

18.3 MODEL ACCURACY DEFINITIONS

A few very different concepts are intertwined in the question: How true are model simulations to real life? Let's first consider the following definitions:

Accuracy	The extent to which a given measurement, or the average of a set of measurements, agrees with the true value for that measurement.
Correlation	How well a set of measurements agrees with a separate set of measurements on the same group of units. One or more attributes may be used.
Dispersion	A distribution. The scattering of values of a measurement around the mean or median of the measurement.
Precision	The extent to which a given set of measurements are repeatable.
Validation	To give *official* conformation or approval. Applied to models, validation means verifying that the model correlates with the standard for the model and will run in simulators available for the purpose. Correlation with the standard for the model implies there is a standard and that the model follows the correct syntax, format, and contains the required data elements.
Verification	A formal assertion of the *truth* of something. Applied to models, verification means confirmation that simulation predictions of device and circuit behavior correlate with laboratory measurements.[1]

---

[1] For IBIS, it is possible to correlate the IBIS simulation with a SPICE simulation if the SPICE simulation was verified by measurements.

The difference between validation and verification is subtle but important. The activity that checks an IBIS file against the requirements of the IBIS Specification is *validation.* When we validate an IBIS model file against the IBIS Spec, we are asserting that it meets the official requirements of the spec. But when we say that simulations with an IBIS model file are true to life, we are asserting that we have *verified* the IBIS file. The activity that checks a simulation using an IBIS file against laboratory measurements is *verification.*

Figure 18-1 illustrates the differences between accuracy and precision.

- The target in the center represents the true value of a measurement.
- The black dots represent observations or simulations.

(a) Accurate but not precise (b) Precise but not accurate (c) Accurate and precise

Figure 18-1. Accuracy and precision

18.4 CONFIDENCE LIMITS IN MEASUREMENTS AND SIMULATIONS

Statistical confidence levels [15, 16, 20, 23, 51, 54, 61, 62, 65, 70, 89, 91, 95] are areas of probabilistic mathematics[2] that seek to convert a deterministic statement into a probability statement. For example, "the mean value of beta is 100" into "the mean value of beta is 100 with a 90% probability."

When establishing simulation-to-measurement correlation, it is more realistic to think in terms of confidence limits. The mathematics and verbiage employed to calculate confidence level, intervals, and limits can be a bit obscure to a circuit design engineer, but the following key concepts are useful:

2 An application of probabilistic modeling and simulation is discussed in "Chapter 23, Using Probability: The Ultimate Future of Simulation."

Confidence interval	A statistical range with a specified probability that a given parameter lies within the range.
Confidence limits	Either of the two numbers that specify the endpoints of a confidence interval.
Confidence level	The probability value, for example 90%, associated with a confidence interval

18.5 HOW MUCH TO GUARD-BAND DESIGN SIMULATION?

It is unrealistic to expect +/-2% correlations between simulations and measurements on a routine basis. When we are verifying a new EDA tool or an important extension to modeling, a +/-2% correlation is reasonable. But when developing a new product, it may be uneconomical to try to achieve such tight correlations routinely.

How much guard-band within our tolerance window is realistic? The answer is, "*It depends*." Simple worst-case design and a 10% guard-band can achieve great success in producing prototypes that work the first time that they are tested in the laboratory.

Simulation errors larger than +/-10% are a cause for concern. When errors are larger than +/-10%, it is best to use the simulations to show what would happen on a *relative* basis. That is, do not use the simulations to make judgments about absolute values; rather use the simulation to indicate relative improvements.

What if our objective is circuitry that works with six-sigma quality in high volume production? That may require working with the semiconductor supplier to get actual population distributions and using them in Monte-Carlo simulations and design of experiments (DOE) simulations. Plus, we will need to take additional steps to desensitize our circuits. An example of desensitizing a circuit is using termination resistors to control reflections instead of relying on modeling and part control. We may also need to work with the PCB fabrication houses to get population distributions on ε_0, Zo, and etch variations, and model those kinds of effects.

What if we are designing with the latest Intel microprocessor and some of our nets have very tight timing windows? In this case, we may need to work with Intel to get information on timing delays due to the bondwires inside the package. We also need to use the [Package Model] model in IBIS, model connectors,[3] and the PCB fabrication houses on modeling non-uniform, dispersive, and lossy dielectrics.

Working with suppliers in such close cooperation requires more than technical skill. It requires tact, patience, trust, and probably long-term relations. More information about this subject is in "Chapter 21, Feedback to the Model Provider Improves Model Accuracy."

Meeting these modeling and simulation challenges could very easily lead to "analysis paralysis." This is especially true if we apply high accuracy considerations across the board. For simple pass and fail decisions on less sensitive circuits, we must design with some prudent guard-banding.

18.6 DIFFERENCES IN ACCURACY, DISPERSION, AND PRECISION FOR SIMULATION AND MEASUREMENT

Table 18-1 compares how the same concept relates to both model simulation and laboratory measurement. The differences are explained in the table.

Table 18-1. Differences between model simulation and measurement

Concept	Model Simulation	Measurement
Accuracy	May be *inaccurate* when compared to the standard of a laboratory measurement.	Considered *real world and therefore accurate.* Mistakes are often made in the test setup. Lack of correlation may be a measurement error.
Correlation	Correlation between two simulators is either not a problem or offset by a predictable amount that can be explained when we compare algorithms.	Correlation between two laboratory measurements can be random unless we control a lot of test variables.
Dispersion	Has zero dispersion of results unless we purposely use a multi-run, statistically variable simulation. If we build in statistical variation, the dispersion will strictly be a function of the variation we have built into the models and our statistical methodology.	Has significant dispersion even without considering a selection of devices or prototypes. Variation arises just from the test setup and random environmental variables.
Precision	Tends to be very *precise.*	Tends to be *imprecise.*
Simplification	Most simulations are very simplified when compared with the physical prototype or its laboratory setup. Usually does not include modeling of laboratory instrumentation and setup.	Most laboratory measurements are complex because all the manufacturing variables are inherently included, plus test variables are added.

---

[3] See the IBIS ICM Specification for connector models [56].

18.7 MODEL LIMITATIONS

Model limitations put a boundary on the level of accuracy that can be expected from a simulation-to-measurement correlation. We may be unaware that we have failed to represent a piece of hardware and its test setup in their full and significant complexity. An outright mistake may occur when we simplify and approximate reality and then use a model in a range where the simplification is not substantiated. For example, simplification of the voltage gain of a BJT leads to:

$$Av \approx Rc/Re \qquad (18\text{-}1)$$

which does not reflect what happens with low or high current beta rolloff. For a more complete description of model extraction, simplification and limitations, refer to "Chapter 20, The Challenge to IBIS." Whatever the model limitations used are, we need to work within these limitations when establishing correlation.

18.8 STANDARDIZATION AND THE COMPACT MODEL COUNCIL

In September 1993, the National Institute of Standards and Technology (NIST) formed a Working Group on Model Validation. The purpose of the Working Group was to establish procedures for the comprehensive evaluation of the circuit simulator models, specifically SPICE.

For the first few years, the undertaking was quite active with membership growing to 160 participants. But then interest waned. The MOS compact model effort was the only really active group. A recommended practice was developed, a number was assigned, and the IEEE evaluated it. A complete re-write was required for submittal. Committee member Britt Brooks got a time extension, but was unable to devote the resources to it. The recommended practice eventually got dropped from consideration. There is still a lot of excellent information on the website. However, there is no formal document for use. Here is the Working Group's charter, which has a worthwhile discussion of the correlation issue:

> The primary objective of the working group is to establish well-defined procedures for the comprehensive evaluation of circuit simulator models. The working group has been established because there is presently no systematic way to determine the

range of applicability of the models provided within circuit simulator component libraries. In addition, the accuracy of component models varies from one circuit simulator to another due to the inclusion of different physical mechanisms in the generic models. The accuracy also varies due to the different methods used to determine the model parameters for specific device part numbers.

The complex behavior of electronic devices prohibits individual users of circuit simulators from evaluating the proper inclusion of model physics and from determining the validity of approximations made to simplify simulator implementation. In addition, software vendors are often reluctant to provide a complete description of their model equations and model parameters to users. Therefore, the goal of the NIST Working Group on Model Validation is to establish experimental test procedures that can be used to comprehensively evaluate circuit simulator models independently of the model equations, the simulator, or the model parameter extraction techniques.

Traditionally, most circuit simulator model evaluation has been performed by comparing simulations with measurements for the easily measured steady-state output characteristics and capacitance-voltage characteristics. However, this type of evaluation is of little use in determining the ability of the models to describe the more important dynamic characteristics. Presently, the only dynamic evaluation performed for circuit simulator models is for narrow ranges of conditions that tend to focus on those physical effects that are included in particular models. For the most part these evaluations are not on comprehensive evaluation of the model's ability to describe the dynamic behavior of the device for the full range of application conditions.

The primary tasks of the working group are to determine the complete range of dynamic conditions that must be described for each device type, then develop well-defined test procedures to evaluate the ability of the models to describe each type of dynamic condition.

It is envisioned that each device type will require different test procedures, and that the circuit parameters for the test procedures will be determined for each part number based upon the device

> manufacturer's ratings and suggested application conditions. Furthermore, the model validation test procedures should evolve to account for new device variations, modeling requirements for new applications, and to prevent the development of models that are designed only to best fit the standard test procedures.
>
> Typical model validation test procedures are expected to consist of test circuits that resemble application conditions, but that are simplified so that the test system is well characterized and is able to isolate the important features of the device characteristics. Ultimately, standard characterization procedures will be established that specify the methods used to construct the test circuits to minimize the influence of parasitics, as well as the procedure for determining the circuit parameters based upon the device manufacturer's ratings. These standard test circuits could then be readily used to compare measured dynamic characteristics with those predicted by different circuit simulator component models. A database of the comparison results could also be maintained.
>
> For the working group to be successful, expertise in the following areas will be essential: electronic component design and manufacturing, model development, software development, component characterization, and circuit and system design. The success of this working group could result in an improved understanding of the validity and limitations of existing circuit simulator component models. This could lead to the development of improved circuit simulator models and an increased confidence in the ability of the CAD tools to aid in the design of electronic systems.

The Compact Model Council (CMC) was formed to promote standard compact models [28]. CMC inherited the efforts and output of NIST's Working Group on Model Validation. Currently the CMC has standardization efforts going for the following models:

- Both MEXTRAM and HICUM as standard BJT models.
- BSIM3V3 as a standard CMOS model.
- BSIMPD as a standard SOI model.
- A next generation standard model effort for CMOS

18.9 SUMMARY

When correlating modeling to laboratory measurements, it is important to balance accuracy and usability. Achieving this balance requires that engineers understand terms like accuracy, precision, usability, correlation, confidence limits, guard-banding, and sensitivity. As well, engineers must ask themselves what the objectives of their correlations are. Is it the proof of something fundamental or is it to make a design decision? If engineers are making a design decision, the circuit can be designed with a wider tolerance window. Guard-banding design margins can be a more economical solution to model accuracy issues than always demanding absolute accuracy.

Chapter 19

DERIVING AN EQUATION-BASED MODEL FROM A MACROMODEL

An example behavioral model is embedded in an RF amplifier circuit and its reverse-transfer functions are derived

Abstract: A 2-Port behavioral Matrix model is simple enough for engineers to derive transfer equations (also called transfer functions). These transfer functions can include the effects of load, generator impedance, and bias circuitry. The derivation of these closed-form equations allows circuit optimization and other modeling insights. When the equations are verified by laboratory measurements, engineers gain confidence in the methodology used to derive the transfer function.

19.1 A "NEW" RF DESIGN CHALLENGE

This topic is an exercise intended to show how transfer (equation-based) functions are derived and thus build confidence in the method. We begin the derivation by formulating what-if questions. Next we apply scientific discipline by analyzing the problem mathematically (modeling). Once the problem is analyzed, we can make predictions that can be verified by measurement. Flexible thinking coupled with analysis can help predict behavior that might otherwise be difficult to predict from observation alone.

19.2 BACKGROUND

This example is about the design of a moderately high-frequency RF *isolation amplifier* using discrete transistors in the 1960s. This example was chosen because it illustrates using macromodels and a bottom-up development of an equation-based model. Also verification measurements were done.

During the initial design, author [Leventhal] asked the question, "What happens when a signal is injected into what is normally the output port of an amplifier circuit?" An isolation amplifier should amplify moderately well in the normal forward direction (at approximately 70MHz in this example), but should provide very good attenuation to signals in the reverse direction (at 68MHz and 72MHz in this example).

At the time this work was done, the available models were primarily 2-Port matrix models. SPICE and IBIS were not widely disseminated or invented, and most suppliers considered S-Parameters too abstract. We used the y-parameter 2-Port model. But this choice of model is fortunate for our example because it facilitated developing closed-form equations from circuit analysis. Closed-form equations lend themselves to simplification and optimization. Simplification and optimization are processes that help develop engineering judgment.

19.3 APPLYING THE RF EXAMPLE TO HIGH-SPEED DIGITAL CIRCUITS

The derivation of equation-based models is used in the latest VHDL-AMS equation-based models being applied to complex IBIS I/O modeling.

A 2-Port, 4-element matrix model, with its bias circuit and coupling elements, is simple enough to be *explicitly* solved with paper and pencil. We use Kirchhoff's Voltage[1] and Current Laws[2] to do this. There are two current loops–input and output–and several nodes in the circuit. After writing the voltage equations around each loop and the current equations at each node, it's just a lot of algebraic equation manipulation to derive expressions for input-output currents and voltages and thence transfer functions and input-output load impedances.

Two-port matrix models (and SPICE models) are forms of macromodels. Because of their simplicity, matrix models work very well for extracting transfer functions and other closed-form equations. Other types of models are harder to solve in closed form. Embedding matrix models in their bias/coupling/signal source/load circuit works well with solving node and mesh equations that include the effects of those elements.

What happens when we try to explicitly solve a physical-level transistor model (SPICE) using Kirchhoff's laws? The task quickly becomes too

---

1 The sum of voltage rises and voltage drops around a closed circuit loop must be zero.

2 The sum of the currents into and out of a node must be zero.

complex. The earliest, simplest such models had about 7 nodes (eight elements) and about 4 loops including bias circuitry. Today a modern CMOS physical level model contains about 30 to 40 nodes and 120 elements. The task of solving such circuits was given to computers to *implicitly* solve by simulation. Hardware description languages (SPICE, MAST, VHDL-AMS, and Verilog-AMS) and computer programs like SPICE were developed to solve circuits. The usual input to the computer is in the form of a structured netlist, SPICE, or a structured set of instructions, AMS. The SPICE approach deals directly with the macromodel. The AMS approach implicitly solves for the circuit equations. The structured netlist approach can have equations embedded in the netlist.

The actual paper-and-pencil process of solving the equation is just a long, algebraic manipulation process that can be found in [75].

19.3.1 Discussion of the What-If Question

These derived equations allowed the author to minimize and maximize the circuit response needed, taking into account the effects of bias networks and loading [73]. The design task presented to the author was to put an amplifier in the paths from a voltage-controlled local oscillator (LO), at 70MHz, to two single-ended down-converting mixers where the sidebands were also injected. These sidebands were at 66-68MHz and 72-74MHz, part of a phase-lock-loop (PLL). The main objective was to keep signals from getting back into the LO–causing spurious lockup, and incidentally, to provide some forward gain. This was before any of the following were readily available, effective, or economical:

- IC isolation amplifiers: OpAmps had barely come on the scene, were pretty low frequency, and were limited in dynamic range.
- Powerful computer programs and the hardware to run them on were not available.
- RF parts characterization with SPICE models was non-existent, and Scattering Parameter models were very rare in the devices then available

Being forced to keep the derivation simple had the benefit of providing insight.

Using a top-down approach on the PLL, we saw that the amplifier would need about 20dB forward gain, and about -60dB reverse isolation. It was obvious that a passive filter with minimal attenuation at 70MHz and 60dB attenuation below 68MHz and above 72MHz could never be practically designed. But, we could readily get y-parameter matrix models for some RF parts. These models lend themselves to mesh-and-node analysis using Kirchhoff's laws.

The "a-hah" was to remember that 2-Port models are, by definition, bi-directional. Their parameters are measured in both the forward and reverse directions. So why could not their gain-impedance equations be derived in both the forward and reverse directions?

The designer can apply a signal source across a transistor's output, to derive formulas for reverse voltage gain, reverse current gain, and "input" and "output" impedances. We derived these closed-form formulas to analyze and design the amplifier. Then we simplified these formulas by considering the usual magnitudes of the y-parameters, the bias resistors, load impedance, and bypassing. Next we examined the simplified formulas to optimize the configuration of a (at minimum) 2-stage amplifier for reverse isolation. He also studied the normal forward-gain equations to verify that the forward-gain objective could be met.

A prototype amplifier was built and measured to verify the analysis and simulations. Units with individually measured model parameters were not available. There was no information about whether the units used were min, typical, or max devices, and thus had to rely on the accuracy of the data sheet.

19.4 PREDICTED AND MEASURED RESULTS

Table 19-1 summarizes the results from analysis and measurement:

Table 19-1. Forward and reverse voltage gain in an RF amplifier: Predicted and measured

Frequency (MHz)	**Predicted**		**Measured**		**Notes**
	Forward Voltage Gain (db)	**Reverse Voltage Isolation (db)**	**Forward Voltage Gain (db)**	**Reverse Voltage Isolation (db)**	
20	24.9	-77.5	33	-73	
70	14.3	-54.5	16.9	-51	

To get closer correlation than this, we would have to measure the y-parameters of the particular units used in the test fixture.

The results in Table 19-1 are certainly good enough for verification. The next step is to find out whether the system designer[3] could accept these results, or whether the design would need to be improved by adding another amplifier stage.

[3] The isolation amplifier was part of a phase-locked loop.

19.5 REVERSE ISOLATION ANALYZED

Reverse isolation is a minor issue in digital circuits, and only a little more of an issue in analog circuits. In addition, BJTs are bit players in today's technologies. However, this example is presented because it previews the process of new model development, verification, and abstraction. Specifically, the process is:

1. Identify the need.
2. Do a thought experiment.
3. Develop the mathematics.
4. Simplify the equations if possible, by looking at relative magnitudes of parameters and what happens when they become very large or very small.
5. Ask what happens when amplifier stages are cascaded. In this case, in various configurations and optimizations.
6. Complete the design-simulation phase by calculating the appropriate parameters that will be measured. In this example forward and reverse voltage and current gains, which can be combined into forward and reverse power gains.

   When doing these calculations, be aware that the input impedance of the next stage loads down the previous stage. Pay close attention to the "input," "output," and direction of signal flow because we analyze the isolation amplifier as a bi-directional circuit. The gain (attenuation) of each stage can be converted to dB and added for easier analysis and plotting. Since this is the frequency domain, complex (real and imaginary parts) number math will be used in the voltage gain times current gain step, and then calculate the absolute value.

7. Prototype the design with due attention to power-supply bypassing, isolation from external signals, instrumentation loading, and other issues.
8. Measure the parameters of interest, and compare them to the predicted simulation results.
9. If the model works, publish the results.

19.5.1 Common Emitter Configuration

Figure 19-1 shows a Common Emitter (CE) circuit and its equivalent CE y-parameter macromodel and bias circuit.

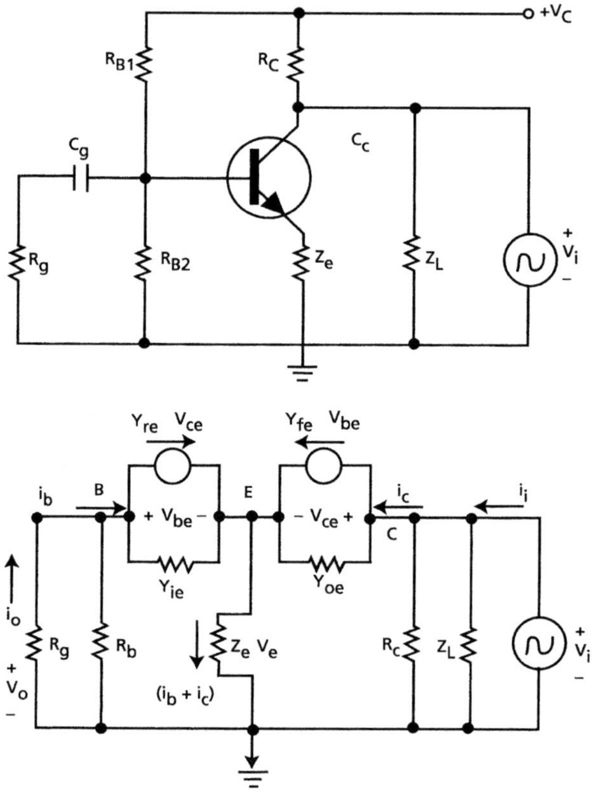

Figure 19-1. Common emitter configuration

LEGEND:
All stages are biased and operating as small-signal class-A amplifiers.
$Vi = V_i * \sin\omega_0 t$
Z_L: a complex load impedance
Z_E: a complex emitter impedance
$1/j\omega_0 Cc <<< Zo$, Rc and Z_L
$1/j\omega_0 C_B <<< R_{B1}$, R_{B2} and Rg
$R_L = RcZ_L /(Rc + Z_L)$
$R_B = R_{B1}R_{B2}/(R_{B1} + R_{B2})$
$Rg` = R_B Rg/(R_B + Rg)$

$\Delta y = y_{ie} y_{oe} - y_{re} y_{fe}$
$\Sigma y = y_{ie} + y_{oe} + y_{re} + y_{fe}$
$\alpha i = (R_L + Zo)/R_L$ - current transfer ratio of the "input" network
$\alpha o = R_B/(R_B + Rg)$ - current transfer ratio of the "output" network

For CE reverse - current gain, voltage gain, output impedance and input impedance - we have:

Reverse Current Gain of the total CE amplifier:

$$A_{i`} = \frac{(y_{re} - \Delta y Z_E)\alpha_o}{[(R_{g`} + Z_E)\Delta y + y_{oe}]\alpha_i} \tag{19-1}$$

Reverse Voltage Gain of the total CE amplifier:

$$A_{v`} = \frac{(y_{re} - \Delta y Z_E)R_{g`}}{1 + y_{ie}R_{g`} + \Sigma y Z_E + \Delta y R_{g`} Z_E} \tag{19-2}$$

CE Reverse Output Impedance

$$Z_o` = \frac{1 + y_{ie}R_{g`} + \Sigma y Z_E + \Delta y R_{g`} Z_E}{y_{oe} + (R_{g`} + Z_E)\Delta y} \tag{19-3}$$

CE Reverse Input Impedance—same as the usual CE Output Impedance

$$Z_i` = \frac{1 + y_{re}R_g` + \sum y Z_e + \Delta y R_g` Z_e}{y_{oe} + \Delta y (R_g` + Z_e)} \tag{19-4}$$

19.5.2 Common Base Configuration

Figure 19-2 shows a Common Base (CB) circuit using an equivalent circuit macromodel in CE y-parameters and its bias circuit.

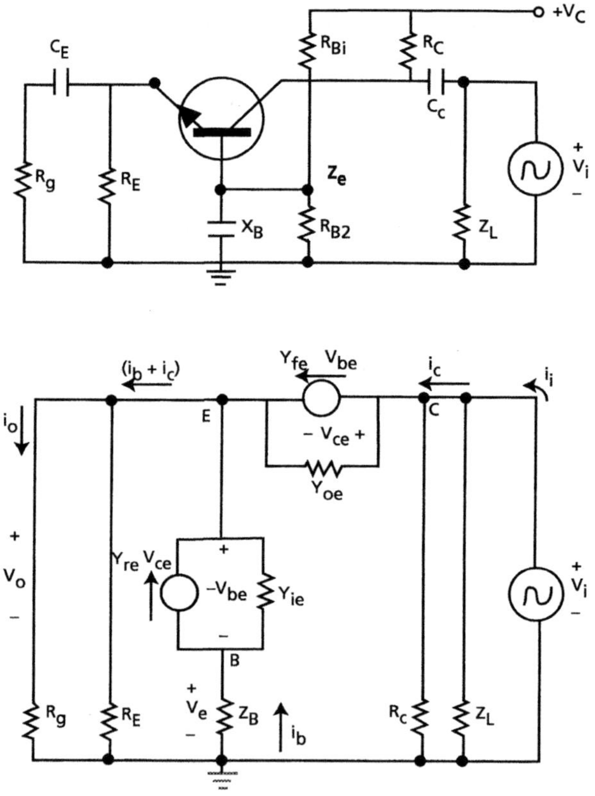

Figure 19-2. Common base configuration

LEGEND:
All stages are biased and operating as small-signal class-A amplifiers.
$Vi = V_i * \sin\omega_0 t$
Z_L: a complex load impedance
Z_B: a complex base impedance $= R_B X_B/(R_B + X_B)$
$1/j\omega_0 Cc \lll Zo$, Rc and Z_L
$1/j\omega_0 C_E \lll R_{B1}$, R_{B2} and Rg
$R_L = RcZ_L /(Rc + Z_L)$

$R_B = R_{B1}R_{B2}/(R_{B1} + R_{B2})$
$Rg` = R_E Rg/(R_E + Rg)$
$\Delta y = y_{ie}y_{oe} - y_{re}y_{fe}$
$\Sigma y = y_{ie} + y_{oe} + y_{re} + y_{fe}$
$\alpha i = (R_L + Zo)/R_L$ - current transfer ratio of the "input" network
$\alpha o = R_E/(R_E + Rg)$ - current transfer ratio of the "output" network

For CB reverse-current gain, voltage gain, output impedance and input impedance, we have:

Reverse Current Gain of the total CB amplifier:

$$A_{i`} = \frac{(y_{oe} + y_{re} + \Delta y Z_B)\alpha_o}{[(R_{g`} + Z_B)\Delta y + y_{oe}]\alpha_i} \tag{19-5}$$

Reverse Voltage Gain of the total CB amplifier:

$$A_{v`} = \frac{y_{oe} + y_{re} + \Delta y Z_B}{1 + y_{ie}Z_B + \Sigma y R_{g`} + \Delta y R_{g`} Z_B} \tag{19-6}$$

CB Reverse Output Impedance

$$Z_{o`} = \frac{1 + y_{ie}Z_B + \Sigma y R_{g`} + \Delta y R_{g`} Z_B}{y_{oe} + (R_{g`} + Z_B)\Delta y} \tag{19-7}$$

CB Reverse Input Impedance—same as the usual CB Output Impedance

$$Zi` = \frac{1 + \sum y R_g` + (y_{ie} + \Delta y R_g`)R_b`}{y_{oe} + \Delta y(R_g` + R_b`)} \tag{19-8}$$

19.5.3 Common Collector Configuration

Figure 19-3 shows a Common Collector (CB) circuit using an equivalent circuit macromodel in CE y-parameters and its bias circuit.

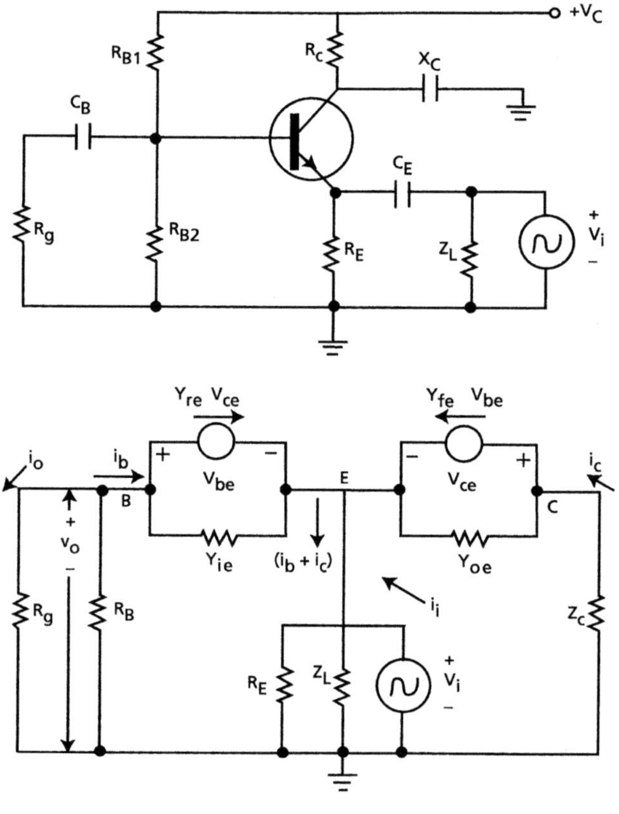

Figure 19-3. Common collector configuration

LEGEND:

All stages are biased and operating as small-signal class-A amplifiers.

$Vi = V_i{*}\sin\omega_0 t$

Z_L: a complex load impedance

Z_C: a complex collector impedance = $R_C X_C/(R_C + X_C)$

$1/j\omega_0 Cc <<< Zo$, Rc and Z_L

$1/j\omega_0 C_E <<< R_{B1}$, R_{B2} and Rg

$R_L = RcZ_L/(Rc + Z_L)$

$R_B = R_{B1}R_{B2}/(R_{B1} + R_{B2})$

$Rg` = R_B Rg/(R_g + Rg)$

$\Delta y = y_{ie}y_{oe} - y_{re}y_{fe}$
$\Sigma y = y_{ie} + y_{oe} + y_{re} + y_{fe}$
$\alpha i = (R_L + Zo)/R_L$ - current transfer ratio of the "input" network
$\alpha o = R_B/(R_B + Rg)$ - current transfer ratio of the "output" network

For CC reverse-current gain, voltage gain, output impedance and input impedance, we have:

Reverse Current Gain of the total CC amplifier:

$$A_{i'} = \frac{(y_{ie} + y_{re} + \Delta y Z_C)\alpha_i}{[(R_{g'} + Z_C)\Delta y + \Sigma y]\alpha_o} \tag{19-9}$$

Reverse Voltage Gain of the total CC amplifier:

$$A_{v'} = \frac{(y_{ie} + y_{re} + \Delta y Z_C)R_{g'}}{1 + y_{ie}R_{g'} + y_{oe}Z_C + \Delta y R_{g'}Z_C} \tag{19-10}$$

CC Reverse Output Impedance

$$Z_{o'} = \frac{1 + y_{ie}R_{g'} + y_{oe}Z_C + \Delta y R_{g'}Z_C}{\Sigma y + (R_{g'} + Z_C)\Delta y} \tag{19-11}$$

CC Reverse Input Impedance—same as the usual CC Output Impedance

$$Z_i` = \frac{1 + y_{oe}Z_c` + y_{ie}R_g` + \Delta y R_g`Z_c`}{\sum y + \Delta y(Z_c` + R_g`)} \tag{19-12}$$

Figure 19-1 showed the y-parameter and bias network schematic, and relevant assumptions used in the analysis of the common-*emitter* configured amplifier. Figure 19-2 showed the y-parameter and bias network schematic, and relevant assumptions used in the analysis of the common-*base* configured amplifier. Figure 19-3 showed the y-parameter and bias network schematic, and relevant assumptions used in the analysis of the common-*collector* configured amplifier.

Each figure, in turn, is followed by the formulas derived for *reverse* current gain, voltage gain, input impedance and output impedance. All formulas are based on common-emitter y-parameters. The common-base and common-collector configurations are converted to the use of common-emitter parameters. All figures and results are from [73, 75].

19.6 OPTIMIZING SINGLE-STAGE REVERSE ISOLATION

A moderately good transistor will have:

y_{re} on the order of $j10^{-6}$
y_{oe} on the order of 10^{-4}
y_{ie} on the order of 10^{-3}
y_{fe} on the order of 10^{-1}
Δy on the order of 10^{-9}

We will use this information in the formulas to optimize reverse isolation and forward gain for each of the three amplifier configurations.

19.6.1 CE Stage

For a CE stage and fair current isolation, we want Z_E small and Rg` relatively large. For a CE stage and good voltage isolation, we want $Z_E$ relatively large and Rg` small. Then $Av` \cong -10^{-7} + j10^{-7}$. We conclude that a CE stage can have fair current isolation and excellent voltage isolation

19.6.2 CB Stage

For a CB stage, current isolation is poor. For a CB stage and good voltage isolation, we want Z_B relatively small and Rg` large. Then: $Av` \cong 10^{-6}$. We conclude that a CB stage can have poor current isolation and excellent voltage isolation.

19.6.3 CC Stage

For a CC stage and excellent current isolation, we want Z_C relatively small and Rg` relatively small. Then: $Ai` \cong 10^{-4}(R_B/(R_B + Rg))(R_L/(R_L + Zo))$. For a CB stage, voltage isolation is poor. We conclude that a CC stage can have excellent current isolation and poor voltage isolation.

To prevent the injection of power (energy spectrum spurs) backward into a LO, a 2-stage amplifier with CC driving CE at the output to a single-ended mixer was chosen.

19.7 COMBINING STAGES FOR POWER ISOLATION

The final configuration after optimization is shown in Figure 19-4:

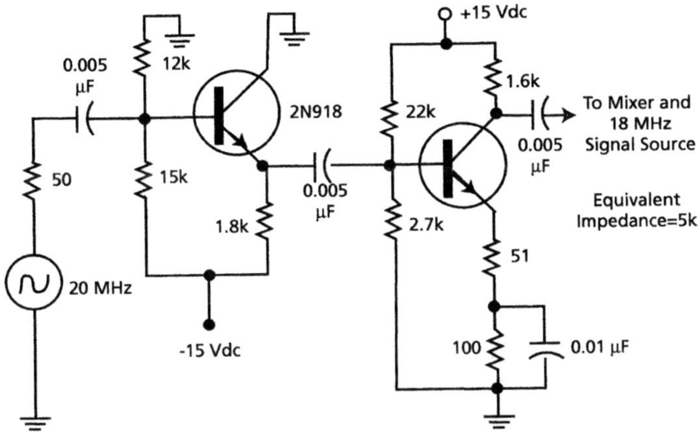

Figure 19-4. A two-stage isolation amplifier

The RF transistor chosen was the 2N918 with the following characteristics.

@ f = 70 MHz
$y_{ie} \cong 4x10^{-3} + j4x10^{-3}$
$y_{oe} \cong 0.25x10^{-3} + j1.0x10^{-3}$
$y_{re} \cong - j0.5x10^{-3}$
$y_{fe} \cong 44x10^{-3} - j39x10^{-3}$

Model properties are displayed as parameter plots in Figure 19-5. Real and Imaginary (reactive) parts are plotted.[4] The source of all the y-parameter data given next is the Fairchild Semiconductor Discrete Products Data Book, July 1973. These plots had not been revised from the ones the author derived in 1966.

---

[4] Example: $y_{ie} = g_{ie} + jb_{ie}$

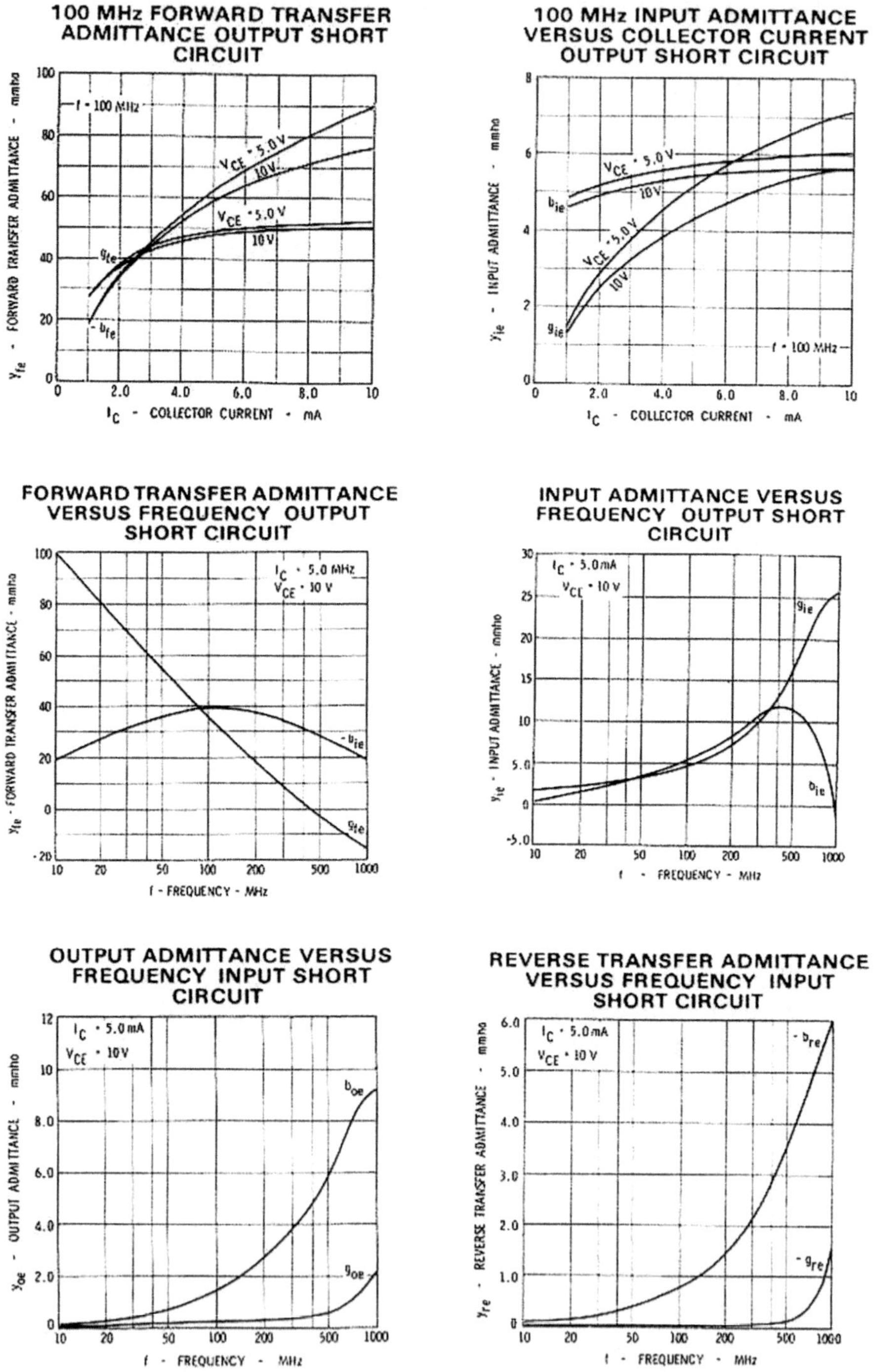

Figure 19-5. Y-parameter plots of the 2N918

19.8 CALCULATIONS VERSUS MEASUREMENTS

Figure 19-6 shows the simulated (solid line curve) versus measured (small circular open dots) of attenuation versus frequency of an inserted signal in the reverse direction of our two-stage isolation amplifier. At 70MHz we have:

$$A_{v(forward)} = -5.18@142.8^{\circ} = 14.3dB$$
$$A_{v`(reverse)} = -1.89x10^{-3}@-17.2^{\circ} = -54.5dB$$

Or a total forward-to-reverse figure-of-merit of 68.8db. This was sufficient for good Phase-Lock-Loop operation without spurious frequency pulling from the carrier sidebands.

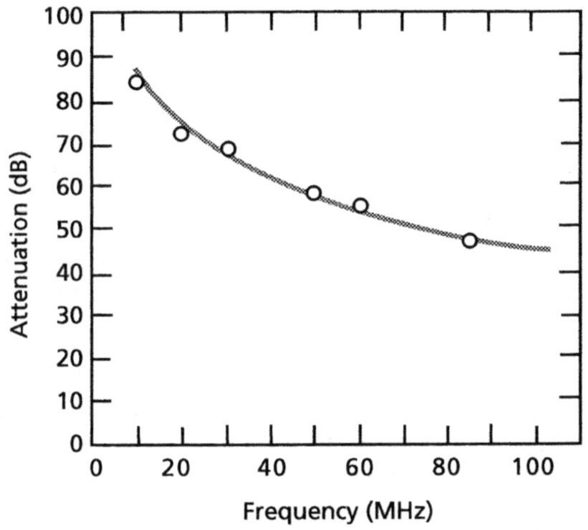

Figure 19-6. Reverse isolation measured (circles) versus calculated results (solid line) in dB

19.9 CONSTRUCTION AND TEST TECHNIQUES

Some care was taken to provide a good RF test fixture, shown in Figure 19-7, for proof-of-concept. This included a solid copper ground plane, a shielded enclosure, input and output RF BNC connectors reasonably isolated from each other, and self-contained bias batteries to avoid coupling across power supply cables.

Still, it appears there was the equivalent 1-2pF of stray coupling capacitance between input and output. That was expected.

Figure 19-7. Isolation amplifier test prototype
Cover removed. Note the batteries to avoid power line coupling to instrumentation.

19.10 SUMMARY

It is very important for engineers to build their confidence in equation-based and macromodeling methods. Macromodels and transfer functions allow engineers to effectively simulate very complex circuits. This example embeds a transistor-level macromodel in an RF amplifier and biasing circuit. Transfer functions are then extracted for the overall circuit. Engineers gain additional confidence when laboratory measurements confirm the equations.

This bottom-up derivation of a circuit transfer function forms an intellectual underpinning for deriving a top-down circuit transfer function and then implementing it in hardware. The example also shows how a transistor level macromodel gets combined with additional circuitry in an overall circuit model and how equations are extracted to form an AMS model.

PART 7: FUTURE DIRECTIONS IN MODELING

Chapter 20

THE CHALLENGE TO IBIS

SPICE as well as IBIS models are being used to simulate today's cutting-edge I/O buffers

Abstract: As semiconductor technology advances, there is a need to balance model complexity with simulation speed. After a decade of dominance in simulating I/O, IBIS is losing ground because of its limitations in simulating complex, programmable I/O. Possible solutions include using physical models (SPICE), behavioral models, macromodels, and the AMS modeling language. The leading contenders for solving the complex I/O modeling issues are SPICE-style macromodels and analog mixed signal equation-based models (AMS). IBIS already incorporates these modeling methods by calling [External Model] and other means. This chapter briefly reviews several modeling methods as they apply to model abstraction.

Everything should be made as simple as possible, but not one bit simpler.
–Albert Einstein

20.1 EMERGING SIMULATION REQUIREMENTS

As semiconductor technology advances, so does the need for more effective semiconductor modeling. As the modeling method of choice in the 1990s, IBIS[1] provided an excellent practical balance between simulation speed and model complexity. But as of this writing, new issues have not been completely solved by improvements to the IBIS model specification.

The new HSDD issues are caused by the following developments:

- *Higher switching speeds* and higher power supply switching currents of today's cutting edge devices.

---

[1] Best embodied in IBIS version 3.1.

In particular, dI/dt power-supply demands of ICs have drastically increased. Modeling these new devices requires considerably more attention to Power Integrity and EMI effects. These concerns include power delivery systems, SSN/SSO, core switching noise, EMI radiation, heat-sinking, and self-heating effects on I/O switching waveforms and behavior.

- *Smaller geometries*, larger number of gates, lower power supply voltages, and lower switching thresholds of today's cutting edge devices.

  Modeling these new devices requires increased attention to heat dissipation, thermal effects, driver loading, threshold voltage shifts, and gate conductivity modulation. No longer can we assume that the interaction between a board and an I/O is negligible. Noise voltage levels that were safe in the past are now no longer safe.

- *More complex I/O* so that some problems can get solved inside the IC chip. This is in response to the problems created by today's faster semiconductor technologies.

  For example, suppliers are adding pre-emphasis, de-emphasis, dynamic clamping, differential pass-through receivers, and other circuitry to I/Os. This increases the needs for more detailed and complex I/O modeling since these IC features must be matched (designed) with signal requirements at the receiver. This is similar to matched filter design in analog circuits.

From its inception in 1993, IBIS has been a work in progress. In response to advancing technology needs, the IBIS Specification continues to evolve. This is part of its strength. V-T curves were first added in IBIS 2.0, followed by driver scheduling, and recently, support for hardware description languages (HDLs), such as SPICE, VHDL-AMS, and macromodeling in IBIS 4.1.

However, IBIS has not entirely kept pace with the needs of EDA tool users. Part of the slowness of the IBIS innovation process is due to the inherent slowness of any standards committee process. By its nature, a standards committee process is deliberative, consensus-building, and needs a time-consuming peer-review. Building consensus has been further slowed because there have been competing approaches to meeting the latest challenges, each with its own champions.

EDA companies—including Cadence, Synopsys, and Mentor—are putting forth solutions that incorporate SPICE and IBIS features that are already built into their simulation tools. There is a push to have either SPICE-based macromodeling or AMS modeling gain favor as the new modeling method. But there is no clear industry consensus on this matter.

The authors ask why does there have to be a winner and loser between these two approaches? Instead, it would be better for IBIS to simply facilitate IC companies to produce models in [External Model] format in whatever language they and their customers want. Language should not be force-fed or dictated by committee.

Meanwhile users, under product development pressures, have asked for transistor-level models of I/Os. Semiconductor companies have increasingly supplied the transistor-level models. These models have been primarily HSPICE. But SPICE modeling has not been an optimum solution for the following reasons:

- First, there are cross-platform and cross-model portability problems.

  Portability between different versions of SPICE models usually doesn't exist. Cadence and Mentor claim HSPICE model support, but the challenge of integrating an end-to-end EDA solutions (outlined in "Chapter 22, Future Trends in Modeling") involves multiple companies, multiple platforms, and many incompatible versions of SPICE.

- Second, there is the intellectual property (IP) problem.

  Obtaining encrypted models and signing NDAs slow down the design process considerably. Manipulating properties for what-if problem solving becomes much more difficult on encrypted models.

- Third, the slowdown in simulation speed using physical models is very significant. In some instances, it is a complete barrier to accomplishing necessary design tasks.

  Users report 10X to 400X speed differences over IBIS versus SPICE simulations. Papers [67] in the literature claim 5 to 10 hours for running 100 bits[2] of a pseudo-random bit sequence (PRBS) for a single SPICE simulation run on a single multi-gigabit serial data net. Determining accurate eye closures may require simulation runs of tens of thousand of bits or up into the millions of bits for a converged eye pattern.

---

[2] Cadence claims 5 to 10 minutes to simulate 100 bits of a pseudo-random bit sequence (PRBS) for a single simulation run on a single multi-gigabit serial data net. This is by using their "channel analysis" methods. These methods incorporate behavioral models in place of gate level physical models [134].

Simulator tool solution convergence of simulation runs, employing very large transistor-level I/O models, is often a very big problem. If a user is successful in getting a high-quality transistor-level model, and clever in getting it to converge to a solution, the user may not have enough time to simulate a large enough bit stream to get meaningful answers. Design skills, usually applied through what-if simulations, are severely limited in being leveraged.

- Fourth, encrypted physical models may not be available in time.

  Models derived directly from silicon layout are not available until *after* the device is completely laid out. For customers designing new devices into their systems in parallel with the IC design, layout-based models are simply not available. This mandates that early behavioral models be utilized for leading edge customers with fast time-to-market requirements.

Therefore, the shift to encrypted physical models (now over 20% of models supplied and growing), while driven by necessity, is viewed as less than optimum, interim solution. A server board can contain more than 1000 components and 10,000 nets. A final verification simulation of all these nets before production sign-off can require over 500,000 simulations. Multi-Gigabit SERDES nets can require simulation of over 10,000,000 bits of PRBS for full eye closure. Simulation run time increases as the cube of the number of nodes in a model. Behavioral model I/Os typically have 4 nodes, while a full physical SPICE I/O typically has 20 to 30. Users want a better simulation speed and more portability and openness than SPICE can provide.

20.2 THE LEADING CONTENDERS TO CHANGE IBIS

There are two leading contenders for providing a new practical balance between simulation speed and model detail. Both have pros and cons, and both are viable approaches for closing the gap on simulation performance. They are:

- SPICE-based macromodeling[3] and
- Analog mixed signal (AMS) equation-based modeling.[4]

---

[3] Macromodeling that employs simplification of an existing physical model.

[4] Use of a model's transfer function.

Both methods are supported in IBIS 4.1 and by major EDA vendors. The EDA vendors' true interests do not lie in delaying IBIS 4.1, but in responding to it effectively with their current tool sets.

There has been much recent discussion on the IBIS email reflector on how to implement SPICE-based macromodeling using the [External Circuit] and [External Model] capability in IBIS 4.1. The [External Circuit] and [External Model] functionality allows combinations of IBIS, SPICE, and AMS devices.

The authors seek to objectively discuss the technical merits and demerits of SPICE macromodeling and AMS. Perhaps because their technical merits and demerits are real, for now the IBIS solution lies in accommodating both approaches. One or the other method may be a clear winner in the long run, but the marketplace will decide that.

20.3 MODELS IN THE CONTEXT OF SIMPLIFICATION

To explain the advantages and disadvantages of different modeling methods, this chapter discusses the progression from a physical model to more abstract modeling. We use the term *abstract* in the sense of *to simplify* a model. The abstraction *process* is explained specifically to show how we go from a detailed to an abstracted model and what simplification occurs along the way.

To find a better balance between model detail and abstraction, we need to understand how abstraction is done. First, recall that a resistor:

can be either a physical element or a behavioral model. The behavioral equation:

$$I = V/R \qquad (20\text{-}1)$$

is abstracted from more fundamental physical properties, such as area, length, and conductivity. Conductivity in turn is derived from electron scattering. Within very wide limits, from microscopic to macroscopic, all models can be thought of as being both physical and behavioral.

When we give a value to a model property, for example 10 ohms, we *parameterize* the model. In the parameterization of the above resistor model, we usually ignore how conductivity changes with temperature or voltage, further simplifying the model. Table 20-1 presents a way to classify models by level of abstraction.

Table 20-1. Abstraction level method of classifying models

Type of Model	**Description**	**Correspondence to Structure of Device Modeled Or Circuit Schematic**	**Abstraction Level**
AMS Model	Linear or non-linear equation models: VHDL-AMS [8, 130] and Verilog-A [37, 129]	Structure usually completely obscured, but circuit synthesis [24, 139] can restore it	From detailed to very simplified
Behavioral Model	Measured characteristic curves, data tables, and combinations of these	Not physically modeled from structure. Observed at the terminals of a device	Completely abstracted, but can be very exact
SPICE-based macromodel	Functional Model: circuit simplification, circuit build-up, symbolic methods, and combinations of these	Building-block circuit primitives incorporating simplified functional elements	Simplified and abstracted
Transistor level circuit model	Primitive Model: devices represented by detailed physical model schematics	Close to one-to-one correspondence with physical devices and structures	Not abstracted

20.4 PHYSICAL MODELING

20.4.1 Discussion

Most engineers consider a transistor-level model (for example the Gummel-Poon model in Figure 20-1) to be a *physical* model of the transistor. However, the Gummel-Poon model is actually a macromodel.[5] It is a (mini) *system* of about a dozen elements (resistors, capacitors, dependent

[5] In this macromodel, many of the elements do not represent any actual physical structure. The macromodel resistors (r_{bb}, r_c, r_e) vary with voltage. Consequently, this model is valid only in certain operating regions. It is not used for large-signal analysis. All so-called physical models (Ebers-Moll [34], Shichman-Hodges [109]) are actually macromodels when compared to the device physics formulations.

signal sources, and other circuit elements) connected in particular way. The Gummel-Poon circuit also displays about a dozen circuit nodes.

Additional non-ideal elements and effects can be added to this simple model [33, 34, 64, 66 and others]. Some additional elements deal with large-signal effects, non-ideal effects, and package effects. Today's more complex, extremely small devices need even more elements to represent newly important physical phenomena. Even the simplified Gummel-Poon model shown in Figure 20-1 is too complex a system for which to easily derive input-output equations.

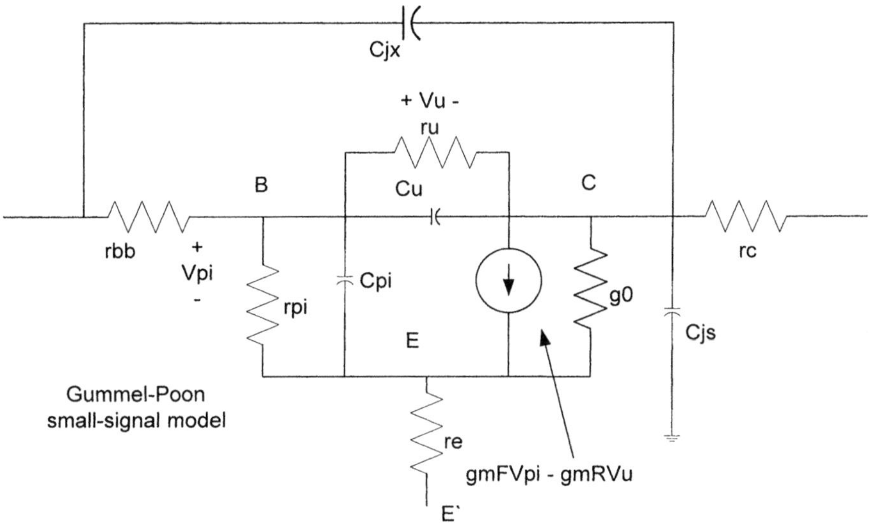

Figure 20-1. Gummel-Poon small-signal BJT model

The SPICE software program was created to simulate linear and non-linear circuits (such as the Gummel-Poon model) and to numerically compute results. The SPICE program uses about 40 *parameterized* elements for a BJT, and about 60 parameterized elements for a level=3 CMOS transistor. A current version BSIM CMOS model has about twice as many (about 120) parameters as a level=3 model. Some parameters are related to physical properties (such as dielectric constant, εr) and geometry (such as gate length), while others are curve-fitting parameters with no physical meaning.

20.4.2 Problems With Physical Models

Even with today's fast computers, the computational burden of simulating an entire board in SPICE is overwhelming. Simulation run time increases as the cube of the number of nodes in a circuit. A modern I/O circuit can contain 50 nodes or more for each I/O. A moderately complex net can contain 25 or more I/Os and 50 or more additional circuit nodes due to stubs, layer changes, and vias. Not all nets are so complex: many are much simpler. When what-if problem solving simulations are needed, each idea investigated requires its own computer run.

Physical models will probably continue to be encrypted and not portable across computing platforms, simulators, and design tasks. Encrypted models require NDAs, which increase project delays. Encrypted models are not particularly flexible. However, they can be used in some what-if simulations, if the data for variations in temperature, voltage, and process corner are embedded in them. But nothing approaching the kinds of simulations outlined in "Chapter 11, Using IBIS Models in What-if Simulations" can be done with encrypted models.

Physical models are not an optimum solution to simulating modern I/O. Remember that one of the challenges before us is to simplify this highly detailed physical modeling, and thus speed up simulation run time.

20.5 BEHAVIORAL MODELING

As a first attempt in the abstraction-simplification process, we place our physical model into a black-box. The black-box hides all the physical detail. What can we tell about device behavior at the input and output ports of the black box?

In IBIS modeling, we describe the device's behavior from the outside looking into the ports. We measure this behavior at the output driver and input receiver pins and say this is how our models *behave*. Traditional IBIS modeling is designed to ignore any pass-through, input-output modeling of behavior. But complex I/O requires some modeling of the buffer internal behavior. A new balance between simulation speed and I/O internal modeling will have to be devised.

20.5.1 Looking into the Black-Box

The black-box model can simplify the physical model of the output stage so that we can model driver and pre-driver at a less detailed level. Due to the abstraction of the model, the simulation runs faster.

We analyze the input-to-output relations of the black box as a first illustration of how abstraction is done. We want to learn:

- The loading effect of the black-box on a signal source driving it. This is the black-box input impedance.
- The strength of the black-box as a driver relative to the load it is driving. This is the black-box output impedance.
- When we drive with a voltage signal generator, what is the voltage signal transferred across the black box to its output? This is the black-box forward voltage transfer function, also called voltage gain.
- When we drive with a current signal generator, what is the current signal transferred across the black box to its output? This is the black-box forward current transfer function, also called current gain.
- The effects of temperature, voltage, and frequency on the behavioral model parameters.

Figure 20-2 shows a 2-loop circuit (Loop 1 and Loop 2) with a 2-Port black-box model forming the common branch between the two loops. One physical realization of the 2-Port model could be a resistor connected across the input and return lines. In that case Loop 1 and Loop 2 currents would both flow through this common resistor.

We now examine a black-box with an input admittance connected across port 1 (loop 1), and an output admittance connected across port 2. There is no direct connection between the two loops. Instead, a port-1 controlled-dependent generator in loop 2 and a port-2 controlled-dependent generator in loop 1 express the coupling between the two ports

We can set our primary signal generator in loop 1 or loop 2. It makes no difference to the fundamental process of solving node and mesh equations using Ohm's and Kirchhoff's Laws.

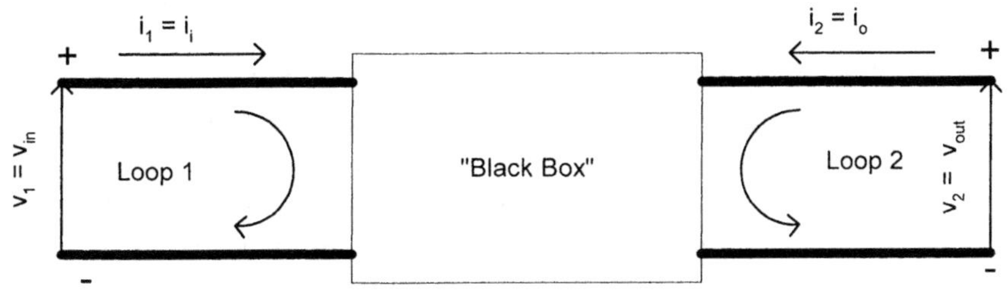

Figure 20-2. Circuit schematic of a 2-Port black-box model

Inside the black-box, we postulate a 4-element model. In this formulation of model elements to represent the black-box, we use input and output admittances, and controlled (dependent) current sources. The admittance parameter set is listed below:

- The input admittance is y_{11}
- The output admittance is y_{22}
- The forward transfer (input-output) admittance is y_{21}
- The reverse transfer (output-input) admittance is y_{12}

20.5.2 Setting up Matrix Parameter Sets

The Admittance Parameter Set Equations

The equations to solve the input and output relations of loop 1 and loop 2 with the intervening black box follow:

$$i_1 = y_{11}v_1 + y_{12}v_2 \tag{20-2}$$

$$i_2 = y_{21}v_1 + y_{22}v_2 \tag{20-3}$$

These admittance parameters are used inside modern EDA tools in matrix operations during simulation. But y-parameters are rarely used in the EDA tool users interface.

The Equations for the S-Parameter Set

S-Parameters are used primarily when modeling passive interconnection or RF and microwave amplifiers. Designers of high-speed digital circuits are increasingly encountering them.

One author [Leventhal] sees a certain amount of convergence coming between the techniques used for RF and microwave circuit design and Gbit serial data bus design. We already know that low voltage, small-signal differential switching tends toward more linearity than traditionally large-signal high-speed switching. We should also become aware that a received switching waveform that is a sine wave, slightly clipped and flattened, may represent the best ones and zeros bit stream that can be transmitted at 10s of GHz. Such a waveform is similar to microwave amplifiers of moderate bandwidth.

Visualizing and solving high-speed serial data bus and EMI problems are aided by thinking in S-Parameters, the frequency domain, and signal wavelengths. Digital designers are going to encounter S-Parameters for modeling connector insertion loss, transmission line losses at GHz frequencies, power plane effects, EMI and more. So, for the sake of completeness and for forecasting the future, we will take a brief look at S-Parameters.

On a transmission line, the incident voltage wave at a point on the line can be notated as $V^+(x)$, and the reflected wave as $V^-(x)$. The variable x is distance.

At some other point on the line, we have:

$$V^+(x) = A\ell^{-\gamma x} \tag{20-4}$$

$$V^-(x) = B\ell^{+\gamma x} \tag{20-5}$$

where $\gamma = \alpha + j\beta$ and for a lossless line $\gamma = j\beta$. The total voltage, V(x), at a given point on the line is:

$$V(x) = V^+(x) + V^-(x) \tag{20-6}$$

and

$$I(x) = I^+(x) - I^-(x) = V^+(x)/Zo - V^-(x)/Zo \tag{20-7}$$

The reflection coefficient, Γ(x), between incident and reflected waves can be written as:

$$\Gamma(x) = \frac{V^-(x)}{V^+(x)} \tag{20-8}$$

Next, we convert from voltage-current waveforms to a power flow notation:

$$\begin{aligned} v(x) &= V(x)/\sqrt{Zo} \\ i(x) &= I(x)\sqrt{Zo} \\ a(x) &= V^+(x)/\sqrt{Zo} \\ b(x) &= V^-(x)/\sqrt{Zo} \end{aligned} \tag{20-9}$$

Although this book does not address power flow, we follow the standard notation for S-Parameters where power flow analysis is used.

We can then write equations (20-6), (20-7), and (20-8) in the form:

$$\begin{aligned} v(x) &= a(x) + b(x) \\ i(x) &= a(x) - b(x) \end{aligned} \tag{20-10}$$

and:

$$b(x) = \Gamma(x)a(x) \tag{20-11}$$

Instead of a 1-Port transmission line, we apply this notation to a 2-Port network with incident wave a_1 and reflected wave b_1 at port 1, and incident wave a_2 and reflected wave b_2 at port 2. Thus, we have:

$$b_1 = s_{11}a_1 + s_{12}a_2 \tag{20-12}$$

$$b_2 = s_{21}a_1 + s_{22}a_2 \tag{20-13}$$

For 2-Port lines, the independent S-Parameter variables a_1 and a_2 are the normalized incident voltages:

$$a_1 = V_{i1}/\sqrt{Zo} \tag{20-14}$$

$$a_2 = V_{i2}/\sqrt{Zo} \tag{20-15}$$

Dependent variables b_1 and b_2 are the normalized reflected voltages:

$$b_1 = V_{r1}/\sqrt{Zo} \tag{20-16}$$

$$b_2 = V_{r2}/\sqrt{Zo} \tag{20-17}$$

The squares of the magnitudes of the a and b parameters represent incident and reflected power flows on ports 1 and 2. S-parameters are further summarized in Table 20-2.

Table 20-2. S-Parameter definitions

Description	**Definition**	**Condition**
Input reflection coefficient	$s_{11}=b_1/a_1$	$a_2=0$ (output properly terminated)
Forward transmission coefficient	$s_{21}=b_2/a_1$	$a_2=0$ (output properly terminated)
Reverse transmission coefficient	$s_{12}=b_1/a_2$	$a_1=0$ (input properly terminated)
Output reflection coefficient	$s_{22}=b_2/a_2$	$a_1=0$ (input properly terminated)

Measuring S-Parameters involves the use of directional couplers for measuring incident and reflected waves. S-Parameters are measured with matched loads at the generator and load ends of the transmission lines connected to the ports of the black-box. Thus there are no re-reflections of the EM waves. Physically, it is much easier to match impedance than to achieve shorts and opens at high frequencies.

20.5.3 Measurement of 2-Port Admittance Parameters

We now return to the admittance parameter set of equations for solving the input-output transfer functions. First, let us define how admittance parameters can be measured. Particularly so, because the black-box is a behavioral model with the observer applying a stimulus to it and measuring how it behaves.

To illustrate the techniques of measurement, let us consider just the admittance set of parameters. Table 20-3 shows how the admittance parameters are measured. For more complete treatments, and other matrix parameter sets, see [2, 3, 65, 124, 126].

Table 20-3. y-Parameter measurement

Description	Definition	Condition
Input admittance	$y_{11} = i_1/v_1$	$v_2 = 0$ (AC short circuit)
Forward transfer admittance	$y_{21} = i_2/v_1$	$v_2 = 0$ (AC short circuit)
Reverse transfer admittance	$y_{12} = i_1/v_2$	$v_1 = 0$ (AC short circuit)
Output admittance	$y_{22} = i_2/v_2$	$v_1 = 0$ (AC short circuit)

y_{21} of a common-emitter amplifier is the same as y_{fe} the common-emitter current gain of the transistor. When measured at low frequencies, it is synonymous with the Beta of the transistor. When measured at high frequencies where the voltage gain of the transistor has rolled off to 1, or zero dB, (the gain-bandwidth of the transistor), then $y_{21} = f_t$.

At high frequencies, the real and imaginary parts of the 2-port parameters are measured to obtain the magnitudes of the currents and voltages, and their phase shift information. The measurements are usually taken with a vector impedance and admittance meter or a network analyzer. For a Class-A amplifier, measurements are taken over a range of frequencies at particular collector-emitter voltages, and collector currents, in the normal class-A amplifier range.

Y-parameter models can be measured accurately out to 1 GHz or so. However, at higher frequencies, component self-resonances, transmission line effects, reflections and other complications make a shift to Scattering Parameters necessary.

So far, our analysis could be applied to any kind of network within the black-box: active, passive, BJT, MOSFET, etc. In the following sections, black-box matrix formulations are applied to a common linear active circuit—the CE BJT amplifier.

20.5.4 A Parameterized Y-Parameter 2-Port Model

Let us assume that within the black box we have a common-emitter (CE) configuration of a BJT amplifier. This is a small-signal circuit where the values of the parameters depend on the transistor's DC bias. The specific realization of the matrix model is shown in Figure 20-3.

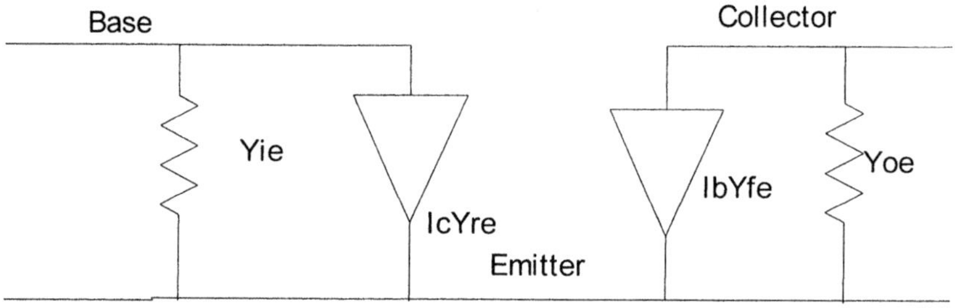

Figure 20-3. Common emitter y-parameter model

The notation change—from generalized matrix parameters to specific CE BJT matrix realization—is as follows:

- Input admittance y_{11} becomes y_{ie}
- Output admittance y_{22} becomes y_{oe}
- Forward transfer admittance y_{21} becomes y_{fe}
- Reverse transfer admittance y_{12} becomes y_{re}

Note that y_{ie}, y_{oe}, y_{fe}, and y_{re} are all bias and frequency dependent now that the general matrix definitions are applied to a specific device (a BJT) in a specific circuit configuration (CE). Figure 19-5, "y-parameter plots of the 2N918," illustrates a set of the admittance parameters for a particular transistor.

Now we have described and defined a particular realization of a behavioral, 2-Port matrix model—the CE BJT amplifier. After measuring and characterizing its parameters, we could embed this model in a computer simulation program. Instead we will solve this circuit in closed-form equations.

A linear small-signal model can be extended to cover large signal operation. In this case, a simulator would evaluate the parameters, calculate I (current) for the current time step, and then calculate Vb and Vc for the next time step. At each time step, the parameters would be updated based on the current values of the voltages.

To cover a wide range of voltage, current and frequency conditions, we would have to measure the model parameters over those ranges and store the data in a series of model tables, or as a set of model equations. To account for process variation of the characterized device, we would further have to characterize the process spread for all of the parameters. The process spread could be measured at three corners (typical, slow and weak, and fast and strong), or over a 16-corner distribution that would separate voltage, temperature, and process corners to describe the corners of the population distribution. An advanced simulator could take advantage of a population distribution function, if enough data is taken on many nearly identical transistors from production runs. For example, the three-corner approach is used in IBIS.

However, since this 4-element model is so simple, why not embed it in its bias, input, and output coupling circuitry? We now develop an overall two-port *macromodel* for that entire circuit block by using the CE BJT amplifier configuration as an illustration.

20.5.5 Applying the Model to a CE Amplifier

Figure 20-4 shows a typical CE BJT amplifier with its bias, input, and output circuitry:

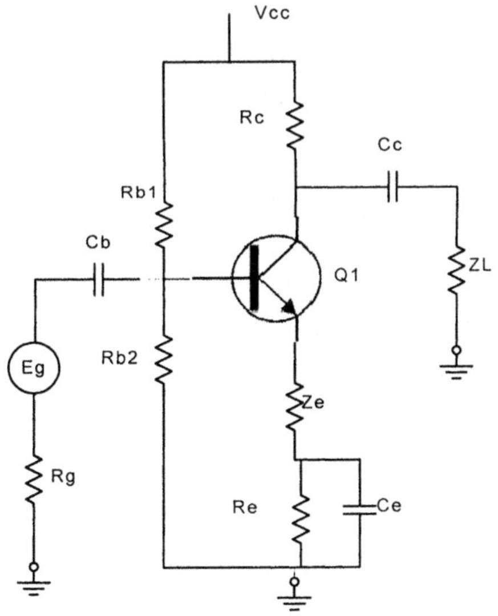

Figure 20-4. A typical common emitter amplifier circuit

Once again, the definitions of the y-parameters and some typical orders of magnitude for small-signal, high-frequency discrete BJT transistors are shown:

- y_{ie} (y_{11}) is the input admittance:
  - typically on the order of 10^{-3} Siemens (mhos)
- y_{oe} (y_{22}) is the output admittance:
  - typically on the order of 10^{-4} Siemens (mhos)
- y_{fe} (y_{21}) is the forward current transfer ratio:
  - typically on the order of 10^{-2} Siemens (mhos)
- y_{re} (y_{12}) is the reverse current transfer ratio:
  - typically on the order of 10^{-5} Siemens (mhos)

For Figure 20-4, Cb, Ce and Cc are considered to be AC short circuits. The values of Rg, Ze and ZL (respectively) are changed to include their effects if their capacitive impedances become significant at frequencies of interest. For example, Rg becomes Zg in that case.

20.6 DEVELOPING A MACROMODEL FROM THE BEHAVIORAL MODEL

Macromodeling is used when we want to reduce the complexity of the original, physical circuit model. The chief advantages are:

- The computer simulations run faster, and
- The user gets a more intuitive grasp of the major behavioral relationships of a circuit

For instance, in the Gummel-Poon model shown in Figure 20-1, it is quite difficult to infer input-output relationships. The Gummel-Poon is simple when compared to a detailed 30+ elements in a typical modern SPICE model. But it is complex compared to the simple 4-element matrix model. In the 4-element matrix model, we almost intuitively understand that there are behaviors that look like:

- Input and output admittances or impedances
- Signals at the input getting transferred to the output
- Signals at the output feeding back to the input

Also, a matrix model, imbedded in some bias circuitry, can be solved for a total closed-form transfer function with reasonable effort. Once derived, these transfer functions may be simplified further to improve the designer's intuitive understanding. Simplifying a model or equation down to their most significant and basic terms often aids understanding.

The macromodel is shown in Figure 20-5:

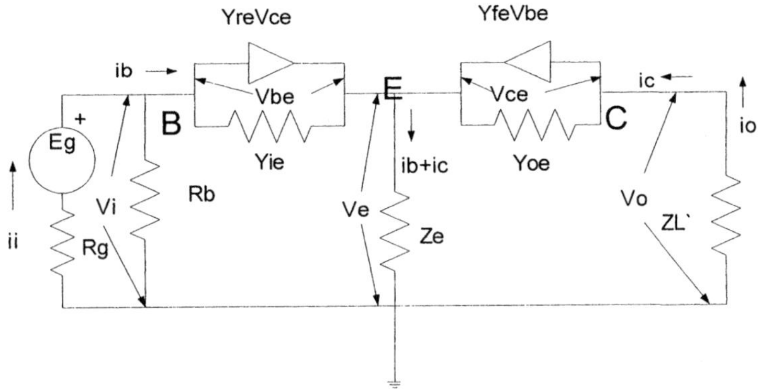

Figure 20-5. Macromodel for the 2-Port common emitter amplifier

Analysis:
Let

$$\Delta y = yie * yoe - yre * yfe \tag{20-18}$$

Let αi be the current transfer ratio of the input network. Rb is the parallel combination of Rb1 and Rb2.

$$\alpha_i = R_b / (R_b + Z_i) \tag{20-19}$$

Let αo be the current transfer ratio of the output network.

$$\alpha_o = R_c / (R_c + Z_L) \tag{20-20}$$

Equivalent load impedance, Zl`

$$Zl` = Rc * Zl / (Rc + Zl) \tag{20-21}$$

Equivalent generator impedance, Rg`

$$Rg` = 1/(1/Rg + 1/Rb1 + 1/Rb2) \tag{20-22}$$

Now, with these substitutions, we have the following macromodel:

Current gain of the 2-Port itself is Ai`

Current transfer function of the macromodel (generator to load), Ai:

$$A_i = \alpha_i * \alpha_o * Ai` \qquad (20\text{-}23)$$

$$Ai` = (\Delta y * Ze - yfe)/(yie + (Ze + Zl`) * \Delta y) \qquad (20\text{-}24)$$

Voltage transfer function of the macromodel, Av:

$$\begin{aligned} Av = &(yie * Zl` * (Ze * \Delta y - yfe))/ \\ &\{[1 + (yie + yre) * Ze] * [yie + (Ze + Zl`) * \Delta y] \\ &+ (yfe - \Delta y * Ze) * [yie * Ze + yre * ((Ze + Zl`)]\} \end{aligned} \qquad (20\text{-}25)$$

Note: Bottom-up behavioral modeling and top-down macromodeling often result in similar equations. This is especially true when models and equations are simplified to increase simulation speed. When we simplify the model equations, we give up detail, subtlety, and accuracy.

Contrast equation (20-25) with the level of abstraction and simplicity shown in equation (20-46). Next, imagine the voltage transfer function that might result from a 100-element BSIM model. If we could derive that equation in a closed form, the equation would be extremely complex. The point is that the level of detail and abstraction of a model is a somewhat flexible concept. The equation's complexity depends on the model maker's skill *and* objectives.

Input impedance (loading of the 2-Port on the signal source) of the 2-Port, Zi:

$$\begin{aligned} Zi = &\{Ze * (yie + yre) * (yie + yfe + \Delta y * Zl`) \\ &+ yie + \Delta y * Ze + yoe * yie * Zl` - yre * \Delta y * Ze * Zl`\} \\ &/\{yie * [yie + \Delta y(Ze + Zl`)]\} \end{aligned} \qquad (20\text{-}26)$$

Output impedance (driver strength) of the 2-Port, Zo:

$$Zo = \{1 + yre * Rg` + (yie + yoe + yre + yfe) * Ze + \Delta y * Rg` * Ze\} / [yoe + \Delta y * (Rg` + Ze)] \tag{20-27}$$

Simplifications:

When Ze is very small or zero, we have the simplified (macro-model) set:

$$Ai`` = -yfe * Zl` / (yie + \Delta y * Zl`)$$

$$Av` = -yfe / (1 + yoe * Zl`)$$

$$Zi = (1 + yoe * Zl`) / (yie + \Delta y * Zl`) \tag{20-28}$$

$$Zo = (1 + yie * Rg`) / (yoe + \Delta y * Rg`)$$

Further simplifications can be made by considering the magnitudes of y_{ie}, y_{oe}, y_{fe}, and y_{re}. These y-parameters are usually measured in steady state, AC, and small-signal transient conditions. The y-parameters are usually *not* measured under large-signal conditions (for example, saturation to cutoff). However, there is no reason they could not be measured large-signal.

One look at the I-V (dynamic, non-linear output impedance) curves of a switching digital I/O shows that the concept of large signal y-parameters has limited practicality.

20.6.1 Limitations in Behavioral Table-Based Macromodels

In behavioral table-based macromodels, parameter data is stored in data table format. Parameter values change markedly over a non-linear current-voltage range. Some of the issues with table-based behavioral models are:

- Linear parameter measurements versus actual non-linear HSDD circuit operation.

- Data tables rapidly become massive when this modeling method is applied to HSDD.[6]
- Accuracy requirements cause a shift to S-Parameter measurements and models as the speed envelope gets pushed.
- Impossibility of probing inside the IC and package to measure the model parameters. We need to be able to access the input to the buffer to measure its model parameters.

There is an even more fundamental reason for abandoning matrix model parameters in our modeling abstraction quest. We could get around the previous four limitations by generating y-parameters, S-Parameters, or any other matrix model set from computer simulations of the I/O SPICE model. In these simulations, SPICE models of the driver, pre-driver, clamps, and other portions of the I/O buffers would be used the same way as when generating most IBIS models. But the penalties in computer simulation time would be again quite large with little payoff to the semiconductor suppliers who provide the models.

Behavioral macromodels don't get derived directly from physical layout; that is the point. Even in SPICE-based macromodeling, the top-level schematic circuit of the I/O is considered, and a parameterized template for this class of I/O is developed, using SPICE and IBIS building blocks. Then the parameters are tweaked to model specific parts, and correlated to whatever is considered "golden" data.

What other techniques could we use to generate a macromodel of an IC? Looking at how operational amplifier (OpAmp) macromodeling was developed is educational. OpAmps (using from approximately 30 individual transistors to several hundred individual transistors) prompted the development of the Boyle OpAmp macromodel and later derivations.

20.7 DEVELOPING A SPICE MACROMODEL FROM A PHYSICAL MODEL

We now return to physical models of transistors. First, we simplify them, which is the approach taken by Boyle and his collaborators.

A full set of SPICE model parameters can tell a device designer a great deal (when aided by a computer) about how to design a device embedded in a circuit. If the designer is given the full SPICE physical model, it is also possible to reverse-engineer the device. However, computer programs often

---

6 Tables are multi-dimensional (bias, frequency), requiring interpolation over more than one independent variable. Using cubic splines (or higher order) generates smoother fits to data, but is slower to calculate than linear interpolation.

do not provide insight about what model properties to control for desired results.

To gain intuitive insight into circuit behavior, we want to use some reasonable model simplifications [80] and, where necessary, derive closed form equations of a circuit. Once we derive the equations, we can also simplify them based on relative parameter magnitudes.

20.7.1 SPICE-Based Macromodels for Amplifiers

Figures 20-6 through 20-9 show the four basic types of amplifiers in simplified, abstracted form. The formulas accompanying each circuit are used to calculate the output quantities from given input quantities.

Linear voltage amplifier with formulas

Figure 20-6 shows a linear voltage amplifier. It can be applied to BJTs, MOSFETs, JFETs, and vacuum tubes.

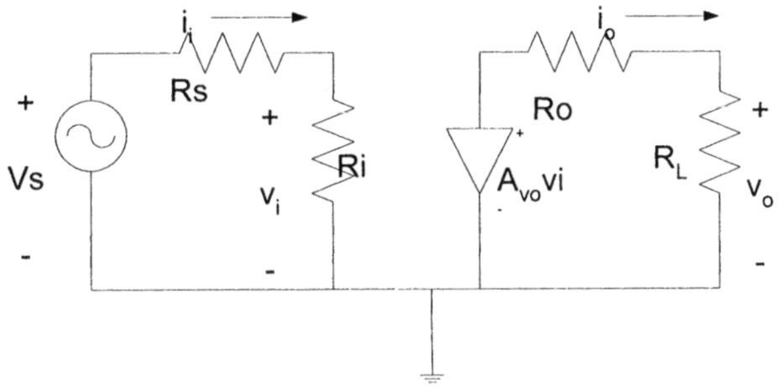

Figure 20-6. Voltage amplifier circuit

$$vo = A_{vo}v_i[R_L/(R_L + Ro)] \quad (20\text{-}29)$$

$$v_i = v_s[Ri/(Ri + Rs)] \quad (20\text{-}30)$$

Linear current amplifier with formulas

Figure 20-7 shows a linear current amplifier. It can be applied to BJTs, MOSFETs, JFETs, and vacuum tubes.

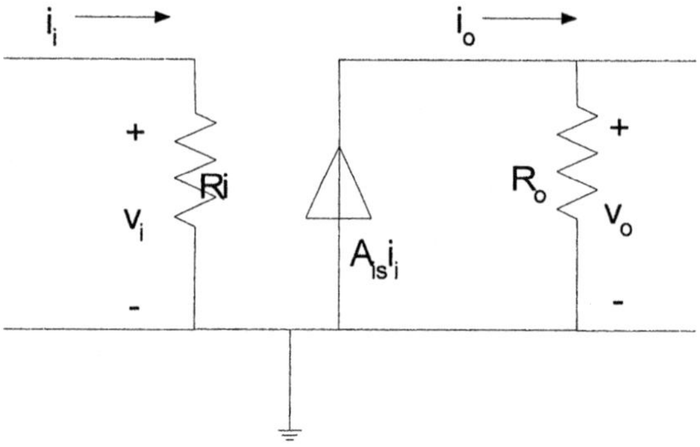

Figure 20-7. Current amplifier circuit

$$i_o = A_{is} i_i [Ro/(Ro + R_L)] \tag{20-31}$$

$$i_i = i_s [Rs/(Rs + Ri)] \tag{20-32}$$

Linear transconductance amplifier with formulas

Figure 20-8 shows a transconductance amplifier circuit.

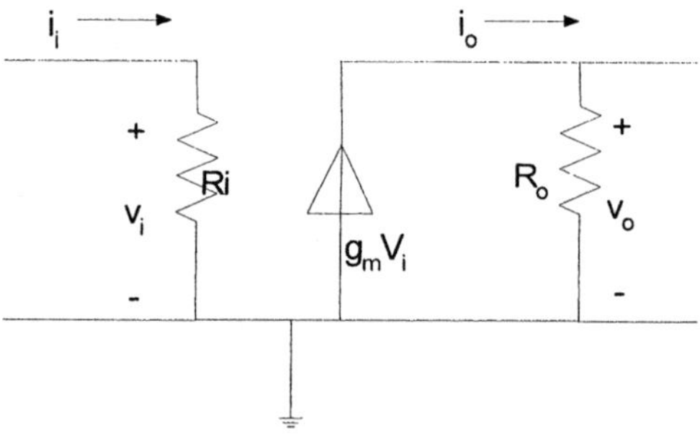

Figure 20-8. Transconductance amplifier circuit

$$i_o = g_m V_i [Ro/(Ro + R_L)] \tag{20-33}$$

$$v_i = v_s [Ri/(Rs + Ri)] \tag{20-34}$$

Linear transresistive amplifier with formulas

Figure 20-9 shows a transresistive amplifier circuit.

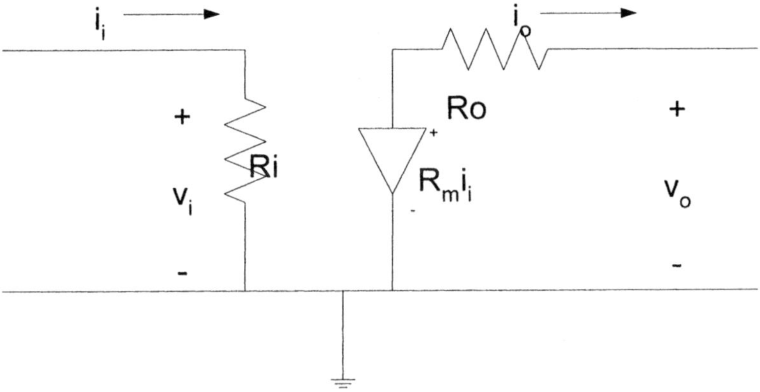

Figure 20-9. Transresistive amplifier circuit

$$v_o = R_m i_i [R_L / (R_L + Ro)] \tag{20-35}$$

$$i_i = i_s [Rs / (Rs + Ri)] \tag{20-36}$$

20.7.2 Equation Relating the Four Simplified Models

$$A_{vo} = A_{is} (R_o / R_i) = g_m R_o = R_m / R_i \tag{20-37}$$

Current amplifier and formulas

We next apply the simplified current amplifier model to our CE amplifier. We look at a current amplifier in Figure 20-10 and Figure 20-11.

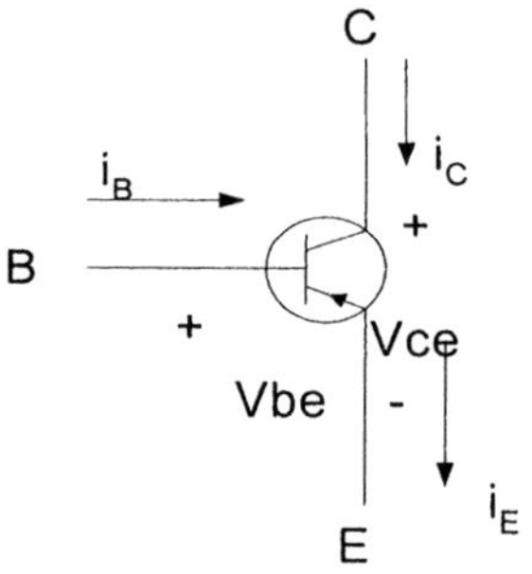

Figure 20-10. BJT common emitter configuration

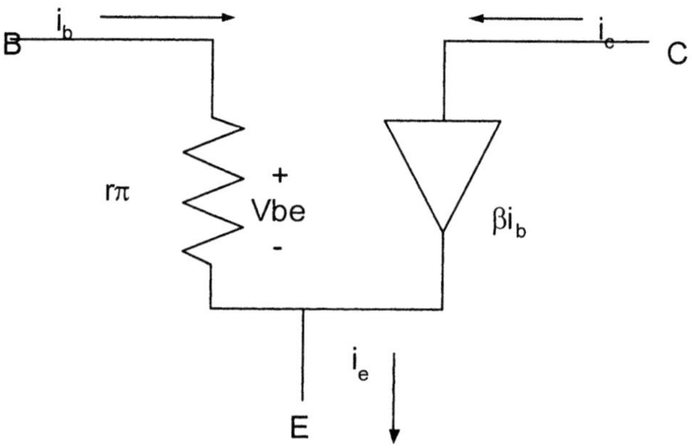

Figure 20-11. Common emitter current amplifier

$$i_b = v_{be} / r_\pi$$
$$i_c = \beta * i_b \qquad (20\text{-}38)$$
$$i_e = (\beta + 1) * i_b$$

For finite Re in the emitter:

$$Av = \frac{\beta R_L}{r_\pi + (\beta + 1) R_e} \qquad (20\text{-}39)$$

20.7.3 SPICE-Based Macromodel Example: OpAmp

An example of a SPICE-based macromodel generated from simplified transistor-level physical models is the Boyle OpAmp SPICE-based macromodel. The abstraction method is to use simplified models (as shown in section 20.7.2) along with simplified circuitry for bias, gain setting, temperature stabilization, and signal clamping. The actual operational amplifier might be composed of several hundred IC gate level elements. Refinements from various contributors have been added to the Boyle [19] model. One such refinement is called the level 2 OpAmp. The SABER listing of the level 2 OpAmp parameters is found in Table 20-4. The equations developed for the OP1 model are as follows:

The transfer function, Vo/Vi, is:

$$Vo / Vi = \frac{A}{(1 + s / P1)(1 + s / P2)} \tag{20-40}$$

where:

A	= open-loop gain
F1	= product of the dominant pole frequency times |A|
P1	= dominant pole frequency
P2	= second pole frequency
SR	= slew rate dVo/dt
V_{os}	= input offset voltage
Ip	= DC average supply current
Rout	= output resistance
s	= the s-plane complex frequency variable that becomes jω in the frequency domain (not the same as S-Parameters)

And:

$$Vo = Vout - (i_{out} * Rout) \tag{20-41}$$

$$Vi = Vin + Vos \tag{20-42}$$

This OpAmp material is presented primarily because it *illustrates* the creation and extraction of SPICE-based macromodels from physical models. Figure 20-12 shows a 7-terminal OpAmp element:

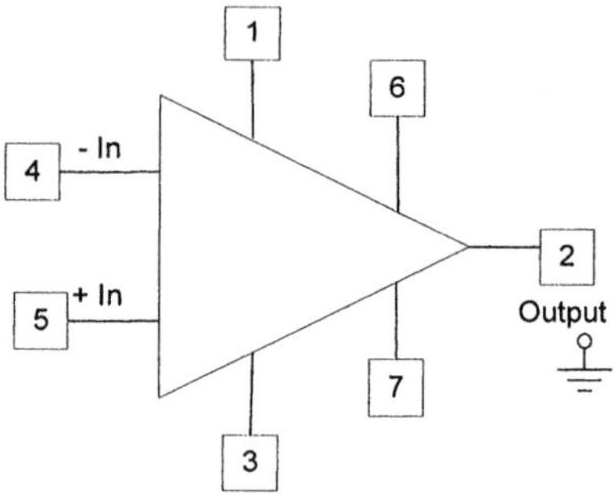

Figure 20-12. OpAmp 7-terminal component

To simplify a macromodel simulation even further, we can ignore terminals 1, 3, 6, and 7. They are for power supplies, DC offset voltage, and compensation. Equation (20-40) is very simplified compared to what we would expect from the circuit in Figure 20-13. Still, equation (20-40) is about as detailed as a closed-form OpAmp transfer function gets. The reality is that even the simplified OpAmp SPICE-based macromodel shown is too much to derive closed-form input-output transfer functions for. Instead, the standard approach is to simulate the OpAmp macromodel in SPICE. For a level 2 OpAmp, added complexity is an even stronger reason for using SPICE computer simulation.

Figure 20-13 is the schematic of the classic Boyle level 1 (OP1) OpAmp model. The SPICE-based macromodel contains differential transistor input stages that establish a high-input impedance and offset effects. Later stages are current sources and filters. Diodes on the output limit output voltage swings.

Using different models for the two transistors in the differential pair captures offset effects caused by transistor mismatches. If those transistors have the same model, they will be *perfectly matched* in SPICE.

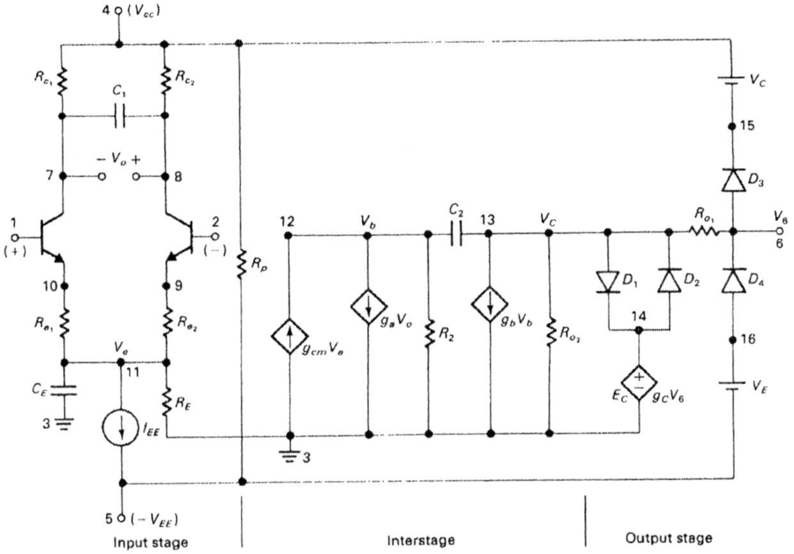

Figure 20-13. Boyle μA741 OpAmp macromodel [100]

A full set of OpAmp parameters follows in Table 20-4 [104]:

Table 20-4. OpAmp parameter set

Parameter	Description	Note
TYPE	Input stage configuration - not device type	
OREF	Reference for output drive signal	
INTECH	Input stage technology	
OUTECH	Output stage technology	
CTECH	Compensation technique	
INBETA	Input stage beta	OP1[7]
INTEMP	Added ib temperature dependence	
OUTZ	Output stage configuration	
MATL	Semiconductor material	
VOS	Input offset voltage	
VOSTC	VOS linear temperature coefficient	
VOSDV	VOS dependence on output voltage	
VOSDIC	VOS dependence on output source current	
VOSDIK	VOS dependence on output sink current	
PSRR	Power supply rejection ratio	
CMRR	Common mode rejection ratio	
IB	Input bias current	OP1
IBTC	IB linear temperature coefficient	
IBOS	Input bias offset current	

[7] OP1 is the same parameter as in the MAST Level 1 OpAmp macromodel.

Table 20-4. (cont.)

Parameter	Description	Note
IBOSTC	IBOS linear temperature coefficient	
IEE	Effective input differential pair tail current	OP1
DVIE	Supply voltage change for which IEE = 0	
CMILL	Compensation capacitance	OP1
EN	Input noise voltage	
EFC	Flicker noise voltage at corner	
EAF	Flicker noise voltage exponent	
IN	Noise current - each input	
IFC	Flicker noise current at corner	
IAF	Flicker noise current exponent	
CIN	Input capacitance - differential	
RIS	Series resistance - each input	
ISD	Input protection diode saturation current	
NPD	Number of input protection diodes	
IS	Input stage saturation current	OP1
BETA	Input stage transconductance	OP1
VTH	Input stage threshold voltage	
DVINP	Max input voltage @ INP	
DVINM	Max input voltage @ INM	
SRP	Positive slew rate	OP1
SRN	Negative slew rate	OP1
AVOL	Open loop gain	OP1
RO1	Input stage output resistance	OP1
F1	Gain bandwidth product	OP1
P1	First real pole -dominant	
P2	Second real pole	
P3	Third real pole	
P4	Fourth real pole	
Z1	First real zero	
Z2	Second real zero	
PHM	Phase margin	
IOQ	Output stage quiescent current	
ROLF	Resistance of drive stage	
RODC	DC output resistance	OP1
CD	Capacitance of drive stage	
ROHF	High frequency output resistance	
OARAT	Area ratio of output transistors to that of their bias diodes	
BFU	Forward beta of upper output stage	
BFL	Forward beta of lower output stage	
FTU	Unity gain frequency of upper output stage	
FTL	Unity gain frequency of lower output stage	
ISCC	Max output short circuit sourcing current	
ISCK	Max output short circuit sinking current	OP1
VXOD	Crossover deadband voltage	
NOD	Number of output clamp diodes	

Table 20-4. (cont.)

Parameter	**Description**	**Note**
NOD	Number of output clamp diodes	
AVMXOD	Gain multiplier in crossover deadband region	
DVOMX	Max output driving voltage	OP1
DVOMN	Max output voltage before ISCC = 0	
DVOIMX	Max output voltage before ISCK = 0	
VCCNOM	VCC at which parameters are defined	
VEENOM	VEE at which parameters are defined	
TNOM	Temperature at which parameters are defined	
SMIN	Minimum slope constant	
IPD	Piecewise linear quiescent current from VCC to VEE	
LIMITS		

The SPICE netlist for Figure 20-13 is shown in Table 20-5:

Table 20-5. uA741 SPICE model netlist

```
* Library file "OPNOM.LIB" for uA741 OpAmp
* connections:
* 1        Non-inverting input
* 2        Inverting input
* 3        ground
* 4        Positive power supply
* 5        Negative power supply
* 6        Output
.SUBCKT UA741 1 2 3 4 5 6
* Vi+ Vi- GND Vp+ Vp- Vout
QI      7    1    10   UA741QA
Q2      8    2    9    UA741QB
RC1     4    7    5.305165D+03
RC2     4    8    5.305165D+03
C1      7    8    5.459553D-12
RE1     10   11   2.151297D+03
RE2     9    11   2.151297D+03
IEE     11   5    1.666000D-05
CE      11   3    3.000000D-12
RE      11   3    1.200480D+07
GCM     3    12   11   3   5.960753D-09
GA      12   3    8    7   1.884955D-04
R2      12   3    1.000000D+05
C2      12   13   3.000000D-11
GB      13   3    12   3   2.357851D+02
RO2     13   3    4.50000D+01
D1      13   14   UA741DA
D2      14   13   UA741DA
EC      14   3    6  3  1.0
RO1     13   6    3.000000D+01
D3      6    15   UA741DB
```

Table 20-5. (cont.)

* Library file "OPNOM.LIB" for uA741 OpAmp			
VC	4	15	2.803238D+00
D4	16	6	UA741DB
VE	16	5	2.803237D+00
RP	4	5	18.16D+03
* End of subcircuit definition			
.ENDS			
* Models for diodes and transistors			
.MODEL	UA741DA	D	(IS=9.762287D-11)
.MODEL	UA741DB	D	(IS=8.000000D-16)
.MODEL	UA741QA	NPN	(IS=8.000000D-16 BF=9.166667D+01)
.MODEL	UA741QB	NPN	(IS=8.309478D-16 BF=1.178571D+02)
* End of library file			

Sedra and Smith [108] devote the following chapters to OpAmp modeling and design:

- Chapter 2 presents OpAmps at the level examined here.
- Chapter 9 gets into the internal design of an OpAmp. Chapter 9 shows how the output stages (class AB) of a BJT power amplifier realization of an OpAmp can be designed at the component level.
- Chapter 10 is devoted to the design and circuit application of analog ICs, particularly OpAmps.

20.7.4 SPICE-Based Macromodeling Templates by Cadence

We turn our attention to SPICE-based macromodeling being done by Cadence Design Systems, Inc. for today's complex digital I/O circuitry. Most of the material in this section is adapted from Cadence literature and is used with their permission.

Viewing transistors as current sources allows the replacement of many stages of transistors. A few strategically placed controlled current sources act as the building blocks of a behavioral SPICE-based macromodel. In the Cadence version of SPICE (tlsim), these sources are called *bdrvrs*. A 7-terminal single-ended driver (see Figure 20-14) and an 8-terminal differential driver are defined for tlsim.

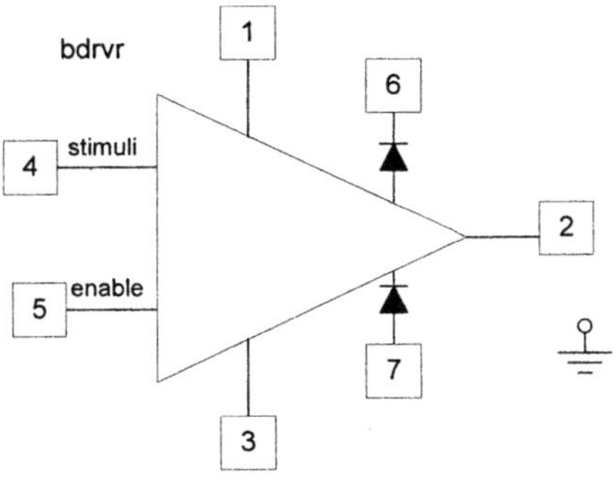

Figure 20-14. Single-ended 7-terminal bdrvr element

The use of these elements in the model of a SPICE-based macromodel pre-emphasis driver with one tap of pre-emphasis is shown in Figure 20-15.

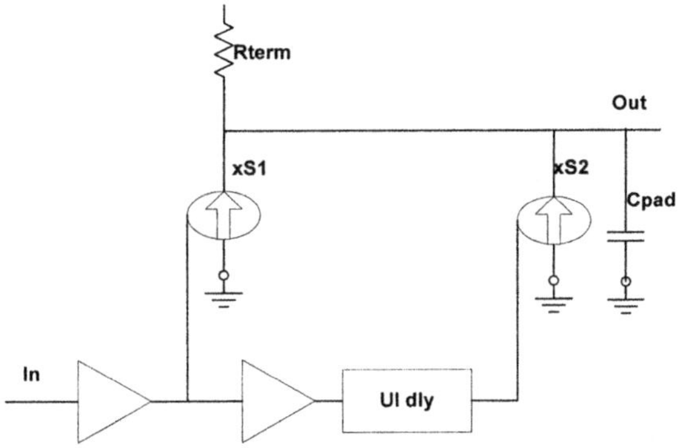

Figure 20-15. Pre-emphasis driver SPICE-based macromodel

In a SPICE-like netlist file (structured model), the model elements include controlled current sources, inverter, delay line, pad capacitance, and on-die termination. S1 and S2 are the scaling factors for the current sources. To speed the simulation, the netlist uses sub-model callouts of behavioral models for the current sources. Additional transistor level effects can be included. Also in place of other sub-model callouts, equation-based models can be embedded in the structured netlist. In the Cadence approach, the design engineer edits a line or two of text in the template file[8] to select different driver characteristics to use in a simulation.

---

[8] The Cadence template is an ASCII text file with instructions for the simulator regarding all pre-emphasis driver characteristics (or other circuit functions) for a given component. An example, pe_2tap.dml, is included in "Appendix B, Sample IBIS Model Files," The file is also saved as "pe_2tap_dml.doc" for readability on the CD. This MacroModel is of MacroType TDiffIO. It is a differential bdrvr. This differs from a single-ended

The schematic in Figure 20-15 shows a data stream signal going into two inverters, or gates. Their current outputs can be steered to aid or oppose through the load. Also, one driver output can be delayed with respect to the other. This is controlled with firmware and software that selects the strength of the driver, the amount of pre-emphasis, and other variables.

By employing a structural, netlist-based model for the pre-emphasis and/or de-emphasis complex driver I/O, a template is created that acts very much like a SPICE-based macromodel, as in the Boyle OpAmp example. Parameter values can be modified and passed into the model. Thus, the model can be tuned to match measured results. Complex driver I/Os with four taps of pre-emphasis and/or de-emphasis as well as other complex I/O configurations are controlled by firmware-software provided by Cadence. The current method for the user to select which tap and how much pre-emphasis and/or de-emphasis is to edit the template file (an ASCII text file). This action selects which pre-emphasis model will be used in the simulation.

The template (structural netlist) enables the simulator to bring in any combination of sub-models embedded in the main netlist. The combination can include IBIS, SPICE, S-Parameter, or equation-based models. Note that designing a driver with pre-emphasis is similar to designing a matched filter for an analog RF circuit. In designing a matched filter, the idea is to optimize the characteristics of the filter to the needs of the signal waveforms that must be delivered to a receiver. Here we are dealing with similar problems in the design of Gigabit high-speed serial nets.

A primary means of evaluating receiver signals is to use an eye diagram like the one shown in Figure 20-16 [134].

MacroModel, which uses a 7-terminal subcircuit. A differential Macromodel such as this uses an 8-terminal subcircuit, the extra terminal being the N-side output. The file pe_2tap.dml is what must be run in the Cadence EDA tools. This file was created and annotated to be an explanatory example and is provided courtesy of Cadence Design Systems, Inc.

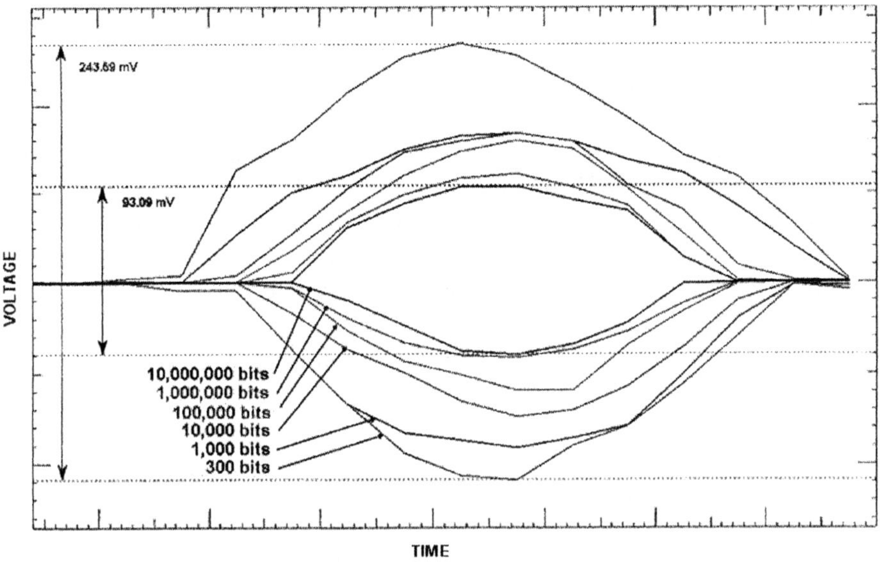

Figure 20-16. Eye-diagram closure versus number of bits simulated.
Used with permission of Cadence Design Systems, Inc.

High-frequency signal losses rise-time roll-off, ringing, and jitter can close down the eye. This means that there is less of a clean, usable bit stream signal at the receiver. Only a limited amount of problem solving can be done by designing the interconnection, termination, and routing.

Design of the interconnection addresses signal-reflection problems. Jitter control must be addressed in the IC design and noise present on the IC power busses. High-frequency signal loss is inherent in transmitting signals across the board. Pre-emphasis compensates for high frequency signal loss by adding extra high frequency signal out of the driver. The right amount of pre-emphasis can open up the eye.

Discontinuities (almost always present) on a transmission line cause reflections. The longer a bit stream runs the more the eye closes down until the line gets fully charged by the bit steam. This phenomenon can be seen in the lab with an oscilloscope's screen persistence set to infinite, and it can be simulated with EDA tools. A microwave circuit design engineer would understand some of these same phenomena in terms of voltage-standing-wave-ratio (VSWR) on the line.

Behavioral SPICE-based macromodeling templates are a Cadence-specific approach. It is a general Cadence strategy to develop specific models by tweaking parameters in the template. It is the same basic philosophy as original IBIS, except the template is not totally "canned" and limited to one circuit. Different templates are required for different general types of I/Os (ex. 2 tap pre-emphasis, etc).

We can see that behavioral SPICE-based macromodeling templates are a viable method for striking a new balance between simulation speed and model detail. Complex I/O can be modeled and run in a reasonable time. Running 100 bits of PRBS using full physical level SPICE models on a net is reported to take 5 to 10 hours. Simulating to get full eye closure can require simulation of 10,000,000 bits. The same 100 bits can be run in 5 to 10 minutes using the Cadence model approach and what they call "channel simulation (simulates 1000 bits per second)."

The templates Cadence provides are available through the following websites:

- 1.5 GBP S-ATA SerDes
  http://www.designcon.com/conference/7-ta3.html
- Differential pass-thru receiver
  http://www.eda.org/pub/ibis/summits/jan00/telian.zip
- Adjustable FPGA SerDes
  http://www.altera.com/corporate/news_room/releases/releases_archive/2004/products/nr-cadence_design_kit.html
- Front-side bus driver, impedance control, SSN, and gate choke effect
  http://www.cadence.com/company/newsroom/press_releases/11_16_98_SQ_Intel_Merced_Processor.doc
- Multi-gigahertz SPICE-based macromodel templates
  http://register.cadence.com/register.nsf/macroModeling?OpenForm

Cadence offers many templates, including:

- Gate-throttle
- Pass-thru Rx
- Multi-tap SerDes
- Rx equalization
- Self-compensating drivers
- DDR2.

To build a SPICE-based macromodel, Cadence recommends collecting some specific data. For example, when pre-emphasis is involved, the data should include: normal IBIS data, Vtt, Rt; pulldown I-V curve, ramp rate, C_comp; and unit interval, pre-emphasis dB or scale factor (x)

Cadence also says that for characterization and final verification simulations, they don't expect behavioral SPICE-based macromodels to replace detailed transistor-level models. They recommend using pre-defined, worst-case bit patterns in final verification simulations.

The shortcoming of traditional IBIS (with its system of Keyword templates) that Cadence seeks to address is lack of flexibility. The Keywords are totally canned and must go through a standard specification approval and vetting process. The Keyword-approval process causes a fairly long delay in responding to advances in IC technology. The Keywords are intended to make data exchange between platforms and software totally portable. By relying on the [External Model], [External Circuit], [Circuit Call] and [Node Declarations] Keywords, Cadence hopes to convert the creation of new templates for rapidly evolving complex I/O to the sphere of an external sub model call. Better flexibility and response time on new template creation is the goal.

20.7.5 Model Abstraction Applied to Top-Down and Bottom-Up Design

As a design progresses from a system specification to working hardware, the design gets decomposed from the top down and verified from the bottom-up. The recommended approach is to verify each progressive step of the decomposed design. After deriving transfer functions that can be realized in a series of individual OpAmp stages, the circuit designer must design the OpAmp stages in realizable hardware. In this chapter we develop several levels of OpAmp modeling detail and abstraction. We will now apply these abstraction levels in the top-down and bottom-up design process in the example listed below:

1. In the first stage of the decomposition process, each block of circuitry can be represented by a very simplified equation. Equations (20-43) and (20-44) are OpAmp models in their most abstracted and simplified form.[9] Abstracted model at this level can be employed as the first step of decomposition. A simulation of the cascaded circuitry blocks would enable the first bottom-up verification of the system transfer function. The simulation would ensure that the cascade of individual circuit blocks could produce the desired response.
2. Next, at the beginning of the individual stage hardware design, the designer will probably account for the additional roll-off and phase shift in the OpAmp due to a secondary pole, as per equation (20-40). We now replace highly abstracted models with more accurate and detailed models.

---

[9] In an example presented in [81], the low frequency gain, input impedance, and corner frequency are first used in a SABER-MAST equation-based model simulation of a single stage of OpAmp.

Each time we do so, we re-verify or adjust the system response. Using the more detailed model provides the designer with more data that is needed to make a component choice and design the entire circuit block.

3. The two-pole simulation can then be followed by a SPICE (Boyle) OpAmp macromodel simulation, where a specific type of commercially available OpAmp is chosen.
4. A simulation that includes the effects of the physical layout of the amplifier block can follow the simulation of the specific OpAmp and its parameters.

During each simulation iteration with increasingly detailed and more accurate models, the hardware design may need to be modified to meet system transfer functions. Conversely, the system transfer functions may need to be modified to conform to what is physically possible. As the design becomes increasingly detailed and evolved, verification should be done at several points in the process. Verification of the model parameters and corresponding implementation from the bottom-up should always be part of the design process. Real hardware and board layout may not duplicate the desired system function exactly.[10]

In Signal Integrity design, a similar simulation process of replacing abstracted models with detailed and physical models is followed. The design starts with a topology schematic and progresses through board layout. Verification is done as physical sections of the board get completed and more accurate models of parts replace more generic models. Digital top-down decomposition of functions into hardware and bottom-up verification of performance proceeds in a parallel manner.

20.8 LIMITATIONS IN MODELS DUE TO SIMPLIFICATION

Models do not always accurately predict device behavior. As we have seen above, we can employ models with various levels of simplification. We believe that more detailed models are more accurate. Let us review some possible sources and reasons of model inaccuracies. They include:

- *Limitations.* If bias and frequency conditions are changed, the model can become invalid. Measurement-based behavioral models are only valid

[10] Mantooth and Fiegenbaum [81] discuss analog top-down design. For example, typical transfer function decomposition for a control system results in a set of OpAmp blocks. The blocks are designed to give a pole-zero and voltage gain at a particular frequency and magnitude respectively.

close to the bias conditions where the data was measured. Limitations to valid measurements are particularly important in semiconductors because they are usually quite non-linear.

- *Assumptions.* One way to simplify a macromodel template is to use behavioral models in place of SPICE physical-level models. But then, for example, a 1.8 Volt logic IBIS model would be inaccurate used at 1.2 Volts even if the same semiconductor process was used. For example, beta (used in the SPICE model) of a transistor measured at room temperature and 5 volts Vce is not the same at 1 volt and an elevated temperature.
- *Simplifications.* Simplified models can be limited in their range of accuracy. A simplification example is the voltage gain transfer function of a BJT transistor. It can be simplified to an equation (20-46) that does not even include beta as a parameter. But that is only true for a mid current beta > approximately 50 and it is not true in the low- and high-current beta rolloff regions.
- *Errors.* Errors are outright mistakes. Models are accurate only when errors have not been made in their measurement or extraction.

What happens when the transistor-level model is just plain wrong? Then *any* SPICE-based macromodel and *any* IBIS-based behavioral model (or any other model) *derived* from the transistor-level physical model would be wrong. Adjusting a template's parameters to match lab results would be much more difficult if important parameters were missing. Missing or wrong parameters can be a result of new or poorly understood physical behavior. As technology advances the underlying physics changes. It would be abnormal for the transistor-level model to stay the same.

In some cases, both the SPICE model and the SPICE-based macromodel are not accurate enough [71]. The inaccuracy may be caused by a wrong assumption made when creating the device physics model. In this case, a SPICE macromodel constructed from the device model will also be wrong. This causes the verification to fail. Therefore, a behavioral model measured on the device should be considered.

Power devices are an example of where the original device physics models were wrong. An example [79] of a wrong device-level model is the SPICE model for a power diode. The problem originally started because the device physics models used to solve the accuracy problem was too computationally expensive. How the device physics equations get simplified is critical. "Normal" SPICE diode equations make simplifying assumptions that are not valid[11] for power devices. Transit time across lightly doped regions, for instance, is significant contrary to the usual assumptions.

Research at the University of Washington led to new equations, with new simplifying assumptions. The research led to new device parameters developed from device physics modeling (structure, doping, and materials properties) using the lumped-charge technique. The group also developed parameter extraction techniques based on measurement.

The new model[12] is an accurate power diode model. The model includes voltage-dependent reverse recovery, forward recovery, conductivity modulation, emitter recombination, and both high- and low-level injection. The new model is more accurate than a SPICE macromodel — with only 13 parameters and 2 pins. Even with 15 parameters, SPICE does not model the effects listed above. This is not a complaint about SPICE, but it does show how important it is to understand several concepts, including:

- Device physics, especially when pushing the technology envelope and the modeling envelope.
- The need to employ several modeling techniques (IBIS, SPICE, S-Parameters, SPICE-based macromodeling, and AMS at minimum) and to be flexible about using them to simulate high-speed digital circuits that use state-of-the-art ICs with complex I/O.
- The need to integrate lab verification into the modeling and design process flow when venturing into new technology areas.

The design of today's high-speed digital and mixed signal circuits I/O would be impossible without modeling and simulation.

20.9 AMS MODELING SIMPLIFIED

20.9.1 Discussion

MAST, VHDL-AMS, and Verilog-AMS are three AHDL modeling languages that differ in capabilities, syntax, and definitions. MAST includes explicit simulator commands for the SABER® simulator. VHDL-AMS

---

[11] A similar case arises with deep sub-micron IC structures. The usual SPICE models assume that junction sidewall capacitance is negligible compared to other junction capacitances. The reverse is true for real, deep submicron structures.

[12] Model source code for the power diode example can be downloaded through a link found on a page of explanation of model parameters at: http://www.ee.washington.edu/research/pemodels/modtext/diode-ac.html

(IEEE 1076.1) and Verilog-AMS (IEEE 1364) models work with simulators from several EDA companies.

In comparison, SPICE allows explicit simulator commands (.OPTIONS), and SPICE 3f5 models are portable across most commercial implementations, even though it is not a published standard.

Conceptually, AMS uses transfer functions of blocks of circuitry to achieve efficiencies in modeling. As a general rule, a model can be expressed as either a set of equations or as structural sub-models, in the form of netlists. Because of how simulators generate system matrices, simulation efficiency is usually improved if the output of the model is a digital value or a current at each pin, rather than a voltage. Unlike a SPICE-based macromodel, an AMS behavioral (equation-based) model does not require the addition of non-physical nodes. Simulation increases in efficiency when there are fewer nodes in the netlist.

As we have seen with OpAmps, the transfer function[13] can be abstracted to a level where it hides most of the internal detail of a block of circuitry. Therefore, when that model is embedded in a larger system, it can be simulated efficiently and quickly. Table 20-6 compares the number of nodes for the 7-pin OpAmp circuit and Level 2 Boyle SPICE-based macromodel (in Figure 20-13) and an AMS implementation.

Table 20-6. Node count for three equivalent OpAmp models

Model Type	**Node Count**
Transistor-level OpAmp	26
Level 2 OpAmp	12
AMS implementation[14]	7

This efficiency is so dependable that transfer functions are used quite often in analog system, top-down design. In this approach a high-level system function equation (such as an amplifier and filter cascaded chain) is decomposed[15] into a set of terms representing blocks of circuitry (such as an OpAmp followed by a filter). The next step is to design the internals of each block of circuitry in increasingly detailed steps, until actual circuitry can be realized. Similar system-level, top-down approaches are followed in digital circuit schematic-capture EDA tools.

[13] Transfer function, as used here, is synonymous with transmission function.

[14] The number of nodes for the AMS model can be reduced to 3. See Figure 20-17 if analysis of power and ground bounce, and thermal drift (only in-out signal gain) are not going to be evaluated.

[15] The Heavyside Partial-Fraction Expansion Theorem is a mathematical technique used for decomposing complex transfer functions into simpler functional blocks.

20.9.2 OpAmp Example

Not only can the transfer function of a block of circuitry be fairly detailed, it can also be highly abstracted. In this section the OpAmp transfer function is abstracted to its most simplified expression. In the case of an ideal (infinite gain-infinite bandwidth), inverting OpAmp, the voltage transfer function [108] is:

$$A_v = -R_2/R_1 \tag{20-43}$$

Even when gain is finite (a real-world case), the equation is still only:

$$A_v = \frac{-R_2/R_1}{1+(1+R_2/R_1)/A} \tag{20-44}$$

where A is the finite open-loop gain of the OpAmp. The inverting OpAmp schematic is shown in Figure 20-17.

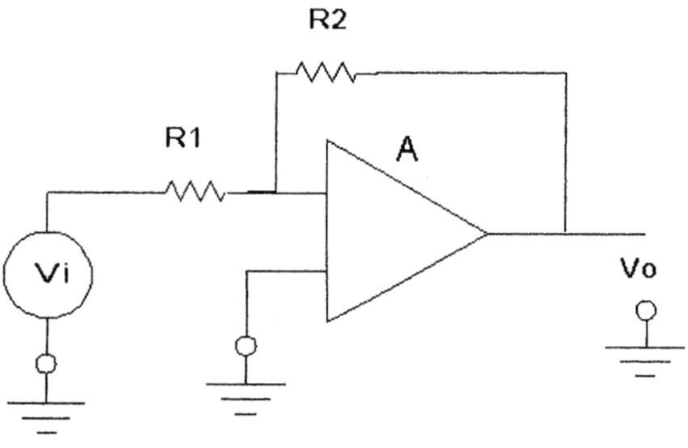

Figure 20-17. OpAmp inverting amplifier schematic

The level of abstraction is variable, and somewhat up to the user and their needs. Model accuracy is affected by how many of the actual physical effects (1st, 2nd, 3rd and higher order terms) are included in the model and how well they are represented. When we construct a transistor-level model and include all effects, we usually believe the model is more accurate than a behavioral model. However, this is not always true. For instance, the transistor-level model may not include certain effects that are readily apparent in a circuit's behavioral data. If care is taken [70] in defining the behavioral equations and extracting the parameters for those behavioral equations, then that equation-based can be as accurate, or even more accurate than the transistor-level model. Usually, it also simulates much faster than the circuit level model.

But there are drawbacks. Two things may be lost in using an equation-based model. First, behavioral models are usually accurate only over a narrow range close to their measurement conditions. If the range of operation is close to the conditions under which the model was derived or measured (temperature, bias, and other conditions), then this is good. The same range limitation may apply to measured SPICE parameters for transistor models. Equations can be made to cover a wider range of Temp, Vcc, etc., if the device behavior is understood and modeled or there is enough measured data.

Second, insight into how a change in the value of any one particular internal model parameter affects behavior is generally lost. In AMS there are often no components or elements–only equations and parameters.

An example of using this equation-based design approach in MAST® follows. A voltage multiplier (mixer) circuit is used in the following top-down design.

20.9.3 Signal Multiplier Example

In this topic, we present an example of an AMS model. This model in the example is a circuit block that up-converts (or modulates) a voice signal for wireless communications onto a RF carrier frequency. Mathematically, this is the same as a times-one multiplication of one signal by the other. The SABER® simulator provided one author [Leventhal] with an opportunity to use an analog behavioral model written in the MAST modeling language.

Writing a SPICE-like behavioral model into a simulator program like SABER was fairly simple. Each model is compiled at run time. This allows for the easy addition of models that are not part of the original simulator. However, this is not possible in SPICE which has all of the models embedded in the simulator code, and no provision for adding external models.[16]

The model for a 4-quadrant multiplier can be written with any of four combinations of input voltage drivers or current drivers and output voltage or current drivers. Figure 20-18 shows how simple a voltage-input-to-current-output template looks in the MAST Analog Hardware Description Language (AHDL):

---

[16] SPICE could have been either interpretive or compiled. It is just a matter of how the programmers implemented the simulator. SPICE's problem is not that is compiled, but that there are no "hooks" for adding a user's compiled model. Other simulators that are netlist-based provide the ability to add compiled models. Compiling one model does not require compiling the entire simulator.

```
element template kmult3 in1 in2 gnd  out = k
electrical in1, in2, gnd out
number k = 1
{
var i iout
val v vin1, vin2, vout
values{
        vin1 = v(in1) - v(gnd)
        vin2 = v(in2) - v(gnd)
        vout = v(out) - v(gnd)
        }
equations{
          i(out->gnd) + = iout
          iout = k*vin1  * vin2
          }
}
```

Figure 20-18. MAST AHDL voltage input to current output multiplier

Next, we can attach this model to a schematic symbol. Figure 20-19 shows a simple schematic circuit with two voltage generator drivers, a black-box symbol for the multiplier, and a resistive load. By attaching property values, such as the resistance of the load, to the symbols in the schematic, the values can be passed into the model and used by the simulator. The schematic symbols' property values can be edited. In Figure 20-19, voltage generator V1 is a ramp function that swings through zero and voltage generator V2 is an exponentially decaying sine wave.

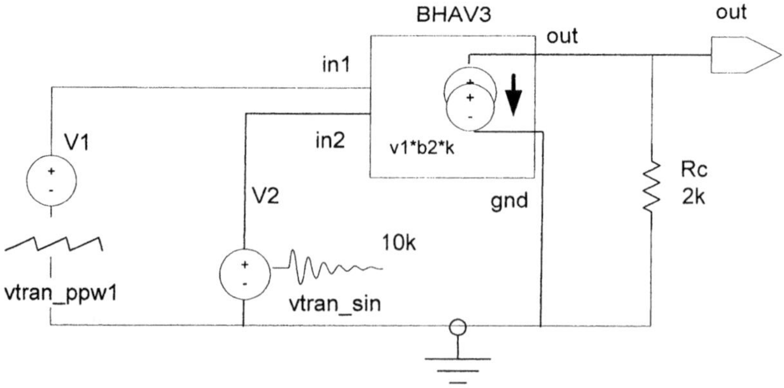

Figure 20-19. Mixer circuit implementing the AMS voltage multiplier

The results of simulating the circuit in Figure 20-19 are shown in Figure 20-20. The multiplication effect of the ramp on the decaying sine wave can be clearly seen. We must look across several cycles in Figure 20-20 to see the sine wave decay. Also, the ramp was offset below zero and we can see that the effect of the 180° phase reversal as the ramp crosses through zero.

Thus, with a mathematical function block (or string of blocks) in a system level circuit, overall system response can be designed. The next step is to implement the optimized mathematical function in physical circuit hardware. If a function block already exists as a parameterized circuit design, a designer would adapt the circuit for his or her own use and generate the layout for each block.

Synthesis is the process of generating a circuit with its components, connections and parameter values from a network transfer function [24, 139]. Circuit and network *analysis* (the inverse) is the process of generating a network transfer function from components, connections and parameter values. Analysis results in a unique transfer function. But synthesis is often not unique and can result in different circuit implementations. These different realizations have been given the name *canonical forms*.

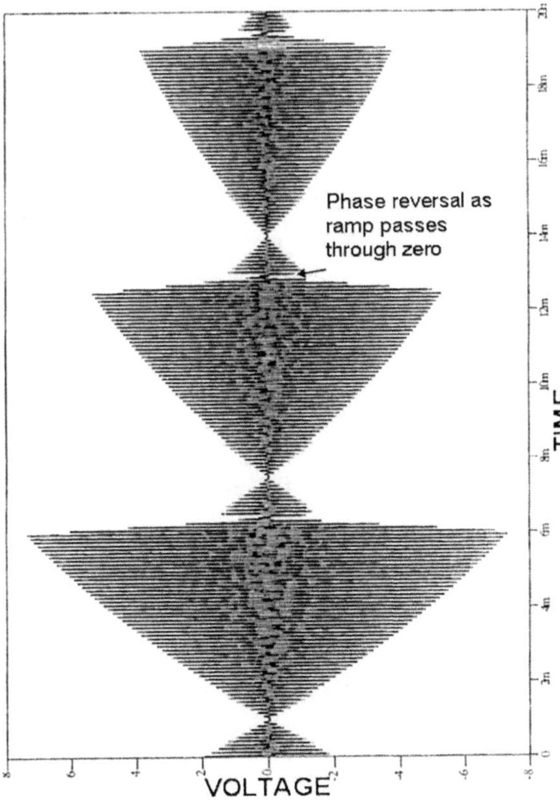

Figure 20-20. Output waveform of the behavioral modulator with the specified voltage generator inputs

In a typical top-down amplifier string, individual blocks are cascaded together to describe the complete function. Total input to output voltage and current gain are represented by multiplying each blocks' transfer function(s) by the next to get the overall response. For example, a 3-stage amplifier with bandpass shaping filters between each stage is easily represented. Typically, the transfer functions are first represented in the Laplace Transform s-plane. Once the correct system level response is derived, these functions are then decomposed[17] and reverse Laplace transformed for hardware implementation.

One way to execute this multiplier in circuitry is to use a simple OpAmp summing amplifier as shown in Figure 20-21. This circuit is presented for illustration only. A 4-quadrant multiplier would probably employ a Gilbert mixer cell.

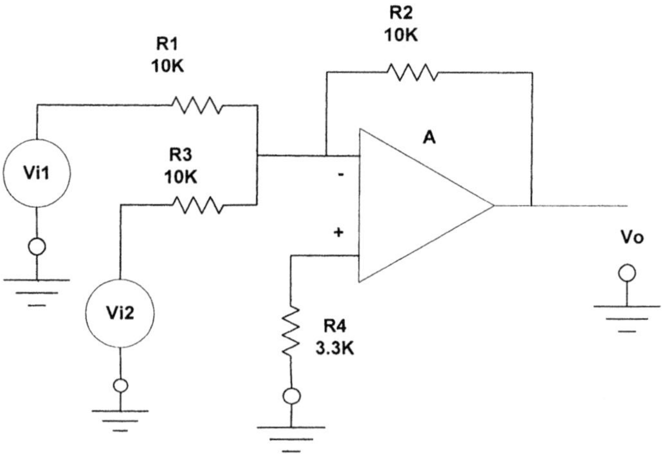

Figure 20-21. Times one, inverting, summing OpAmp amplifier (multiplier)

This amplifier has a gain of –1 for each input signal (Vi1 and Vi2) to output. Therefore:

$$Vo = -(Vi1 + Vi2) \qquad (20\text{-}45)$$

R4 is chosen as the parallel combination of the input and feedback resistors. This stage can be followed by an additional inverting OpAmp stage, with gain multiplication K set where desired.

[17] Heavyside's partial traction expansion theorem.

20.9.4 Limitations in AMS Modeling

The AMS equation-based approach to modeling is ideal for many uses. AMS has much to offer. But there are several important barriers to implementation:

- *Lack of familiarity in writing equations.* Someone must write the equations for the IC I/O buffers. The best sources of knowledge regarding what to write are the IC chip designers. Semiconductor companies already deeply committed to mixed-signal IC design and complex I/O already use AMS heavily. If AMS models are needed, many companies already have experts to get chip level I/O modelers started, but there is a learning curve.

  However, *many* semiconductor companies are not dedicated to providing good circuit-level models. They may see little payoff in providing circuit-level models, or penalties for not providing good circuit models. HSPICE and TCAD are used to control semiconductor processes, and the model parameters of interest are only those that directly track process control. Some advanced semiconductor companies supplying state-of-the-art microprocessors and processor boards see things differently. They strongly support providing good circuit level models.

- *Simulator convergence is often a problem.* It takes skill to write equations that converge for simulators.
- *Analog equations are different than digital equations.* Digital designers who know VHDL or Verilog will be able to use AMS, but probably will not be skilled analog model makers. Digital designers rarely use analog equations.

The first bullet applies to AMS and SPICE-based macromodeling along the lines, for example, of an OpAmp SPICE-based macromodel. The second and third bullets mean a significant learning curve issue if AMS is to be a clear winner for striking a new balance in IBIS modeling.

Analog *systems* designers are most familiar with transfer functions, s-planes, and similar modeling approaches. Analog *hardware* designers are less familiar with these tools, and digital designers least familiar. The positive expectation is that AMS models will be portable and will tend to hide the (IP) internal structure of the I/O buffer and the transistor parameters. Two motivations for IBIS are portability and IP protection. A third motivation is a practical balance between model detail and simulation speed.

That balance depends on the details of how to create an AMS model for an I/O device.

20.10 LIMITATIONS BECAUSE OF PARAMETER VARIATION

By whatever method we use to abstract and simplify a model, we need to be aware of the limitations imposed on the simplified results. To arrive at a simplified model, we can start with physical-level modeling, SPICE-based macromodeling, or equation-based modeling. As an illustration of how model accuracy and circuit design application can interact, a CE-BJT amplifier is discussed next.

20.10.1 Range of Validity

Given a moderately high β (50 to 100), we can simplify a CE amplifier voltage gain equation of a transistor as [80]:

$$A_v \approx -Rc/Re \qquad (20\text{-}46)$$

This equation is the gain of a single stage of CE BJT amplifier at its most abstract form. Note that transistor parameters do not appear in this equation! As an example of a transfer function, this equation is very simple. Equation (20-46) is also desirable because it shows that we can design an amplifier to be independent of its variables. Transistor parameters are usually much more variable than the ratio of 2 passive resistors. Even in an IC realization of an amplifier, where the absolute values of Rc and Re are not that easy to control, their ratio Rc/Re can be tightly controlled.

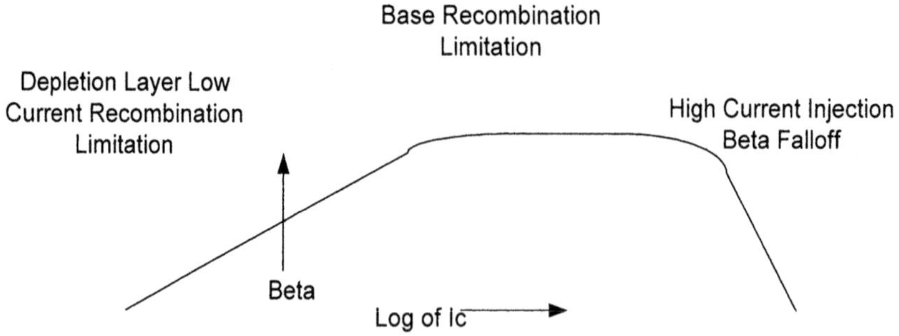

Figure 20-22. Beta versus Ic: The beta turn-on region, mid-current region, and high-current injection falloff regions

However, we immediately see that equation (20-46) gives us no information regarding low- or high-current beta roll off. There are many other details we also lose when we abstract (simplify) a model. Thus, the abstracted and simplified model is valid only in the flat mid-current beta range and below the frequencies where gain starts to roll off significantly due to f_t limitations. If we use the simplified model outside of those *limitations,* it is no longer valid. For illustration, the behavior of beta with collector current is shown in Figure 20-22.

Also consider the Early effect where the collector-emitter voltage affects beta, which is shown in the collector characteristics in Figure 20-23.

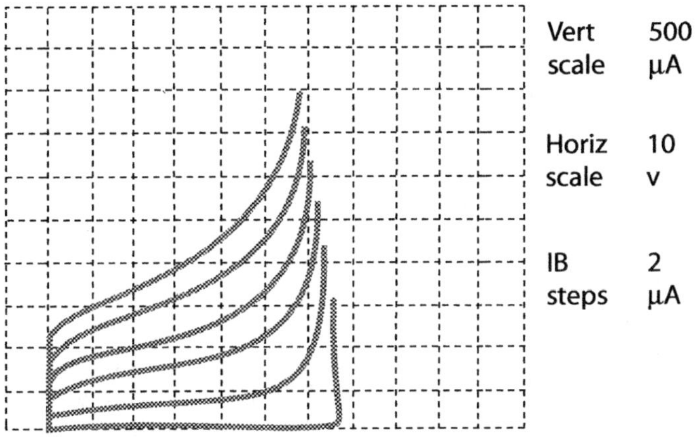

Figure 20-23. Collector characteristics of a narrow base silicon planar device Photo from a Tektronix 576 curve tracer. [74]

Such pronounced Early effect is typical in a high frequency BJT. Deep submicron low-voltage MOSFETs are subject to a similar effect–gate conductivity modulation.

The designer of an analog amplifier or a high-speed switch is expected to exercise skill in applying a device whose characteristics vary with bias (Vce, Ic and Ib). Designers are expected to try to bias it in an operating region where those variations are small. A place to start is where collector is in a range where beta is flat, and greater than 50 to 100.

Next, given a beta of 100, the Ib would be Ic/100. The current through a simple resistor divider string to bias up the base would be set at about 10*Ib for bias stability. DC bias voltage across the output and the collector (pullup) resistor would be set—depending on the voltage swing required at the output and the peak current at the load. Given these rough estimates, the designer can then start looking at transistors that can handle the signal swings, power dissipation, gain, gain-bandwidth (alternatively, switching speed and swing), and other performance factors. Power dissipated by the transistor and in the bias network are factors that have to be traded off.

The intended use of the device also influences how it is modeled and applied. Collector current tends to be low on an input stage, mid-range on an intermediate stage, and high on an output stage. The same device operated in low, mid and high currents is then modeled differently based on its use in the circuit. In mid-range Rc/Re is an adequate model. At the low and high ends of the beta range, different models involving beta rolloff should be used.

Alternatively, the input, intermediate, and output stages can be sized differently based on signal swings and power delivery requirements. Differently sized devices generally have different values for their SPICE model parameters. I/O buffer sizing and modeling is standard practice upon a process technology base.

Non-constant beta might be handled adequately by the model found in equation (20-47) with data on beta variation called as part of a simulation. Alternatively, beta in equation 20-47 could call a model of how it varies with properties such as collector current.

$$Av = \frac{\beta R_L}{r_\pi + (\beta + 1)R_e} \tag{20-47}$$

The frequency or temperature variation of beta has not yet been discussed in this section. Accounting for these additional concerns adds more detail, accuracy, *and simulation time.* Therefore, we can say that model complexity versus simplification is a function of the effects that *need* to be modeled. A model should be as simple as possible, but no simpler. That need also depends on how the device is being applied, the performance it has to deliver, the skills of the designer, and the design of the device itself. Model simplification is an art form that is not easily separated from the reality it has to represent. Another way of saying this is that the model abstraction process should not be approached as a standard process–that is, one method fits all.

20.10.2 Device Variation

Multiple units of a device vary in their manufacture. Setting aside the question of defectively produced devices, a population of units exhibit a range of behavior that should, in most cases, be reflected in their model. Sometimes device variation may not matter. Indeed that is usually the desired outcome. But once again, a highly simplified model such as gain = Rc/Re cannot account for statistical variation, whereas equation (20-47) may answer such needs.

Given an equation similar to equation (20-47) and population data, we can apply the methods of statistical design to account for device variation. Statistical design can be computationally expensive, but extremely effective in producing a manufacturable product. Many engineers are familiar with worst-case, and Monte-Carlo methods. Some engineers know about the methods used in Design-Of-Experiments (DOE) and Analysis-Of-Variance (ANOVA). Still, we must examine some of our usual assumptions about device variation and how we model that variation using statistics and statistical design.

The most common assumption is that multiple properties on a device (and its model) vary in a linearly independent way. The simulator may change one variable at a time or several at a time as in DOE and ANOVA. But the variables are treated as independent and this is often not true for an active semiconductor device. We can still go about our deterministic simulation of models and systems by invoking the methods and software of Multivariate-Statistical-Analysis [155]. To do this, there will be a simulation run time price to pay. But the problems of deterministic design can escalate beyond common-sense control, as we shall next discuss.

20.11 LIMITATIONS OF DETERMINISTIC MODELING AND DESIGN

The first step in design, modeling and simulation is usually to calculate and/or test to see if a circuit or product built from typical parts will work. Since we typically do not have all typical parts, the next reasonable step is to calculate and/or test to see if a circuit/product built from worst-case parts will work. It is even less probable that we will build a unit from exclusively minimum or maximum parts. Absolute worst-case design should only be used where human safety is at risk and then in concert with the methods of HALT and HASS.

If we are wise, and have available time and management support, we should very quickly progress to using statistical design methods like Monte-Carlo and DOE. The authors advocate using statistical design–particularly, when the goal is products that will be manufactured with very high quality, reliability, and low cost.

However, statistical design is basically still a deterministic approach to computation and design. In very complex systems, especially those that will be repaired and maintained, we need to ask what is the probability of a particular combination of parts being built into any particular unit? With the addition of the mathematics of probability to the methods of statistical design we get probabilistic [23, 153, 154] design. Concepts like confidence interval and confidence level (probability) come into use by the designer.

20.11.1 Transfer Functions

Both bottom-up, behavioral SPICE-based macromodeling and AMS modeling result in explicit or implicit transfer functions. Likewise, AMS top-down modeling starts with transfer function equations. Transfer function equations can be expressed using a wide variety of mathematical tools.

An easy way of trying out some of these other functions is to follow the above multiplier example and substitute new equations in the multiplier model. Defining pins, variables, constants, parameters, and equations would have to be done appropriately, depending on the modeling language.

Among the types of mathematical functions encountered in transfer functions are:

- Rational polynomials
- Pole-Zero functions
- Non-Linear functions
- Fourier Transforms and frequency responses
- Control system functions
- Laplace S-Domain transform functions
- Z-Domain transform functions
- W-Domain transform functions

Transfer functions were developed in the 1920s by the founders [Bode, Brune, Butterworth, Chebyscheff, Cauer, Darlington, Foster, Norton, Nyquist, and others] of modern network synthesis [24, 125, 139]. The objective of network synthesis is to design one or more networks that realize a given network function. Again, this is top-down design. These methods were first applied only to passive R-L-C networks because of the unreliability of the active elements of the day—vacuum tubes.

Top-down methods and transform methods are now being actively applied to networks containing semiconductors. In many cases, the resulting voltage transfer function of cascading a set of circuits can be found with the chain rule. The chain rule states that the overall transfer function can be found by simply multiplying the individual voltage transfer functions together.

The chain rule is permissible when the loading of the succeeding stage does not significantly affect the transfer function of the preceding stage. And the source impedance of the driving stage does not significantly affect the voltage transferred to the receiving stage. This occurs when the load impedance is substantially larger than the driving impedance. A common implementation of a cascaded amplifier chain is the use of op-amp voltage[18] follower circuits. These follower circuits are in between wave-shaping circuits to isolate them and provide necessary gain.

In HSDD the load impedance of a receiver is usually matched to that of a routed PCB transmission line. The line is usually less than 100-ohms. Also, the capacitive input impedance, $Xc=1/j\omega C$ of the receiver tends to dominate at very high frequencies. That is, Xc becomes very low at very high frequencies, much lower than 100 Ohms.

Even when working into an unmatched high-impedance receiver, the driver, due to propagation delay, is usually working into the impedance of the transmission line. This makes application of the chain rule problematical in high-speed digital circuits. The output impedance of the driver is commonly in the range of 15- to 25-ohms. In the case of PCI technology (reflective wave switching), it is 50-ohms.[19]

The way to get correct results from the voltage transfer chain rule for low impedance loads is to derive the transfer function, taking into account source and load impedances. Then to achieve the desired transfer function, be sure to implement the cascaded chain in that way. In the derivation of the SPICE-based macromodel of section 20.6, source and load impedances are accounted for.

It is very common to design to match 50-ohm input and output impedances in RF and microwave work. 50-ohms was chosen because early coaxial cable design could be optimized for low loss and size at the frequencies then used. Today, 50-ohms is a common high speed digital and RF and microwave PC board characteristic impedance for similar and historic reasons.

20.11.2 IBIS 4.1 Support for AMS and Macromodeling

IBIS 3.2, 4.0 and 4.1[20] added a set of features to support AMS modeling and macromodeling. IBIS 3.2 added support for submodels. IBIS 4.1 added support for external models in SPICE 3f5, VHDL-AMS, and Verilog-AMS as well as for "External Models." Some details are still being worked out for these features. Thus IBIS now supports hardware description languages (HDLs). The IBIS ICM Specification for connectors, packages, and interconnections supports S-Parameter models.

---

[18] High input impedance and low output impedance.

[19] Typically receivers have an input impedance of a few pF in parallel with several M-ohms. Receivers are typically not matched to any PCB Zo. Matching is usually done at the PBC or package level (termination resistor) for these parts.

[20] 4.1 is the current version as of this writing.

The IBIS Specification itself contains good introductions about how it handles support and integration for special I/O functions, macromodeling, and analog mixed signal (AMS) modeling. Therefore, considerable material from IBIS 4.1 is used here, with permission from the IBIS Committee Chair.

New features get added to IBIS primarily through the [Keyword] mechanism. Prior versions of IBIS included keywords such as [Series Switch Groups] and [Model Selector]. For example, [Model Selector] is used to pick a [Model] from a list of [Model]s for a pin which uses a programmable buffer.

The [Driver Schedule] keyword that existed in version 3.1 is particularly interesting. From the current version of IBIS we have:

```
 [Driver Schedule] Describes the relative model switching
sequence for referenced models to produce a multi-staged
driver.

Required: No, Description & Usage Rules:
The [Driver schedule] keyword establishes a hierarchical
order between models and should be placed under the [Model],
which acts as the top-level model. The scheduled models are
then referenced from the top-level model by the [Driver
Schedule] keyword.
When a multi-staged buffer is modeled using the [Driver
Schedule] keyword, all of its stages (including the first
stage, or normal driver) have to be modeled as scheduled
models. If there is support for this feature in an EDA tool,
the [Driver Schedule] keyword will cause it to use the
[Pulldown], [Pulldown Reference], [Pullup], [Pullup
Reference], [Voltage Range], [Ramp], [Rising Waveform] and
[Falling Waveform] keywords from the scheduled models instead
of the top-level model, according to the timing relationships
described in the [Driver Schedule] keyword.
Consequently, the keywords in the above list will be ignored
in the top-level model. All of the remaining keywords not
shown in the above list, and all of the subparameters will be
used from the top-level model and should be ignored in the
scheduled model(s).
However, both the top-level and the scheduled model(s) have
to be complete models, i.e., all of the required keywords
must be present and follow the syntactical rules. . .
```

Thus we can see that considerable effort was made prior to IBIS 4.1 to grow the IBIS Specification to include the behavior of more complex I/O and advanced technologies. One of the next steps forward in extending the IBIS Specification was the addition of the [Add Submodel] keyword. [Add

Submodel] references a submodel to be added to an existing model. The [Add Submodel] keyword is invoked within a model to add the functionality that is contained in the submodel or list of submodels in each line that follows. From IBIS 4.1:

```
The [Add Submodel] keyword can be used under a top-level
[Model] keyword to add special-purpose functionality to the
existing top-level model. This section describes the
structure of the top-level model and the submodel.

TOP-LEVEL MODEL:
When special-purpose functional detail is needed, the top-
level model can call one or more submodels. The [Add
Submodel] keyword is positioned after the initial set of
required and optional subparameters of the [Model] keyword
and among the keywords under [Model]. The [Add Submodel]
keyword lists of name of each submodel and the permitted mode
(Driving, Non-Driving or All) under which each added submodel
is used.

SUBMODEL:
A submodel is defined using the [Submodel] keyword. It
contains a subset of keywords and subparameters used for the
[Model] keyword along with other keywords and subparameters
that are needed for the added functionality. The [Submodel]
and [Submodel Spec] keywords are defined first since they are
used for all submodels. The only required subparameter in
[Submodel] is Submodel_type to define the list of submodel
types. No subparameters under [Model] are permitted under the
[Submodel] keyword.
The following sets of keywords that are defined under the
[Model] keyword are supported by the [Submodel] keyword:
[Pulldown] [Pullup] [GND Clamp] [POWER Clamp] [Ramp] [Rising
Waveform] [Falling Waveform]
The [Voltage Range], [Pullup Reference], [Pulldown
Reference], [GND Clamp Reference], and [POWER Clamp
Reference] keywords are not permitted. The voltage settings
are inherited from the top-level model.
These additional keywords are used only for the [Submodel]
are documented in this section: [Submodel Spec] [GND Pulse
Table] [POWER Pulse Table]. The application of these keywords
depends upon the Submodel_type entries listed: Dynamic_clamp,
Bus_hold, Fall_back. Permitted keywords that are not defined
for any of these submodel types are ignored. The rules for
what set of keywords are required are found under the Dynamic
Clamp, Bus Hold, and Fall Back headings of this section. . .
```

Multi-lingual model support adds support for SPICE[21] 3f5, VHDL-AMS and Verilog-AMS models under IBIS. This is accomplished under the new keywords [External Model], [External Circuit], [Circuit Call] and [Node Declarations].

One limitation of the existing [Model] keyword is that it matches exactly one I/O pin to a model. Differential pins always assign the same model to both differential pins. This inherently voids any interaction between the two halves of the differential circuit on the IC. Using [External Model] and [External Circuit] gets around this by allowing multiple pins from the IC to be connected to the same "model" for the differential pair.

The beginning portion of the IBIS 4.1 section on external models is quoted here:

```
Section 6b
Multi - Lingual Model Extensions
INTRODUCTION:
IBIS supports the SPICE, VHDL-AMS and Verilog-AMS languages.
This chapter describes how models written in these languages
can be referenced and used by IBIS files. The language
extensions use the following keywords within the IBIS
framework:
[External Circuit], [End External Circuit]
References enhanced descriptions of structures on the die,
including digital and/or analog, active and/or passive
circuits.
[External Model], [End External Model]
Same as [External Circuit], except limited to the connection
format and usage of the [Model] keyword, with one additional
feature added: support for true differential buffers.
[Node Declarations], [End Node Declarations]
Lists on-die connection points related to the [Circuit Call]
keyword
[Circuit Call], [End Circuit Call]
Instantiates [External Circuit]s and connects them to each
other and/or die pads
```

[21] An implementation of SPICE 3f5 is available as open source from the University of California at Berkeley.

```
The placement of these keywords within the hierarchy of IBIS
is shown in the following diagram (Figure 1):

|-- [Component]
|    | ...
|    |-- [Node Declarations]
|    |-- [End Node Declarations]
|    |    | ...
|    |    | ...
|    |    |-- [Circuit Call]
|    |    |-- [End Circuit Call]
|    |    | ...
|    | ...
|    |-- [Model]
|    |    | ...
|    |    |-- [External Model]
|    |    |-- [End External Model]
|    |    | ...
|    | ...
|    |-- [External Circuit]
|    |-- [End External Circuit]
|    | ...
|
LANGUAGES SUPPORTED:
IBIS files can reference other files, which are written using
the SPICE, VHDL-AMS, or Verilog-AMS languages. In this
document, these languages are defined as follows:
"SPICE" refers to SPICE 3, Version 3F5 developed by the
University of California at Berkeley, California. Many
vendor-specific EDA tools are compatible with most or this
entire version. "VHDL-AMS" refers to "IEEE Standard VHDL
Analog and Mixed-Signal Extensions", approved March 18, 1999
by the IEEE-SA Standards Board and designated IEEE Std.
1076.1-1999. "Verilog-AMS" refers to the Analog and Mixed-
Signal Extensions to Verilog-HDL as documented in the
Verilog-AMS Language Reference, Version 2.0. This document is
maintained by Accellera (formerly Open Verilog
International), an independent organization. Verilog-AMS is a
superset that includes Verilog-A and the Verilog Hardware
Description Language IEEE 1364-2001.
In addition the "IEEE Standard Multivalue Logic System for
VHDL Model Interoperability (Std_logic_1164)" designated IEEE
Std. 1164-1993 is required to promote common digital data
types.
```

```
Note that, for the purposes of this section, keywords,
subparameters and other data used without reference to the
external languages just described are referred to
collectively as "native" IBIS.
OVERVIEW:
The four keyword pairs discussed in this chapter can be
separated into two groups based on their functionalities.
The [External Model], [End External Model], [External
Circuit] and [End External Circuit] keywords are used as
pointers to the models described by one of the external
languages.
The [Node Declaration], [End Node Declaration], [Circuit
Call], and [End Circuit Call] keywords are used to describe
how [External Circuit]s are  connected to each other and/or
to the die pads.
The [External Model] and [External Circuit] keywords are very
similar in  that they both support the same external
languages, and they can both be used to describe passive
and/or active circuitry. The key difference between the two
keywords is that [External Model] can only be placed under
the [Model] keyword, while [External Circuit] can only be
placed outside the  [Model] keyword. This is illustrated in
(Figure 1) above.
The intent behind [External Model] is to provide an upgrade
path from native IBIS [Model]s to the external languages (one
exception to this is the  support for true differential
buffers). Thus, the [External Model] keyword can be used to
replace the usual I-V and V-T tables, C_comp, C_comp_pullup,
C_comp_pulldown, C_comp_power_clamp, C_comp_gnd_clamp
subparameters, [Ramp], [Driver Schedule], [Submodel]
keywords, etc. of a [Model] by any modeling technique that
the external languages allow.
For [External Model]s, the connectivity, test load and
specification parameters (such as Vinh and Vinl) are
preserved from the [Model] keyword and the simulator is
expected to carry out the same type of connections and
measurements as is usually done with the [Model] keyword. The
only difference is that the model itself is described by an
external language.
In the case of the [External Circuit], however, one can model
a circuit having any number of ports (see definitions below).
For example, the ports may include impedance or buffer
strength selection controls in addition to the usual signal
and supply connections. The connectivity of an [External
Circuit] is defined by the [Node Declaration] and [Circuit
Call] keywords.
```

```
Currently, the test loads and measurement parameters for an
[External Circuit] can only be defined inside the model
description itself. The results of measurements can be
reported to the user or tool via other means.
The [Circuit Call] keyword acts similarly to subcircuit calls
in SPICE, instantiating the various [External Circuit]s and
connecting them together. Please note that models described
by the [External Model] keyword are connected according to
the rules and assumptions of the [Model] keyword. [Circuit
Call] is not necessary for these cases and must not be used
. . .
```

20.12 SUMMARY

Two emerging approaches to modeling complex I/O are macromodels and AMS languages. These approaches allow IBIS to keep up with circuit design evolution of complex I/O. EDA vendors are already supporting the IBIS 4.1 language features. These language features include circuit calls for [External Model] and [External Circuit]. These submodel calls incorporate the macromodels and equation-based models in an IBIS simulation. Now model developers need to explore these new options and decide which ones are most effective in terms of portability and design requirements. Time will tell which approach will be most widely adopted.

Chapter 21

FEEDBACK TO THE MODEL PROVIDER IMPROVES MODEL ACCURACY

Sharing the validation and verification results with a model provider improves the design process

Abstract: Communication between semiconductor suppliers and OEM users should be interactive. Today, cutting edge ICs, net topology, routing, and termination are developed by a team of logic design engineers, signal integrity engineers, and PCB designers. The chip developer must be brought into the board development cycle.

21.1 CONTINUING NEED FOR BETTER MODELS

As design moves from concept to hardware, the design choices become more constrained. During this process, the places in the design where tolerance and safety margins are tight become apparent. Default models and other makeshift models are then no longer sufficiently accurate for running simulations. The designer must find more accurate models.

The objective of prudent design is to simulate all circuits before release to production. Therefore, components that initially did not have models will now need them. The approximately 35% of components that may initially have had part specific IBIS models at the start of a development must be increased in coverage to more like 80%. IBIS models that may have had only dV/dt and typical data must now include V-T and typ-min-max data.

Working very closely with suppliers on the increasing accuracy needs and requesting more models are part of an ongoing process. In spite of the many problems with models, cooperation with the model provider is desirable. When calling attention to model problems, it is important to be tactful. Problem resolution is another instance in which people skills are just as important as technical skills.

Adding to the stress between supplier and user is the rapid advancement of technology, which pushes everyone to constantly improve. One place where true belief in the effectiveness of modeling and simulation is growing is the automobile industry. There, ICs are being bought and pass incoming inspection on the basis of their VHDL models. Data sheets are not being used.

21.2 HOW FAR WE HAVE COME

Much of the current success of applying EDA tools to Signal Integrity problems belongs to the improvement in process control in the semiconductor industry. Indeed, without that improved process control, no modeling and simulation activity could be successful.

Businesses that create measured IBIS models report that the usual measured +/-3 sigma variation around a process average value for a model parameter is now typically +/-5%. The few exceptions usually exhibit a big jump in parameter values from those expected. This is probably caused by outright surprises, as opposed to a lack of ongoing process control. Imposing Quality Control discipline on a supplier is now recognized as second best to working together with them to resolve problems.

Achieving six-sigma quality is impossible by testing because that level of quality cannot be tested in.[1] Even when the model is accurate and everyone is supportive, achieving a six-sigma design is a matter of very close cooperation between supplier and user. Six-sigma quality has to be designed in. Modeling, simulation, and circuit desensitization are part of designing in six-sigma quality.

In an effective design team, a Signal Integrity engineer works with a Product (Logic) Design engineer to examine the benefits of a few different termination schemes on a net that shows too much ringing. Likewise, working with a PCB Designer calls for the same team approach and sharing of data.

Not only is the team approach good inside a company, it is also good when working with suppliers. The preferred way of resolving problems and design challenges between suppliers and users involve the sharing of data, and information. This includes sharing data and information that illuminate

---

[1] At less than 10 PPM, every time the parts are handled and tested, the potential for creating more than 10 PPM defects has to be considered. Plus the potential for "test noise," causing a mis-test or test escape is very likely to result in greater than 10-PPM errors. Mis-tests will cause good parts to be counted as bad, and bad parts to be counted as good.

the problem, the need, the opportunity, and alternative solutions. Sharing data, information, and user feedback with suppliers is part of "Specs that Swim Upstream."[2]

21.3 FOUR-STEP UNIVERSAL PROCESS FOR IMPROVEMENT

Working constructively with model makers requires several things, among them:

- Use of objective data as the primary content of communications
- Technical competence
- Interpersonal communication, and team skills
- A good process for problem solving

The four-step universal process for process and product improvement is:

1. Detect and acknowledge the problem
2. Diagnose the problem's root cause
3. Design a fix based on the root cause
4. Verify the fix

21.4 SPECS THAT SWIM UPSTREAM: A NEW APPROACH

To work in a close, symbiotic relationship with suppliers, the user should have some purchasing power. The team approach is a good way to work with both suppliers and customers. Table 21-1 suggests how to go about this:

Table 21-1. Steps in working with model suppliers

Step	Activity
Communicate	Communicate, as early as possible, what is really needed—in a *Purchase Spec* if necessary. Be sure it clearly spells out bias conditions, measurement definitions (sometimes in a companion boilerplate document), and anything else needed for clear understanding.
Get Buy-In	Get buy-in to what needs to be done. Build trust and understanding —especially technical understanding— with the supplier. Negotiate and be sure to listen carefully to each other. Remember, this is primarily an exchange of technical information.

[2] We are indebted to DC Sessions for the phrase "Specs that swim upstream."

Table 21-1. (cont.)

Step	Activity
Testing	If testing is involved, the supplier may have some standard test setups close to what the customer proposes, which will correlate to the proposed test setup. This requires the seller and buyer to come into alignment with each other.
Bottom-Up and What-if	Do early what-if simulations. The earlier the better. Even if a model is preliminary the simulations will get discussion started. A much better approach is to base the model on a real device, with perhaps, some tweaks. Typical tweaks would be a little tighter limit on slew rate for the fast-switching device. What-if simulations function as a bottom-up verification design process.
Top-Down	Do top-down design with early efforts to desensitize the design to device variations and characteristics. An example would be placement based on timing constraints. Another example would be a more liberal use of termination. Share the design approach with the device supplier.
Listen	Listen to the supplier and respond to their suggestions, until the most mutually beneficial agreement or specification is arrived at.
Take Action	Take steps to deal with those things that cannot be solved effectively at the device level. Do not believe that predicted problems magically cure themselves at layout, or on the shop floor. In other words, right the first time has to be designed in, and action must be taken based on what the simulations are predicting.
Feedback and Share	Bottom-up feedback of the lab and simulation results should bring out a response to clear up model discrepancies from the supplier. Verify to them when a correct model is being provided. Do this even when the model was right from the start. Suppliers will appreciate it.

21.5 WARNINGS ABOUT DOING WHAT-IF MODEL SIMULATIONS

Sometimes suppliers and users seem like adversaries because their objectives may not always coincide. But that is not logical; they need each other in order to succeed. Model [Disclaimers], [Copyrights], and [Notes] can be a particular instance of this problem. The advocacy of experimenting with models and specs that swim upstream can run afoul of copyrights, and legal constraints.

Disclaimers are put in to protect suppliers from any legal liability arising from a misuse of their parts, or from incompetent user design. In a court of law such disclaimers never protect suppliers from a failure to exercise due diligence. But, some disclaimers contain language that can be particularly irksome. George Opsahl of Clearbrook Systems provided the following summary of current issues about disclaimers:

"Frequently statements are included with models and relate to ownership of the model and the use of the model. The statements can impact the ownership of designs and changes to the models. The statements are usually found in the Notes, Disclaimer, and Copyright sections of the model. In addition, the statements are sometime found as part of the Comments that are scattered throughout the model. Also, the statements might be attached to a model in separate documents such as a License or Confidential Disclosure document.

Reading and understanding the statements are very important. The statements might restrict the use of the model, making changes to the model, and providing the model to others. The statements could result in ownership of a design that uses the models by another party. The other party might be the component provider that works with the design's competitors. Another type of party is a third party that would distribute the model changes or designed product to others that are competitors.

The best models are provided without any restrictions placed on the model's usage. If there are restrictive statements, the statements need to be reviewed by an attorney. If the model provider is unwilling to remove restrictive statements, then careful consideration needs to be given before moving forward with the related component in a design. . ."

Users should be able to tweak models to get them to work successfully with simulation. If the restrictive disclaimer issue arises, the issues will need to be resolved before the user can fix or experiment with a model.

21.6 SELLING THE IDEA OF BETTER MODELS AND SIMULATION

To make progress and improve processes or products, selling new ideas is always involved. For example, designers may need to sell the following ideas for improvements:

- *Getting better models.* This starts with getting supplier buy-in regarding what needs to be done to a model.
- *Doing more what-if simulations.* This sometimes starts with getting management's buy-in to doing modeling and simulation at all. Given that modeling issues (such as lack of models and their poor quality) persist,

the question might arise "Why do modeling at all?" The answer is: because in spite of all the problems, the results are hugely beneficial.

Let's look at the benefits. Big benefits usually involve large, long term cost savings. However, to get the necessary organizational turn-around, reaping big benefits takes effort and time.

Along the way to big successes, designers must give management a steady stream of small successes to assure them that they are getting value for their investment. It is in the day-to-day specifics of problem solving that the positive feedback loop is short enough for people to first notice improvements. Without the steady support of management, those big, long-term successes will not happen.

21.6.1 Selling the Big Successes

One author [Leventhal] was part of the design team at 3Com that realized a big, long-term improvement in the design process for high-speed digital designs. Over a three-year period, we applied modeling and simulation to Signal Integrity problems and reduced the length of time it took to get a first working prototype, on average, from three-and-a-half prototype cycles to a little more than one prototype cycle.

The designs simulated were dense, high-speed, complex boards that were very comparable to high-speed computer server boards and backplanes. It was quite common for the designs to have over 1500 components and 3000 nets on a board. Components often included microprocessors like the Pentium III, and FPGAs with pin counts of 500 and more. Clock speeds at the beginning exceeded 33MHz, increasing to 266MHz, and approaching GHz speeds by the end of the 3-year period.

Scrap and rework cost per prototype cycle could easily exceed $100,000 per design.[3] During the three years we worked on about 36 such designs. Naturally, we made mistakes and there was a learning curve for the organization. But in the end, we saved the company a few million dollars over and above labor, materials, and software license costs. Plus, the improvements in engineering development costs were insignificant when compared to the profit leverage gained by improved time-to-market.

21.6.2 Day-to-Day Success Example

The following example illustrates some important ideas about cooperation and bringing practical insight to the act of running the EDA

---

[3] US dollars, 2002.

tool. The example showed that when the model launched less high-frequency signal content on a backplane bus, the signal showed less noise. That result should be of no great surprise. The example involved a GTLP bus on a backplane. Simulation results, before and after a change was made to the driver IBIS model, show a startling improvement in Signal Integrity performance.

The backplane was simulated using a model that, at first, did not include enough detail. The backplane bus was an 18-slot, daisy chain, bi-directional[4] bus using a GTLP technology part. Data rate was 8.162 MHz. Pitch between slots was one inch, and daughter-card stub length was 1-1/2 inches. The customer equipment configuration could have two to eighteen applications cards plugged into the backplane.

Most digital-logic design books show this type of bus with what amounts to zero-length daughter card connection paths (stubs), as if no reflections will occur at the point where the cards are plugged into the backplane. This is permissible for logic design. But it is nonsense for Signal Integrity design.

The reflections were much worse as simulated with one supplier's model file (shown in Figure 21-1) versus a competitor's model file for the same device. The cause turned out to be the use of dV/dt in simulating this kind of bus.

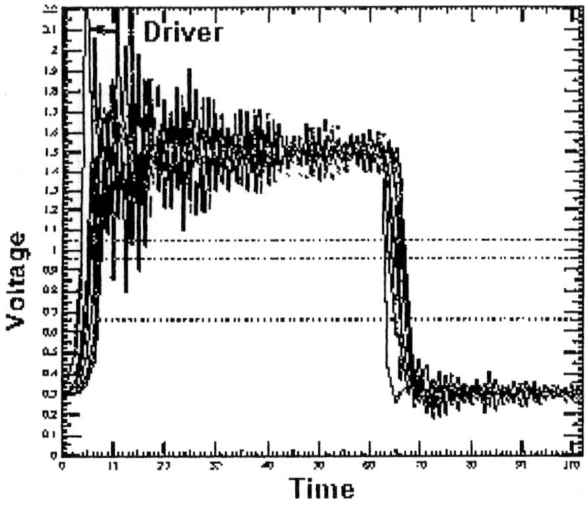

Figure 21-1. Simulation using slew rate data only

In reality (analog effects of high-speed logic signals) this topology is complex enough, and frequency content is high enough, to set up many

4 Meaning that any card on the bus can become the driver.

noisy switching reflections on the bus. Reflections will occur everywhere there is an impedance discontinuity. There is no feasible termination strategy that can fix the reflection problems on a bus like this. Instead, if possible, we must avoid launching high-frequency content signals from a driver onto the bus. The product designer asked us to take a closer look at the poorer-performing part. He wanted to utilize the more convenient packaging available from the supplier of the poorer-performing part.

Looking inside the model file that gave poor results revealed that switching, voltage-time data was given only as dV/dt data, which is all that the IBIS Spec requires. IBIS permits including a V-T table at the model supplier's option. GTLP technology is specifically designed to handle complex backplane topologies, by having soft turn-on and turn-off corners in its switching characteristics. Such characteristics cannot be adequately modeled with dV/dt data alone. The simulation engine is being told to include too much high-frequency content by using a dV/dt model for this part. A V-T table is needed that correctly models the soft corners for the part in question. The supplier was asked to include the added detail, which they provided about three months later. Figure 21-2 is the graphical plot of the V-T table data after its addition to the IBIS model file.

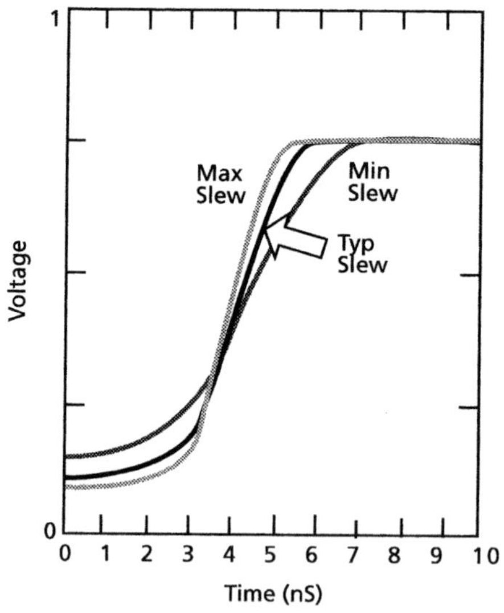

Figure 21-2. New V-T data curve

Simulations with the corrected model showed reflection results that were even better than with the competitor's model—as shown in Figure 21-3 below:

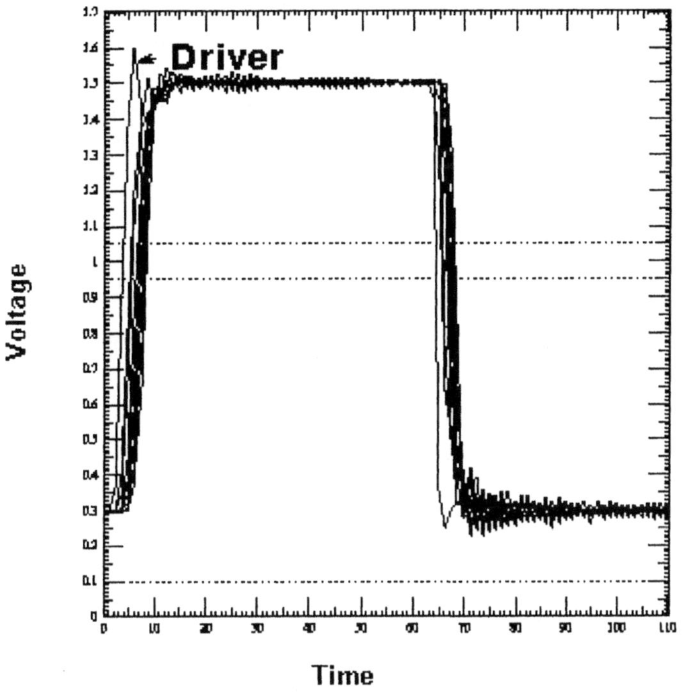

Figure 21-3. New simulation using V-T table data

The improved results were not just due to model improvements. 3Com also studied limits on stub length, optimization of backplane pullup resistors, and optimization of target backplane-daughter card characteristic impedance, connector modeling, and various termination schemes. The ideas reinforced by this example are:

- Simulate early and often
- Bring engineering insight into the application of EDA tools
- Feedback simulation results to the semiconductor supplier, in a cooperative frame of mind
- Hope that the supplier has enough time to respond with model improvements, if that is indicated
- Keep an open mind

21.7 SUMMARY

Most engineers understand the benefits of what-if modeling, especially at the level of manipulating board layout, routing, and termination. However, they may not appreciate the benefits of what-if model manipulation with the I/O model itself—even though it may be very important. In the case of components with programmable and controllable I/O, that what-if manipulation is done routinely. When doing what-if simulations, engineers can achieve similar benefits with non-programmable I/O. For example, changing V-T switching waveform frequency content and scaling up drive capability.

Chapter 22

FUTURE TRENDS IN MODELING

Modeling and simulation are being successfully applied to solve Power Integrity and EMI/EMC problems

Abstract: The rapid advance of technology has opened new opportunities and challenges in modeling and simulation. Signal Integrity, Power Integrity, EMI/EMC, and simulation are becoming more necessary and more challenging. New ideas being applied to high-speed digital design include the use of macromodeling, VHDL-AMS, and S-Parameter models. Advances in EDA tools and simulation models are helping engineers meet the new challenges of SI, PI, and EMI/EMC in Gigabit circuits.

22.1 BRIDGES TO THE FUTURE

Predicting the future is a risky business. If the authors claimed to know what the next exciting new electronic products or applications would be, they probably would be quite wealthy. Even so, the authors are confident that modeling and simulation will enable innovative future developments and give engineers with these skills a competitive edge in their career advancement. The themes presented in this book are a bridge into the future. What are those themes and how do they point ahead?

- Modeling and simulation involve *skilled specialists* in a wide variety of disciplines working together as a team.
- Modeling and simulation requires that individuals possess a *multi-disciplined skill set.* No one model or simulation method handles all of today's design tasks, but a combination of them does.
- Modeling and simulation requires a *can-do attitude, with a realistic approach.* Accurate and abundant models are not a given. Knowing how

to design within limitations has, and will continue to be, a key design skill.

- *RF modeling and simulation techniques will be used* on very high-speed digital design challenges. High-speed digital nets are operating at microwave speeds and RF skills will be used to design those nets.
- Modeling and simulation will be systematically and consistently applied in the design *process* for maximum benefits. It is better to have a simulation that is 90% accurate and timely, than one that is 100% accurate and too late. *Virtual design* will ensure that new product prototypes are right by design.
- Modeling and simulation *for EMI/EMC control* is one of the next big areas for competitive advantage. Early in the design process, numerical computational electromagnetics will be used on EMI/EMC problems to lower expenses and meet development schedules.

What are the future trends for modeling and simulation? Those future trends will address the above issues in an improved and successful manner. The exciting news is that these predictions are already happening.

22.2 CHALLENGE OF HSDD

Semiconductor process technology shows no sign of slowing down in its quest for ever-smaller geometries, faster devices, and more functions (gates) on ICs. The industry is well beyond million-gate ICs and is headed towards billion-gate devices.[1] As feature sizes shrink, so have power supply voltages and threshold voltages, while edge rates get ever faster. Devices continue to grow in complexity and size, and so do their requirements for power-supply switching currents. Signal Integrity is now an issue on the IC die itself [113].

These trends make device I/O electrical performance much more subject to loading and interaction with signal reflections on the PCB interconnections. The strength of IBIS modeling and simulation was and continues to be that it intentional ignores the behavior inside the driver and receiver I/O. Now I/O and load interact. Complex I/O designed to provide adjustable pre- and de-emphasis on output waveforms requires that I/O behavior and interconnection behavior be modeled and simulated *together*.

This is similar to solving a SPICE model amplifier circuit. Also, the modeling and simulation of I/Os using pre- and de-emphasis to improve eye-opening is similar to matched filter design familiar to analog engineers.

---

[1] Billion-gate ICs may exist by the time this is published.

Table 22-1 shows a trend of semiconductor technology over the past twenty years and projected to 2007.

Table 22-1. Semiconductor technology trends since 1985 [13]

CMOS Technology (μm)	Year	Core Power Supply (Volt)	Density (gates /mm²)	Current peak per gate (mA)	Clock frequency (MHz)	Switching Delay
1.2	1985	5	8 K	1.1	4 - 50	1 ns
0.8	1990	5	15 K	0.9	4 – 90	0.5 ns
0.5	1993	5	28 K	0.75	8 – 120	0.3 ns
0.35	1995	5 - 3.3	50 K	0.6	16 – 300	0.2 ns
0.25	1997	5 - 2.5	90 K	0.4	40 – 450	0.12 ns
0.18	1999	3.3 – 1.8	160 K	0.3	100 – 900	0.1 ns
0.12	2001	2.5 -1.2	240 K	0.2	150 – 1200	70 ps
0.09	2004	2.5 - 1	480 K	0.15	300 – 2000	50 ps
0.065	2007	2.5-0.7	900 K	0.1	500 – 3000	30 ps

Used with permission of the authors

Noise signals on the power deliver and return interconnections and SSN/SSO switching noises are now much more significant and interact with data and clock signals. Shifts in threshold voltages due to self and signal heating are also much more significant and interact with data and clock signals. Higher frequency components of faster switching waveforms excite more electromagnetic coupling and resonance modes in the board, enclosure, and cables, thus further interacting with signals. Extremely small dDevices are more sensitive to device physics effects like gate conductivity modulation. Hence, the old IBIS approach of ignoring internal device behavior does not work so well for cutting edge, complex I/O.

In the past, IBIS worked well because it struck a good balance between speed and complexity. Signal behavior on the interconnection dominated and internal I/O behavior was secondary. To accommodate more I/O behavior, a new balance between speed and complexity needs to be found.

These issues were discussed in detail in "Chapter 20, The Challenge to IBIS." The authors concluded that a combination of accommodation to HDLs (SPICE, SPICE macromodeling, AMS equation-based modeling, and S-Parameters) and external models promises to provide that new balance and is a future trend in modeling.

The context of the discussion in "Chapter 20, The Challenge to IBIS" is primarily that of modeling what happens on the signal interconnections and in the I/Os that drive and receive the signals. Other issues including Power Integrity, SSN/SSO, lossy transmission lines, and EMI/EMC also need to be addressed. In this chapter, future trends and suggestions on these issues are presented. This topic discusses how well the various types of models help us in addressing these challenges. But before new and innovative solutions can be implemented, some attitudes must be changed, particularly in EMI/EMC. Attitude can often be the toughest barrier to progress.

22.3 HOW DESIGN METHODS HAVE CHANGED

Historically, circuits were constructed first, and then measured to verify that they performed the intended functions. Computations were sparse because they were done on paper, using slide rules, and later with hand calculators. Computations were time consuming and tedious. It was easier and faster to go to the laboratory, construct something, and find an answer. Most development schedules were not as compressed as they are today. In order to measure and figure out how to make a prototype work, we usually waited until the actual hardware was available. Today, it's often faster to go to the computer, simulate something, and find an answer.

We can no longer afford to wait until the actual hardware is available to make a new product work. This is for many reasons, which are discussed throughout this book. But waiting for a prototype to debug is still the way that most EMI/EMC design is done today.

Today, for Signal Integrity, we run our virtual design on computers and use modeling and simulation to assist in a proactive design process. We employ mathematical models and simulation to study the behavior of the circuit. Based on the computations, we may change the design of the circuit until we get the predicted behavior we want. The predictions must be done early enough in the design cycle to head off expected problems. Once we are satisfied with this "paper" design, we can then give a good set of directions on how to build, test, and verify it for the intended behavior. The phrase "right-by-design" applies here.

However, critics often say, "Wait. Simulation does not work because reality is too complex to be represented by modeling, and the simulations would run too long to be able to use the results. Besides, it takes too big of an investment to get into that game."

Critics have always made these objections throughout the advancement of computation technology–do not believe them!

One author [Leventhal] heard them in the early 1960s when working on 3-stage RF amplifier circuits. Those objections were not believable then, and they still are not believable. Instead, we were taught that the best engineers successfully combined rules of thumb, modeling, simulation, physical layout smarts, and measurement skills.

Often, the critics of EMI/EMC modeling and simulation have valid concerns that should be listened to. But remarkable progress has been made in the field of modeling and simulation. Indeed today's technological progress would be nearly impossible without modeling and simulation.

Breadboarding—that is, building the circuit and debugging it on the bench (or test and verify)—is still an important part of any engineer's skills. Indeed, the physical act of laboratory debug and measurement enhances cognitive skills [99]. But feature sizes on PCBs are shrinking to a point where they cannot be probed and measured. Therefore, bench test and debug are becoming very difficult to do and we are becoming more dependent on modeling and simulation.

In terms of complexity, speed, shrinking sizes, and increasing functionality demands, it is impossible to meet schedule, cost, and performance requirements without modeling and simulation. Many companies will not sign off on a design unless a successful Signal Integrity simulation has been run on it. Chip designs always require computer validation before being released to production.

Modeling and simulation can take a considerable investment to get into and to develop the necessary skills and knowledge base. But return-on-investment (ROI) has been proven in analog, RF and microwave, and high-speed Signal Integrity at the chip, PCB, and system level.

22.4 ATTITUDES IN EMI/EMC ABOUT MODELING AND SIMULATION

The one area where reluctance to model and simulate still dominates is in EMI/EMC simulation. Historically, opponents of EDA were right when they said that the task was too complex. New software tools and computers are becoming much more powerful, and the ability to solve EMI problems is rapidly increasing. The opponents sometimes claim that the tried and true methods of design rules, build, and debug have always worked. So why change? The truth is that 70% to 90% [43] of new designs fail first-time EMC Regulatory testing. That is unacceptable.

The iterative cycle of design→build→test→find→fix (redesign) adds to cost and missed schedules. It also adds to missed chances for EMI engineers to favorably affect the final product. If we need to wait until we have the hardware before we can apply engineering knowledge, we are usually in deep trouble. A common management answer[2] to problems that arise at that stage is: *Keep testing it until it passes*. Not: *Find the problem, fix it, and we*

---

[2] As often commented on by EMI/EMC engineers and Regulatory test houses.

will change the design or slip the schedule if necessary. Since Regulatory testing is an inexact science, this approach often works, but it usually means producing a second-class product.

Often EMI engineering is viewed as a mysterious activity performed by wizards and curmudgeons after disaster strikes.[3] With today's short market window of opportunity, we need products that are built-right-the-first-time. Not redesigned and fixed after several prototype cycles.

At 3Com, after the design team came up a learning curve on Signal Integrity, all PCB products went, on average, from 3.5 prototype runs to 1.1 prototype runs. Without fail, however, when it came to passing Regulatory requirements on EMI/EMC, it was *run in circles, scream, and shout.* 3Com had very good EMI/EMC engineers and they tried to exercise their skills through careful application of design rules. But EMI/EMC knowledge needs to be applied early and often and with the aid of computers to keep track of complexity–not by waiting until hardware is in hand before getting a chance to affect outcomes.

In addition, as we push the envelope on cutting-edge designs, EMI/EMC rules-of-thumb are getting much harder to sort out and apply. Today's designs are being driven to consider phenomena and problems that could have been ignored in the past. Signal Integrity, crosstalk, Power Integrity, electromagnetic radiation, thermal, and various reliability issues extend down into the die and individual-gate realm and outward to the system level at all scale factors. In any given design task, several different effects have to be accounted for. The *only* ways to sort, combine, and weigh these multiple effects is through numerical computation, that is, modeling and simulation.[4]

22.5 HIGH-SPEED DESIGN IS BECOMING MORE CHALLENGING

The next two tables highlight recent trends that affect the tasks of designing for good performance, reliability, low cost, and timely market introduction. Table 22-2 shows what happens as edge rates (*1/Tr*) increase to bandwidth requirements and wavelength (λ). Today's deep-submicron

---

[3] Often, with the added remark from the EMI expert: *Any dummy could have avoided that mistake.* But this is typical. In the days before this author used much simulation, he often cursed himself when discovering problems in a prototype.

[4] For example, automotive and aeronautic customers ask for drastically low emission and high immunity-level requirements. Automobile and aeronautic products now include more on-board electronics and computer power than was contained in the spacecraft that got us to the moon.

CMOS technologies are now switching at GaAs edge rates and faster. So high-frequency effects are very pronounced.

Table 22-2. Trends in switching speeds

Basic Technology	Type	Tr (Ns)	Bandwidth Requirements (MHz)	1/2 λ in stripline (inch)
TTL	LS	6	53	16.67
	ALS	3	106	8.33
	FAST	2	159	5.55
CMOS	HC	4	79	11.11
	FACT	2	159	5.55
	ALVT	0.8	398	2.22
ECL	10H	1	318	2.78
	100K	0.7	455	1.94
	ECLinPS	0.5	637	1.39
GaAs		0.2	1590	0.56

Table 22-3 shows how the demands for clean power delivery have become much harder to satisfy. Larger ICs require more total current and *more dI/dt.* This requirement is in spite of the individual gate current shrinking by more than an order of magnitude in the past twenty years.

Because *dI/dt* current demand has grown by more than two orders of magnitude, the effective power supply impedance, *Ztarget* seen at the IC supply pins has decreased, drastically in order to deliver clean power.

Ztarget is the required transfer impedance at the IC power pins looking back towards the power supply.

Table 22-3. Trends in power delivery requirements

Year	Voltage[5] (Volts)	Power (Watts)	Current (Amps)	Z_{target} (m-Ohms)	Frequency (MHz)
1990	5	5	1	250	16
1993	3.3	10	3	54	66
1996	2.5	30	12	10	200
1999	1.8	90	50	1.8	600
2002	1.2	180	150	0.4	1200

Instantaneous power-supply switching-current demands have become very pronounced. Note in particular how the needs of bandwidth, current, and power delivery continue to grow, while signal wavelengths and inherent noise-immunity-threshold windows shrink as power-supply voltages shrink.

Passives used on PCBs are shrinking also. Resistors and capacitors, used to terminate and bypass the ICs, are now the size of grains of pepper! Still,

[5] The power-supply corresponds to the internal power-supply. IO supply has not scaled down this fast. It is expected that the voltage decrease will not continue as fast for the next technologies. For example, 65nm at Intel is still 1.2V, as well as 90nm and for 130nm.

they consume too much room, and many will migrate into the layers (stackup) of the PCB as printed circuit elements. These elements will then become mostly inaccessible for measurement or manipulation.

ICs must deliver more and more functionality, even though feature sizes shrink into the nanometers. Along with their gate sizes, chips keep getting bigger instead of shrinking. Million-gate ICs are now commonplace, and billion-gate chips will soon be here if they are not already. This leads to the design method trends in Table 22-4.

Table 22-4. Past and future technique trends used in circuit design

Past	Future
Test and fix	Boards will be built to be correct by design
Debug board on bench	Debug virtual board in computer
What-if analysis done by soldering in resistors, etc.	What-if analysis done by manipulating computer variables
Current and voltage generator instruments	I, V, E, and H generator models
Generator internal impedance, R, L, and C, etc.	Generator internal impedance, R, L, C, I-V/V-T tables for I/Os, I-T tables for power pins, etc.
SPICE and IBIS modeling techniques	SPICE and IBIS modeling techniques will continue. S-Parameters will be added for in-package parasitics. Macromodeling and AMS modeling will be added for complex I/O
IC, Hybrid, MCM and (SIP) SystemInPackage manufacturing, test and trim techniques. Example: Current mirrors, resistor laser trim, inked resistors etc.	IC, Hybrid, MCM and SystemInPackage manufacturing, test and trim techniques will be applied to boards where possible

Progress in modeling and simulation is driven by need and measured in capability. As technology evolves, the need for improved analysis tools and improved models also evolve. Need for more advanced analysis almost always marginally exceeds capability.

22.6 ADVANTAGES OF SPICE, S-PARAMETERS, AND IBIS

SPICE, S-Parameters, and IBIS modeling methods have different capabilities for solving the following design tasks:

- Simulation of signals on the interconnection and in the I/Os that drive and receive those signals, including lossy transmission lines.
- Simulation of noise on the power delivery system and its interaction with the signal system.

- Simulation of EMI/EMC on the product as a whole and its interaction with other electronic equipments.

Let's take a closer look at the advantages and disadvantages of SPICE, S-Parameters, and IBIS modeling methods.

22.6.1 SPICE Advantages

As an illustration of the type of problems SPICE models are designed to handle, consider the issue of thermal shift and shifts in switching thresholds. Figure 22-1 shows how Cadence Design Systems illustrated threshold shift due to heating at a receiver.

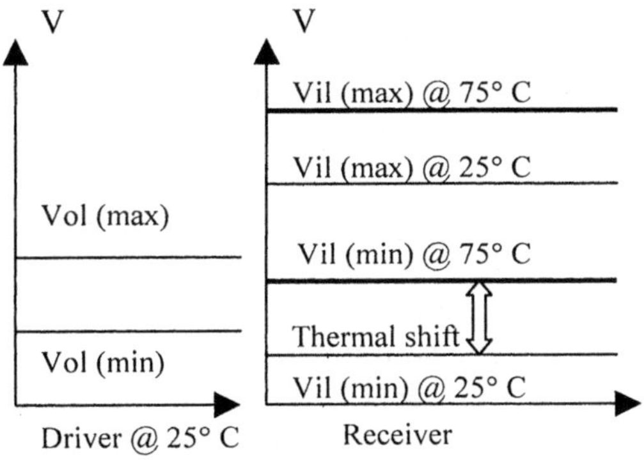

Figure 22-1. Threshold thermal shift example

This figure illustrates the idea that Vol min and max of a driver at 25°C cannot drive past Vil min and max at a receiver at 75°C. With today's lower threshold voltages, we would expect that relative thermal shift due to temperature to be even more sensitive. Gate conductivity modulation is another issue that can be studied efficiently with SPICE models. For studying these types of issues, SPICE is excellent.

Additionally, SPICE retains an advantage in emission and susceptibility analysis with a non-ideal supply, and an almost immediate simulation of impedance versus frequency.

22.6.2 S-Parameter Advantages

We are witnessing convergence between the methods employed by RF and microwave engineers and digital engineers addressing GBIT serial data busses. The following list summarizes the reasons for discussing S-Parameters in this book.

- Users may have already encountered S-Parameters in very high frequency digital circuit analysis involving interconnection.
- S-Parameters efficiently handle multi-port coupling analysis and conceptualization problems, as encountered in IC packaging design. S-Parameters do so without revealing much information about actual physical construction. This is because they are black-box models. That matrix formulation does not even postulate internal elements in the S-Parameter black box. S-Parameters are only transmission and reflection coefficients.
- Radiation propagation through space can be visualized better with a multi-port S-Parameter approach as in the TLM method.
- S-Parameters handle issues like dispersion and loss well and these issues are now faced in designing the fastest digital nets.
- Some simplifying assumptions that, until recently, allowed software to simulate PCBs quickly and accurately are no longer always true. Those assumptions were that:
  - Transmission lines were *lossless* and frequencies were below waveguide cutoff. In this situation, waves propagate in the transverse electromagnetic wave mode (TEM). Under those assumptions, quasi-static solution methods, often in 2D, work well.
  - Transmission lines were *low loss* and frequencies were close to waveguide cutoff. This case can still be approximated by quasi-TEM methods. Under those assumptions, quasi-static solution methods, usually in 2D with 3D at vias, pin fields, and other complex structures still work well.

  Today, these assumptions are no longer true for the fastest digital nets. Frequencies are higher and signal losses in the direction of propagation are significant. Therefore, voltage fields in the direction of propagation become significant due to these higher losses and the assumption of TEM mode propagation no longer is valid. Quasi-static solvers become inaccurate and full-wave solvers must be used. Lumped element R, L, G and C models for the transmission line become invalid and S-Parameter models that can better describe loss, dispersion, distributed network, and high-frequency related effects must be used.

- In "Chapter 8, Selecting the Best Model for a Simulation," we discussed transform methods. These methods enable:

  - Transferring a problem back and forth between the time and frequency domains to aid solution.
  - Performing convolution and conformal mapping of S-Parameter models into the time domain. The mapped model is then used in a time domain transient simulation that can account for high-frequency losses and dispersion.
- High-speed serial data busses operate at 10 GBITs or 10GHz and beyond. They need to be simulated from die pad to die pad. In-package time delay and modeling of wire bond, solder bump, and package interconnections are best handled by S-Parameter analysis.

22.6.3 Using Microwave Design Techniques in Digital Circuits

One challenging aspect of deep sub-sub-micron CMOS technology is that driver behavior is becoming sensitive to loading effects seen at the output terminals of the driver. This is a wholly new phenomenon for the digital circuit designer, but it is very familiar to RF and Microwave amplifier designers. For them, it is common practice to compute the effects of receiver loading at the *reference plane* of a driver. S-Parameter models are commonly used in computing the effects of network loading as seen at the driver. Either the Smith Chart[6] or the transmission line equations are used to compute the driver loading effect at a particular frequency.

How exactly are S-Parameter models used in such a process? Once S-Parameters are available and extracted for the components and structures on a board, they can be *transformed or translated* [45, 78, 88, 94] to the driver output pins of the IC.

In a *frequency domain* steady-state analysis, the impedance effect at a driver input to a transmission line of a load at the line's receiving end can be modeled. This is done by shifting reference planes [88, page 274] from the receiver to the driver. The effects of loading at the driver can then be computed.

If the length of the line between these two ports is ℓ, and s_{11} (input reflection coefficient at the receiver) is the only S-Parameter we need to reference to the driver pins, then the propagation constant, γ, is:

---

[6] The process of using the Smith Chart to compute the receiver load, as seen at the driver, is called *rotating* the load (by x-number of wavelengths) to the plane of the driver. The effect of multiple stubs (impedance discontinuities) might be handled by matrix model combinations at the discontinuities. Or perhaps, superposition would allow them to all be rotated to the driver and combined there. Wavelength at a given frequency translates to electric distance on the transmission line. RF and Microwave engineers have used Smith Charts since 1939 [115, 116].

$$\gamma = (\alpha + j\beta) \tag{22-1}$$

where α = attenuation constant, for a lossless line, γ reduces to

$$\gamma = j\beta \tag{22-2}$$

where β = phase constant. If $s`_{11}$ is the load reflection coefficient seen at the input to the line, we have:

$$s`_{11} = s_{11} e^{-j2\beta\ell} \tag{22-3}$$

Whenever working with S-Parameters and reflection coefficients, we need to "renormalize" them if the impedance of the line is different from the impedance for which their values were originally computed. To renormalize:

- Convert S-Parameters to Z- or Y-parameters using the original normalizing Zo
- Re-convert back to S-Parameters using the new normalizing Zo
- If Zo is complex or the line is lossy, be careful to use complex variables math. For example, transmission lines over lossy or conductive substrates.

A second way of computing the effect is by using the transmission line equation, plus the knowledge that $Z_{in} = Z_L$ transformed to the driving point input to the transmission line. In the following equations,[7] we can account for line losses:

$$Zin = Zo\left[\frac{Z_L + Zo(\tanh(\gamma\ell))}{Zo + Z_L(\tanh(\gamma\ell))}\right] \tag{22-4}$$

And the propagation constant of the transmission line, γ, is:

$$\gamma = \alpha + j\beta = \sqrt{(R = j\varpi L) * (G + j\varpi C)} \tag{22-5}$$

Therefore, both attenuation and dispersion can be accounted for in these equations. Also, the characteristic impedance of a lossy, dispersive transmission line is:

---

[7] The transmission line equations also have the historical label of *telegrapher's equations*. That indicates how long they have been in use.

$$Zo = \sqrt{(R + j\varpi L)/(G + j\varpi C)} \tag{22-6}$$

For both γ and Zo, the RLGC parameters are extracted in an electromagnetic field solver, with the appropriate material properties for the dielectric and the copper traces.

With the aid of S-Parameter techniques, RF and Microwave engineers have long solved for receiver-loading effects on the driver. The authors speculate that the same methods can be used in HSDD. The first step is to solve for driver-receiver loading effects on the waveform, which is launched by a high-speed digital driver. Then, a "standard" IBIS approach of not solving for internal driver behavior could proceed within some limits. Or, as in the case of designing the amount of pre-emphasis to use, the simulation of driver and equivalent load could be iterated to refine the launch waveform.

A semi-graphical technique using normalized impedance, the *ZY* Smith Chart [115, 116] is another approach to computing the effect of a load as seen at other points in a distributed circuit. Graphical techniques can aid in the visualization of referencing load impedance to different locations on a transmission line. The receiver S-Parameter model can be re-normalized to the working characteristic impedance, Zo, as necessary. To combine the effects of several loads and impedance discontinuities, we should use matrix model transformations as required. For details on matrix transformations, see "Chapter 8, Selecting the Best Model for a Simulation."

A complex net on a PCB can be decomposed into individual loading effects at the driver and recombined there. Once the effects of all loads are transformed to the driving pins, they can be combined using matrix mathematics. The y-matrix parameters of loads in *parallel* can be added directly, and the z-matrix parameters of loads in *series* can be added directly. All the modeling and layout information can be incorporated into an equivalent load at the IC pins. At this point, the circuit can be solved using Ohm's Law or transmission line methods if the equivalent independent generator plus generator internal impedance model of the IC are available.[8]

To summarize, the effects of loading at the driver output terminals in a distributed network (transmission line) can be computed.

S-Parameters are a frequency domain model. To solve the problem in that domain, we would simulate the impulse response and incorporate the spectrum of the IC signal into that response. Then, given a Fast-Fourier-Transform method, we can derive the time-domain waveform. To solve the

---

8 We might even speculate that other classic microwave PCB techniques might one day be used in very high-speed digital boards. For example, co-planar waveguide transmission lines, stub tuning, and couplers of various types.

problem in the time domain, we would first convolve[9] the equivalent impulse response into the time domain.

22.6.4 IBIS Advantages

IBIS is used for the majority of PCB simulations where the problems are simpler and likely to remain so for quite some time. Once a better IBIS model of complex I/O behavior is achieved, there is nothing inherently wrong with the behavioral driver and bounce diagram method used by IBIS. This method can be used to scan multiple nets on complex boards for reflection issues, crosstalk issues and more. We recommend a first order modeling and simulation strategy employed early in product design. SPICE and other more complex modeling methods can then be employed at final verification or for very critical nets.

Internal effects at driver and receiver will probably remain a semiconductor supplier modeling responsibility. Models supplied by the model makers could be a more detailed form of IBIS. That more detailed form of IBIS would include more complete guidelines for loading conditions, power input limits for receivers, and other modeling assumptions. Even thermal shift data could be incorporated in the expanded model. With these adjustments, an IBIS model approach at the board level would still work, with its advantages of simulation speed. If a user needed to operate outside of the model's assumptions and limitations, that user would need to acquire an adjusted model.

22.7 COMBINING MODELS AND EDA TOOLS TO DESIGN HIGH-SPEED SERIAL BUSSES

As of this writing, designing high-speed serial, point-to-point data busses is a challenging task. The authors indicate throughout this book that a combination of modeling methods and simulation tools are needed to address the tasks of designing such data busses. Today, those SERDES-type busses are being designed routinely to run at 2.5 to 10 GBITs per second. The design tasks for such high-speed nets now include issues of Signal Integrity, Power Integrity, and EMI/EMC. The techniques of routing, termination, impedance control, and topology design do not quite solve all these complex issues [17, 46, 49, 60] by themselves. The techniques of

---

[9] Convolution is a mathematical method that converts (maps) quantities and properties between different domains such as the frequency and time domains.

designing complex buffer I/O to compensate for high-frequency signal losses must also be applied.

EDA tools are available to apply to all high-speed serial data bus design issues, but they come from many different providers. A lack of integration between these different EDA tools raises design flow problems [4]. Unfortunately, combining models and EDA tools from separate sources produces a design flow that is error prone. How one EDA vendor, Ansoft, has teamed with various semiconductor suppliers and other EDA suppliers to address the challenge is found at the following websites:

- http://www.gigabitbackplanedesign.com and,
- http://www.ansoft.com/membership/login.cfm

In addition to the white paper authored by [4], other white papers can be downloaded from those URLs or received via Ansoft's email newsletter.

22.8 IBIS: PAST, PRESENT, AND FUTURE SPECIFICATION ADDITIONS

This topic presents a brief history of improvements, additions, and clarifications to the IBIS Specification. Past history is one indication of what might be expected in the future. From the start of electronics design, technological evolution has been pushing the modeling and simulation envelope.

Previously, additions and expansions to the Specification have been addressed by additions and expansions to a system of keywords. In the future, major improvements are more likely to come from support for HDL extensions and calling external models. What remains constant is the IBIS Committee's constructive attitude that the Specification is a work-in-progress that needs to keep up with advances in technology.

In June 1993, IBIS V1.0 was ratified. At that time, there were some issues that IBIS V1.0 model did not address well. From the start, a process of revision and re-ratification has been ongoing. The following extensions, additions, and clarifications have been added to the Spec:

- Improvements to modeling pin parasitics, including mutual coupling
- Addition of minimum and maximum values to parameters
- Ability to handle series switches, such as analog switches
- Driver scheduling for soft turn-on and turn-off devices
- V-T waveform tables
- Differential pins

- Behavioral descriptions (.ebd model) of PCBs
- Enhanced characterization of receiver thresholds
- Tree diagram of Specification keywords to aid readability
- Support for Berkeley SPICE, VHDL-AMS, and Verilog-AMS hardware description languages

Practical Issues With IBIS Models, written by Bob Ross, long the Chairman of the Committee, explains the reasons for changing the IBIS Specification, and the change implementation up to v3.2 [102]. Later versions of IBIS tend to include a readme file that documents the changes and additions. Table 22-5 summarizes the major additions of the different versions of IBIS [72].

Table 22-5. Major additions in IBIS Specification versions

IBIS 1.0 1993	IBIS 2.1 1995	IBIS 3.2 1999	IBIS 4.0 2002	IBIS 4.1 2004
• Ramp • V-I Curves • C_comp	• Model_types • Waveforms • Package model • Pin mapping	• Model_spec • Submodel • Driver schedule • Z0 package • Dimm modules	• Receiver thresholds • Submodel extensions	• External models SPICE • External models AMS

Simulator companies are continually expanding their capabilities to model and simulate new technologies and new issues. Consider the following developments shown in Table 22-6.

Table 22-6. Recent major developments in circuits

1990	2005
High-speed digital boards averaged 5 volts.	High-speed digital boards are targeted at 1.2 volts and below.
Clock speeds averaged 16 MHz.	Clock speeds are targeted at 2000 MHz and above.
Ground and power-plane transfer impedances (looking back towards the power supply) averaged 250 milliohms.	Ground and power-plane transfer impedances are targeted to be at 0.4 milliohms and below.

Future Developments

As of this writing, the IBIS Committee is working on expanding the IBIS Specification to include:

- Improved quality-checking procedures, explanations, and an improved parser to make the job of producing a correct IBIS model easier.

- Discussions of EMI/EMC modeling, and how to incorporate it in or as a companion specification to IBIS.
- Discussions of expanded [External Circuit] keyword capability to handle the modeling of complex I/O.

Table 22-7 presents a preview of how the [External Circuit] call in IBIS will handle SPICE and AMS model sub-program calls. Table 22-7 and the implementation information on the [External circuit] keyword are courtesy of the IBIS Committee.

Table 22-7 [External circuit] keyword and sub-parameters

Keyword:	[External Circuit], [End External Circuit]
Sub-Params:	Language, Corner, Ports
Usage Rules:	Each [External Circuit] keyword is followed by a unique name that differs from any other name used for [Model] or [Submodel]. The [External Circuit] name is referred to as 'Ext_name', and a [Model] name is referred to as 'Model_name'. Each [External Circuit] is referenced in a unique manner by [Circuit Call] keywords that are part of the [Component] keyword.
Language:	Accepts "Spice", "VHDL-AMS", or "Verilog-AMS". The Language subparameter is required and can appear only once.
Corner:	Three entries follow the Corner subparameter on each line: corner_name, file_name and circuit_name.
corner_name	The corner_name entry is "typ", "min", or "max".
file_name	The file_name entry points to the referenced file in the same directory as the .ibs file. It is recommended that lower-case file_name entries be used to avoid possible conflicts with file naming conventions in different operating systems.
circuit_name	The circuit_name entry provides the name of the circuit within the referenced file to be executed by the EDA tool. For Spice files, this is normally a ".subckt" name. For VHDL-AMS files, this is normally an "entity", "architecture" name pair with the architecture name enclosed in parenthesis. For Verilog-AMS files, this is normally a "module" name.
Ports:	Gives port names to the terminals as "ports" that are used to connect the external model. The port assignment is by position, and the port names do not have to match exactly the terminal names of the external model. The list of port names can be presented over several lines by using the Ports subparameter for each line.

SPICE Implementation

```
[External Circuit]  BUS_SPI
Language   Spice
| Corner       corner_name   file_name     circuit_name (.subckt name)
Corner         Typ           bus_typ.spi   Bus_typ
Corner         Min           bus_min.spi   Bus_min
Corner         Max           bus_max.spi   Bus_max
|
| Ports are in same order as defined in Spice
Ports     vcc gnd io1 io2
Ports     int_ioa vcca1 vcca2 vssa1 vssa2
Ports     int_iob vccb1 vccb2 vssb1 vssb2
|
[End External Circuit]
```

Vhdl_AMS Implementation

```
| VHDL_AMS Example:
|
[External Circuit]  BUS_VHD
Language   VHDL-AMS
| Corner       corner_name   file_name     circuit_name
entity(architecture)
Corner         Typ           bus.vhd       Bus(Bus_typ)
Corner         Min           bus.vhd       Bus(Bus_min)
Corner         Max           bus.vhd       Bus(Bus_max)
|
| Ports are in the same order as defined in VHDL-AMS
Ports     vcc gnd io1 io2
Ports     int_ioa vcca1 vcca2 vssa1 vssa2
Ports     int_iob vccb1 vccb2 vssb1 vssb2
```

Verilog_AMS Implementation

```
| Verilog_AMS Example:
|
[External Circuit]  BUS_V
Language   Verilog-AMS
| Corner       corner_name   file_name     circuit_name (module)
Corner         Typ           bus.v         Bus_typ
Corner         Min           bus.v         Bus_min
Corner         Max           bus.v         Bus_max
|
| Ports are in the same order as defined in Verilog-AMS
Ports     vcc gnd io1 io2
Ports     int_ioa vcca1 vcca2 vssa1 vssa2
Ports     int_iob vccb1 vccb2 vssb1 vssb2
|
[End External Circuit]
```

The [External Circuit] keyword is used to reference an external code file for a more detailed die interconnection description. The [External Circuit] keyword and sub-parameters can be positioned as a group before or after any [Component] keyword group.

22.9 ADVANTAGES OF PRE-LAYOUT SIMULATION FOR EMI/EMC

In Signal Integrity, it is standard practice to simulate nets from schematic topologies prior to their actual layout and routing. Generating a topology is referred to as "mapping" a proposed net or circuit.

Design constraints on routing, characteristic impedance, termination, time delay, and more are then passed into the CAD department where they guide the PCB designer. Critical nets and devices are placed and routed first, followed by the remaining nets. After routing, nets are extracted from the physical board and re-simulated. A partially routed net can be simulated with the mapped and extracted portions intermixed. An iterative feedback and feed-forward loop gets exercised until the PCB design goals are met and the board is routed. During this process, the logic design engineer, signal integrity engineer, and PCB designer continually communicate with each other.

During this iterative process, un-routed sections get modeled as idealized transmission lines. This transmission line modeling starts with a postulated characteristic impedance (Z_o), length (t_{pd}), and propagation velocity (v_p). Or the simulator can be instructed to extract the necessary parameters from a postulated board stackup, such as dielectric, etch width, and route length. Supplier provided models (extracted) of connectors and I/Os can be inserted in the network. Vias and other features can be similarly handled.

As connections get routed, actual transmission line parameters, vias, and other structures can be extracted and modeled with the simulator's EM field solver. The actual board stackup, dielectric, etch width, and route length can be used along with the effects of routing through pin fields. 2D or 3D field solvers can be used as needed. As these structures get modeled, they replace the postulated structures in the simulation. The result is a more accurate simulation. In a prudent design *all* nets on the entire board eventually[10] get completely routed and simulated for min-typ-max conditions before final release to board fabrication.

---

[10] Remember that Murphy's Law of the Unexpected is always operating.

Power Integrity issues can be simulated from an early stage. The placement and sizing of bypass capacitors can be set by design.[11] Non-ideal power distribution system effects can be modeled. The placement and dI/dt waveform behavior of power pins can be simulated in an EMI/EMC simulator.

If all this postulated and virtual prototyping can be done for Signal Integrity and Power Integrity simulations, why not do it for EMI/EMC simulations? Cannot antenna characteristics be attached to the virtual prototype transmission line sections, other structures, and the location of image planes be postulated? Near field simulations can be run, and radiation hot-spots reduced. E-H vectors can be passed up to a far field simulator along with source characteristics, such as antenna resistance [85].

The virtual design objective is *problem avoidance*. The sooner potential problems are identified, the more effective the proposed solution. The later a fix to a problem is applied, the more chance that it will not be implemented properly. Late fixes tend to be band-aids; the fixes are often less effective and may easily fail.

It is in this problem-avoidance paradigm (as opposed to find-and-fix) that rules-of-thumb (and rules-checking software) are effective. If good design rules are followed from the outset, fewer simulations need to be run, and the simulations that are run are more effective.

22.10 INTERCONNECTION DESIGN APPLIED TO EMI/EMC

In Signal Integrity, we already do waveform manipulation when we match terminate a line, use soft turn-on and turn-off, pre-emphasis, and a host of other things. Waveform manipulation is another area where convergence between methods employed by RF and microwave engineers and digital signal integrity engineers is likely to occur.

Using a trapezoid-like driving waveform has advantages in some situations. It limits the frequency spectrum of the signal and eliminates higher frequencies that cause more reflection problems. Termination helps reduce reflections. The mismatching effects of C_comp at both driver and receiver are hard to ignore at very high frequencies. This mismatching contributes to reflected energy on the line (VSWR) and it affects eye closure for GBIT serial data busses.

In RF and microwave circuits, transmission line matching stubs are often used to improve the matching on a line, lower reflections, and improve

---

[11] PowerIntegrity® for instance is a tool available from Cadence.

VSWR. Until now, they have been ignored on digital lines. This is because they are narrow band techniques. But perhaps on GBIT serial data busses, matching stubs could be used to tweak out the effects of C_comp.

22.11 MODELING FOR POWER INTEGRITY AND EMI/EMC

22.11.1 Introduction

To get good EMI/EMC results, we must start with good Signal Integrity (SI) and Power Integrity (PI) performance. The challenge for the designer is to control and manage signals to accomplish this.

The primary energy input supplying the signals on the board is the power supply and the power distribution system (PDS). The PDS must be designed so that it does not become a pathway for the addition of noise to our desired signals. Amplifiers and switches use and process the energy of the PDS to operate logic or process useful analog signals.

Undesired signal interference, losses, and a number of noise generation phenomena interfere with the desired signals and cause problems. It helps in the analysis to separate out SI, PI, and EMI/EMC problems. In reality, all of these phenomena interact. The problems, generally referred to as *noise problems,* arise in three general ways:

- Directly by the signals traveling in the signal circuits. Reflections, losses, jitter and other phenomena cause noise. We can also get crosstalk noise when signal lines are close enough for near field coupling to be significant. These noise problems are collectively referred to as SI problems.
- Directly in the PDS system by transient switching, amplification current demands, and the finite impedance of the PDS system.[12] Noise generated by the switching of neighbors of a particular switch is referred to as simultaneous switching noise (SSN). Power supply noise problems include noise due to simultaneously switching outputs (SSO). These noise problems are collectively referred to as PI problems.
- Received from far field sources or conducted from unwanted signal sources. These noise problems are collectively referred to as EMI/EMC emissions interference problems.

---

[12] Additional noise can be coupled into the PDS system from the signal circuitry.

Noise is any unwanted signal. For the unwanted signal to create a problem, three things must exist: A source of the noise, a signal path, and a receiver (victim) of the noise. Noise is undesired, uncontrolled electromagnetic energy flowing around inside electrical equipment until it gets dissipated. What can we do with this undesired electromagnetic energy to minimize problems? We can:

- Minimize the unwanted signal, referred to as *source minimization*. This is usually the most economical approach, but is not always possible. This is especially true when the signal is desired in another part of the equipment.
- Interrupt or attenuate the interference path.
- Make the receiver less susceptible to the noise signal that arrives at it. For example, filtering an incoming PDS line at an amplifier or switch.

22.11.2 Techniques to Minimize PI and EMI/EMC Problems

To minimize the noise on the PDS, various design techniques are applied to PI problems. These techniques include proper bypassing and filtering of power supply pins on the IC package where it is mounted to the power planes [92, 93, 106, 136, 137, 138]. To minimize EMI/EMC interference problems, various design techniques are applied. For example, shields and enclosures are transmission path interruption and attenuation techniques.

Very detailed design information is required to completely solve PI and EMI/EMC issues. Effects including interaction between image-plane and IC-package, radiation, crosstalk, and emission often have to be accounted for. For example, to completely model SSN/SSO behavior would require modeling the following interconnections and components:

- Die pad to pin-to-pin-escape connection parasitics in the power and ground circuits
- Any on-die decoupling
- Connections and electrical behavior of I/Os sharing a common IC package power and ground bus structure
- Pin escape connection parasitics to the power supply planes
- Power plane decoupling
- Transfer impedance in the power and return planes back to the ultimate power supply source

22.11.3 When and How to Apply PI and EMI/EMC Design Techniques

The earlier in a design development that EMI/EMC design techniques are applied the more effective they are. Next, early application needs to be followed up with continual adjustment and attention to detail. But early in a product development, detail is lacking–especially in layout and power plane design.

The detail to fully solve PI/EMI/EMC problems usually does not exist early in a design development. This creates a dilemma. Sometimes the detailed information will not be available in an humanly-acceptable timeframe. In today's world, it is common for the PCB to be designed and prototyped before the first working silicon (much less verified model) is produced. This state of affairs is commonly referred to as IC-PCB *co-design* and is very common today. The authors suggest the following methods for moving ahead with an IC-PCB co-design.

During design development, the designer must get as much information flowing between semiconductor user and semiconductor supplier as possible. An analog signal generator and equivalent generator impedance will work well as a drive source. The IC can be represented as a first-order noise or signal generator model. The board can be represented as a first-order equivalent-load. The first models can be adapted, inferred, and cross-referenced from existing process models as cover in-depth in "Chapter 14, Sources of IBIS Models." As the design progresses, partial solutions can be produced and detail added later. Modifications are made to the design of the IC and/or the PCB as seem prudent. This process continues until at least enough design margin in meeting goals is indicated, or alternatives must be considered. Alternatives may include:

- Using a different design that ensures greater success when simulated.
- Building and testing a prototype before knowing if the design will work.

Very close cooperation between semiconductor supplier and OEM parts user require NDAs, considerable negotiating skill, and trust. Information sharing includes very proprietary information between both. Also, even with the best of supplier-user relationships the new, custom, state-of-the-art IC may not physically exist until after the PC board has been laid out.

This suggested design flow method parallels the refinement of models approach already recommended by the authors for board design. In that design flow, it is common to start with approximate models and refine them as the board gets laid out. Simulations are run, routing is adjusted, and bypassing is added, and so forth in an iterative loop. To get started designing a board, we do not need the completed, finalized board design in great detail. The earlier the simulation is started (when strategic choices are being made), the better the result.

Next, we already know that if we do source minimization, attenuate an interference path, and make a receiver less susceptible, we are applying prudent design. We know that representing a noise or signal source as an equivalent generator with an equivalent source impedance has been an acceptable electronics engineering method for generations.[13] We can refine models as we progress, and as the requirements of the design indicate. If necessary, we can eventually replace some of the abstracted modeling with physical modeling.

Equivalent generator modeling for EMI/EMC sources can be represented as E and H vector generators and antenna impedances. PI effects can be represented as dI/dt waveforms and source admittances. This equivalent source approach has additional advantages. It hides construction details and speeds simulations by passing behavioral models up to the next level of system integration. Using these IBIS-like methods for the IC and the modeling of the power planes, proprietary information can be protected, and first-order behavioral models can be exchanged in a timely fashion.

Modeling by reasonable inference from similar devices[14] can be done for new devices being developed in parallel with board layout. Virtual placement of the prototype IC on the PCB can then be done early and intelligently. The extraction of equivalent-load models from the virtual PCB, and some information on image-plane placement, can be provided to the IC supplier. The IC supplier can then use this information for re-simulating and re-estimating IC-package radiation and refining early first-order models.

Noise sources,[15] on a board act like energy pumps. They inject noise into other I/Os and ICs, much like heat generated in one IC can get pumped into a victim IC. This occurs via the source-pathway-receiver connection. Providing behavioral models of multiple noise sources on the PCB can be used to simulate possible effects on the subject IC. One could even begin to consider enclosure and shielding design.

During early component placement, thermal dissipation design is often examined and may be simulated. A decision to add a heat sink or PCB metal layer heat spreader can affect EMI behavior. The possibilities of that interaction can, and should, be simulated.

Noise sources studied can include equivalent source noise generators and parasitic elements for core switching and power plane interactions. Likewise,

---

[13] In IBIS modeling, we reduce our signal generator to an equivalent source plus an equivalent source impedance.

[14] Assuming the same process technology base.

[15] Examples are switching I/Os and ICs.

simultaneous switching noise (SSN/SSO) coupling from power and ground pins can be modeled by including a coupling impedance to the I/O pins. Noise pathways can be modeled and simulated before the board is routed.

We already do early simulations for SI purposes by creating and simulating a topology with virtual prototype sections of transmission line. Early models include Zo and delay. These early models are followed by extraction of the transmission line parameters from a proposed stackup, etch cross section, dielectric constant, and so forth.

These models are used to develop and manage design constraints that guide the PCB designer's efforts in the CAD department. These later models are still virtual, and still not the actual, routed PCB. Once the actual, routed PCB database is available, SI is re-simulated and verified, or adjusted. The usual practice is to design critical nets first, and then the entire board, then simulate and verify the completed board.

This methodology works extremely well for SI and PI. Why not for EMI/EMC? One would first use some simplified models of possible radiating structures. Models extracted from a proposed stackup and route would replace these simplified models. Modeling of the virtual board would follow this. Early, strategic decisions about running a high-speed signal as microstrip, or stripline, would be better informed. Likewise, placement of image planes, slots, and much more can be modeled early on.

A source[16] and coupling path model is being proposed in the IEC and IBIS Committees to provide first-order behavioral models of the noise generated by the logic-switching cores of ICs. This noise can appear on the I/O pins. We examine this topic next.

22.11.4 Concepts Proposed for Solving Power Integrity and EMI

Solving Power Integrity and EMI problems include modeling and simulating noise from adjacent pins and noise from core switching [13, 18, 133]. For modeling adjacent I/O pins and their SSN/SSO in the current IBIS package model, a conceptual framework exists. This framework is presented in the proposed Specification IEC 62014-3 from which most of the material in this section was derived [58]. This framework includes parasitics due to the connections between the package pins and the IC die-pads. The companion IBIS Connector Specification expands this conceptual framework for pin-pin coupling. To complete this pin-pin conceptual framework, an equivalent noise generator must be placed inside the package at the die pads. The equivalent generator approach is behavioral—it can be

16 Modeled as a core switching current generator dI/dt and a coupling network.

measured and characterized externally, or can be generated by the supplier by simulating the proprietary SPICE deck and chip layout model. For a first-order model, a simplified approach can be taken.

Internal IC switching activities generate noise, which gets injected into the system that the IC is part of. These mechanisms are sketched in Figures 22-2 through 22-7.

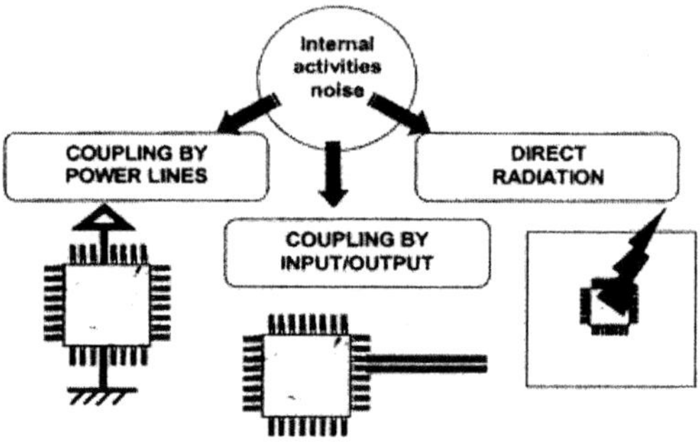

Figure 22-2. Mechanisms for emissions

Courtesy of Etienne Sicard, INSA-Toulouse, France.

Figure 22-3 shows simultaneous switching gate activity (numbers of) versus time of a typical currently marketed IC.

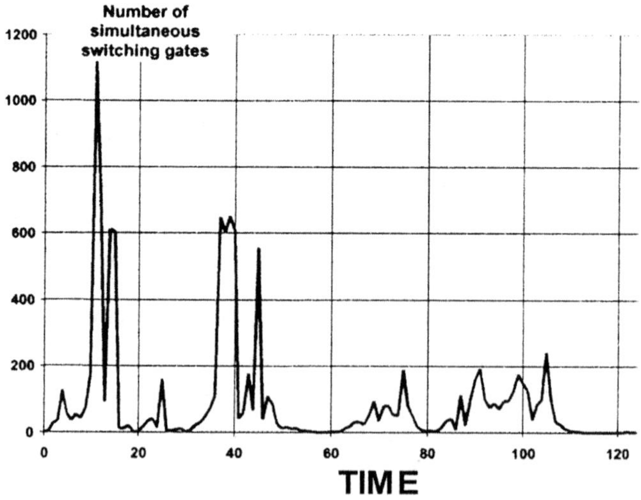

Figure 22-3. Typical number of switching gates versus time [157]

The equivalent circuit and noise generator looking into the power supply pins proposed in the IEC model is shown in Figure 22-4.

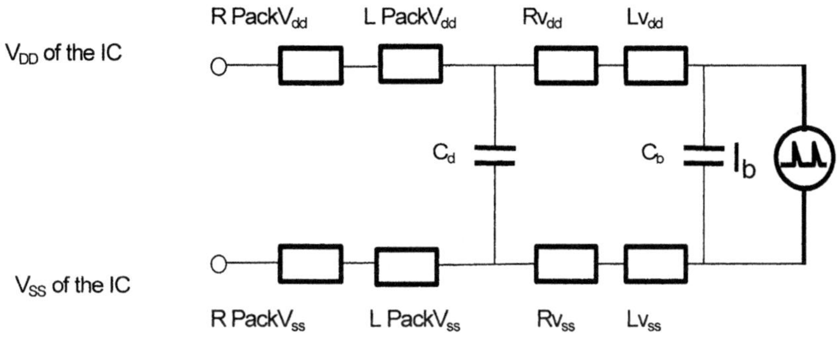

Figure 22-4. Model of the IC supply lines [156]

LEGEND:
Ib = noise-current generator
LpackVdd = package inductance of the positive supply line, Vdd
LpackVss = package inductance of the negative supply line, Vss
RpackVdd = package resistance of the positive supply line, Vdd
RpackVss = package resistance of the negative supply line, Vss
Cd = parasitic capacitance between Vdd and Vss package pins
(note that Cd, Cb, and Ci/o could be designed to provide some bypassing)
Cb = internal die capacitor in parallel with the equivalent local noise-current generator on the power supply lines
Ci/o =internal die capacitor in parallel with the equivalent local noise-current generator on the I/O lines
Rvdd = series resistance of Vdd bonding and internal die metalization
Rvss = series resistance of Vss bonding and internal die metalization
Lvdd = series inductance of Vdd bonding and internal die metalization
Lvss = series inductance of Vss bonding and internal die metalization
Zsub coupling impedance = Valid for most CMOS technologies with p-type substrate. It accounts for the substrate coupling path between the core Vss and the I/O Vss

A major pathway for noise interference is crosstalk and mutual-capacitive coupling. But as a first order approximation, we suspect that this model is a good start for the discussion of modeling power-supply noise. Next, a characterization of the internal noise-current generator for a single gate is shown in Figure 22-5.

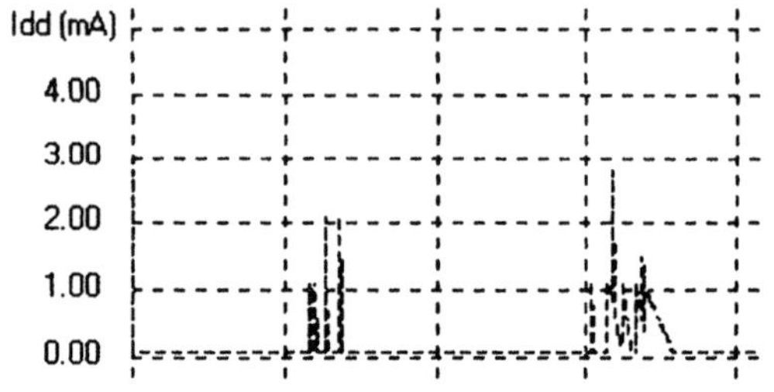

Figure 22-5. Idd defined as a PWL description versus time [158]

Figure 22-6 shows typical current source magnitude for Ib(t) for a typical 8-16 bit microprocessor. This behavioral characterization of Ib(t) and Ii/o(t) represents the coupled and conducted power-supply noise from the switching of the logic core. At this stage in the development of methods to analyze this source of switching noise, and its effect on a system, it probably makes sense to characterize the current generators worst-case.

Worst-case design sometimes results in overly conservative and costly product design. But it probably makes sense the way this first-order modeling fits in the design flow. The first step in a SI-PI-EMI simulation of a board is normally to analyze for voltages and currents given the IC-equivalent generator and impedances presented here. The following step would be to analyze for near- and far-field vectors, taking into account factors such as board stackup, routing structure, and image planes.

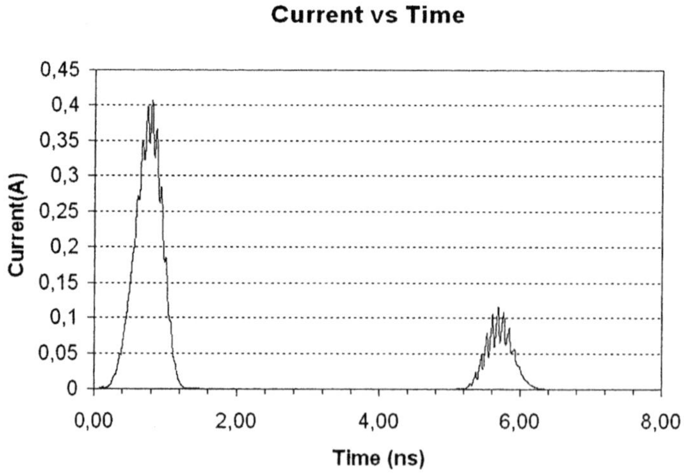

Figure 22-6. Typical current source defined as a PWL versus time
Data courtesy of Etienne Sicard, INSA-Toulouse, France.

Once the IC and board are tentatively placed and routed, we can examine second-order effects, and best-case and worst-case. After initial place and route, we can also present data on equivalent loads and location of image planes to the IC packaging engineer. The IC packaging engineer can refine the previously supplied first-order models if radiation off the package and other noise signal issues should be analyzed further. In the end, if this kind of close co-operation and data sharing is called for, we may want to negotiate appropriate non-disclosure agreements (NDAs) between supplier and user.

Next, Table 22-8 shows some typical parasitic-element values for our typical million-gate IC.

Table 22-8. Range of values for the model parasitic elements.
Used with permission [18, 133].

Parasitic Element	Min	Max
LpackVdd, LpackVss	1nH	10nH
RpackVdd, RpackVss	1.0 ohm	1.0 ohm
Cd	10pF	100nF
Cb	10pF	100nF
Lvdd, Lvss	1nH	20nH
Rvdd, Rvss	0.1Ω	10Ω
Ci/o	10pF	100nF
Zsub DC value	0Ω	100Ω

This IEC model seems conceptually straightforward, and a good place to start a discussion of modeling for SSN/SSO analysis.

Next, let us consider the possibility of conducted emissions[17] on the I/O lines of the IC. The proposed model is shown in Figures 22-7 and 22-8.

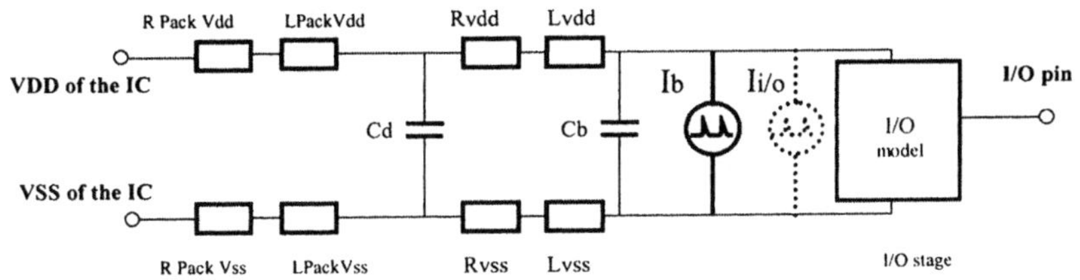

Figure 22-7. Coupling between the core and the I/O [156]

17 Conducted emission is noise superimposed on the I/O ports due to switching of the logic core.

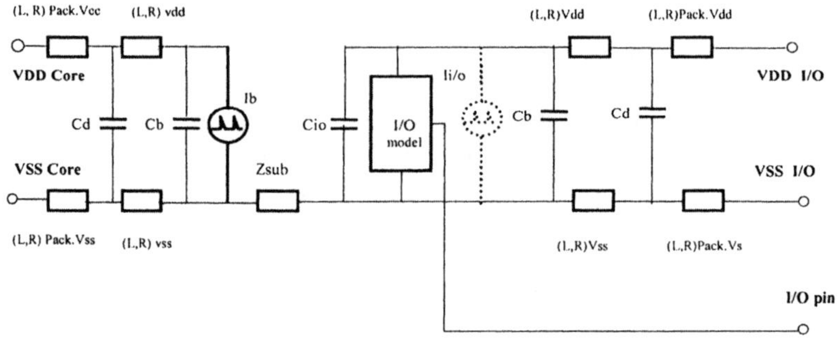

Figure 22-8. Model of the IC I/O lines—separate supplies [157]

Note the following in Figures 22-7 and 22-8:

- The I/O pin has its own associated parasitic Rs, Ls, and Cs. These are already part of the IBIS model.
- The I/O block is a black-box. We may need to derive the transfer function of the noise-current generator, impressed across its power supply connections, to the I/O pin itself. That transfer function probably depends on the state that the I/O is in: pullup, pulldown, open-collector, and receiver input.

22.11.5 EMI/EMC Design Process Changes

Specialists who warn of possible problems just before a product is released to production are at an extreme disadvantage. By that time, the development money may have run out, the project may be late, and the smallest changes are major reworks. Whether such specialists work in quality, reliability, safety, or EMI/EMC, they are often looked upon as obstructionists when such late-project pressures are present in full effect. At that point, specialists have much less influence, and only the most blatant product failures will receive attention and remedial measures. Late design-cycle problems are *very unwelcome*.

Early modeling and simulation have the potential to change that entire paradigm of late input being ignored and change the image of specialist from obstructionist to *enabler*. Modeling and simulation can occur early in a design cycle, and continuously thereafter as a product moves closer to physical realization. Modeling and simulation of what-ifs, combined with clever, experience-informed suggestions to head off potential problems, can be time saving and cost-effective when applied early in a project.

Why has not more EMI/EMC simulation been done if it is so wonderful? Part of the answer is cultural resistance. Another part of the answer is that it is often a notoriously difficult and complex modeling and simulation problem. The modeling and simulation difficulties are also compounded by lack of EDA use and experience.

For instance, what should EDA providers concentrate on for dealing with questions of potential radiation problems from an IC package? Should they concentrate on radiation from the supply ring loop, or radiation from vias? That question remains open although the indications are that the vias are the bigger problem.

EMI/EMC modeling and simulation are still in their youth compared to the application of modeling and simulation for other purposes. What to model, how to model it (engineering methodology and computational methods), and issues of scale are still poorly understood.

22.12 COMPUTATIONAL ELECTROMAGNETICS

For today's ever-faster designs and their power supply current requirements, precise solutions to very complex electromagnetic simulation problems are required for good PI/EMI/EMC and even SI. Equivalent generators and first-order models may help us get started, but we often need more exact methods.

Numerical methods now play a major role in the field of applied computational electromagnetics (CEM). CEM is used for both impedance modeling and emissions modeling. Computational numerical methods, employing high-speed digital computers, are applied when the problem gets too tough to analyze by ordinary means. The theoretical tools behind these methods have been available for many years. The birth of computational electromagnetics was in the 1960s, shortly after the appearance of the first scientific mainframe computers.

22.12.1 Finite Element Methods (FEM)

What are the mathematical methods of computational electromagnetics? For details, see [6, 25, 50, 96, 123]. In 1942, Courant [29] first outlined the finite element method (FEM). At that time, he published a two-page appendix on the use of variational methods in potential theory. Choosing a piecewise linear approximation on a set of triangles, which he called elements, he constructed two two-dimensional examples, marking the birth of the finite element method (FEA and FEM).

Historically, numerical methods for field solvers came from microwave and RF modeling. The methods were adapted to package modeling and board-level interconnection modeling. Current applications include semiconductor material characterization, but these are still quite rare. Finite elements when combined with the Method of Moments (MoM) are useful to characterize arbitrarily shaped samples of material.

22.12.2 CAD-Oriented Finite Element-based Simulators

Analyzing microwave circuits with finite elements is a fairly well established technical discipline. But mesh optimization is still largely an unsolved problem. For example, the mesh used close to where a via penetrates a layer must be much finer than the mesh used far away from the transition. A fine mesh will impede the simulator if used throughout, and making variable-gridding work is a challenging programming problem.

Time domain applications of finite elements are still quite rare. Only a few applications have been reported, concerning scattering problems, planar transmission line characterization, and axisymmetric as well as aperture radiators. Variable-gridding approximation methods are used in the Finite Difference Time Domain (FDTD) technique. FEA-FEM is further improved by its hybridization with the FDTD technique. This achieves a flexible and computationally efficient time domain field simulator that is able to accurately take into account the presence of curved interfaces.

22.12.3 Models from Boundary Element (BEM) Analysis

One of the oldest field solver approaches is the boundary element method (BEM). This method analyzes the charge distribution on all surfaces, and relates that to potentials and impedances.

The BEM approach is still widely used today for solving the impedance of PCB traces over a ground plane. Because the method requires a plane boundary, it is not suitable for many IC packages, or for most interconnections inside ICs. The BEM also does not handle large slots or holes in the reference plane. The impedance of a trace over a slot must be calculated through other methods.

22.12.4 Models from Finite Difference Time Domain (FDTD) Analysis

FDTD time-domain device and field simulation are applied for a rigorous simultaneous solution of Maxwell's and semiconductor transport equations. FDTD tools could be useful for analyzing today's very high-speed digital devices. But it is sometimes unclear whether the wavelength of signals is small enough (in the case of sub-micron devices) to cause them to act as traveling wave structures. So it is primarily at the chip-to-chip board-level scale that FDTD simulators are seeing application.

Advanced methods have been developed for analysis of SI and EMI in the absence of a ground plane. These methods include FDTD, Asymptotic Waveform Evaluation (AWE) using FEA techniques, as well as many other techniques. In the absence of a ground plane, FDTD continues to be a hot research topic, both commercially and at universities.

Commercial full-wave, non-linear electromagnetic SPICE simulators for SI, PI, and EMI analysis are available. FDTD analysis is being applied at the PCB, IC, and package level for the analysis of very high-speed digital nets.

22.12.5 Recommendations on the CEM Method to Use

Which of the following Computational Electromagnetics Methods (CEM) should we use?

- Integral and Green's Function methods [27, 123]
- Method-Of-Moments (MoM) [50]
- Finite-Element-Analysis (FEA/FEM) [29, 96]
- Finite-Difference-Time-Domain methods (FDTD) [120]
- Transmission-Line-Modeling (TLM) [25]
- Partial-Element-Equivalent-Circuit (PEEC) [6]

Or some combination of the above?

Tables 22-9 through 22-13 list the advantages and disadvantages of four CEM methods. The tables are adapted from [147]; a very good discussion of the current numerical techniques for solving CEM problems is in [150, chapter 3].

Table 22-9. MoM advantages and disadvantages

MoM Advantages	MoM Disadvantages
• Effective modeling of antennas, wires, surface structures and wires attached to large surfaces • Detailed modeling of current distributions on structures of any shape • Radiation condition allows modeling E and H fields anywhere outside radiating structure • Calculates antenna parameters: input impedance, gain, and radiation	• Great difficulty in modeling interior regions • Difficulty in modeling inhomogeneities • Large computational resources in terms of memory, hence useful only at low frequency

Table 22-10. FEA/FEM advantages and disadvantages

FEA/FEM Advantages	FEA/FEM Disadvantages
• Electrical and geometric properties can be defined separately • Does not need staircase modeling space as in FDTD • Effective for modeling interior problems • Can model inhomogeneous media	• Unbounded problems require modeling the space around the scatterer • Computationally intensive • Difficult to model open problems where fields are not known at points on a closed boundary

Table 22-11. FDTD advantages and disadvantages

FDTD Advantages	FDTD Disadvantages
• Simple approach for solving Maxwell's equations • Does not require storage for modeling space • Can model inhomogeneous media easily • Effective for modeling interior problems. Explicit two-step procedure.	• Unbounded problems require modeling the space around the scatterer • Prone to dispersion errors • Computationally intensive • Basically calculates field propagation: other parameters are much more difficult to calculate, for example current distribution.

Table 22-12. TLM advantages and disadvantages

TLM Advantages	TLM Disadvantages
• All E and H field components can be calculated at the same point • Can model inhomogeneous media • Effective for modeling interior regions • Less dispersion error than FDTD	• Requires more computational resources than FDTD • Bandwidth limited by dispersion error • Unbounded problems require large space around the scatterer • Basically calculates field propagation: other parameters are more difficult.

Table 22-13. PEEC advantages and disadvantages

PEEC Advantages	PEEC Disadvantages
• The usual MoM advantages, but: • PEEC converts physical structure to an equivalent circuit. Very easy to add lumped circuit elements to model. • Well suited to PCB analysis. • Time and frequency domain analysis available. • E-H fields found at any point. Fields between plates easy. • DC, skin effect and dielectric dispersion handled well.	• The usual MoM disadvantages, but: • Increasingly difficult to satisfy the condition that $d \ll \lambda$ as frequencies continue to increase.

PEEC derives from MoM with a key assumption. The assumption is that the dimensions of the object modeled are much smaller than λ (wavelength). Therefore, given a retardation of time-dependent change, a quasi-static condition exists. Then the equivalent partial inductance and capacitance matrix can be found for the object modeled for all grid points

When we do the initial step of solving for currents and voltages–are we in TEM mode, or do we need to use full-wave simulation? At what point do power-plane resonances and the effects of enclosures have to be accounted for in our initial simulation of currents and voltages?

Integral and Green's Function methods have been used successfully to model near-field a few millimeters off a PCB. Solving the near field at an imaginary surface a few millimeters above the PCB surface inverts the usual problem formulation. Instead of trying to use the method to solve for fields in the far radiative region, we use it to solve for the near-field. Likewise, using the TLM method combined with three-dimensional S-Parameters allows us to solve for far-field EMI propagation once the PCB signal voltage and currents are known. This also inverts the usual approach of using TLM for near field solutions. This technique of inverting the usual way a CEM method is employed can be very productive.

The next questions are about what to model, and at what scale. For instance, a via in an IC, a pin on an IC package, and a via on a PCB are all very important EMI monopole antenna problems. Each creates current flowing on a pin perpendicular to an image plane (if there is one) and creates no counterbalancing image currents in the plane.

In this problem what to solve for is primarily the radiation pattern and the radiation impedance of the structure. Of the two, radiation impedance is more difficult, and the scaling question complicates the issue. The scaling involves mixed scales because modeling the structure close to where the monopole pierces the plane requires a fine FEA mesh that will bog down the simulator further away. Mixed scale problems are one of the knottiest issues facing EMI/EMC modeling and simulation.

Nevertheless, progress in EMI/EMC modeling and simulation has been made, and practical and useable tools exist. Much remains to be done to verify the limits, accuracy, and applicability of these tools. Enough progress has been made for standards committees to get into the act. For example the following organizations are active in this area:

- The International Electrotechnical Commission drafted a Specification, IEC 62014-3, which addresses some aspects of Power Integrity and EMI modeling. IEC 62014-3 addresses modeling of power- and ground-pin parasitics, the dI/dt table of supply-current-demand and emitted energy in a TEM test cell for ICs. The IEC is a worldwide organization for standardization, comprising all national electrotechnical committees.
- IBIS: Consideration is being given to making the 62014-3 Spec a companion specification for addressing Power Integrity and EMI modeling of ICs in IBIS. Refer to:
  http://domino.iec.ch/webstore/webstore.nsf/artnum/029517
  http://intrage.insa-tlse.fr/~etienne/emccourse/macro_model.html
- The University of Missouri – Rolla (UMR) has a number of projects in the modeling of EMI/EMC effects underway. Refer to http://www.emclab.umr.edu/

22.13 EDA TOOL SUPPLIER SURVEY

In the discussion just presented, we have not fully addressed the role that EDA tools play. The tools from a single EDA provider company do not cover the entire design process outlined. Consequently, to get the necessary analysis capabilities, we must purchase tools from multiple EDA tool providers. This can cause problems in the integration of the design process. For an overview of EDA tool involvement in the design process, refer to Figures 2-10 and 2-11.

Some tool providers claim their tools handle everything and that their tools are deeply integrated, but that remains to be demonstrated. Despite advertising claims, there is no single, numerical technique that is fast and accurate in all cases. Each approach has advantages and disadvantages [148].

It is difficult to do the many simulations required when scaling up from inside an IC to a large system–and do them all well. The authors, however, believe that if we want to do simulation from the TCAD level to simulating a fiber-optic link across town, it is likely that we need to work with software from several companies. In that case, it is unlikely that we will ever get such a suite of tools so well integrated that data and models flow seamlessly between them. Software companies do not have a good track record of such cooperation, even when they cannot do a particular simulation job by themselves. This leaves the users doing end-to-end simulations as a series of isolated activities with manual passing of data across the interfaces. In such a situation, the design process is prone to mistakes. The formation of a standards committee for data exchange on EMI/EMC models is suggested.

Nevertheless, it is worthwhile to review the tools provided by EDA suppliers (see Tables 22-14 through 22-18). All tables have three columns. The Task column lists the type of extraction, modeling, and simulation task the company's EDA tool software can perform. The two right columns indicate the scale size of the system being analyzed. The abbreviations in the two right columns are for the EDA tool providers, which are listed in Table 22-19.

Disclaimer: The tables in this chapter are mostly the authors' assessments, which are based on Internet and marketing information. Most assessments have neither been confirmed nor corrected, despite inquiries to the indicated companies.

Table 22-14. Model extraction: Die and package

Task	**IC Die Scale**	**IC Package Scale**
TCAD active cell simulation	S, SYN	
Active cell SPICE extraction	AST, OEA, OP, S, T	AST
Interconnection SPICE extraction	AN, AST, C, M, OEA, OP, S, SIG, SON, SYN, T, ZE	AN, AST, B, C, CST, OEA, OP, OPT, SIG, SIS, ZE
IBIS generation and/or validation	AST, OP, SIS, ZU	AST, C, SIS
S, Y, Z matrix extraction	AN, AST, CST, OP, SON, ZE	AN, AST, B, OP, SIG, SIS, ZE

Table 22-15. Model extraction: Topology, stackup, components, cables, connectors and .ebd

Task	Virtual: Topology, Etch and Stackup Scale	Components, Cables, Connectors, Boards Scale
Interconnection SPICE extraction	AN, AST, C, OP, Q, S, SIG, SIM, SIS	AN, AST, B, C, CST, EMCS, OEA, OP, OPT, Q, SIG, SIM, SON, W, ZE
IEC 62014-3 variables extract	S	SYN
S, Y, Z matrix extraction	AN, AST, OP, SIG, SIS,	AN, AST, B, CST, EMCS, OP, SIG, SIM, SYN, ZE
E and H vector extraction	AN, AST	AN, AST, B, CST, EMCS, Q, SIM, ZE

Table 22-16. Simulation: Die and package

Task	IC Die Scale	IC Package Scale
Signal Integrity	AN, AST, B, M, OEA, OP, OPT, SYN, T, ZE, ZU	AN, AST, B, C, CST, OEA, OP, OPT, SIG, SOS, SON, SYN, ZE, ZU
Power Integrity	AN, B, M, OEA, OPT, SYN, ZE, ZU	AN, AST, B, C, CST, OEA, OPT, SIG, ZE, ZU
Crosstalk Coupling	AN, AST, B, M, OEA, OP, OPT, SYN, T, ZE, ZU	AN, AST, B, C, CST, OEA, OP, OPT, SIG, SOS, SON, ZE, ZU
EMI	AN, AST, B, S, SYN, ZE, ZU	AN, AST, B, C, CST, S, SIG, SOS, ZE, ZU

Table 22-17. Simulation providers at the circuit and PCB level

Task	Topology Scale	PCB Scale
Signal Integrity	AL AN AST C OP Q SIG SIS SOS	AL, AN, AST, B, C, EMCS, FU, OEA, OP, OPT, Q, SIM, SIS, SOS, SON, W, ZE
Power Integrity	AN AST SIG	AN, AST, B, C, EMCS, FU, OEA, OPT, SIG, SIM, ZE
Crosstalk Coupling	AN AST C OP Q SIG SIS	AN, AST, B, C, EMCS, FU, OEA, OP, OPT, Q, SIG, SIM, SIS, SOS, SON, W, ZE
EMI	AN AST C SIG	AN, AST, B, C, EMCS, FLO, FU, Q, SIG, SIM, SOS, ZE

Table 22-18. Simulation providers at the multi-PCB and system level

Task	PCB to PCB Scale	System and Enclosures Scale
Signal Integrity	AN, AST, C, CST, EMCS, FU, OEA, OP, Q, SIG, SIM, SIS, SOS, W, ZE	AN, C, EMCS, FU, OEA, OP, SIG
Power Integrity	AN, AST, CST, EMCS, FLO, FU, OEA, SIG, SIM, ZE	EMCS, FLO, FU, SIG
Crosstalk Coupling	AN, AST, C, CST, EMCS, FU, OEA, OP, Q, SIG, SIM, SOS, W, ZE	C, EMCS, FU, OP, SIG
EMI	AN, AST, CST, EMCS, FU, Q, SIG, SIM, SIS, ZE	AN, EMCS, FLO, FU, SIG

Table 22-19 identifies the companies named by the abbreviations in Tables 22-14 to 22-18.

Table 22-19. Company abbreviations for Tables 22-14 through 22-18

Company Abbreviation	Company	URL
ACS	Andro Computational Solutions	http://www.androcs.com/frame_services.html
AEMC	Advanced EMC Solutions	http://www.aemcs.com/
AG	Agilent	http://www.home.agilent.com/USeng/home.html
AL	Altium	http://www.altium.com/
AN	Ansoft	http://www.ansoft.com
ANS	AnSys	http://www.ansys.com/
AST	Applied Simulation Technology	http://www.apsimtech.com/
B	Bay Technology	http://www.bay-technology.com/
C	Cadence Design Systems	http://www.cadence.com/
CST	Computer Simulation Technology	http://www.cst.de/
EE	e*ECAD	http://www.eecad.com
EMCS	EMCoS	http://www.emcos.com/
FLO	Flomerics/Flo EMC	http://www.floemc.com/

Table 22-19. (cont.)

Company Abbreviation	Company	URL
FU	Fujitsu	http://translate.google.com/translate?hl=en&sl=ja&u=http://salesgroup.fujitsu.com/plm/eda/html/ACCUFIELD.html&prev=/search%3Fq%3Daccufield%2Beda%26hl%3Den%26hs%3Df3e%26lr%3D%26client%3Dfirefox-a%26rls%3Dorg.mozilla:en-US:official_s%26sa%3DG
I	Interactive Products Corp	http://www.interactive-products.com
IN	Intercept	http://www.intercept.com (CAD only)
M	Magma Design Automation	http://www.magma-da.com/c/@saAXEynp3_muw/Pages/index.html
MEN	Mentor Graphics	http://www.mentor.com
OEA	OEA International	http://www.oea.com/
OP	OptEM	http://www.optem.com
OPT	Optimal	http://www.optimalcorp.com/
Q	Quantic EMC	http://www.quantic-emc.com
S	Silvaco	http://www.silvaco.com/homepage.html
SIG	Sigrity	http://www.sigrity.com
SIM	SimLab	http://www.simlab-emc.com
SIS	SiSoft	http://www.sisoft.com
SOS	SoftSim	http://www.softsim.com/tsl/TSL_softwarenew.asp
SON	Sonnet	http://www.sonnetsoftware.com
SP	SuperMax ECAD	http://www.dde-eda.com/
SYN	Synopsys	http://www.synopsys.com/
T	Tanner EDA	http://www.tanner.com/EDA
W	Wavecrest	http://www.wavecrest.com/
ZE	Zeland	http://www.zeland.com/
ZU	Zuken	http://www.zuken.com

Tables 22-14 to 22-18 show that all electromagnetic issues can be modeled and simulated from the individual gate and semiconductor junction level to an entire system such as an automobile. That is, provided we are very persistent, prepared to deal with many different EDA companies, and extract models out of suppliers. Further, the models and simulations would not necessarily flow seamlessly between the different product integration scales, and simulators.

It would be a big improvement if there were some kind of data exchange format and standard similar to IBIS. The standard should enable EMI/EMC models and simulations to flow more seamlessly between the different product integration scales and providers. A good place to start would be a data exchange format that covered the following capabilities:

- S-Parameters for parasitics and passive structure models.
- I and V tables and spectrum versus frequency
- E and H vectors for passing field information up to the next level of integration.
- A data exchange format for mechanical structure information.

22.14 RISK MANAGEMENT AND THE LIMITATIONS OF SIMULATION

As of this writing, practical EMI simulation is a final frontier of EDA analysis. The sheer complexity of modeling and simulating an EMI far-field that radiates from a complex PCB, out through an enclosure, and onto cabling has limited the use of EMI modeling. The radiation field changes with the data pattern running on the board.

Accurate correlation of EMI predictions with measurements gives engineers confidence in their modeling and simulation. However, very little verification work has been published. We should start with simple boards set up specifically for correlation studies.

Beyond correlation studies and exercises to build trust in EDA tools, we have the task of making good decisions in everyday product development. When we get to the Regulatory test laboratory and test site, the complexity of the EMI/EMC testing task is recognized in the "80-80 rule." That is, we seek an 80% confidence level that 80% of our product units will pass Regulatory testing. Our Regulatory test community is wise enough to use probabilistic testing in so complex a task. We, on the modeling and simulation side, should be wise enough to use probabilistic design [23] in so complex a task.

22.15 SUMMARY

The electromagnetic interaction among IC, PCB, and equipment enclosure is now more pronounced than ever. The concerns and techniques of Logic Design, Signal Integrity, Power Integrity, and EMI/EMC are completely intertwined and interdependent. Modeling semiconductors for simulating high-speed designs requires engineers to be knowledgeable in IBIS, SPICE and S-Parameter modeling. Convergence between the analysis techniques of the Time Domain and the Frequency Domain is occurring.

Electromagnetic signaling issues arise at the sub-atomic level and extend to radiation from large systems.

Engineers are applying their experience and knowledge at almost every level of assembly and area of analysis using EDA tools. But most of these EDA tools operate in isolation from each other and are not integrated into a cohesive design flow. This lack of integration can cause design flows that are prone to mistakes. Data and models must be ported manually from tool to tool. Moreover, the tools require specialized knowledge to operate. To master any one tool, engineers require specialized expert knowledge. More than at any time in the past, there is a need for vision, team skills, and management facilitation for high technology to continually progress.

Chapter 23

USING PROBABILITY: THE ULTIMATE FUTURE OF SIMULATION

The end-game for design

Contributing author: Darren J. Carpenter, BT Exact

Abstract: This chapter examines the nature of modeling and simulation over the long term and speculates on the ultimate state that can be expected. The view reached is of a discipline very different from that of today.

This chapter presents modeling as a deductive exercise in the mapping of a parameter space. Each single-valued model parameter is replaced by a statistical parameter that exists within a physical range with *a defined probability function*. To be economical, most designs must be produced within the probability of particular combinations of variables occurring at any one time. An absolute worst-case combination of variables is usually unlikely. For the target quantity, simulation yields both a physical range and a confidence level by mapping of the model parameter space. Examples are presented for very simple problems of practical interest today.

23.1 INTRODUCTION

As computing resources increase, so does the complexity of the simulation that can be performed within a humanly-acceptable timeframe. However, just because more complex calculations can be run does not mean that the simulation has been formulated correctly. The linkage that must be forged in simulations of the future is the connection between any particular value of a parameter in a range, and the *probability* of its occurrence.

Almost all modeling and simulation work performed today is classified by the authors as *deterministic* modeling.[1] The essential feature leading to this classification is the fact that each of the known model parameters is

---

1 Even statistical design techniques can be viewed as deterministic unless the *probabilities of the occurrence* of particular values of the variable are accounted for.

given a single, finite value and the results of the simulation (the solved unknowns) are a set of single, finite values.

However, many parameters considered within engineering problems exist naturally within ranges (sometimes wide) defined by the underlying physics. Hence, our problems may be thought of as existing within large, multi-dimensional parameter spaces. Deterministic modeling, by considering a single value for each parameter, therefore restricts the simulation to consider a single point in this parameter space. The validity of its results is therefore critically dependent upon the accuracy with which parameter values are specified. However, information on parameter values is often not known with great confidence, and best-guess approaches are taken. Therefore, the selection of parameter values is part of the *black art* of simulation.

An alternative approach is known as *probabilistic* modeling[2]. This approach sees the deductive acceptance that each known model parameter exists within some defined range, with an associated probability distribution. The approach taken in probabilistic modeling is to systematically map the associated multi-dimensional parameter space, considering not just a single point within it but rather the entirety of the space. From this approach, the range of the problem unknowns is determined, as is the probability distribution that describes the likelihood that the unknowns shall reside within the range.

The authors believe that the current convention of *deterministic* modeling will inevitably evolve into *probabilistic* modeling over the long term due to a number of fundamental issues that are confronted as the complexity of attempted simulation increases.

What is probabilistic modeling? Let us start by defining probabilistic.

PROBABILISTIC DEFINITION:[3]

1. Of, based on, or affected by probability, randomness, or chance: "The Big Bang universe is... exemplified in the probabilistic and indeterminate interactions of the smallest known physical properties." (Frederick Turner).
2. Relating to, or governed by, probability. The behavior of a probabilistic system cannot be predicted exactly but the probability of certain behaviors is known. Such systems may be simulated using pseudo-random numbers. Evolutionary computation uses probabilistic processes to generate new (potential) solutions to a problem.

---

[2] The methods in this section were developed under US patent **6,597,184** and are protected under it [23].

[3] From www.dictionary.com

23.2 LIMITATIONS OF DETERMINISTIC MODELING AND DESIGN

When dealing with the future of modeling and simulation, many observers point towards the ongoing advancement of Moore's Law.[4] These observers argue that the associated increases in computer resources (processing-power and addressable memory space) will provide simulations that consider progressively more and more parameters. Such observers claim that the inclusion of progressively more parameters leads to the consideration of models that accurately represent the problem of interest. These observers see simulation as gradually progressing towards its natural end game, being the point when it is possible to model reality.

The authors have a major problem with this general approach, as it is little more than a linear extrapolation of the present onto the very-near future. Such extrapolations are notoriously dangerous. When the development of modeling and simulation is considered on a longer term, a number of fundamental issues are encountered that demand that this view be revisited.

The problem is as follows. As more complex codes are developed in response to the increased computer resources delivered by Moore's Law, we are able to consider more complex physical descriptions of our problems, involving the consideration of more and more parameters and still have them run within a decent timeframe. However, a quite basic trap emerges: as the ability to consider more and more parameters in our simulations arrives, this capability demands that we specify values for the known parameters such that the simulation may yield the value for the unknown parameter(s). This is the First Trap of deterministic modeling.

This presents its own challenges, as many of the known parameters of interest can exist within a range of values (sometimes quite wide) and exact information on the actual values for a given problem may not be available. In this instance, what is the modeler to do? It is possible to measure the exact values of "knowns," but such measurements are very expensive to make, requiring skilled people to do accurately (and the skills demanded are normally very different to those held by the modeler); sophisticated and calibrated measurement equipment and take considerable time. Now, the whole point of simulation is to minimize measurement time, so if it takes days to measure the parameters required for the simulation, it may be as well

---

4 An observation, in 1965, by Intel co-founder Gordon Moore. He noticed the number of transistors per square inch on integrated circuits had doubled every year since their invention. Moore predicted the trend would continue for the foreseeable future. Although the pace has slowed, the number of transistors per square inch has since doubled approximately every 18 months. This is used as the current definition of Moore's Law.

to abandon the simulation altogether and simply measure whatever it is was going to be simulated. This is the knowledge that many simulation specialists secretly know, but will not publicly admit!

Antenna modeling is a fine example of the above argument. The earliest (and hence simplest) modeling codes considered perfectly conducting ground planes that were perfectly flat and infinite in extent. The next generation of modeling codes supported the consideration of more realistic ground planes (in fact referred to within the community as *real grounds*), as perfectly homogeneous semi-infinite dielectric slabs (semi-infinite in that the dielectric slab occupies half of the problem volume) with finite electrical conductivity and electrical permittivity. Once available, users were then faced with the challenge of having to define these two parameters for their problems. The measurement of these parameters for real grounds over the wide frequency ranges considered within antenna modeling is a very difficult task, taking great time and expertise. Add to this the fact that soil parameters can change over time due to the effect of soil water and ground water content (both being generally highest during the winter months and lowest during the summer months) further complicates the issue. In the author's view, limited progress has been made on this issue within the antenna modeling community, with the result that the selection of the real ground parameters remains the biggest black art of current antenna modeling practice.

When a problem of interest has a large number of known parameters that can exist over some range of values, the problem exists within a parameter space defined by these known parameters' ranges. By defining single values for each known parameter, only one point in that problem space is considered during the simulation.

For example, if our problem had two known parameters (and is hence two-dimensional), then the range of each known parameter can be thought to define a planar parameter space. Selection of single values for each of the two known parameters places the problem at a single point on that plane. Similarly, if our problem had three known parameters (and is hence three-dimensional), then the range of each known parameter can be thought to define a volumic parameter space. But again, the selection of single values for each known parameter places the problem at a single point within that volume.

As the number of known parameters in the model increases, the parameter space defined by the physical ranges of each parameter becomes larger, but our choice of single parameter values means that our simulation will cover a smaller and smaller part of that parameter space.

Hence the solution obtained from the simulation should be delivered with a health warning, in that it applies to a specific set of known parameter values and that solutions for other known parameter values cannot be inferred. This why the simulation of very complex systems using models that consider very many single-valued parameters conjures up the term "Determinism's Greatest Folly" – progressively more and more accurate answers are delivered for a progressively smaller region of the problem's parameter space *without considering the significance of the answer obtained.*

There is a second trap of deterministic modeling: the worst-case scenario. When faced with a situation in which the exact values of the known parameters are unknown, values may be selected to maximize the unknown parameters to be obtained through simulation. While such an approach is useful in that it provides an upper-limit to the range of the unknown parameters (i.e. it does answer the question – what is the worst that can happen?) it encourages decision making to be based on an upper limit boundary without considering the likelihood with which this worst-case value may occur. This issue is discussed in more detail later in this chapter.

23.3 A NEW APPROACH: PROBABILISTIC MODELING

The authors sought to approach simulation from a completely new perspective. This would consider problems with a large number of parameters, but avoid the conventional approach of demanding single values of each parameter (both known and unknown). Instead, it would consider - from the very start - the physical range of all parameters and treat the entire problem space. This means abandoning the complex solution engines that require hours of solution time for each single-value answer and instead the consideration of more generalized solution engines and models that can be executed at high speeds so that the complete problem space can be mapped methodically within a human-significant time frame.

By sacrificing accuracy to the nth decimal place, we gain a view of the big picture - that is, the complete space over which the problem can exist. We can see that while the worst-case can indeed exist, for problems of a suitable complexity, it is actually highly unlikely to do so. As the number of unknowns in the problem increases, the probability of the worst-case occurring also decreases.

The rest of the chapter is concerned with the principals of probabilistic modeling and presents the results of its application to a practical problem: the management of EMI within a product or system design.

23.4 COMPLEXITY OF THE EMI CHAIN OF CAUSE AND EFFECT

The chain of causation for EMI radiation begins deep inside an IC and extends outward though the PCB, enclosure, cables, and reflections off of other system cabinets. There are many models, structures, and interactions to account for. Each factor has uncertainties and limits on accuracy. This produces predictions with wide margins of tolerance. The usual approach to design simulation and prediction in this case is to use a worst-case design methodology to predict the typical, min, and max of system performance. But worst-case design leads to very stringent limits on individual EMI radiators that are very difficult and expensive to meet.

In HSDD there is an additional factor that widens the min-max distribution of system EMI predictions. That factor is the vector-addition of many emitting sources on a PCB at a particular point in space and time to result in a particular field strength. To have the data to do vector addition requires the magnitude and phase of the individual field emissions, of the individual sources for each particular emitting frequency of the energy spectrum of the sources on the PCB.

Consider two sources emitting at the same frequency and strength on the PCB. Second, assume a point equidistant from the two emitters. Third, assume no intervening reflection structures and no reflection structures behind the measurement apparatus of consequence.

Then the min-max results at a single far field point ranges from zero to twice the individual field emission strengths because the vectors add either in phase or out of phase. The results would have to be modified for unequal path lengths, unequal field strengths, and reflecting and shielding structures nearby. But beyond adding a bit to system complexity, nothing fundamental has changed to make our min-max predictions more complex or dispersed.

Next consider that the numbers and locations of emitters active on a board are driven by the digital data pattern and processing on the board. This adds an extra dimension of randomness and tolerance spread to the min-max field emission results. Add to this the randomness of the IC component driver strengths and speeds on a given board and we have a very difficult problem in making deterministic modeling and simulation predictions of field emissions from a PCB.

Thus we are faced with quite a challenge in verifying the accuracy of EMI predictions, in a deterministic fashion, on a complex production PCB. Even predicting the statistical envelope of likely responses from a simplified, and controlled, proof-of-prediction lab board is no easy task.

23.5 RISK MANAGEMENT MATHEMATICS

In the world of practical risk management for new product design we are primarily concerned with staying below some maximum field strengths over a test spectrum and in a defined test setup.

Modeling and simulation aside, even the complexity of *measuring* the emissions from a piece of equipment are recognized by regulatory requirements. The requirements of CISPR 22 [26] call out that we need to show that 80% of a population of equipment can be expected to fall below some emission limit, L, with an 80% statistical confidence limit. This is known as the 80-80 rule. This may be written:

$$L \geq X + S^*k \tag{23-1}$$

where:

L = The limit level
X = The arithmetic mean of the measured emission levels at a specific frequency
S = The standard deviation
K = A defined factor corresponding to the 80% confidence level (i.e. that 80% of the area of the probability distribution lies below the value X + Sk)

Therefore, the nature of our difficulties in making accurate, deterministic predictions of EMI field strength begins to moderate with the following two steps:

- First, we will use the principles of statistical design to predict a mean and standard deviation for field strengths.
- Second, we will use the principles of probability, specifically that 80% of the population of manufactured equipments should fall below a limit on emissions, L, with 80% statistical confidence.

In summary, we will not concern ourselves with minimum emission levels, nor with the absolute worst-case maximum emission level. Absolute maximum worst-case emission level leads to an overly conservative, overly constrained design. Also, mathematical modeling [15] can be used to generate both the probability and cumulative probability distributions that describe a corresponding system emission level for some number of emitters operating at a given frequency and known maximum field strengths.

Therefore, a mathematical "net" is cast around a number of EMI emitters to capture the maximum *likely* combined field strength that will be generated given a number of emitters operating at a particular frequency. Designers can then respond to a more realistic limit on the emission level allowed from individual sources to meet the 80-80 rule.

Using mathematical modeling and simulation to design products likely to pass regulatory requirements with a defined statistical confidence level becomes the new, more realistic objective. The approach avoids over-constrained and under-constrained product designs.

The statistical net (envelope) that captures the maximum *likely* combined field strength at a measurement frequency that will be generated by a number of emitters operating at that frequency, can be applied to any set of data. The data can be modeling and simulation predictions of the individual emitters, or measurement results. The scale of the emission sources can run from individual ICs (or smaller) on a board to the cumulative effects of various equipments inside a telephone exchange.

23.5.1 Basic Mathematics

In this topic, we consider the general problem of the unintentional radiated emissions from a system that is composed of a number of emissions sources that emit at common frequencies. It is assumed that the emissions levels produced by the individual sources are known, whereas the system-level emission is not. Hence, our general problem may represent a variety of practical problems, ranging in scale and complexity from:

a) The emissions of a complex PCB, containing a number of circuits that independently generate radiated emissions at common frequencies – if the emissions from each circuit are known, what is the emission of the PCB as a whole?
b) The emissions of a product, containing a number of components that independently generate radiated emissions at common frequencies – if the emissions from each component are known, what is the emission of the product as a whole?
c) The emissions of a composite system, containing a number of equipment items that independently generate radiated emissions at common frequencies – if the emissions from each equipment item are known, what is the emission of the system as a whole?

When a system contains a number of emission sources that each individually emit at some known level (at some standard measurement distance) at one or more common frequencies, mathematical tools exist to predict the combined, system-level radiated emission level at each common frequency.

Imagine we have a number, N, of independently radiated RF emissions at some common frequency, f, incident at some point of interest. Assume that each independently radiated emission is represented as a simple cosine function. Let the *i*th independently radiated RF emission be written as:

$$E_i(t) = E_{oi}\cos(\alpha_i \pm \omega t) \tag{23-2}$$

where:

$E_i(t)$	=	The radiated RF field level due to the *i*th independently radiated RF emission at time, *t*
E_{oi}	=	The amplitude of the *i*th independently radiated RF emission
α_i	=	The phase delay with respect to some agreed reference of the *i*th independently radiated RF emission
ω	=	$2\pi f$

The sum of these emissions at the point of interest can also be expressed as a simple cosine function at the same frequency, viz:

$$E_0(t) = E_o\cos(\alpha \pm \omega t) \tag{23-3}$$

where:

$E_0(t)$	=	The radiated RF field level of the sum field at time, t
E_0	=	The amplitude of the sum field
α	=	The phase delay with respect to some agreed reference of the sum field

$$E_0^2 = \sum_{i=1}^{N} E_{0i}^2 + 2\sum_{j>i}^{N}\sum_{i=1}^{N} E_{oi}E_{oj}\cos(\alpha_i - \alpha_j) \tag{23-4}$$

Careful examination of this equation indicates that, to know the sum radiated RF emission, two pieces of information are required for each independently radiated RF emission:

- The amplitude, E_i
- The phase, α_i, with respect to some reference.

While it is reasonable to assume knowledge of (a), it is not reasonable to assume knowledge of (b): only half the information required to use this equation is available.

23.5.2 Basing Certification on Worst-Case Emission

The equation in (23-4) does allow the extremes of emission performance to be calculated. The worst-case scenario is generated when all of the individually radiated RF emissions arrive at the measurement point in phase with one another and hence undergoes constructive interference. In this instance, the amplitude of the system-level emissions, E_0, becomes:

$$E_0 = \sum_{i=1}^{N} E_{0i} \tag{23-5}$$

That is, the simple arithmetic sum of the amplitudes to the individually radiated emission levels.

23.6 IDENTICAL EQUIPMENTS CASE

If it is assumed that the amplitude of the system-level emissions, E_{System}, are the absolute worst-case combined field generated by some number, n, of identical equipment, each independently radiating at some amplitude, $E_{Equipment}$, at some common frequency, then it is possible to write:

$$E_{System} = E_{Equipment} + 20\ log_{10}\{n\} \tag{23-6}$$

That is, that the amplitude of the system-level emissions is greater than that from the individual equipment by a 20 $\log_{10}\{n\}$ margin.

To prevent the system-level emissions exceeding some limit level, E_{limit}, it is therefore necessary to restrict the individual equipment emissions to a level below the limit by the 20 $\log_{10}\{n\}$ margin, i.e.:

$$E_{Equipment} = E_{limit} - 20\ log_{10}\{n\} \tag{23-7}$$

Adoption of this approach would lead the Systems Integrator to define a new, tighter emissions limit as part of its procurement process.

However, adoption of this approach has a number of implications to the Systems Integrator. As the number of identical equipment items within the system increases, the restricted emissions are unlikely to be met by commercial-off-the-shelf (COTS) apparatus. As a result, the Integrator is likely to have to organise the customized reengineering of an existing COTS product. This will generally cause delay to system deployment and significantly increase the unit costs (as economies of scale in manufacturing collapse and the need to recoup redevelopment cost). Fundamentally, this approach may lead to the specification of an emissions requirement that is simply not physically possible.

23.6.1 Example

Let us assume that a composite system contains ten items of identical equipment. Assume also that each item emits identically at a number of common frequencies and that the highest emission occurs at the frequency of 100 MHz and at a level of 25 dBμV/m when measured at the 10-meter distance. Such Apparatus would be considered as extremely well engineered, complying with the CISPR 22 Class A emissions limit of 40 dBμV/m by a margin of 15 dB.

The worst-case emission level generated would be $(25 + 20 \log_{10}\{10\}) =$ 45 dBμV/m, i.e. 5 dB in excess of the limit. Hence, consideration of the worst-case would lead the Systems Integrator to reengineer or seek alternate equipment that emits at or below a level of $(40 - 20 \log_{10}\{10\}) = 20$ dBμV/m, i.e. for the Systems Integrator to reduce the highest emission by another 5 dB.

23.7 NON-IDENTICAL EQUIPMENTS CASE

This argument has so far considered a system containing a number of identical equipment, that each emits at a similar level at a common frequency. The same argument can equally be applied to a system containing a number of different equipment items that each emit at different levels at a common frequency (the mathematics is slightly less elegant).

23.8 RISK ASSESSMENT

23.8.1 Overview

Basing system-level certification upon the absolute worst-case is fundamentally flawed, since a second question must be considered: how likely is this to occur? Consideration of this question leads naturally to the development of an alternative approach to the system-level certification issue that is now described.

23.8.2 Statistical Approach

The worst-case approach assumes that each equipment's individually radiated RF emissions are in phase at the measurement point and hence interfere constructively. To determine the likelihood of this occurring, it is necessary to have knowledge on the likelihood of the unknown phase values adopting values.

It is assumed that the phase relationships between the individual equipment emissions are both unknown and uncontrolled: in principal, any value is possible. Given the *a priori* lack of specific knowledge, the individual phase terms are assumed to be random: capable and equally likely to adopt any of the possible values (i.e. between 0 and 2π radians).

Monte Carlo processing of the basic equation allows a probability distribution (PD) and a cumulative probability distribution (CPD) of the system-level RF emissions amplitude to be determined.

The PD indicates the relative frequency (i.e. probability) or likelihood that the system emissions amplitude will be found at a given level and hence measured.

The CPD indicates the relative frequency with which the systems emission amplitude will be found to be equal to or below a specified level. If the level in question is an emissions limit, the CPD displays the compliance probability with that limit.

23.9 DISTRIBUTION EXAMPLES

23.9.1 N=2

When the system contains two items that emit at a common frequency, the PD and CPD display a characteristic form. Examples are displayed on Figures 23-1 and 23-2 for the case of common emissions amplitudes (in this case E1 = E2 = 40 dBμV/m).

Examination of Figures 23-1 and 23-2 indicate that the amplitude of the combined, system-level emissions in this case occur between the worst-case limit of (40 dBμV/m + 20 $\log_{10}\{2\}$) = 46 dBμV/m and a best-case limit of zero.

Figure 23-1 indicates that the PD displays a clear maximum at the worst-case field level: the worst-case system-level amplitude is, therefore, the most likely to occur and the Systems Integrator is best to adopt the approach discussed earlier within "23.7, Non-Identical Equipments Case."

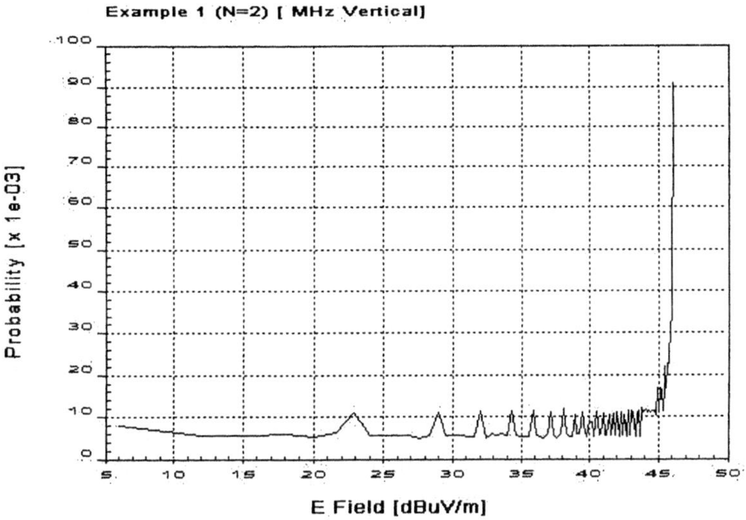

Figure 23-1. Example PD for N = 2

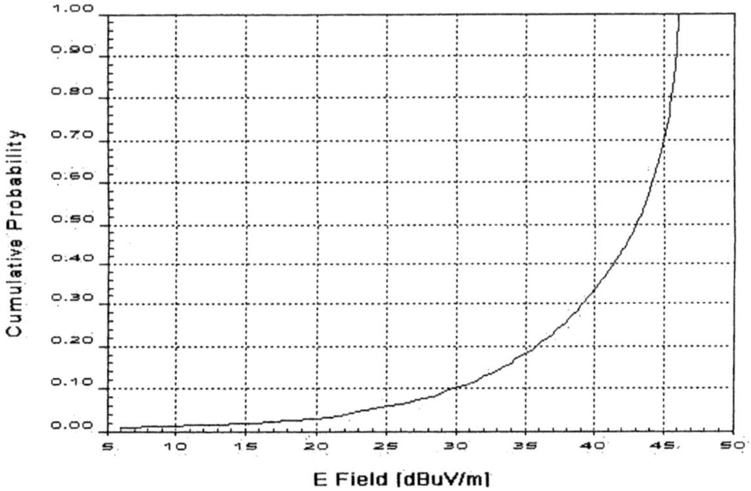

Figure 23-2. Example CPD for N = 2

Formal mathematical proof of this characteristic form exists [153, 154].

23.9.2 N=3

When the system contains three items that emit at a common frequency, the PD and CPD also display a distinct form. Examples are displayed on

Figures 23-3 and 23-4 for the case of common emissions amplitudes (in this case E1 = E2 = E3 = 40 dBμV/m).

Examination of Figures 23-3 and 23-4 indicate that the amplitude of the combined, system-level emissions in this case occur between the worst-case level of (40 dBμV/m + 20 $\log_{10}\{3\}$) = 49 dBμV/m and a best-case limit of zero.

Figure 23-3 indicates that the PD displays a clear maximum at the common emissions amplitude, i.e. when Ec = E1. This means that the amplitude of the composite system that is most likely to occur is the common amplitude of the constituent equipment: i.e. that the Integrator is most likely to measure no change in the emissions amplitude from considering one equipment item to considering three. This result disagrees with the arguments presented in "23.7, Non-Identical Equipments Case."

Formal mathematical proof of this characteristic form does not currently exist. However, this is still under study by the author and any progress made will be presented in a future reference work.

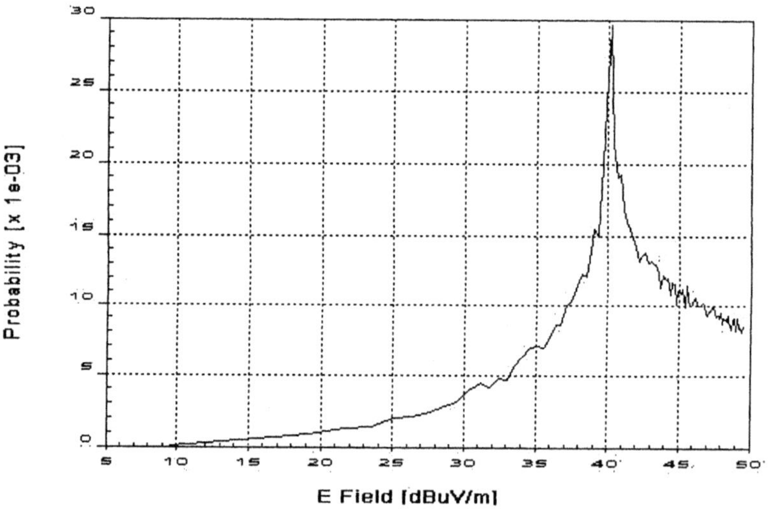

Figure 23-3. Example PD for N = 3

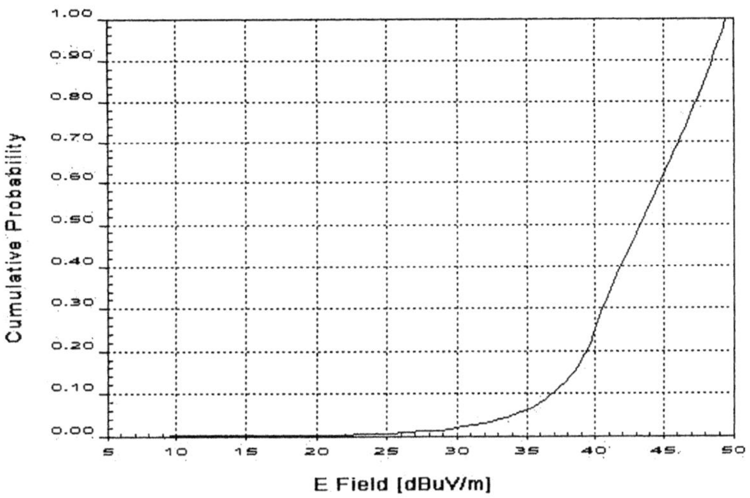

Figure 23-4. Example CPD for N = 3

23.9.3 N=5

When the system contains five items that emit at a common frequency, the PD and CPD display a characteristic form. Examples are displayed on Figures 23-5 and 23-6 for the case of common emissions amplitudes (in this case E1 = E2 = E3 = E4 = 40 dBμV/m).

Figure 23-5 indicates that the PD displays a maximum at a systems emissions amplitude of ~ 45 dBμV/m. This is noted to be some ~ 9 dB below the worst-case value.

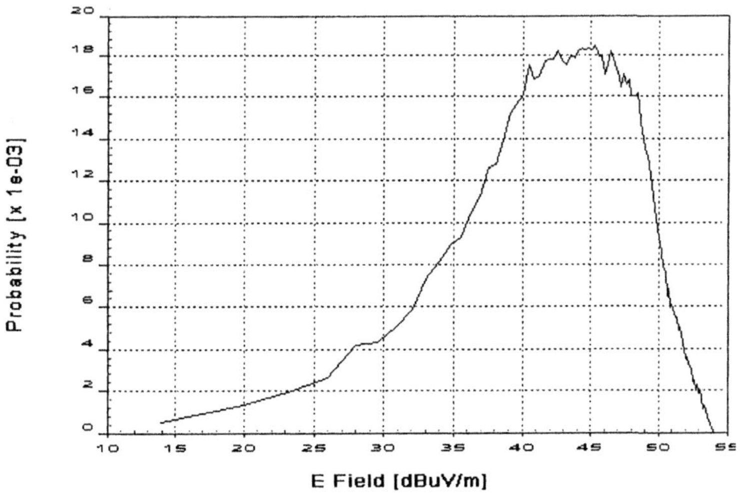

Figure 23-5. Example PD for N = 5

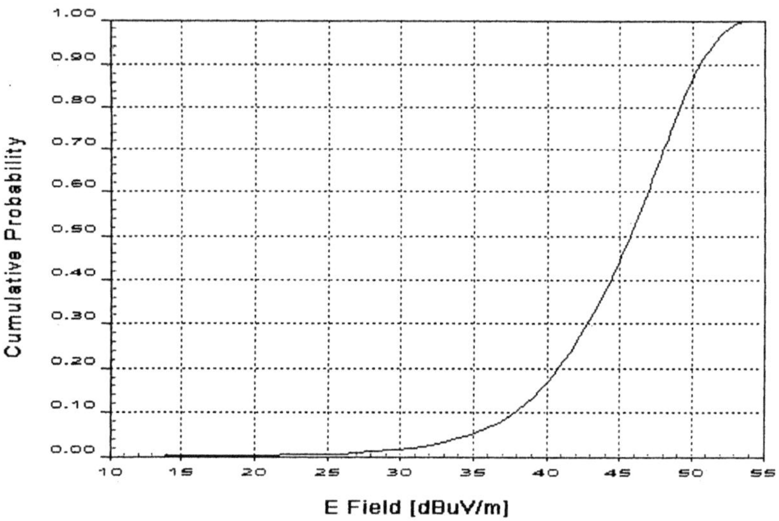

Figure 23-6. Example CPD for N = 5

Examination of Figures 23-5 and 23-6 indicate that the amplitude of the combined, system-level emissions in this case occur between the worst-case limit of (40 dBμV/m + 20 $\log_{10}\{5\}$) = 53.97 dBμV/m and a best-case limit of zero.

23.9.4 N=10

When the system contains ten items that emit at a common frequency, the PD and CPD display a characteristic form. Examples are displayed on Figures 23-7 and 23-8 for the case of common emissions amplitudes (in this case 40 dBμV/m).

Examination of Figures 23-7 and 23-8 indicate that the amplitude of the combined, system-level emissions in this case occur between the worst-case limit of (40 dBμV/m + 20 $\log_{10}\{10\}$) = 60 dBμV/m and a best-case limit of zero.

Figure 23-7 indicates that the PD displays a maximum at a system emissions amplitude of ~ 48 dBμV/m. This is noted to be some 12 dB below the worst-case value.

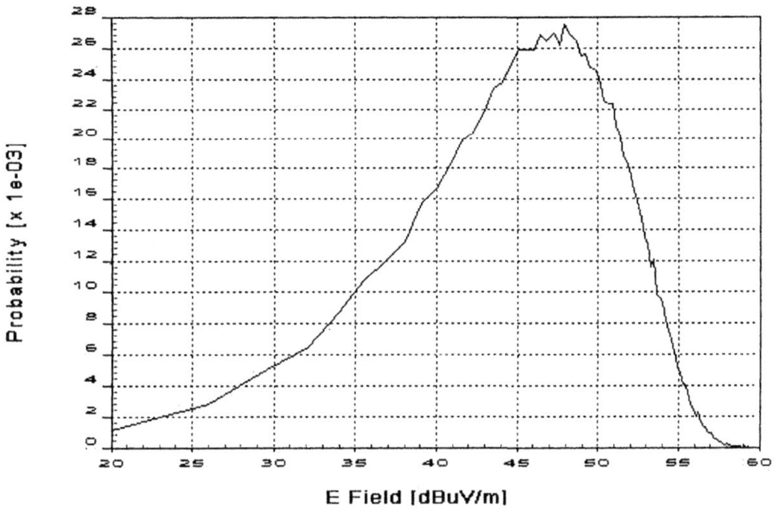

Figure 23-7. Example PD for N = 10

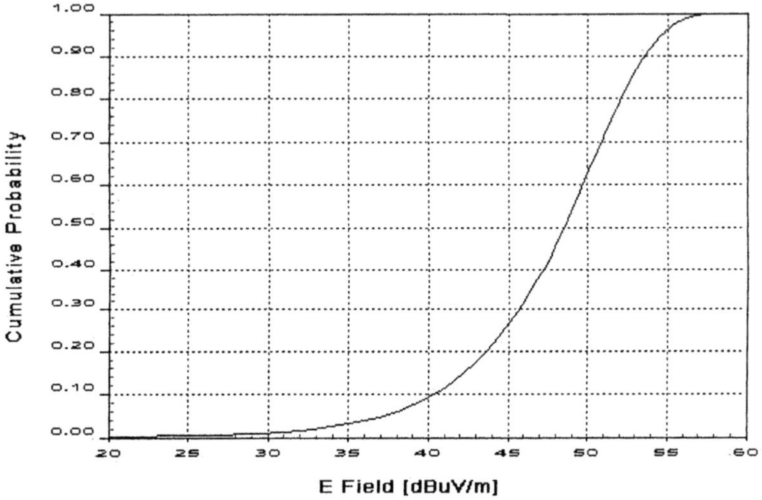

Figure 23-8. Example CPD for N = 10

23.9.5 N=100

When the system contains one hundred items that emit at a common frequency, the PD and CPD display a characteristic form. Examples are displayed on Figures 23-9 and 23-10 for the case of common emissions amplitudes (in this case 40 dBμV/m).

Examination of Figures 23-9 and 23-10 indicate that the amplitude of the combined, system-level emissions in this case occur between the worst-case limit of (40 dBμV/m + 20 $\log_{10}\{100\}$) = 80 dBμV/m and a best-case limit of zero.

Figure 23-9 indicates that the PD displays a maximum at a system emissions amplitude of ~ 58 dBμV/m. This is noted to be some 22 dB below the worst-case value.

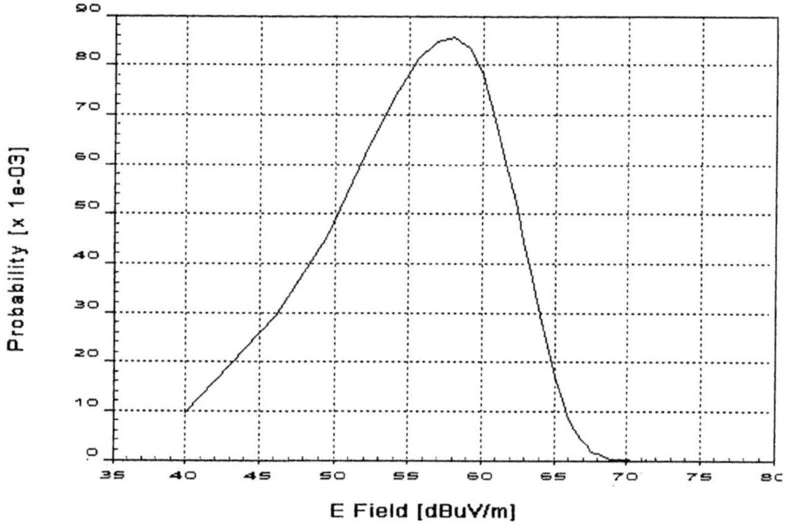

Figure 23-9. Example PD for N = 100

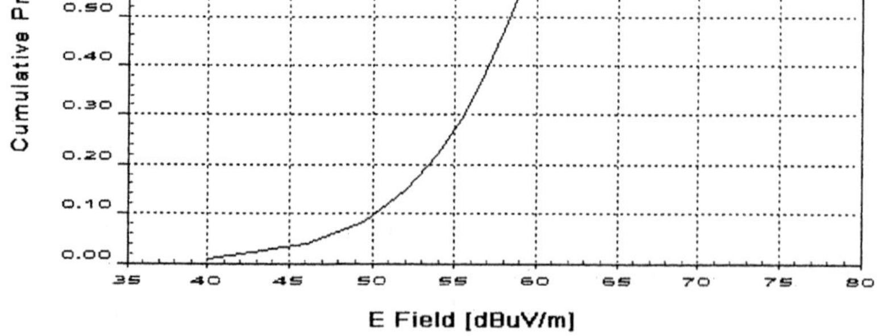

Figure 23-10. Example CPD for N = 100

23.10 REVIEW OF PROBABILITY DISTRIBUTIONS

Review of the PD plots presented in the previous examples indicates a pattern: as the number of independent emissions at a common frequency increases, the likelihood of the worst-case level occurring decreases. For example:

- When only two independently radiated field terms are being added, the PD reaches its maxima when the system emissions amplitude approaches the worst-case.
- When ten independently radiated field terms are being added, the PD reaches a minimum when the systems emissions amplitude approaches the worst-case. For the example displayed, the worst-case amplitude is 60 dBμV/m, but the PD approaches zero values above 58 dBμV/m.
- When one hundred independently radiated field terms are being added, the PD reaches a minimum when the systems emissions amplitude approaches the worst-case. For the example displayed, the worst-case amplitude is 80 dBμV/m, but the PD approaches zero values above ~70 dBμV/m.

Hence, the worst-case amplitude is extremely unlikely to occur and should not be used as the basis of certifying the system. Also, as the number of independent emissions at a common frequency increases, the margin between the most likely amplitude and the worst-case amplitude increases. For example:

- When only two independently radiated field terms are being added, the PD reaches its maxima when the system emissions amplitude approaches the worst-case. Hence the margin between the most likely and worst-case levels is 0 dB.
- When ten independently radiated field terms are being added, the PD reaches a maximum at ~48 dBμV/m while the worst-case is 60 dBμV/m. Hence the margin between the most likely and worst-case levels is ~12 dBμV.
- When one hundred independently radiated field terms are being added, the PD reaches a maximum at ~58 dBμV/m while the worst-case is 80 dBμV/m. Hence the margin between the most likely and worst-case levels is ~ 22 dBμV.

An intuitive explanation for this behavior exists. The worst-case system emissions amplitude is associated with a single, specific combination of events: when *all* of the independently radiated field terms arrive at the measurement point in-phase with one another. As more independently radiated field terms are added, the probability of this singular event occurring becomes progressively smaller. For other lower emission amplitudes, there are generally many different combinations of phase values (i.e. many different events) among the independently radiated field terms that generate the amplitudes. Hence, the probability of these amplitudes occurring is higher.

23.11 FOLLOW UP SIMULATION WITH PRODUCT ASSURANCE

The first step in producing a good design is to use modeling and simulation, both deterministic and probabilistic. But there are limitations to modeling and simulation. Consider what happens when we must produce a design where failure is unacceptable and service is not an option. We would want our design to work with 100% confidence under absolute worst-case conditions. However, even deterministic modeling and simulation should never be counted on exclusively to reach that level of certainty.

In high-risk, high-consequence cases, designers should apply a spectrum of techniques for resolving risk, and mitigating the effects of possible failure. This spectrum of techniques follows modeling and simulation with reliability engineering, burn-in and more. Quality control, and the techniques of HALT and HASS should be applied in proportion to the risks and consequences of failure.

Early and effective application of modeling and simulation has been shown to improve the chances of first prototype success from less than 25% to greater than 90%. Late or no application of modeling and simulation usually has dire consequences even when followed by reliability engineering and additional product assurance steps. There are two basic phenomena that make this assertion so: 1) Problems discovered late in a development cycle tend to "escape" into the finished product, and; 2) It is very difficult to build and test enough prototypes to discover all possible combinations of problems.

By screening out most potential design flaws early on, modeling and simulation prevents multiple late problems from overwhelming additional product assurance activities. Compared to building hardware prototypes, building and testing multiple virtual prototypes on a computer is easy, economical, and provides greater design insight.

23.12 SUMMARY

Probabilistic modeling and simulation produces a more *economical* design than deterministic modeling and simulation. Probabilistic design is a statistical method that takes the probability of a variable assuming particular values in a range into account.

The radiated RF emissions of a system were selected as a practical example of the application of the technique. Such design tasks are often complex, and are often characterized by a level of uncertainty regarding some model parameters. For instance, the radiated emissions on a board are driven, both in strength and phase, by the data patterns on the board. Designing for absolute worst-case is usually unwise and uncalled for. Rather, model parameter values have a certain probability of occurring that should be accounted for in the simulation.

The method generates a probability distribution (PD) and cumulative probability distribution (CPD) that describes the system's radiated RF emissions amplitude. The PD indicates the amplitude that is *most likely* to occur (and hence be measured). The CPD can be used to quantify the *compliance probability* of the systems' radiated RF emissions with an emissions limit.

Use of this method converts the emissions certification process from a simple pass or fail to a risk management approach that parallels the methods of measuring and certifying compliance in a test cell. A number of examples of the application of this method have been presented. These examples indicate that, as the number of equipment items radiating at a common frequency increases, the *most likely* amplitude of the radiated RF emissions is generally:

- Much lower than the worst-case
- Does not increase at the same rate as the worst-case with equipment number

As systems grow ever more complex, managing the probability of success using modeling, simulation, testing and other techniques will be the coming state of design.

PART 8: GLOSSARY, BIBLIOGRAPHY, INDEX, AND CD-ROM

GLOSSARY

Use the following link for acronyms not found in this glossary: http://www.acronymfinder.com/

Another excellent link is: http://www.answers.com/

A

A	Symbol for ampere.
abstract	To simplify.
ABT	Acronym for Advanced BiCMOS Technology, as in 74ABT245.
ABTE	Acronym for Advanced BiCMOS Technology/Enhanced logic, as in 74ABTE245.
ABTH	Acronym for Advanced BiCMOS Technology logic with bus-Hold, as in 74ABTH245.
AC	Acronym for Alternating Current.
AC	Acronym for Advanced Cmos logic, as in 74AC245.
accuracy	1. The extent to which a given measurement, or the average of a set of measurements, agrees with the true value for that measurement. 2. Agreement between simulation results and lab measurements.

ACL	Acronym for Advanced CMOS Logic, as in 74ACL245.
ACT	Acronym for Advanced CMOS logic with TTL compatible inputs.
AEA	Acronym for American Electronic Association.
AEA	Acronym for American Engineering Association.
AHDL	Acronym for Analog Hardware Description Language.
AMS	Acronym for Analog Mixed-Signal (hardware description language).
analog	Information represented as continuously varying voltage or current rather than in discrete levels as opposed to digital data varying between two discrete levels.
ANOVA	Acronym for Analysis Of Variance
ANSI	Acronym for American National Standards Institute.
AQL	Acronym for Acceptable Quality Level.
ASCII	Acronym for American Standard Code for Information Interchange.
ASIC	Acronym for Application Specific Integrated Circuit.
ASTM	Acronym for American Society for Testing and Materials.
AWE	Acronym for Asymptotic Waveform Evaluation. Model order reduction. Such as reducing a large RC network to a smaller RC network for faster simulation. This modeling method has known frequency limits for accuracy.

B

behavioral model	Provides data on behavior at the input and output ports of a device. What the device consists of internally remains a *black (opaque) box*. Usually used when scanning a large design for the first time. At that point speed is desired over detail.
behavioral modeling	System-level modeling consisting of a functional specification plus modeling of the timing of an implementation. A behavioral model consists of an HDL description of a device or component, which is expressed at a relatively high level of abstraction (higher than the register-transfer level or gate level). It uses underlying mathematical equations to represent the functional behavior of the component. See also *functional modeling*.
BEM	Acronym for older Boundary Element Method. Method of numerical computational electromagnetics. Dates from before the 1980s. Requires ground plane.
BER	Acronym for Bit Error Rate.
BGA	Acronym for Ball Grid Array.
BiCMOS	Acronym for Bipolar Complementary Metal Oxide Semiconductor.
BIRD	Acronym for Buffer Issue Resolution Document – IBIS Committee.
BJT	Acronym for Bipolar Junction Transistor.
BNC	Acronym for Bayonet Neill Concelman Connector.
BOM	Acronym for Bill Of Materials.
BSIM	Acronym for Berkeley Short-channel IGFET Model.
BTL	Acronym for Backplane Transceiver Logic, as in 74BTL245.

C

C	1. Symbol for capacitance. 2. Abbreviation for degrees Celsius/Centigrade.
CAD	Acronym for Computer Aided Design.
CAE	Acronym for Computer Aided Engineering.
CAM	Acronym for Computer Aided Manufacturing.
CB	Acronym for Complementary Bipolar, as in 74CB245.
CEM	Acronym for Computational Electromagnetics.
characterization model	Another term for a device's data sheet. Displays the behavior of various properties over current, frequency, temperature, population spread, etc. Characterize means to describe and depict.
circuit	Interconnection of components to provide an electrical path between two or more components.
CISPR	Acronym for Comité Internationale Spécial des Perturbations Radioelectrotechnique.
CMC	Acronym for Compact Model Council.
CML	Acronym for Current Mode Logic.
CMOS	Acronym for Complimentary Metal Oxide Semiconductor.
CMRR	Acronym for Common Mode Rejection Ratio.
Confidence interval	A statistical range with a specified probability that a given parameter lies within the range.

Confidence level	The confidence level is the probability value (1-α) associated with a confidence interval. It is often expressed as a percentage. For example, say α=0.05=5%, then confidence level = (1-0.05) = 0.95, that is, a 95% confidence level. Example: Suppose an opinion poll predicted that, if the election were held today, the Conservative party would win 60% of the vote. The pollster might attach a 95% confidence level to the interval 60% plus or minus 3%. That is, he thinks it very likely that the Conservative party would get between 57% and 63% of the total vote.
Confidence limit	Either of the two numbers that specify the endpoints of a confidence interval.
correlation	1. How well a set of measurements agrees with a separate set of measurements on the same group of units. One or more attributes may be used. 2. Process of making a quantitative comparison between two sets of data. Also, from the word roots co: together and relation.
correlation metric	A means of quantifying agreement between two sets of data.
COTS	Acronym for Commercial Off-The-Shelf apparatus.
CPD	Acronym for Cumulative Probability Distribution (function).
crosstalk	The amount of signal from one conductor that gets coupled onto an adjacent conductor through the mutual capacitance and inductance between them. The magnitude of the effect is always a fraction of the active, or aggressor, net upon the passive, or victim, net. The amount of coupling depends on the proximity of the nets, the proximity of any ground/power plane, the length of any parallel runs, the risetime of the driving signal and the dielectric medium between them.
current carrier	A current carrier is either a conduction band electron or a conduction band hole.

D

DARPA	Acronym for Defense Advanced Research Projects Agency.
data sheet	Documentation created after the properties and behavior of a population of devices, from a statistically *in-control* process, are characterized.
dB	Acronym for decibel. The number of decibels denoting the ratio of the two amounts of power being ten times the logarithm to the base 10 of this ratio. With P1 and P2 designating two amounts of power and n the number of decibels denoting their ratio.(ANSI C63.14 – 1992) N = 10log10 (P1/P2) dB
dBµV/m	Symbol for decibels-micro-Volts per meter. where dBµv = 20 log [Signal (µV)/1µV] Decibels relative to one microvolt across same resistance.
DC	Acronym for Direct Current.
detailed physical model	A device model wherein its internal detail is described as closely as possible. There is usually a close correlation with the physical construction of the device.
device	A component or part.
DFM	Acronym for Design For Manufacturability.
DFT	Acronym for Design For Test.
DIE	Acronym for Die Information Exchange.
dielectric	Dielectric materials are poor conductors of electricity, and are insulators that are used to provide separation between conductors. Dielectric materials can be made to hold an electrostatic charge while dissipating minimal energy in the form of heat. Glass, porcelain, mica, rubber, plastics dry air, vacuums and some liquids and gases are dielectric. For example: air, FR4, and GETEK.

dielectric constant (relative)	An inherent property of dielectric materials that determines the amount of electric charge that can be stored. The higher the constant, ε, the higher the energy stored in the material's capacitance. ε is usually measured and reported in formulas relative to air (ε_r). Air then gets an ε_r, = 1. FR4 often has a low frequency $\varepsilon_r = 4.3$ depending on starting materials and processing. In solid dielectrics ε_r will fall off starting with high frequencies while signal attenuation increases. ε_r affects transmission line behavior, particularly characteristic impedance (Zo) and propagation delay (t_d). The changes in t_d with frequency cause decreased delay leading to dispersion, or spreading distortion, of a time domain pulse that is composed of many frequencies.
digital	Data varying between two discrete levels. Electronic signals or switches based on discrete, binary electrical levels (ones and zeros) found in such products as touch-tone telephones and audio compact disk players. These signals are either ON or OFF, HIGH or LOW, YES or NO. The mathematical description in digital products is simple, since it is either ON or OFF.
DIP	Acronym for Dual InLine Package.
discrete	As in discrete component. Package containing only a single component as opposed to an integrated circuit containing many components in a single package.
dispersion	1. The scattering of values of a measurement around the mean or median of the measurement. 2. A distribution.
dispersion, dielectric	A high frequency effect in dielectric materials. At high frequencies the dielectric constant decreases causing an increase in propagation velocity and a dispersion in arrival times of wave frequency components.
DML	Acronym for Device Modeling Language (Cadence) and associated .dml files.
documentation	Reports such as data sheets, process control, product characterization, process characterization, and models. May be produced by either the supplier or customer to describe or depict the semiconductor process.
DOE	Acronym for Design Of Experiments.

doping	The process of adding impurity atoms to intrinsic (pure) silicon or germanium to improve the conductivity of the semiconductor material.
DSP	Acronym for Digital Signal Processing.
DTL	Acronym for Diode-Transistor Logic.
DUT	Acronym for Device Under Test.

E

Early effect	A behavior in narrow base BJT transistors named for Dr. J. Early who explained it. The effect is an increase in β due to base narrowing caused by increasing base-collector bias voltage.
EBD	Acronym for Electronic Board Description (IBIS) and the associated .ebd files.
ECL	Acronym for Emitter Coupled Logic.
EDA	Acronym for Electronic Design Automation. EDA software tools or tool suites may display simulator output for analysis (as in waveform analyzers) or which may analyze the reliability, electromagnetic interference, metal migration, signal integrity, or thermal characteristics of a design. The tools in this category may work at any level of abstraction — behavioral, register-transfer-level (RTL), gate-level, or with the physical layout of an IC device or electronic system.
EDIF	Acronym for Electronic Design Interchange Format.
EIA	Acronym for Electronic Industries Association.
EIAJ	Acronym for Electronics Industries Association of Japan.
EKV	Acronym for Enz-Krummenacher-Vittoz model.

ElectroMagnetic Integrity	The technical discipline of designing for ElectroMagnetic Control (EMC) and low ElectroMagnetic Interference (EMI), so that one circuit (or equipment) does not interfere with another circuit (or equipment). The design task begins with designing for good Signal and Power Integrity and adds the concerns of coupling and radiation. The noise energy available to the coupling and radiation mechanisms can be suppressed with good Signal and Power Integrity. The next step is good design control over coupling and radiation mechanisms. These additional actions include controlling: shielding; cabling; leakage; board resonances; enclosure resonances; structural behavior that operates as an antenna.
EM	Acronym for ElectroMagnetic.
EMC	Acronym for ElectroMagnetic Compatibility. The capability of Electrical and Electronic Systems, equipment and devices to operate in their intended electromagnetic environment within a defined margin of safety, and at design levels of performance, without suffering or causing unacceptable degradation as a result of electromagnetic interference (ANSI C63.14 – 1992).
EMI	Acronym Electromagnetic Interference. The electromagnetic disturbances (electronic noise) in an environment that can affect an electronic device, or that which is being produced by an electronic device, or both. EMI analysis tools are used to verify EMC compliance during the design of high-speed PCBs and IC packages. The traditional EMI remedies involve the addition of extra components, metal shields, metal plans, or even redesigning the entire system. Synonym: radio-frequency interference.
EMI/EMC	Acronym for Electromagnetic Interference/Electromagnetic Control.
equation-based model	A model based on equations that describe the behavior of the device, or circuit, modeled. Most importantly, the output as a function of the input.
ESD	Acronym for ElectroStatic Discharge. A transfer of electric charge between bodies of different electrostatic potential in close proximity or through direct contact (ANSI C63.14 – 1992).
ESS	Acronym for Environmental Stress Screening.

eV Abbreviation for electron-volt.

F

F Symbol for Farad, measure of capacitance.

f Abbreviation for frequency.

FACT Acronym for Fairchild Advanced Cmos Technology, as in 74FACT245.

FAST Acronym for Fairchild Advanced Schottky TTL, as in 74F245.

FCT Acronym for Fast Cmos Technology, as in 74FCT245.

FDTD Acronym for Finite Difference Time Domain modeling.

FEA Acronym for Finite Element Analysis.

FEM Acronym for Finite Element Method. Method of numerical computational electromagnetics. Essentially everybody does some form of this for field solvers, such as T-lines.

FET Acronym for Field Effect Transistor.

FFT Acronym for Fast Fourier Transform.

FOM Acronym for Figure Of Merit.

FPGA	Acronym for field programmable gate array.
functional modeling	Modeling by the use of mathematical functions, algorithms and formulas.

G

G	Symbol for giga, as in10^{+12}
GaAs	Acronym for Gallium Arsinide.
GB or Gbit or GBIT	Acronym for GigaBit.
glue logic	Small ICs that level shift and otherwise perform simple functions that enable large blocks of logic (ASICs, microprocessors, memories) to work together. Thus *gluing* them together.
GND	Acronym for GrouND. A reference connection commonly connected to Earth, whose electric potential is usually equal to zero.
ground bounce	The transient rise or fall in voltage on a ground plane or ground pin from its ideal quiescent value of zero due switching currents on and off through impedance (mostly inductance) in the ground path. A similar effect on the power plane causes power bounce. The result is noise on the signal that can decrease signal to noise ratio in analog circuits or lead to false switching in digital circuits.
GTL	Acronym for Gunning Transceiver Logic, as in 74GTL245.
GTLP	Acronym for Gunning Transceiver Logic Plus, as in 74GTLP245.
guard-banding	The practice of adding safety margin, or extra safety margin, to specification limits or population distributions.

H

H	Abbreviation for Henry, the unit of inductance.
HALT	Acronym for Highly Accelerated Life Test.
HASS	Acronym for Highly Accelerated Stress Screening.
HC	Acronym for High speed Cmos, as in 74HC245.
HCT	Acronym for High speed Cmos with TTL thresholds, as in 74HCT245.
HDL	Acronym Hardware Description Language. One of several specialized high-level languages used by semiconductor designers to describe the features and functionality of chips and systems prior to handoff to the IC layout process. HDL descriptions are used in both the design implementation and verification flows. Currently, the two standard HDLs in use worldwide are Verilog-HDL and VHDL. Several proprietary HDLs also exist, mainly for describing logic that is targeted for vendor-specific programmable logic devices.
high-speed digital design	The design of digital circuits relative to their analog behavior. Particularly, where the fast switching edge rates used cause transmission line effects to become significant.
hole	A gap left in the covalent bond when a valence electron gains sufficient energy to jump to the conduction band.
HSDD	Acronym for High-Speed Digital Design. HSDD, as used in this book, applies to both active and passive circuit elements.
HSTL	Acronym for High Speed Transceiver Logic.
HTL	Acronym for High Threshold Logic.
HTTL	Acronym for High power Transistor-to-Transistor Logic.
Hz	Symbol for Hertz, as in cycles per second.

I

I/O	Symbol for Input/Output.
IBIS	Acronym for Input/output Buffer Information Specification.
IBIS model	A data file produced in conformance with the IBIS Specification. The IBIS file, or model, provides necessary data for a simulator to predict the analog behavior of digital circuits.
IBIS quality levels	Four levels of increasing quality as defined in the IBIS quality checklist.
IBIS rise time	Rise time based on the IBIS [Ramp] specification of 20% to 80% of 0 Volts to Vcc switching.
IC	Acronym for Integrated Circuit.
ICEM	Acronym for Integrated Circuits Electromagnetic Model, specification 62014-3 from IEC.
ICM	Acronym for IBIS Interconnect Modeling Information (specification). Connectors, packages, and interconnections.
ideal-generic models	Used early in the product development when part-specific models have not yet been obtained and only estimates are needed. Models that have been simplified, primarily to save simulation time, by ignoring 2nd and 3rd order effects. Ideal models also describe ideal components, such as capacitors with no internal inductance that are unrealizable in real life. In spite of this they can be very useful in analysis. Examples: ideal current source, ideal operational amplifier.
IEC	1. Acronym for International Engineering Consortium 2. Acronym for International Electrotechnical Commission.
IEEE	Acronym for Institute of Electrical and Electronics Engineers.
IGBT	Acronym for Insulated Gate Bipolar Transistor.
IGFET	Acronym for Insulated Gate Field Effect Transistor.

impedance	Measured in ohms, it is the total opposition to the flow of current offered by a circuit element. Impedance consists of the vector sum of resistance and reactance. The symbol, or term, for impedance is Z.
in-control (SPC)	A statistical process control (SPC) term that means that only random variation is seen in the process and that there are no *special cause* reasons for variation.
IP	Acronym for Intellectual Property.
ISO	Acronym for International Standards Organization.

J

JEDEC	Acronym for Joint Electron Device Engineering Council.
JFET	Acronym for Junction Field Effect Transistor.
junction	Contact or connection between two or more wires or cables. The area where the *p*-type material and *n*-type material meet in a semiconductor.

K

K	Symbol for kilo, as in 1000.
Keyword	Special words, as used in the IBIS Specification, that are reserved and set off by square brackets. Example: [File Name]. These terms are recognized and parsed by simulation software programs.

L

L	Symbol for inductance.
LSL	Acronym for Lower Spec Limit
LSTTL	Acronym for Low power Schottky Transistor-to-Transistor Logic, as in 74LSTTL245.
LVC	Acronym for Low Voltage Cmos, as in 74LVC245.
LVDS	Acronym for Low Voltage Differential Signaling, as in 74LVDS245.
LVT	Acronym for Low Voltage Transceiver, as in 74LVT245.
LVTTL	Acronym for Low Voltage Transistor-to-Transistor Logic, as in 74LVTTL245.

M

M	Symbol for million, 1,000,000.
macromodel	A model composed of several sub-elements. Used early on in top-down design approaches or to simplify standard blocks of circuitry. For example an OpAmp macromodel. Still possesses an ability to be decomposed into smaller, simpler blocks of circuitry.
majority carrier	The conduction band electrons in an n-type material and the valence band holes in a p-type material. Produced by pentavalent impurities in n-type material and trivalent impurities in p-type material.
MAST	Not an acronym, but the abbreviation of MAST AHDL (language), a registered trademark of Analogy, Inc.
matched	Condition that occurs when the output impedance of a source is equal to the input impedance of a load so that maximum power is transferred between the two.

Matrix model	A black-box N by N model in the form of a square matrix where the behavior inside the black box can be characterized by measurements at its input and output ports.
maximum	The usual maximum of a property of a population of devices. Usually defined to be three sigma standard deviations greater than the mean.
mean (and typical)	The average of a property of a population of devices.
micron	A unit of measure equivalent to one-millionth of a meter, synonymous with micrometer. 1 mil = 1 thousandth of an inch = 25.4 microns.
minimum	The usual minimum of a property of a population of devices. Usually defined to be three sigma standard deviations less than the mean.
minority carrier	The conduction band holes in n-type material and valence band electrons in p-type material. Most minority carriers are produced by temperature rather than by doping with impurities.
model	A mathematical prediction of the behavior of a physical system. Also, a functional representation of a device or system. This representation contains the basic structure and characteristics of a design object, which is used to perform design verification. During the development of an electronic system, models are exercised along with signals entering from the outside environment to simulate the behavior of the system in software and ensure that it will operate properly before being manufactured in hardware.
modeling	The process of using models for analysis and simulation.
MoM	Acronym for Method of Moments method of numerical computational electromagnetics. Dates from at least the 1980s. Does not require ground plane.
MOS	Acronym for Metal Oxide Semiconductor.
MOSFET	Acronym for Metal Oxide Semiconductor Field Effect Transistor.
MTBF	Acronym for Mean Time Between Failures.

MTTF	Acronym for Mean Time To Failure.
MTTR	Acronym for Mean Time To Repair.

N

n-type semiconductor	A semiconductor compound formed by doping an intrinsic semiconductor with a pentavalent element. An n-type material contains an excess of conduction band electrons.
NA	1. Acronym for Not Applicable. 2. Acronym for Not Available.
NASA	Acronym for National Aeronautics and Space Administration.
NC	Acronym for Not Connected.
NDA	Acronym for Non Disclosure Agreement.
net, or network or topology	Combination of interconnected components, circuits or systems.
NIST	Acronym for National Institute for Standards and Technology.
NMOS	Acronym for N-channel Metal Oxide Semiconductor.
Normal distribution	Population distribution. Synonymous with Gaussian distribution. Called *normal* because its shape is the most common normally seen in nature.
npn	A bipolar junction transistor in which a p-type base element is sandwiched between an n-type emitter and an n-type collector.

O

OEM	Acronym for Original Equipment Manufacturer.

P

p	Symbol for pico, as in 10^{-12}
parameter	A variable. A means by which an application or user can customize the behavior or characteristics of a model instance when it is created. A parameter is set to a constant value during design entry.
parameterize	A variable that is assigned a specific value or is otherwise made into a physical *thing*, as in a *resistor*.
parasitic extraction	Software tools that translate IC and PCB layout data into networks of electrical circuit elements (transistors, resistors and capacitors) and parasitic elements (interconnect capacitance and resistance).
PCB	Acronym Printed Circuit Board. An electronic interconnect product, which is the foundation of most electronic systems. PCBs are used to mount and interconnect chips, capacitors, resistors, and other discrete components required in a piece of electronic equipment. The base material of a PCB is called a dielectric and is generally made of rigid fiberglass, rigid paper, or flexible thin plastic laminates. Those dielectric substrates are then coated with copper and may be fabricated into rigid single- or double-sided, multilayer, or flexible circuits. Also referred to as printed wiring board.
PCI	Acronym for Peripheral Component Interconnect bus, as in reflected wave switching.
PD	Acronym for Probability Distribution (function).
PDS	Acronym for Power Distribution System.
PECL	Acronym for Positive/pseudo Emitter Coupled Logic.

PEEC	Acronym for Partial Element Equivalent Circuit.
physical model	See *detailed physical model.*
PI	Acronym for Power Integrity.
PKG	Acronym for IBIS Package file denoted .pkg.
PMOS	Acronym for P-type Metal Oxide Semiconductor.
pnp transistor	A bipolar junction transistor with an n-type base and p-type emitter and collector.
Power Integrity	The technical discipline of maintaining clean power distribution in electronic equipment. One where the power distribution system does not transmit noise, and voltages do not sag, bounce or droop under the demands for power.
PPM or ppm	Acronym for parts-per-million.
PRBS	Acronym for Pseudo-Random Bit Sequence.
precision	The extent to which a given set of measurements are repeatable.
probabilistic	Relating to, or governed by, probability. The behavior of a probabilistic system cannot be predicted exactly but the probability of certain behaviors is known. Such systems may be simulated using pseudo-random numbers.
process characterization	Documentation that describes and depicts a process.
process model	Model that applies to an entire process from which more than one device may be manufactured.
product characterization	Documentation that describes and depicts a product.
PWB	Acronym for Printed Wiring Board. See *PCB.*
PWR	Abbreviation for PoWeR. The rate of generating, transferring or using energy.

Q

Q	Abbreviation for Quality factor of a resonant circuit.
QA	Acronym for Quality Assurance.
QC	Acronym for Quality Control.
quality	The measure of a device's, or population of devices', fitness for use.

R

R	Symbol for resistance.
RASSP	Acronym for Rapid Prototyping of Application Specific Signal Processors (DARPA project).
reflection coefficient, ρ	The fraction of a signal that gets reflected in a distributed circuit.
RF	Acronym for Radio Frequency.
RFIC	Acronym for RF Integrated Circuit.
rise and fall time	The time for a signal to rise, or fall, from a defined percentage of output swing to a second defined percentage of output swing.
RMS	Acronym for Root Mean Square.
ROI	Acronym for Return On Investment.
RTL	Acronym for Resistor Transistor Logic.

RTL	Acronym for Register Transfer Level. A type of HDL description in which a circuit is modeled by specifying the data flowing between a set of registers, which are elements of a design that transition between states based on an event (a high or low edge) occurring on a clock signal. The register-transfer level of abstraction is above the gate level and below the behavioral level.

S

SABER	SABER is not an acronym. It is a registered trademark of Analogy, Inc., for their SABER analog simulator.
safety margin	An extra margin of tolerance on a spec limit thus providing an extra margin of safety.
saturation	Condition in which a further increase in one variable produces no further increase in the resultant effect. In a bipolar junction transistor, the condition when the emitter to collector voltage is less than the emitter to base voltage. This condition puts forward bias on the base to collector junction.
Scattering-parameter model	See *S-Parameter model.*
schematic diagram	Illustration of an electrical or electronic circuit with the components represented by their symbols and wire connections represented by connecting lines.
semi	Abbreviation for semiconductor.
semiconductor	An element that is neither a good conductor nor a good insulator, but rather lies somewhere between the two. Characterized by a valence shell containing four electrons. Silicon, germanium and carbon are the semiconductors most frequently used in electronics.
SERDES	Acronym for Serializer/ De-serializer. High-speed I/O for Backplane Ethernet, PCI E & XAUI
Si	Abbreviation for Silicon.

SI	Acronym for Signal Integrity.
SiGe	Acronym for Silicon-Germanium.
sigma, σ	A standard deviation in a population distribution. A statistic used as a measure of the dispersion or variation in a distribution, equal to the square root of the arithmetic mean of the squares of the deviations from the arithmetic mean.
Signal Integrity	The technical discipline of maintaining the waveform integrity of a signal so that the digital logic can work properly. Actions include eliminating reflections, crosstalk, and noise on the signal. The 1's and 0's must be kept clean enough so that they are still recognized as 1's and 0's.
simulation	The process of verifying an electronic design using EDA software, which reads in models and input/output vectors, exercises the device under test. The software records the resulting electrical behavior and timing information for the purpose of identifying and debugging any incorrect or unexpected behavior.
skin effect	A physical effect on metallic conductors where, as frequency becomes high enough, electric fields cannot penetrate into the conductor. All current then gets conducted on the outer surface (i.e., skin) of the conductor. The thickness of the skin varies as $1/\sqrt{\text{frequency}}$. This raises the per unit length resistance of the conductor and its losses. This is the reason that a hollow metallic waveguide is used, or even just a dielectric rod, at microwave frequencies to guide signals.
SMD	Acronym for Surface Mount Device.
SMT	Acronym for Surface Mount Technology.
S-Parameter model	A type of Matrix model where the port properties are characterized by forward and reflected wave scattering coefficients. Measured into transmission line matched loads. Also know as Scattering-parameter model.
SPC	Acronym for Statistical Process Control.
specsmanship	The practice of guard-banding specification limits with more, or less, than sufficient margin, while maintaining the perception of a well-performing part.

SPICE model	Acronym for Simulation Program with Integrated Circuit Emphasis. An industry-standard analog simulation language, which contains many models for most circuit elements and can handle complex nonlinear circuits. Also refers to a freely distributed simulation tool, which simulates circuitry described in the SPICE language. By long usage has become synonymous with *SPICE model.*
SSTL	Acronym for Stub Series Terminated Logic, as in 74SSTL245.
STTL	Acronym for Schottky Transistor-to-Transistor Logic, as in 74STTL245.
SWR	Acronym for Standing Wave Ratio.
system	A group of objects operating together.

T

TCAD	Acronym for Technology Computer Aided Design
TDR	Acronym for Time Domain Reflectometry.
TEM	Acronym for Transverse ElectroMagnetic wave.
TLM	Acronym for Transmission Line Modeling.

transmission line — Any circuit connection where time delay of a signal propagating down it is large enough so that signal at the receiving end does not instantaneously follow the signal at the sending end. Transmission lines are characterized by inherent impedance, Zo, per unit length that is determined by structure geometry, dielectric and conductor properties. Not all of the signal can be transferred into the load if the input impedance at the receiving end does not match the Zo of the transmission line. A *reflection* of signal (positive or negative) will be set up to maintain conservation of energy. In a zero time delay circuit this means that the signal level at the sending end would instantaneously adjust to a level consistent with the load. This tracking of send-receive can get out of step if the round trip time delay is greater than the rise time of the signal. Then, the conditions at the sending end, current sent = (generator voltage)/(generator impedance + transmission line Zo) may not match conditions at the receiving end, current accepted = (generator voltage)/(transmission line Zo + receiving end input impedance). If so, signal reflections will be set up until signal power gets absorbed into the transmit-receive ends and/or dissipated in the transmission line structure.

TTL — Acronym for Transistor-Transistor Logic.

U

UL — Acronym for Underwriters Laboratory.

USL — Acronym for Upper Spec Limit

V

V — Symbol for volt.

validation	To give official confirmation or approval. Applied to models, it means verifying that the model correlates with the standard for the model and will run in simulators available for the purpose. Correlation with the standard for the model implies there is one and that the model follows the correct syntax, format, and contains the required data elements. For IBIS it's possible to also include an *eyeball reality check* of things like clamp and pullup curves.
VBIC	Acronym for Vertical Bipolar Inter-Company.
verification	A formal assertion of the truth of something. Applied to models, it means confirmation that simulation predictions of device and circuit behavior correlate with lab measurements. For IBIS it's possible to correlate the IBIS simulation with a SPICE simulation if the SPICE simulation was verified by measurements.
Verilog	A hardware description language. An industry-accepted standard language used by electronic designers to describe and design their chips and systems prior to fabrication.
Verilog-AMS	Acronym for Verilog Analog Mixed-Signal (hardware description language) extension to Verilog.
VHDL	Acronym for Very-high-speed-integrated-circuit Hardware Description Language. An IEEE-standard hardware description language originally developed by the U.S. Department of Defense as a common means of documenting electronic systems. Specified in the IEEE 1076 standard and used by electronic designers to describe and simulate their chips and systems prior to fabrication, an alternative language to Verilog. Also, referred to as VHSIC.
VHSIC	See *VHDL*.
VLSI	Acronym for Very Large Scale Integration.
VNA	Acronym for Vector Noise Analyzer.
VRM	Acronym for Voltage Regulator Module.
VSWR	Acronym for Voltage Standing Wave Ratio.

W

W	Symbol for watts.

X

Xtalk	Acronym for crosstalk.

Y

y-parameter model	A type of matrix model where the port properties are characterized by input, output, forward and reverse admittances. Measured into AC short circuits. Also known as admittance parameter model.

Z

z-parameter model	A type of matrix model where the port properties are characterized by input, output, forward and reverse impedances. Measured into AC open circuits. Also known as impedance parameter model.

BIBLIOGRAPHY

[1] Agilent, *In-Fixture Measurements Using Vector Network Analyzers (AN 1287-9)*, Agilent Technologies, 2000. http://cp.literature.agilent.com/litweb/pdf/5968-5329E.pdf

[2] Agilent, *Pulsed-RF S-Parameter Measurements Using a Vector Network Analyzer*, Agilent Technologies, 2005. http://cp.literature.agilent.com

[3] R. W. Anderson, "S-Parameter Techniques for Faster, More Accurate Network Design," *Hewlett Packard Application Note 95-1, Hewlett Packard Journal,* February 1967

[4] Ansoft Staff, *Reference Flow for High-Speed Serial Interconnect Design,* Ansoft, Inc., and Intel, Inc. 2005

[5] G. Antonini, C. Ciccomancini Scogna, A. Orlandi, C. Ritota and A. Duffy, "Applications of FSV to EMC and SI Data," *IEEE International Symposium on EMC,* Chicago, 2005

[6] B. Archambeault, C. Brench, O. Ramahi, *EMI/EMC Computational Modeling Handbook 2nd Ed.*, Kluwer Academic Publishers, 2001.

[7] B. Archambeault, S. Connor and A. Duffy, "Comparing FSV and Human Responses to Data Comparisons," *IEEE International Symposium on EMC,* Chicago, 2005

[8] P. Ashenden, G. Peterson, and D. Teegarden, *The System Designer's Guide to VHDL-AMS,* Morgan Kaufman, 2003.

[9] D. B. Atkins, *Correlation Analysis of Simulated Voltage Responses in Printed Circuit Board Transmission Lines,* Motorola, The Six Sigma Research Institute Series, Addison-Wesley, 1992.

[10] Avanti, *HSPICE User's manual* (Avant! was acquired by Synopsys), 2001. http://www.ece.uci.edu/eceware/hspice_docs/hspice_and_qrg/hspice_2001_2-142.html.

[11] R. J. Baker, H. W. Li, and D.E. Boyce, *CMOS Circuit Design, Layout, and Simulation*, IEEE Press, 1998.

[12] J. Barnes, *Robust Electronic Design Reference Book, Volumes 1 and 2,* Kluwer Academic Publishers, 2004.

[13] S. Bendhia, M. Ramdani, and E. Sicard, *Electromagnetic Compatibility of Integrated Circuits: Techniques for Low Emission and Susceptibility,* Springer, 2005.

[14] Bergeron, L. J. B., *Water Hammer in Hydraulics and Wave Surges in Electricity*, Wiley, 1961.

[15] D. H. Besterfield, Quality Control, Prentice-Hall, 1972.

[16] P. R. Bevington, *Data Reduction and Error Analysis for the Physical Sciences.* McGraw Hill, 1969.

[17] E. Bogatin, Signal Integrity – Simplified, Prentice-Hall PTR, 2004.

[18] A. Boyer, S. Bendhia, B. Vrignon, J. L. Levant, and M. Ramdani, "Modélisation d'un Boîtier TQFP144 par Mesures et Simulation," *Telecom 2005, 4th JFMMA*, 2005.

[19] G.R. Boyle, B.M. Cohen, D.O. Pederson and J.E. Solomon, "Macromodeling of Integrated Circuit Operational Amplifiers," *IEEE Journal of Solid-State Circuits*, Vol. SC9, No. 6, December 1974.

[20] J. T. Burr, *SPC Tools for Operators*, Quality Press, ASQC, 1989.

[21] Cadence Design Systems, Inc., Staff, *PSPICE A/D Reference Guide, Product Version 10.2*, Cadence Design Systems, Inc. June 2004.

[22] Cadence Design Systems, Inc., Staff, *Cadence Circuit Components and Device Models Manual, Product Version 6.0 Spectre*, Cadence Design Systems, Inc. November 2004.

[23] D. Carpenter, US patent **6,597,184** Method and apparatus for electromagnetic emissions testing, BT Exact Technologies, 2003.

[24] D. K. Cheng, *Analysis of Linear Systems*, Addison-Wesley, 1959.

[25] C. Christopoulos, *The Transmission-Line Modeling Method*, IEEE Press & Oxford University Press, 1995.

[26] CISPR: International Special Committee on Radio Interference (CISPR), Publication 22: *Information technology equipment — Radio Disturbance characteristics — Limits and methods of measurement,* 1993.

[27] R. E. Collin, *Field Theory of Guided Waves, 2nd Ed.*, IEEE Press, 1991.

[28] Compact Model Council web page: http://www.eigroup.org/cmc/

[29] R. Courant & D. Hilbert, *Volume 1, Methods of Mathematical Physics*, Wiley-Interscience, 1989

[30] A. L. Drozd, "Selected Methods for Validating Computational Electromagnetic Modeling Techniques," *IEEE International Symposium on EMC,* Chicago, 2005

[31] A. Duffy, A. Martin, G. Antonini, A. Orlandi and C. Ritota, "The Feature Selective Validation (FSV) Method," *IEEE International Symposium on EMC,* Chicago, 2005.

[32] R.W. Dutton & Z. Yu, *Technology CAD – Computer Simulation of IC Processes and Devices,* Kluwer Academic Publishers, 1993.

[33] J.M. Early, "Effects of Space-Charge Layer Widening in Junction Transistors." *Proceedings of the IRE*, Vol 40, pp. 1401-1406, November 1952.

[34] J. J. Ebers & J. L. Moll, "Large Signal Behavior of Junction Transistors," *Proceedings of the IRE*; pp. 1761 - 1771+, December 1954.

[35] C. C. Enz, F. Krummenacher, & E. A. Vittoz, "An Analytical MOS Transistor Model Valid in All Regions of Operation and Dedicated to Low-Voltage and Low-Current Applications," *Analog Integrated Circuits and Signal-Processing,* pp 83-114, July 1995. Also, EKV Model home page: http://legwww.epfl.ch/ekv/

[36] F. C. Fitchen, *Transistor Circuit Analysis & Design*, D. Van Nostrand & Co., 1960.

[37] D. Fitzpatrick and I. Miller, *Analog Behavioral Modeling with the Verilog-A Language,* Kluwer, 1998.

[38] T. A. Fjeldly, T. Ytterdal & M. Shur, *Introduction to Device Modeling and Circuit Simulation*, J. Wiley & Sons, 1998.

[39] Fluke Instrument Corp., Staff, *Calibration: Philosophy and Practice*, Fluke Instrument Corp, 1994.

[40] D. Foty, *MOSFET Modeling With SPICE: Principles and Practices,* Prentice Hall, 1997.

[41] Y. W. Fowlkes, C. L. Creveling, *Engineering Methods for Robust Product Design*, Addison-Wesley, 1995.

[42] D. Frohman-Bentchkowsky, A.S. Grove, "On the Effect of Mobility Variation on MOS Device Characteristics," *Proceedings of the IEEE*, 56, 1968.

[43] F. German, Flowmerics, Inc., "Design Early for EMC" *EDN http://www.edn.com* October 28, 2004.

[44] I. Getreu, *Modeling the Bipolar Transistor*, Tektronix, Inc., 1976.

[45] G. Gonzalez, *Microwave Transistor Amplifiers – Analysis and Design*, Prentice-Hall, 1984.

[46] T. Granberg, *Handbook of Digital Techniques for High-Speed Design: Design Examples, Signaling and Memory Technologies, Fiber Optics, Modeling and Simulation to Ensure Signal Integrity*, Prentice Hall PTR, 2004.

[47] A. S. Grove, *Physics and Technology of Semiconductor Devices,* John Wiley & Sons, 1967.

[48] K. Gummel & H. C. Poon, "An Integral Charge Control Model of Bipolar Transistors," *Bell System Technical Journal,* pp. 827-852, May 1970.

[49] S. Hall, G. Hall and J. McCall, *High Speed Digital System Design: A Handbook of Interconnect Theory and Design Practices*, Wiley Interscience, 2000.

[50] R. F. Harrington, *Field Computation by Moment Methods*, IEEE Press, 1993.

[51] G. E. Hayes and H. G. Romig, *Modern Quality Control,* Glenco Publishing, co., Inc., Revised Edition 1982.

[52] G. K. Hobbs, *Accelerated Reliability Engineering: HALT and HASS,* John Wiley & Sons, 2000.

[53] J.A. Hoerni, "Planar Silicon Transistors and Diodes," *Institute of Radio Engineers Electron Devices Meeting, Washington, D.C.,* 1960.

[54] J. Hoskins, B. Stuart, & J. Taylor, *SPC – Statistical Process Control,* BR392/D, Motorola, Inc., 1989.

[55] IBIS I/O Buffer Accuracy Handbook, http://www.eda.org/pub/ibis/accuracy/

[56] *IBIS Interconnect Model Spec,* http://www.eda.org/pub/ibis/icm_ver1.0/

[57] IBIS Quality Checklist: http://www.sisoft.com/ibis-quality/

[58] "Electromagnetic Compatibility Part 3 Integrated Circuits Electrical Modeling," *ICEM #62014-3 IEC Standard Proposal,* 2002.

[59] International Organization for Standardization, *ISO9000 Quality Management and Quality System Elements,* http://www.iso.org/iso/en/ISOOnline.openerpage

[60] H. Johnson and M. Graham, *High Speed Digital Design: A Handbook of Black Magic,* Prentice Hall, 1993.

[61] R. A. Johnson, *Miller and Freund's Probability and Statistics for Engineers*, Prentice-Hall, 1994.

[62] J. M. Juran and F. M. Gryna, Jr., *Quality Planning and Analysis: From Product Development Through Use,* McGraw-Hill, 2nd Edition 1980.

[63] R. Kielkowski, *SPICE – Practical Device Modeling,* McGraw-Hill, 1995.

[64] C.T. Kirk, Jr., "A Theory of Transistor Cutoff Frequency (fT) Falloff at High Current Densities," *Institute of Radio Engineers Transaction on Electron Devices*, p 164 – 174, March 1962.

[65] E. Kreyszig, *Advanced Engineering Mathematics, 6th Ed.,* J. Wiley & Sons, 1988.

[66] G. M. Kull, L. W. Nagel, S.W. Lee, P.Lloyd, E.J. Pendergast, and H.K. Dirks, "A Unified Circuit Model for Bipolar Transistors Including Quasi-Saturation Effects," *IEEE Transactions on Electron Devices,* ED-32, p1103-1113, 1985.

[67] C. Kumar, K. Willis, D. Telian, & M. Ali, "Behavioral MacroModeling for Multi-GigaHertz Applications, *DesignCon East*, 2004.

[68] K. S. Kundert, *The Designers Guide to SPICE & SPECTRE,* Kluwer Academic Publishers, 1995.

[69] K. Kundert, O. Zinke, *The Designer's Guide to Verilog-AMS,* Kluwer Academic Publishers, 2004.

[70] R. Kurshan - Editor, *Computer-Aided Verification*, Kluwer Academic Publishers, 1992.

[71] P. O. Lauritzen, Selected publications and recent conference papers, http://www.ee.washington.edu/people/faculty/lauritzen/

[72] E. Lenski, "Can We Stop the Growing Disparity Between the Potential of IBIS Model Parameters and the Reality of Delivered Model Parameters?" *EIA IBIS Open Forum Summit,* Munich, 2005

[73] R. Leventhal, "Block Reverse Gain in Your Transistor Amplifier," *Electronic Design,* August 16, 1969, pp205-210.

[74] R. Leventhal, *Device Physics Diagnostics by Curve Tracer,* 1977.

[75] R. Leventhal, *Reverse Transfer Isolation of Transistor Amplifiers*, Lawrence Radiation Laboratory, University of California, Livermore, UCRL 70888, 1968.

[76] R. Leventhal, *Switching Techniques and Equipment for Small Signal Discrete Devices*, Fairchild Semiconductor Research & Development Technical Report #569, 1974.

[77] W. Liu, *MOSFET Models for Spice Simulation, Including BSIM3v3 and BSIM4*, J. Wiley & Sons, 2001.

[78] R. Ludwig & P. Bretchko, *RF Circuit Design - Theory and Applications*, Prentice-Hall, 2000.

[79] C. L. Ma, P. O. Lauritzen, J. Sigg, "Modeling of Power Diodes with the Lumped-Charge Modeling Technique, *IEEE Transactions on Power Electronics,* May 1997. Vol. 12, No. 3, pp. 398-405.

[80] A. P. Malvino, *Transistor Circuit Approximations,* McGraw-Hill, 1968.

[81] H. A. Mantooth, M. Fiegenbaum, *Modeling With an Analog Hardware Description Language,* Kluwer Academic Publishers, 1995.

[82] A. J. M. Martin, A. R. Ruddle, & A. P. Duffy, "Comparison of Measured and Computed Local Electric Field Distributions Due to Vehicle-Mounted Antennas Using 2D Feature Selective Validation," *IEEE International Symposium on EMC,* Chicago, 2005.

[83] W. J. McCalla and W. G. Howard, Jr., "BIAS-3-A Program for the Nonlinear DC Analysis of Bipolar Transistor Circuits," *IEEE Journal of Solid-State Circuits, Vol. SC6,* p 14 – 19, February 1971.

[84] L. G. Meares & C. E. Hymowitz, *Simulating With SPICE,* IntuSoft, 1988.

[85] S. Mee, R. Leventhal, M. Ventham, A. Wexler, "Simulation for the Suppression of PCB Emissions in Digital Telematics Devices," *ITEM™ Update,* pp 56-66, 2002.

[86] Meta-Software, Inc., *HSPICE User's Manual: Vol. 2, Elements and Models, HSPICE Version H92,* Meta-Software, Inc., 1992.

[87] J. W. Miller, "Dependence of the Input Impedance of a Three-Electrode Vacuum Tube upon the Load in the Plate Circuit," *National Bureau of Standards Science Papers,* 15 (1919-1920), No. 351, 367-385.

[88] D. K. Misra, *Radio Frequency and Microwave Communications Circuits - Analysis and Design,* J. Wiley & Sons 2001.

[89] Motorola, Inc., Training materials on "Six Sigma Process Control," Motorola, Inc, 1983.

[90] L. W. Nagel, *SPICE2: A Computer Program to Simulate Semiconductor Circuits,* Electronics Research Laboratory, College of Engineering University of California, Berkeley, Memorandum # ERL-M520, 9 May 1975.

[91] W. Nelson, *How to Analyze Data With Simple Plots,* American Society for Quality Control, 1986.

[92] I. Novak, "Frequency-domain Power Distribution Measurements - An Overview," *DesignCon East,* http://home.att.net/~inovak/papers/DC03-East_HP-TF1_SUN.PDF 2003.

[93] I. Novak, "Frequency Dependent Characterization of Bulk and Ceramic Bypass Capacitors," *EPEP, 2003,* http://home.att.net/~istvan.novak/papers/EPEP2003_cap_models.PDF

[94] J. Ogrodzki, *Circuit Simulation Methods and Algorithms*, CRC Press, 1995.

[95] M. S. Phadke, *Quality Engineering Using Robust Design*, Prentice-Hall, 1989.

[96] A. Peterson, S. Ray & R. Mittra, *Computational Methods for Electromagnetics*, IEEE Press & Oxford University Press, 1998.

[97] T. Quarles, A. R. Newton, D. O. Pederson, A. Sangiovanni-Vincentelli, *SPICE3 Version 3f3 User's Manual,* Department of Electrical Engineering & Computer Sciences, University of California-Berkeley, 1993.

[98] J. M. Rabaey, *The Devices,* Digital Integrated Circuits & Prentice-Hall, 1995.

[99] V. S. Ramachandran, *A Brief Tour of Human Consciousness,* PI Press, 2001.

[100] M. H. Rashid, *SPICE for Circuits and Electronics Using PSPICE*, Prentice-Hall, 1990.

[101] L. F. Root, *Radio Frequency/Microwave Robust Design Techniques Applied to a Transistor Amplifier Test Fixture,* Addison-Wesley, 1993.

[102] R. Ross, "Practical Issues With IBIS Models," *PCB Design Conference East,* **EIA IBIS Open Forum Summit,** 1997: http://www.eigroup.org/ibis/pcb_cvr.htm

[103] SABER® home page. http://www.synopsys.com/cgi-bin/ces/detail.cgi?id=1200

[104] Staff, SABER® User Manuals Release 3.3c, Analogy, Inc. 1994.

[105] SABER-MAST modeling course: Class notes. http://www.synopsys.com/cgi-bin/ces/detail.cgi?id=1201

[106] D. Sanders and J. Muccioli, T. North, K. Slattery, "Quantitative Measurement of the Effectiveness of Decoupling Capacitors in Controlling Switch Transients from Microprocessors." *Capacitor and Resistor Technology Symposium (CARTS),* March 22,23 of 2005.

[107] S. R. Schmidt & R. G. Launsby, *Understanding Industrial Designed Experiments" 4th Ed.,* Air Academy Press, 1994.

[108] A. S. Sedra & K. C. Smith, *Microelectronic Circuits 3rd Ed,* and *Additional Problems With Solutions 3rd Ed.,* Saunders College Publishing, 1991.

[109] H. Shichman and D.A. Hodges, "Modeling and Simulation of Insulated-Gate Field-Effect Transistor Switching Circuits," *IEEE Journal of Solid-State Circuits,* SC-3 1968.

[110] W. Shockley, J. Bardeen and W. Brattain, "The transistor—a semiconductor triode," Physical *Review.* Vol. 74:1948 p 230W, Shockley, Patent Number(s) 2,502,488; 2,524,035, 1949.

[111] W. Shockley, J. Bardeen and W. Brattain, "The Theory of P-N Junctions in Semiconductors and P-N Junction Transistors," *Bell System Technical Journal, Vol 28: 1949, p 435 Semiconductive Materials Transistor.*

[112] W. R. Simpson & J. W. Sheppard, *System Test and Diagnosis*, Kluwer Academic Publishers, 1994.

[113] R. Singh - Editor, *Signal Integrity Effects in Custom IC and ASIC Design,* IEEE Press & J. Wiley & Sons, Inc. 2002.

[114] D. C. Smith, *High Frequency Measurements and Noise in Electronic Circuits*, Van Nostrand-Reinhold, 1993.

[115] P. H. Smith, "Transmission- Line Calculator," *Electronics,* 1939. Vol 12, pp 29-31.

[116] P. H. Smith, "An Improved Transmission- Line Calculator," *Electronics,* 1944. Vol 17, p 130.

[117] SPICE: SPICE3: http://newton.ex.ac.uk/teaching/CDHW/Electronics2/userguide/
SPICE3: http://infopad.eecs.berkeley.edu/~icdesign/SPICE/
BSIM3: http://www-device.eecs.berkeley.edu/~bsim3/get.html
BSIM4: http://www-device.eecs.berkeley.edu/~bsim3/bsim4.html
BSIMSOI: http://www-device.eecs.berkeley.edu/~bsimsoi/get2.html

[118] B. G. Streetman, *Solid State Electronic Devices*, 2nd Ed Prentice-Hall, 1980.

[119] T. Sudo, H. Sasaki, N. Masuda, J. L. Drewniak, "Electromagnetic Interference (EMI) of System-on-Package (SOP)," *IEEE Transactions on Advanced Packaging Vol 27,* #2 ITAPFZ pp 304-314.

[120] D. Sullivan, *Electromagnetic Simulation Using the FDTD Method*, IEEE Press & John Wiley & Sons, Inc., 2000.

[121] S. M. Sze, *Physics of Semiconductor Devices 2nd Ed.*, J. Wiley & Sons, 1981.

[122] G. Taguchi, *Introduction to Quality Engineering*, Asian Productivity Organization, 1986.

[123] C. T. Tai, Dyadic Green's Function in Electromagnetic Theory, 2nd Ed, IEEE Press, 1993.

[124] TDA Staff, *Effects of Measurement Fixtures of S-Parameters Compliance Testing of Infiniband Cable Assemblies*, TDA Systems, Inc., 2005. http://www.tdasystems.com/library/appnotes/CBLS1204.pdf

[125] R. E. Thomas & A. J. Rosa, *Analysis and Design of Linear Circuits*, Prentice-Hall, 1994.

[126] TI Staff, Editors: J. A. Walston & J. R. Miller, *Transistor Circuit Design*, McGraw-Hill, 1963.

[127] Y. Tsividis, *Operation and Modeling of the MOS Transistor*, Oxford University Press, 1999.

[128] R. R. Tummala, *Fundamentals of Microsystems Packaging*, McGraw-Hill, 2001.

[129] Verilog-AMS Home Page: http://www.eda.org/verilog-ams/

[130] VHDL-AMS Home Page: http://www.modelica.org/links/document_view

[131] J. Vlach and K. Singhal, *Computer Methods for Circuit Analysis and Design, 2nd Ed*, Van Nostrand Reinhold, 1994.

[132] A. Vladimirescu, *The SPICE Book*, J. Wiley & Sons, 1969.

[133] B. Vrignon, S. Bendhia, E. Lamoureux, and E. Sicard, "Characterization and Modeling of Parasitic Emission in Deep Submicron CMOS," IEEE Transactions on Electromagnetic Compatibility, vol. 47, n° 2, IEEE Publishers (May 2005) 382-387.

[134] M. Wang, H. Maramis, D. Telian, K. Chung, "New Techniques for Designing and Analyzing Multi-GigaHertz Serial Links," *DesignCon 2005,* Intel Corporation and Cadence Design Systems, 2005.

[135] T. Westerhoff, "Things You Can Learn From an IBIS Model," *CDNLive2005! Santa Clara,* 2005.

[136] S. Weir, "Measurement and Comparative s21 Performance of Raw and Mounted Decoupling Capacitors," *Interference Technology, EMC Test and Design Guide, ITEM™ Publications,* Weir Design Engineering, *2004.*

[137] S. Weir, "Does position matter? Locating bypass capacitors for effective power distribution and EMC control," *Meeting, Santa Clara Valley Chapter of the IEEE EMC Society,* Teraspeed Consulting Group, January, 2005.

[138] S. Weir, "Considerations for Capacitor Selection in FPGA Designs," *Capacitor and Resistor Technology Symposium (CARTS),* Weir Design Engineering, March 22,23 of 2005.

[139] A. B. Williams & F. J. Taylor, *Electronic Filter Design Handbook, 3rd Ed.*, McGraw-Hill, 1995.

[140] T. Ytterdal, Y. Cheng, and T. A. Fjeldly, *Device Modeling for Analog and RF CMOS Circuit Design,* J. Wiley & Sons, 2003.

[141] Staff, Fairchild Semiconductor, *Discrete Products Databook, Fairchild Semiconductor Components Group,* 1973.

[142] ANSI/EIA Standard 656 – A, *IBIS (I/O Buffer Information Specification)* http://www.eigroup.org/ibis/specs.htm

[143] IBIS Cookbook: http://www.eda.org/pub/ibis/wip/ckbkwfw3.doc

[144] L. Green, "IBIS Modeling for High Speed Designs," JEDEX, April 2004.

[145] N. Drakos, R. Moore, "Transmission and Delay Lines," Computer Based Learning Unit, Univ. of Leeds, Mathematics Dept. Macquarie Univ., 2002 http://fourier.eng.hmc.edu/e85/lectures/transmission_line/transmission_line.html

[146] J. Olah, S. Gupta, C. Chavez-Dagostino, "Signal Integrity Analysis and Simulation Tools Include IBIS Models," *High Frequency Electronics*, 2004.

[147] R. Perez, Editor, *Handbook of Electromagnetic Compatibility,* Academic Press, Inc. 1995.

[148] Staff, "Electromagnetic Field Computation," Advanced^{EM} of Caligate Technology Corporation, 2005 http://advancedem.com/methods.html

[149] J. Sifri, N. Kanaglekar, "Simulation Tools Converge on Large RFICs," CommsDesign, June 2002. http://www.commsdesign.com/showArticle.jhtml?articleID=16505209

[150] D. G. Swanson, Jr., W. J. R. Hoefer, *Microwave Circuit Modeling Using Electromagnetic Field Simulation,* Artech House, 2003.

[151] L. DeForest, *Space Telegraphy,* Patent US879532, 1908.

[152] A. Vladimirescu, S. Liu, *The Simulation of MOS Integrated Circuits Using SPICE2, Memo No. UCB/ERL M80/7*, Electronics Research Laboratory, University of California, Berkeley, 1980.

[153] D. Carpenter, "A De-Mystification of the U-Shaped Probability Distribution," *IEEE International Symposium on Electromagnetic Compatibility,* Boston, 2003.

[154] D. Carpenter, "A Further De-Mystification of the U-Shaped Probability Distribution," *IEEE International Symposium on Electromagnetic Compatibility,* Chicago, 2005.

[155] Multivariate Statistical Analysis: IBM Data Explorer: http://www.research.ibm.com/dx/

[156] S. Baffreau, C. Huet, C. Marot, S. Calvet, and E. Sicard, "A Standard Model for Predicting the Parasitic Emission of Micro-Controllers," *Proceedings of EMC Europe 2002,* Italy 2002.

[157] S. Calvet, C. Huet, J. Levant, C. Marot, J. Perrin, and E. Sicard, "Integrated Circuit Electromagnetic Model (ICEM) Cookbook," Release 1.c 2001 LESIA-INSA Toulouse, ESEO-Angers, IUT-Tarbes. Available at: http://www.ic-emc.org

[158] C. Xi, "Design and Characterization of ASIC for Medical Application," Ph.D. Dissertation, 2002. Available at http://www.lesia.insa-toulouse.fr

INDEX

Note: Terms set off by brackets [ ] are IBIS keywords.

USING THE COMPANION CD-ROM

This book comes with a CD that includes the following software and book information:

- *Sonnet® Lite from Sonnet Software, Inc.*
- *Visual IBIS Editor® from Mentor Graphics Corp.*
- *The Appendixes* include example data sheets, sample model files, IBIS syntax guides, presentations with practical model-building laboratory exercises (including input, intermediate, and output files), and other documentation. For a description of the various Appendix files, go to the Appendix directories and open the readme files.

Sonnet® Lite from Sonnet Software, Inc.

Sonnet Lite is a free introductory version of Sonnet's Professional Suite for high frequency 3D Planar electromagnetic (EM) analysis. Sonnet provides EM analysis to hundreds of companies across the world. Sonnet Lite can be used to analyze planar structures such as transmission lines, matching networks, microstrip or stripline filters, vias, crosstalk, package resonances, and much more.

This software runs on Windows platforms. The installed program requires a minimum of 75 Meg of storage. The memory requirements needed for the creation and analysis of design problems increase memory storage requirements.

Installation instructions:

1. With the CD in the CD-ROM drive, browse to the SonnetSoftware directory.
2. Double-click **setup.exe** and follow the setup instructions.
3. After installation, go to Start→Programs→Sonnet 10.51→Register and follow the registration instructions that appear.
   You will receive two emails in response to registering. One email will contain a set of license keys. A second email will contain information on where to get updated software and how to join a Sonnet web-based Community Forum.

4. Open the email titled "Sonnet Lite Key Codes from Sonnet Software" and follow the instructions for license installation.
   As of this writing, the license installation instructions ask you to save the email as a text file on your C: drive and to start the Sonnet program.

5. From the task bar, select Admin→Make License. Follow the program's instructions.

6. When done, you are ready to start learning the program. Nothing happens until you select a task from the task bar. The authors strongly recommend that you follow the Tutorials and User Manuals as you begin your learning journey.

7. If you have further questions, contact Sonnet at:

   Sonnet Software, Inc.
   100 Elwood Davis Rd.
   North Syracuse, NY 13212 USA
   Voice: 315-453-3096
   Toll Free: 877-776-6638 (USA & Canada only)
   FAX: 315-451-1694
   Web: http://www.sonnetsoftware.com
   Email: info@sonnetsoftware.com

Visual IBIS Editor® from Mentor Graphics

Visual IBIS Editor is a free tool from Mentor Graphics for creating IBIS (I/O Buffer Information Specification) model files. This tool validates syntax, creates model templates, graphically displays V/I and V-t (waveform) tables, edits model text, and provides a hierarchical view of IBIS models. Anyone who uses or creates IBIS models will benefit from using this editor, regardless of whether or not they own any of Mentor Graphics' other products.

This software runs on Windows platforms. The installed program requires 4.3 MB on your hard drive.

Installation instructions:

1. With the CD in the CD-ROM drive, browse to the MentorSoftware directory, and double-click **visibis32.exe**.
2. To complete the installation, follow the on-screen instructions.
3. When complete, you are ready to start learning the program. You may want to read or scan the Visual IBIS Editor documentation (Help menu→ Help) as you begin your learning journey.

A more advanced version of the Visual IBIS Editor is included with Mentor Graphics HyperLynx, ICX, and ICX Pro products. These products offer technical support as well. To make sure that you have the latest version of the Visual IBIS Editor, go to http://www.mentor.com/hyperlynx. Visual IBIS Editor is provided by Mentor Graphics Corporation, Systems Design Division.

GPSR Compliance

The European Union's (EU) General Product Safety Regulation (GPSR) is a set of rules that requires consumer products to be safe and our obligations to ensure this.

If you have any concerns about our products, you can contact us on ProductSafety@springernature.com

In case Publisher is established outside the EU, the EU authorized representative is:

Springer Nature Customer Service Center GmbH
Europaplatz 3
69115 Heidelberg, Germany

Batch number: 10755320

Printed by Printforce, the Netherlands